Advances in Experimental Medicine and Biology

Volume 997

Advances in Experimental Medicine and Biology presents multidisciplinary and dynamic findings in the broad fields of experimental medicine and biology. The wide variety in topics it presents offers readers multiple perspectives on a variety of disciplines including neuroscience, microbiology, immunology, biochemistry, biomedical engineering and cancer research. *Advances in Experimental Medicine and Biology* has been publishing exceptional works in the field for over 30 years and is indexed in Medline, Scopus, EMBASE, BIOSIS, Biological Abstracts, CSA, Biological Sciences and Living Resources (ASFA-1), and Biological Sciences. The series also provides scientists with up to date information on emerging topics and techniques. 2015 Impact Factor: 1.953

More information about this series at http://www.springer.com/series/5584

Mitsuo Tagaya • Thomas Simmen
Editors

Organelle Contact Sites

From Molecular Mechanism to Disease

Springer

Editors
Mitsuo Tagaya
School of Life Sciences
Tokyo University of Pharmacy and Life Sciences
Hachioji, Tokyo, Japan

Thomas Simmen
Department of Cell Biology
University of Alberta
Edmonton, Alberta, Canada

ISSN 0065-2598 ISSN 2214-8019 (electronic)
Advances in Experimental Medicine and Biology
ISBN 978-981-10-4566-0 ISBN 978-981-10-4567-7 (eBook)
DOI 10.1007/978-981-10-4567-7

Library of Congress Control Number: 2017947694

Printed on acid-free paper

This Springer imprint is published by Springer Nature
The registered company is Springer Nature Singapore Pte Ltd.
The registered company address is: 152 Beach Road, #21-01/04 Gateway East, Singapore 189721, Singapore

Preface

Eukaryotic cells have a variety of membrane-enclosed compartments termed organelles. Each organelle has unique functions, which are based on proteins that are specifically transported into each organelle according to their intrinsic "address tags". Organelles can produce cellular building blocks as well as provide energy, both of which are required for cell survival, proliferation, and differentiation. Research from several decades, assembled in this book, shows that the specific functions of organelles are not only geographically confined within membranous compartments, but also depend on rather intimate inter-organellar communication. Historically, such communication was thought to occur via vesicular trafficking. However, studies starting in the 1970s on lipid synthesis and transport have led to the discovery of another way that organelles can communicate, generating the concept of "membrane contact sites" (MCS). The importance of MCS truly emerged by the finding in the late 1990s that the endoplasmic reticulum (ER) releases Ca^{2+} in hotspots directed towards mitochondria. Recent studies have discovered not only major MCS regulatory proteins, but also new roles of MCS, including their involvement in a number of diseases, including neurodegenerative diseases, type 2 diabetes, infections, inflammation, and cancer. Today, MCS research is one of the most rapidly developing and exciting fields in cell biology and basic medicine.

This book aims to provide a comprehensive coverage of our quickly evolving knowledge of organelle communication at MCS and the significance of MCS for disease. The book is organized into two parts in addition to the overview (Chap. 1). The first part consists of nine chapters that set forth the organization and roles of MCS. The ER plays a principal role in organelle contacts, because it forms a wide network in cells and has several unique sub-domains with distinct functions. This allows the organelle to form contacts with other organelles in a variety of ways. Four chapters deal with the mitochondria-associated membrane (MAM), a domain of the ER that faces mitochondria. Other chapters describe the contact of the ER with the Golgi apparatus, endosomes, the plasma membrane, and lipid droplets. While the Golgi apparatus at the *cis*-side accepts transport vesicles from the ER, at the *trans*-side it interacts with the ER through MCS. Lipid droplets are unique among organelles in that they are surrounded by a phospholipid monolayer. Enzymes that are responsible for the biogenesis of lipid droplets are concentrated at the MAM, highlighting that these organelles are highly dependent on MCS formation. The second part (seven chapters) concerns viruses and pathogens (e.g., hepatitis C virus, *Chlamydia*, and *Coxiella*) whose targets are MCS and diseases that are caused by malfunctions of MCS, including Alzheimer's disease, Parkinson's disease, type 2 diabetes, and inflammation.

Given the rapidly evolving knowledge about MCS, it is inevitable that there are many controversies and conflicting reports in this field, which often prevents non-expert researchers from a deeper understanding of MCS and their significance. This book nicely summarizes the present status of MCS research for researchers who are not familiar with this field. MCS has been attracting much interest from young, talented researchers, which is reflected by the fact that several authors of this book have recently started their own laboratories. As editors, we hope that this book will stimulate graduate students and postdocs, who have the potential to energize, drive, and develop this field in the near future, and clinical fellows who may extend disease-related MCS research. The book will also be a valuable reference for established researchers who want to view their own research data and projects from the standpoint of MCS and who have so far focused on a type of cell biology that is based on "isolated organelles".

We thank the authors who have taken the time to write each chapter and Dr. Misato Kochi at Springer Japan for providing us with the opportunity to edit this book.

Hachioji, Tokyo, Japan — Mitsuo Tagaya
Edmonton, Alberta, Canada — Thomas Simmen

Contents

1 Organelle Communication at Membrane Contact Sites (MCS): From Curiosity to Center Stage in Cell Biology and Biomedical Research

Thomas Simmen and Mitsuo Tagaya

Abstract

Cell biology has long recognized that organelles can communicate with each other. Initially, such communication was thought to occur primarily via vesicular trafficking between biochemically distinct organelles. However, studies starting in the 1970s on lipid metabolism have unearthed another way how organelles can communicate and have spawned the field of membrane contact sites (MCS). While, initially, MCS had been recognized as fluid entities that mediate lipid and ion transport in an ad hoc manner, more recently MCS have been found to depend on protein–protein interactions that control themselves a variety of MCS functions. As a result, the cell biological definition of an intracellular organelle as an isolated membrane compartment is now being revised. Accordingly, the organelle definition now describes organelles as dynamic membrane compartments that function in a milieu of coordinated contacts with other organelles. Through these mercurial functions, MCS dictate the function of organelles to a large extent but also play important roles in a number of diseases, including type 2 diabetes, neurodegenerative diseases, infections, and cancer. This book assembles reviews that describe our quickly evolving knowledge about organellar communication on MCS and the significance of MCS for disease.

Keywords

Membrane contact sites • Endoplasmic reticulum • Golgi apparatus • Mitochondria • Endosomes • Lysosomes • Peroxisomes • Plasma membrane • Ca^{2+} • Lipid synthesis

T. Simmen (✉)
Faculty of Medicine and Dentistry, Department of Cell Biology, University of Alberta, Edmonton T6G2H7, Canada
e-mail: Thomas.Simmen@ualberta.ca

M. Tagaya
School of Life Sciences, Tokyo University of Pharmacy and Life Sciences, Hachioji, Tokyo 192-0392, Japan
e-mail: tagaya@toyaku.ac.jp

M. Tagaya, T. Simmen (eds.), *Organelle Contact Sites*, Advances in Experimental Medicine and Biology 997, DOI 10.1007/978-981-10-4567-7_1

1.1 Introduction

Historically, cell biology has attempted to assign specific functions to the various intracellular organelles of the eukaryotic cell, based on their precise protein as well as lipid composition. These efforts have led to the impression that organelles are defined, separate entities that exert clear roles for cellular survival, proliferation, and differentiation. Following the discovery of clathrin-coated vesicle invaginations in endocytosis (Anderson et al. 1977), it was postulated that vesicular trafficking would mediate interorganellar communication between these defined compartments (Goldstein et al. 1979; Palade 1975). By doing so, vesicular trafficking along the cytoskeleton could enable the cell to achieve communication even between fairly distant organelles. Indeed, the study of the intracellular trafficking of vesicular stomatitis virus (VSV) G protein (Rothman et al. 1980; Rothman and Fine 1980) and of yeast mutants unable to secrete invertase and acid phosphatase showed that organelles receive material and signals from rather distant intracellular locations via vesicular transport (Novick et al. 1980; Novick and Schekman 1979). The 2013 Nobel Prize has recognized the importance of these discoveries (Bonifacino 2014). That Nobel for intracellular trafficking included, however, also the discovery of SNAP receptors (SNAREs), which mediate the fusion of synaptic vesicles within neurons with target membranes (Söllner et al. 1993).

Unbeknownst to researchers at the time, the discovery of this group of proteins anticipates a connection between vesicular trafficking and membrane contact sites (MCS), where organelles interact directly with each other, rather than via vesicles. This connection arises from synaptotagmin, a type 1 transmembrane protein that accelerates the profusion activity of SNAREs in a Ca^{2+}-dependent manner (Perin et al. 1990). Synaptotagmin promotes vesicular fusion with its cytosolic C2 domain, which mediates phospholipid binding (Nalefski and Falke 1996). Through this activity, synaptotagmin promotes the fusion of docked synaptic vesicles with the plasma membrane in a Ca^{2+}-dependent manner and thus triggers neurotransmitter release following the arrival of an action potential (Geppert et al. 1994). However, synaptotagmin also uses its C2B domain to interact with phosphatidylinositol 4,5-bisphosphate (PIP_2) with the intent to form synaptic vesicle-plasma membrane docking sites (de Wit et al. 2009; Park et al. 2015). A lipid microdomain of about 73 nanometers in size subsequently attracts syntaxin 1A, a SNARE protein (van den Bogaart et al. 2011a), while synaptotagmin itself functions as a distance regulator for SNARE nucleation and thus tethers the liposomes close enough for membrane fusion (van den Bogaart et al. 2011b). By forming apposed membrane structures, synaptotagmin mediates the formation of a prototype MCS.

Synaptotagmin is related to extended synaptotagmins (E-Syts), which contain synaptotagmin-like mitochondrial lipid-binding protein domains (SMP) in addition to C2 domains (Herdman and Moss 2016; Perez-Lara and Jahn 2015). SMP domains allow for the formation of protein–protein complexes, as well as the formation of tubular paths for lipid transfer (Schauder et al. 2014). With these characteristics, E-Syts are recognized today as the classic paradigm for organelle tethers (Perez-Lara and Jahn 2015), in particular at MCS between the endoplasmic reticulum (ER) and the plasma membrane (Fernandez-Busnadiego et al. 2015). Here, SMP-containing E-Syts dimerize in a head-to-head fashion to mediate MCS architecture (Reinisch and De Camilli 2016). It is currently unclear, whether E-Syts are alone in this function, since they do not significantly influence STIM1/Orai-mediated store-operated Ca^{2+} entry (SOCE), a plasma-membrane-localized flow of Ca^{2+} ions directed to the ER (Giordano et al. 2013). More recently, SMP domains have also been detected within proteins that form another MCS, the ER–mitochondria encounter structure (ERMES) in yeast, where they similarly mediate ER–mitochondria tethering as well as lipid transfer (Kornmann and Walter 2010). Despite the absence of ERMES from mammalian cells (Wideman et al. 2013), this finding nevertheless indicates a central role of SMP domains in the formation of MCS.

One of the earliest incarnations of the concept of MCS is the formation of GERL contacts between the ER, mitochondria, and additional organelles (Novikoff 1976) that can bring Golgi, ER, lysosomes (GERL), and mitochondria into close proximity (Holtzman et al. 1967). The term MCS itself stems from observations on the ultrastructure of mitochondria, where physical contact between inner and outer membranes occurs, a prototype of apposed membranes (Hackenbrock 1968). Early studies on this type of MCS determined that intramitochondrial MCS accommodate protein translocation activity across these apposed membranes (Pon et al. 1989; Schwaiger et al. 1987), as well as cholesterol transfer to the mitochondrial inner membrane (Cherradi et al. 1996; Thomson 2003). Moreover, mitochondrial MCS also show enrichment for hexokinase, Ca^{2+} (Kottke et al. 1988), proteins of permeability transition pore (Brdiczka et al. 1990), and phosphatidylserine (Simbeni et al. 1991), highlighting their central role for the cellular power plant. In yeast, formation of the mitochondrial MCS requires the GTPase Fzo1p (Fritz et al. 2001) and furthermore the mitochondrial contact site complex (MICOS), also present in mammalian cells (Harner et al. 2011). On mitochondrial contact sites of mammalian cells, pro-apoptotic Bcl2 family proteins such as tBid regulate the release of cytochrome c (Kim et al. 2004). Similar to yeast, the distribution of these outer mitochondrial membrane (OMM) proteins depends on Fzo1p orthologs called mitofusins that also catalyze the fusion of the OMM (Weaver et al. 2014).

However, the mitochondrial MCS is but one of many examples of membrane contacts. In fact, the typical MCS occurs not within an organelle, but rather between two or more organelles, as outlined in several hallmark review articles on the topic (Levine 2004; Levine and Patel 2016). As a result, and due to these extensive interorganellar contacts, our understanding of an organelle has moved from one of an isolated membrane compartment with a defined function to a concept where organelles dynamically coordinate their membrane composition with their function.

1.2 The Basics on MCS Architecture and Types

Most known MCS correspond to interorganellar contacts formed between the ER and virtually all other organelles, including the mitochondria, cell membrane, endosomes, lysosomes, Golgi apparatus, lipid droplets, peroxisomes, and chloroplasts (Fig. 1.1). The ER as the MCS centerpiece was first proposed in a groundbreaking review article by Tim Levine about a dozen years ago (Levine 2004). However, while the ER is still recognized today as being involved in many MCS types, its participation is no longer seen as a requirement for MCS formation that may lack the ER as a participant altogether (Gatta and Levine 2017).

The ER is the largest organelle in most cell types and spreads out along microtubules to the cellular periphery, but originates at the nuclear membrane. Throughout its structure, the ER forms subdomains that have distinct functions as well as characteristic lipid and protein composition. On these subdomains, the ER segregates enzymes accelerating the folding and degradation of proteins, catalyzing the synthesis of phospholipids and cholesterol, enabling cellular detoxification, and mediating Ca^{2+} flux as well as storage (Lynes and Simmen 2011). Structurally, ER membranes have been observed to form tubular or sheet-like structures (Voeltz et al. 2006). More recently, using super-resolution microscopy, the exact structure of the sheet-like ER subtype has been refined as densely packed tubules forming structures called ER "matrices" or clusters of tubular ER with heterogeneous, highly irregular structures (Nixon-Abell et al. 2016). Such irregular tubules are also the prototype ER structure found in proximity of the plasma membrane, where they form one of the best-characterized MCS (Fernandez-Busnadiego et al. 2015), sometimes called the plasma membrane-associated membrane (PAM) (Pichler et al. 2001). At this location, microtubule-associated remodeling of the ER takes place in a Ca^{2+}-dependent manner, leading to sprouting of plastic tubules (Grigoriev et al. 2008) that undergo a process called ER

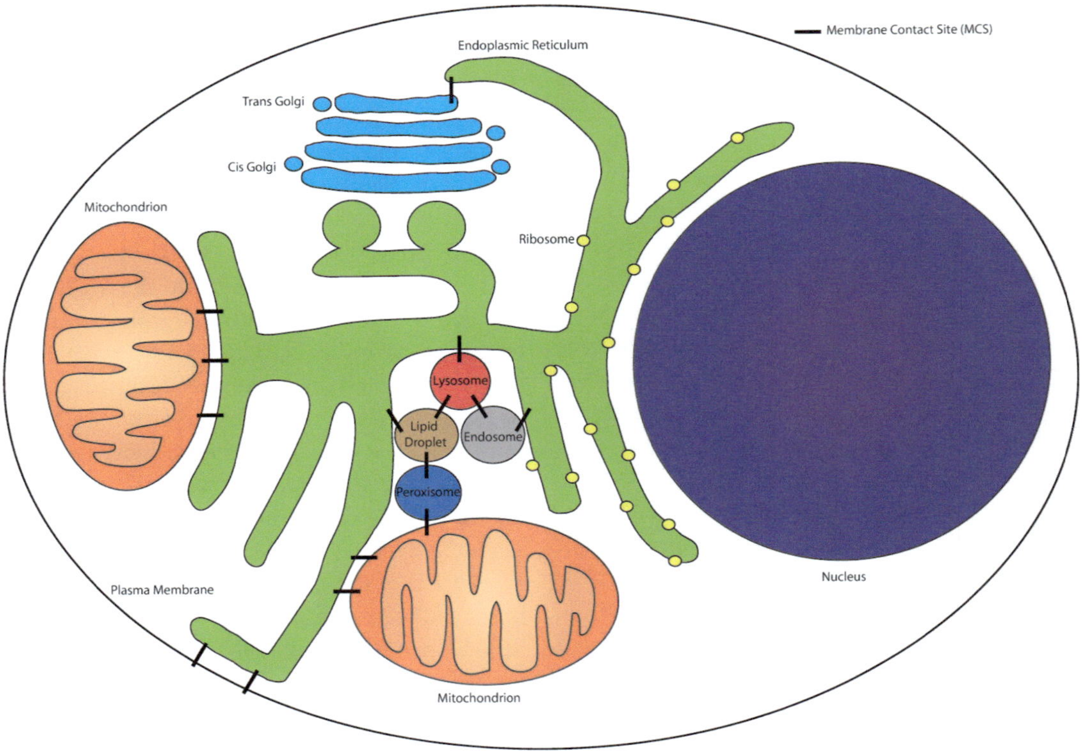

Fig. 1.1 Representation of major membrane contact sites (MCS), as described in the text. Organelles form a dense meshwork with multiple contacts between each other. At MCS, protein tethers anchor the organelles to each other. The sites allow for lipid and ion flux between partner organelles without the involvement of vesicular trafficking

sliding to interact with their partner membrane to create novel MCS (Wozniak et al. 2009). Given the drastic reduction in the number of ER tubules upon their depletion, proteins of the reticulon and DP1/YOP1 family are implicated in these events that ultimately lead to MCS formation (English and Voeltz 2013). At the plasma membrane, as well as in the case of other MCS, the ER approaches other organellar membranes to a distance of up to about 3–15 nm (Phillips and Voeltz 2016). Such tight interactions can occur on puncta where the ER touches another organelle (Varnai et al. 2007), or as outright wrapping of an ER tubule around the partner organelle, leading to mitochondrial or endosomal constriction (Friedman et al. 2011; Rowland et al. 2014). Due to the lack of space between the ER and partner membranes, ribosomes are often excluded from contact sites, although contacts of lesser proximity can form with ribosome-studded ER (Giacomello and Pellegrini 2016).

Other MCS have been discovered in the past decade that are in part based on organelles whose biogenesis is functionally tied to the ER, such as lipid droplets or peroxisomes. For instance, lipid droplets formed from the local accumulation of lipids at the ER (Wilfling et al. 2014) can interact with lysosomes during the process of lipophagy (Singh et al. 2009; van Zutphen et al. 2014). Likewise, peroxisomes can form from ER-derived membranes (Smith and Aitchison 2013) and interact with mitochondria (Neuspiel et al. 2008), lipid droplets (Schrader 2001), or lysosomes (Chu et al. 2015). Another large group of MCS involves membraneous contacts between mitochondria and partner organelles, for example, with the plasma membrane (Westermann 2015). An important physiological role for these contacts is revealed in T cells, where mitochondria–plasma membrane contacts control T cell activation via Ca^{2+} signaling through Orai (Quintana et al. 2011). In most cases, little if any structural information is available about these MCS.

1.3 MCS Function as a Conduit for Lipids and Ions

While MCS have been detected first by electron microscopists such as Bernhard in the 1950s (Bernhard et al. 1952; Porter and Palade 1957) or Novikoff in the 1960s (Holtzman et al. 1967), no function was assigned to these intracellular structures for a long time. Starting in the early 1970s, it was postulated that intracellular lipid synthesis requires the interaction between organelles, in particular between the ER and mitochondria (McMurray and Dawson 1969; Sauner and Levy 1971; Wirtz and Zilversmit 1968). However, it took until the early 1990s to biochemically prove and fully develop this hypothesis (Rusinol et al. 1994; Shiao et al. 1995; Vance 1990, 1991). This scientific breakthrough led to the coining of the first official term for an MCS in the form of the "mitochondria-associated membrane (MAM)." At the MAM, an ER membrane close to mitochondria, newly synthesized phosphatidylserine (PS) is transported in a vesicle-independent manner to mitochondria. Here, a decarboxylase that is present in the inner mitochondrial membrane catalyzes its transformation to phosphatidylethanolamine (PE). Thereafter, PE is distributed to other organelles by currently unknown mechanisms. In particular, a moiety of PE is returned to the ER to yield phosphatidylcholine (PC) in the liver (Vance 2015).

Another important function of MCS is the production and distribution of cholesterol in human cells or ergosterol in yeast cells. The production site of cholesterol is found within the ER (Reinhart et al. 1987). However, inside this organelle, cholesterol levels are kept low on most subdomains. This exclusion of cholesterol from most ER membrane domains occurs via vesicular export of excess cholesterol from the transitional ER to the Golgi complex, endosomes, and the plasma membrane, which all contain much more cholesterol than the ER itself (Maxfield and Wustner 2002). Conversely, under conditions of low ER cholesterol, the export of the transcription factor sterol regulatory element-binding protein-2 (SREBP-2) from ER exit sites to the Golgi complex (Radhakrishnan et al. 2008) allows for its proteolytic cleavage of SREBP-2 and subsequent activation of cholesterol metabolism genes (Goldstein et al. 2006). Despite a general scarcity of cholesterol on ER membranes, it has been surprisingly shown that ER MCS are enriched in cholesterol (Hayashi and Su 2003) and caveolin (Sala-Vila et al. 2016), in particular MCS based on smooth ER (Reinhart et al. 1987). The enrichment of cholesterol on MCS results in their biochemical properties resembling lipid rafts (Helle et al. 2013). These cholesterol and sphingolipid-enriched lipid microdomains are insoluble in nonionic surfactants such as Triton X-100 (Pike 2003). Through these properties, lipid rafts exhibit low membrane fluidity and are often enriched in lipid-modified (acylated) proteins such as palmitoylated or myristoylated proteins (Levental et al. 2010), for example, the palmitoylated ER chaperones calnexin and TMX1 (Lynes et al. 2012).

Aside from the biosynthetic route that originates at the ER, organelles of the later secretory pathway mostly receive cholesterol due to low-density lipoprotein (LDL) endocytosis from the plasma membrane to the lysosome (Goldstein et al. 1979; Goldstein and Brown 2015). MCS might play an important role in the redistribution of this cholesterol moiety as well, since from lysosomes, cholesterol reaches peroxisomes via MCS that require synaptotagmin-7 (Chu et al. 2015).

Another important function of MCS is ion transport, in particular Ca^{2+} transport. This mechanism operates prominently in the context of muscle fibers, where the sarcoplasmic reticulum (SR), a specialized form of the ER, makes contacts with t-tubules through the protein junctophilin (Barone et al. 2015; Takeshima et al. 2000). Interestingly, caveolin-3 mediates the development of t-tubules from the sarcolemma (Munoz et al. 1995), suggesting classic MCS contact formation processes are involved in the spatial organization of muscle cells. Junctophilins are critical for the muscle-contracting Ca^{2+} release via ryanodine receptors (RyRs) that localize to t-tubule/SR contact sites (Phimister et al. 2007). Here, released Ca^{2+} can migrate over to closely apposed mitochondria to activate the citric acid cycle and replenish ATP that has been lost due to muscle contraction (Eisner et al. 2014; Rudolf et al. 2004).

In other cell types, ER Ca^{2+} release more typically occurs through inositol 1,4,5-trisphosphate receptors (IP3R) (Mikoshiba 2007). MCS between the ER and mitochondria as well as the plasma membrane then determine the consequences of this release: on the one hand, this Ca^{2+} migrates across the MAM over to mitochondria, where it boosts respiration (Cardenas et al. 2010). This ion migration requires MCS formation between the ER and mitochondria (Patergnani et al. 2011), as well as the formation of cytosolic Ca^{2+} microdomains (Rizzuto et al. 1993). On the other hand, when the ER Ca^{2+} concentration decreases, $[Ca^{2+}]_{ER}$ is replenished from the extracellular milieu via store-operated Ca^{2+} entry (SOCE) that requires the formation of ER contacts with the plasma membrane (Stathopulos et al. 2008). Upon Ca^{2+} release through IP3Rs, the ER Ca^{2+} channel STIM1 migrates onto puncta (Liou et al. 2005), which align with the plasma membrane Ca^{2+} pore Orai1 (Luik et al. 2006; Varnai et al. 2007; Wu et al. 2006; Xu et al. 2006). This alignment then leads to Ca^{2+} influx via CRAC channels (Mercer et al. 2006; Peinelt et al. 2006; Prakriya et al. 2006). Also part of MCS Ca^{2+} handling is the SERCA Ca^{2+} pump. Increased SERCA activity can reduce ER-mitochondria Ca^{2+} flux (Lynes et al. 2013) and decrease MAM formation (Raturi et al. 2016). Consistent with this finding, under conditions when MAM formation is increased, like with increased concentration of cholesterol at the MAM, SERCA activity decreases (Li et al. 2004).

Other intracellular Ca^{2+} fluxes that occur on MCS have been recently discovered between late endosomes, lysosomes, and the ER. These mechanisms are sensitive to the third messenger nicotinic acid adenine dinucleotide phosphate (NAADP) that leads to Ca^{2+} release from the endolysosomal system (Calcraft et al. 2009). Combined with IP_3R and RyR-released Ca^{2+} from the ER (using IP_3 and cADPR, respectively), contacts between the ER and the endolysosomal system create a Ca^{2+} flux similar to the one between the ER and mitochondria (Gerasimenko et al. 2015; Parrington et al. 2015). This flux requires proteinaceous contact formation nucleated around oxysterol-binding proteins of the Osh (yeast) and oxysterol-related protein (ORP, mammalian) protein families (Honscher and Ungermann 2014) and is critical for insulin secretion (Kim et al. 2008).

1.4 Current and Future Challenges of MCS Research

Historically, the best evidence for MCS formation has been at a biochemical and electron microscopic level (see above). However, electron microscopic immunolocalization has so far only rarely been successful to pinpoint the localization of a given MCS protein to membraneous contacts (Lynes et al. 2012; Missiroli et al. 2016; Wang et al. 2000). Moreover, biochemical approaches to demonstrate the existence and characteristics of MCS face multiple challenges. On the one hand, they must distinguish between association artifacts, but on the other hand, they could miss existing associations due to incorrect incubation methods that disrupt delicate tethers. Unfortunately, the small distance between two organelles forming an MCS (normally within 30 nm) precludes the usage of conventional immunofluorescence confocal microscopy that has a resolution of 200 nm. This variety of technical difficulties has often led to erroneous localization of MCS proteins to exclusively either one or the other partner membrane, when they actually localize at least in part to both. A good example for these issues is seen with the MAM, where proteins such as calnexin or the voltage-dependent anion channel (VDAC) have been previously found exclusively on the ER (Wang et al. 2000) or mitochondria (Colombini et al. 1996), respectively, when they are actually at least in part MCS localized (Lynes et al. 2012; Shoshan-Barmatz et al. 2004). To resolve this issue, superresolution microscopy has now been introduced to the canon of techniques used for MCS research, but results are for the moment limited (Bhuvanendran et al. 2014; Colberg-Poley et al. 2015).

Current and future technology developments will aid in the further characterization of MCS. Besides the promise of superresolution microscopy, proximity ligation assays (PLA) have been recently used in the study of MCS tethers (Tubbs et al. 2014). A similar approach is based on split GFP that is reconstituted upon close apposition (Alford et al. 2012). To address the multiple functions of MCS in Ca^{2+} and ROS signaling, GFP-based indicators serve to detect local accumulations of these signaling molecules and indicate their role for membrane apposition (Booth et al. 2016; Csordas et al. 2010). The closeness of membranes in MCS and their three-dimensional arrangement have led to the introduction of electron tomography to the field as early as 1997 (Perkins et al. 1997). However, the full power of this technology is being harnessed only now and has led to the insight that MCS play an important role in organelle dynamics and morphology (Friedman et al. 2011; Rowland et al. 2014).

Another field of intense research is the significance of MCS formation and functioning for human disease. Recent results implicate the malfunctioning of MCS, in particular between the ER and mitochondria, in cancer (Marchi et al. 2014), neurodegeneration (Krols et al. 2016), diabetes (Rutter and Pinton 2014), cardiovascular disease (Eisner et al. 2013), and infection (Jacobs and Coyne 2013). While these *potential* connections are currently a hot topic of biomedical research, less *direct* connections between specific MCS proteins and disease are currently known. One example of such *bona fide* connections is Niemann–Pick type C disease, a lipid trafficking disorder where unesterified intracellular cholesterol accumulates (Schultz et al. 2016). It has been hypothesized that one reason that this lipid disorder arises is because of a loss of interaction between NPC1 on late endosomes and the ER-resident ORP5 (Du et al. 2011). Another example is Charcot–Marie–Tooth disease type 2A (CMT2A), where mutations in the MAM protein mitofusin-2 lead to neuropathy (Züchner et al. 2004). However, in both the NPC and CMT examples, it is currently unclear whether an MCS malfunctioning is at the basis of the disorder.

Given the rapid progress in the field, as evidenced in the present book, we predict that the coming years will lead to rapid further insight into MCS-based disease mechanisms that will shift from phenomenology to protein-associated disorders. There is therefore no doubt that this progress will lead to increased knowledge about MCS in the next few years, but also increased understanding of the biological and medical significance of MCS.

Acknowledgments We thank Maria Sol Herrera for the design of the figure and critical reading of the manuscript.

References

Alford SC, Ding Y, Simmen T, Campbell RE (2012) Dimerization-dependent green and yellow fluorescent proteins. ACS Synth Biol 1:569–575

Anderson RG, Brown MS, Goldstein JL (1977) Role of the coated endocytic vesicle in the uptake of receptor-bound low density lipoprotein in human fibroblasts. Cell 10:351–364

Barone V, Randazzo D, Del Re V, Sorrentino V, Rossi D (2015) Organization of junctional sarcoplasmic reticulum proteins in skeletal muscle fibers. J Muscle Res Cell Motil 36:501–515

Bernhard W, Haguenau F, Gautier A, Oberling C (1952) Submicroscopical structure of cytoplasmic basophils in the liver, pancreas and salivary gland; study of ultrafine slices by electron microscope. Z Zellforsch Mikrosk Anat 37:281–300

Bhuvanendran S, Salka K, Rainey K, Sreetama SC, Williams E, Leeker M, Prasad V, Boyd J, Patterson GH, Jaiswal JK et al (2014) Superresolution imaging of human cytomegalovirus vMIA localization in sub-mitochondrial compartments. Virus 6:1612–1636

Bonifacino JS (2014) Vesicular transport earns a Nobel. Trends Cell Biol 24:3–5

Booth DM, Enyedi B, Geiszt M, Varnai P, Hajnoczky G (2016) Redox nanodomains are induced by and

control calcium signaling at the ER-mitochondrial interface. Mol Cell 63:240–248
Brdiczka D, Bucheler K, Kottke M, Adams V, Nalam VK (1990) Characterization and metabolic function of mitochondrial contact sites. Biochim Biophys Acta 1018:234–238
Calcraft PJ, Ruas M, Pan Z, Cheng X, Arredouani A, Hao X, Tang J, Rietdorf K, Teboul L, Chuang KT et al (2009) NAADP mobilizes calcium from acidic organelles through two-pore channels. Nature 459:596–600
Cardenas C, Miller RA, Smith I, Bui T, Molgo J, Muller M, Vais H, Cheung KH, Yang J, Parker I et al (2010) Essential regulation of cell bioenergetics by constitutive $InsP_3$ receptor Ca^{2+} transfer to mitochondria. Cell 142:270–283
Cherradi N, Rossier MF, Vallotton MB, Capponi AM (1996) Calcium stimulates intramitochondrial cholesterol transfer in bovine adrenal glomerulosa cells. J Biol Chem 271:25971–25975
Chu BB, Liao YC, Qi W, Xie C, Du X, Wang J, Yang H, Miao HH, Li BL, Song BL (2015) Cholesterol transport through lysosome-peroxisome membrane contacts. Cell 161:291–306
Colberg-Poley AM, Patterson GH, Salka K, Bhuvanendran S, Yang D, Jaiswal JK (2015) Superresolution imaging of viral protein trafficking. Med Microbiol Immunol 204:449–460
Colombini M, Blachly-Dyson E, Forte M (1996) VDAC, a channel in the outer mitochondrial membrane. Ion Channels 4:169–202
Csordas G, Varnai P, Golenar T, Roy S, Purkins G, Schneider TG, Balla T, Hajnoczky G (2010) Imaging interorganelle contacts and local calcium dynamics at the ER-mitochondrial interface. Mol Cell 39:121–132
de Wit H, Walter AM, Milosevic I, Gulyas-Kovacs A, Riedel D, Sorensen JB, Verhage M (2009) Synaptotagmin-1 docks secretory vesicles to syntaxin-1/SNAP-25 acceptor complexes. Cell 138:935–946
Du X, Kumar J, Ferguson C, Schulz TA, Ong YS, Hong W, Prinz WA, Parton RG, Brown AJ, Yang H (2011) A role for oxysterol-binding protein-related protein 5 in endosomal cholesterol trafficking. J Cell Biol 192:121–135
Eisner V, Csordas G, Hajnoczky G (2013) Interactions between sarco-endoplasmic reticulum and mitochondria in cardiac and skeletal muscle – pivotal roles in Ca^{2+} and reactive oxygen species signaling. J Cell Sci 126:2965–2978
Eisner V, Lenaers G, Hajnoczky G (2014) Mitochondrial fusion is frequent in skeletal muscle and supports excitation-contraction coupling. J Cell Biol 205:179–195
English AR, Voeltz GK (2013) Endoplasmic reticulum structure and interconnections with other organelles. Cold Spring Harb Perspect Biol 5:a013227
Fernandez-Busnadiego R, Saheki Y, De Camilli P (2015) Three-dimensional architecture of extended synaptotagmin-mediated endoplasmic reticulum-plasma membrane contact sites. Proc Natl Acad Sci U S A 112:E2004–E2013
Friedman JR, Lackner LL, West M, DiBenedetto JR, Nunnari J, Voeltz GK (2011) ER tubules mark sites of mitochondrial division. Science 334:358–362
Fritz S, Rapaport D, Klanner E, Neupert W, Westermann B (2001) Connection of the mitochondrial outer and inner membranes by Fzo1 is critical for organellar fusion. J Cell Biol 152:683–692
Gatta AT, Levine TP (2017) Piecing together the patchwork of contact sites. Trends Cell Biol 27(3):214–229
Geppert M, Goda Y, Hammer RE, Li C, Rosahl TW, Stevens CF, Südhof TC (1994) Synaptotagmin I: a major Ca^{2+} sensor for transmitter release at a central synapse. Cell 79:717–727
Gerasimenko JV, Charlesworth RM, Sherwood MW, Ferdek PE, Mikoshiba K, Parrington J, Petersen OH, Gerasimenko OV (2015) Both RyRs and TPCs are required for NAADP-induced intracellular Ca^{2+} release. Cell Calcium 58:237–245
Giacomello M, Pellegrini L (2016) The coming of age of the mitochondria-ER contact: a matter of thickness. Cell Death Differ 23:1417–1427
Giordano F, Saheki Y, Idevall-Hagren O, Colombo SF, Pirruccello M, Milosevic I, Gracheva EO, Bagriantsev SN, Borgese N, De Camilli P (2013) $PI(4,5)P_2$-dependent and Ca^{2+}-regulated ER-PM interactions mediated by the extended synaptotagmins. Cell 153:1494–1509
Goldstein JL, Brown MS (2015) A century of cholesterol and coronaries: from plaques to genes to statins. Cell 161:161–172
Goldstein JL, Anderson RG, Brown MS (1979) Coated pits, coated vesicles, and receptor-mediated endocytosis. Nature 279:679–685
Goldstein JL, DeBose-Boyd RA, Brown MS (2006) Protein sensors for membrane sterols. Cell 124:35–46
Grigoriev I, Gouveia SM, van der Vaart B, Demmers J, Smyth JT, Honnappa S, Splinter D, Steinmetz MO, Putney JW Jr, Hoogenraad CC et al (2008) STIM1 is a MT-plus-end-tracking protein involved in remodeling of the ER. Curr Biol 18:177–182
Hackenbrock CR (1968) Chemical and physical fixation of isolated mitochondria in low-energy and high-energy states. Proc Natl Acad Sci U S A 61:598–605
Harner M, Korner C, Walther D, Mokranjac D, Kaesmacher J, Welsch U, Griffith J, Mann M, Reggiori F, Neupert W (2011) The mitochondrial contact site complex, a determinant of mitochondrial architecture. EMBO J 30:4356–4370
Hayashi T, Su TP (2003) Sigma-1 receptors (sigma (1) binding sites) form raft-like microdomains and target lipid droplets on the endoplasmic reticulum: roles in endoplasmic reticulum lipid compartmentalization and export. J Pharmacol Exp Ther 306:718–725
Helle SC, Kanfer G, Kolar K, Lang A, Michel AH, Kornmann B (2013) Organization and function of membrane contact sites. Biochim Biophys Acta 1833:2526–2541

Herdman C, Moss T (2016) Extended-Synaptotagmins (E-Syts); the extended story. Pharmacol Res 107:48–56

Holtzman E, Novikoff AB, Villaverde H (1967) Lysosomes and GERL in normal and chromatolytic neurons of the rat ganglion nodosum. J Cell Biol 33:419–435

Honscher C, Ungermann C (2014) A close-up view of membrane contact sites between the endoplasmic reticulum and the endolysosomal system: from yeast to man. Crit Rev Biochem Mol Biol 49:262–268

Jacobs JL, Coyne CB (2013) Mechanisms of MAVS regulation at the mitochondrial membrane. J Mol Biol 425:5009–5019

Kim TH, Zhao Y, Ding WX, Shin JN, He X, Seo YW, Chen J, Rabinowich H, Amoscato AA, Yin XM (2004) Bid-cardiolipin interaction at mitochondrial contact site contributes to mitochondrial cristae reorganization and cytochrome C release. Mol Biol Cell 15:3061–3072

Kim BJ, Park KH, Yim CY, Takasawa S, Okamoto H, Im MJ, Kim UH (2008) Generation of nicotinic acid adenine dinucleotide phosphate and cyclic ADP-ribose by glucagon-like peptide-1 evokes Ca^{2+} signal that is essential for insulin secretion in mouse pancreatic islets. Diabetes 57:868–878

Kornmann B, Walter P (2010) ERMES-mediated ER-mitochondria contacts: molecular hubs for the regulation of mitochondrial biology. J Cell Sci 123:1389–1393

Kottke M, Adam V, Riesinger I, Bremm G, Bosch W, Brdiczka D, Sandri G, Panfili E (1988) Mitochondrial boundary membrane contact sites in brain: points of hexokinase and creatine kinase location, and control of Ca^{2+} transport. Biochim Biophys Acta 935:87–102

Krols M, van Isterdael G, Asselbergh B, Kremer A, Lippens S, Timmerman V, Janssens S (2016) Mitochondria-associated membranes as hubs for neurodegeneration. Acta Neuropathol 131:505–523

Levental I, Lingwood D, Grzybek M, Coskun U, Simons K (2010) Palmitoylation regulates raft affinity for the majority of integral raft proteins. Proc Natl Acad Sci U S A 107:22050–22054

Levine T (2004) Short-range intracellular trafficking of small molecules across endoplasmic reticulum junctions. Trends Cell Biol 14:483–490

Levine TP, Patel S (2016) Signalling at membrane contact sites: two membranes come together to handle second messengers. Curr Opin Cell Biol 39:77–83

Li Y, Ge M, Ciani L, Kuriakose G, Westover EJ, Dura M, Covey DF, Freed JH, Maxfield FR, Lytton J et al (2004) Enrichment of endoplasmic reticulum with cholesterol inhibits sarcoplasmic-endoplasmic reticulum calcium ATPase-2b activity in parallel with increased order of membrane lipids: implications for depletion of endoplasmic reticulum calcium stores and apoptosis in cholesterol-loaded macrophages. J Biol Chem 279:37030–37039

Liou J, Kim ML, Heo WD, Jones JT, Myers JW, Ferrell JE Jr, Meyer T (2005) STIM is a Ca^{2+} sensor essential for Ca^{2+}-store-depletion-triggered Ca^{2+} influx. Curr Biol 15:1235–1241

Luik RM, Wu MM, Buchanan J, Lewis RS (2006) The elementary unit of store-operated Ca^{2+} entry: local activation of CRAC channels by STIM1 at ER-plasma membrane junctions. J Cell Biol 174:815–825

Lynes EM, Simmen T (2011) Urban planning of the endoplasmic reticulum (ER): how diverse mechanisms segregate the many functions of the ER. Biochim Biophys Acta 1813:1893–1905

Lynes EM, Bui M, Yap MC, Benson MD, Schneider B, Ellgaard L, Berthiaume LG, Simmen T (2012) Palmitoylated TMX and calnexin target to the mitochondria-associated membrane. EMBO J 31:457–470

Lynes EM, Raturi A, Shenkman M, Ortiz Sandoval C, Yap MC, Wu J, Janowicz A, Myhill N, Benson MD, Campbell RE et al (2013) Palmitoylation is the switch that assigns calnexin to quality control or ER Ca^{2+} signaling. J Cell Sci 126:3893–3903

Marchi S, Giorgi C, Oparka M, Duszynski J, Wieckowski MR, Pinton P (2014) Oncogenic and oncosuppressive signal transduction at mitochondria-associated endoplasmic reticulum membranes. Mol Cell Oncol 1: e956469

Maxfield FR, Wustner D (2002) Intracellular cholesterol transport. J Clin Invest 110:891–898

McMurray WC, Dawson RM (1969) Phospholipid exchange reactions within the liver cell. Biochem J 112:91–108

Mercer JC, Dehaven WI, Smyth JT, Wedel B, Boyles RR, Bird GS, Putney JW Jr (2006) Large store-operated calcium selective currents due to co-expression of Orai1 or Orai2 with the intracellular calcium sensor, Stim1. J Biol Chem 281:24979–24990

Mikoshiba K (2007) The IP_3 receptor/Ca^{2+} channel and its cellular function. Biochem Soc Symp 74:9–22

Missiroli S, Bonora M, Patergnani S, Poletti F, Perrone M, Gafa R, Magri E, Raimondi A, Lanza G, Tacchetti C, Kroemer G, Pandolfi PP, Pinton P, Giorgi C (2016) PML at mitochondria-associated membranes is critical for the repression of autophagy and cancer development. Cell Rep 16:2415–2427

Munoz P, Rosemblatt M, Testar X, Palacin M, Zorzano A (1995) Isolation and characterization of distinct domains of sarcolemma and T-tubules from rat skeletal muscle. Biochem J 307:273–280

Nalefski EA, Falke JJ (1996) The C2 domain calcium-binding motif: structural and functional diversity. Protein Sci 5:2375–2390

Neuspiel M, Schauss AC, Braschi E, Zunino R, Rippstein P, Rachubinski RA, Andrade-Navarro MA, McBride HM (2008) Cargo-selected transport from the mitochondria to peroxisomes is mediated by vesicular carriers. Curr Biol 18:102–108

Nixon-Abell J, Obara CJ, Weigel AV, Li D, Legant WR, Xu CS, Pasolli HA, Harvey K, Hess HF, Betzig E et al (2016) Increased spatiotemporal resolution reveals highly dynamic dense tubular matrices in the peripheral ER. Science 354:aaf3928

Novick P, Schekman R (1979) Secretion and cell-surface growth are blocked in a temperature-sensitive mutant of *Saccharomyces cerevisiae*. Proc Natl Acad Sci U S A 76:1858–1862

Novick P, Field C, Schekman R (1980) Identification of 23 complementation groups required for post-translational events in the yeast secretory pathway. Cell 21:205–215

Novikoff AB (1976) The endoplasmic reticulum: a cytochemist's view (a review). Proc Natl Acad Sci U S A 73:2781–2787

Palade G (1975) Intracellular aspects of the process of protein synthesis. Science 189:347–358

Park Y, Seo JB, Fraind A, Perez-Lara A, Yavuz H, Han K, Jung SR, Kattan I, Walla PJ, Choi M et al (2015) Synaptotagmin-1 binds to PIP_2-containing membrane but not to SNAREs at physiological ionic strength. Nat Struct Mol Biol 22:815–823

Parrington J, Lear P, Hachem A (2015) Calcium signals regulated by NAADP and two-pore channels-their role in development, differentiation and cancer. Int J Dev Biol 59:341–535

Patergnani S, Suski JM, Agnoletto C, Bononi A, Bonora M, De Marchi E, Giorgi C, Marchi S, Missiroli S, Poletti F et al (2011) Calcium signaling around mitochondria associated membranes (MAMs). Cell Commun Signal 9:19

Peinelt C, Vig M, Koomoa DL, Beck A, Nadler MJ, Koblan-Huberson M, Lis A, Fleig A, Penner R, Kinet JP (2006) Amplification of CRAC current by STIM1 and CRACM1 (Orai1). Nat Cell Biol 8:771–773

Perez-Lara A, Jahn R (2015) Extended synaptotagmins (E-Syts): architecture and dynamics of membrane contact sites revealed. Proc Natl Acad Sci U S A 112:4837–4838

Perin MS, Fried VA, Mignery GA, Jahn R, Südhof TC (1990) Phospholipid binding by a synaptic vesicle protein homologous to the regulatory region of protein kinase C. Nature 345:260–263

Perkins G, Renken C, Martone ME, Young SJ, Ellisman M, Frey T (1997) Electron tomography of neuronal mitochondria: three-dimensional structure and organization of cristae and membrane contacts. J Struct Biol 119:260–272

Phillips MJ, Voeltz GK (2016) Structure and function of ER membrane contact sites with other organelles. Nat Rev Mol Cell Biol 17:69–82

Phimister AJ, Lango J, Lee EH, Ernst-Russell MA, Takeshima H, Ma J, Allen PD, Pessah IN (2007) Conformation-dependent stability of junctophilin 1 (JP1) and ryanodine receptor type 1 (RyR1) channel complex is mediated by their hyper-reactive thiols. J Biol Chem 282:8667–8677

Pichler H, Gaigg B, Hrastnik C, Achleitner G, Kohlwein SD, Zellnig G, Perktold A, Daum G (2001) A subfraction of the yeast endoplasmic reticulum associates with the plasma membrane and has a high capacity to synthesize lipids. Eur J Biochem 268:2351–2361

Pike LJ (2003) Lipid rafts: bringing order to chaos. J Lipid Res 44:655–667

Pon L, Moll T, Vestweber D, Marshallsay B, Schatz G (1989) Protein import into mitochondria: ATP-dependent protein translocation activity in a submitochondrial fraction enriched in membrane contact sites and specific proteins. J Cell Biol 109:2603–2616

Porter KR, Palade GE (1957) Studies on the endoplasmic reticulum. III Its form and distribution in striated muscle cells J Biophys Biochem Cytol 3:269–300

Prakriya M, Feske S, Gwack Y, Srikanth S, Rao A, Hogan PG (2006) Orai1 is an essential pore subunit of the CRAC channel. Nature 443:230–233

Quintana A, Pasche M, Junker C, Al-Ansary D, Rieger H, Kummerow C, Nunez L, Villalobos C, Meraner P, Becherer U et al (2011) Calcium microdomains at the immunological synapse: how ORAI channels, mitochondria and calcium pumps generate local calcium signals for efficient T-cell activation. EMBO J 30:3895–3912

Radhakrishnan A, Goldstein JL, McDonald JG, Brown MS (2008) Switch-like control of SREBP-2 transport triggered by small changes in ER cholesterol: a delicate balance. Cell Metab 8:512–521

Raturi A, Gutierrez T, Ortiz-Sandoval C, Ruangkittisakul A, Herrera-Cruz MS, Rockley JP, Gesson K, Ourdev D, Lou PH, Lucchinetti E et al (2016) TMX1 determines cancer cell metabolism as a thiol-based modulator of ER-mitochondria Ca^{2+} flux. J Cell Biol 214:433–444

Reinhart MP, Billheimer JT, Faust JR, Gaylor JL (1987) Subcellular localization of the enzymes of cholesterol biosynthesis and metabolism in rat liver. J Biol Chem 262:9649–9655

Reinisch KM, De Camilli P (2016) SMP-domain proteins at membrane contact sites: structure and function. Biochim Biophys Acta 1861:924–927

Rizzuto R, Brini M, Murgia M, Pozzan T (1993) Microdomains with high Ca^{2+} close to IP_3-sensitive channels that are sensed by neighboring mitochondria. Science 262:744–747

Rothman JE, Fine RE (1980) Coated vesicles transport newly synthesized membrane glycoproteins from endoplasmic reticulum to plasma membrane in two successive stages. Proc Natl Acad Sci U S A 77:780–784

Rothman JE, Bursztyn-Pettegrew H, Fine RE (1980) Transport of the membrane glycoprotein of vesicular stomatitis virus to the cell surface in two stages by clathrin-coated vesicles. J Cell Biol 86:162–171

Rowland AA, Chitwood PJ, Phillips MJ, Voeltz GK (2014) ER contact sites define the position and timing of endosome fission. Cell 159:1027–1041

Rudolf R, Mongillo M, Magalhaes PJ, Pozzan T (2004) In vivo monitoring of Ca^{2+} uptake into mitochondria of mouse skeletal muscle during contraction. J Cell Biol 166:527–536

Rusinol AE, Cui Z, Chen MH, Vance JE (1994) A unique mitochondria-associated membrane fraction from rat liver has a high capacity for lipid synthesis and contains pre-Golgi secretory proteins including nascent lipoproteins. J Biol Chem 269:27494–27502

Rutter GA, Pinton P (2014) Mitochondria-associated endoplasmic reticulum membranes in insulin signaling. Diabetes 63:3163–3165

Sala-Vila A, Navarro-Lerida I, Sanchez-Alvarez M, Bosch M, Calvo C, Lopez JA, Calvo E, Ferguson C, Giacomello M, Serafini A et al (2016) Interplay between hepatic mitochondria-associated membranes, lipid metabolism and caveolin-1 in mice. Sci Rep 6:27351

Sauner MT, Levy M (1971) Study of the transfer of phospholipids from the endoplasmic reticulum to the outer and inner mitochondrial membranes. J Lipid Res 12:71–75

Schauder CM, Wu X, Saheki Y, Narayanaswamy P, Torta F, Wenk MR, De Camilli P, Reinisch KM (2014) Structure of a lipid-bound extended synaptotagmin indicates a role in lipid transfer. Nature 510:552–555

Schrader M (2001) Tubulo-reticular clusters of peroxisomes in living COS-7 cells: dynamic behavior and association with lipid droplets. J Histochem Cytochem 49:1421–1429

Schultz ML, Krus KL, Lieberman AP (2016) Lysosome and endoplasmic reticulum quality control pathways in Niemann-Pick type C disease. Brain Res 1649:181–188

Schwaiger M, Herzog V, Neupert W (1987) Characterization of translocation contact sites involved in the import of mitochondrial proteins. J Cell Biol 105:235–246

Shiao YJ, Lupo G, Vance JE (1995) Evidence that phosphatidylserine is imported into mitochondria via a mitochondria-associated membrane and that the majority of mitochondrial phosphatidylethanolamine is derived from decarboxylation of phosphatidylserine. J Biol Chem 270:11190–11198

Shoshan-Barmatz V, Zalk R, Gincel D, Vardi N (2004) Subcellular localization of VDAC in mitochondria and ER in the cerebellum. Biochim Biophys Acta 1657:105–114

Simbeni R, Pon L, Zinser E, Paltauf F, Daum G (1991) Mitochondrial membrane contact sites of yeast. Characterization of lipid components and possible involvement in intramitochondrial translocation of phospholipids. J Biol Chem 266:10047–10049

Singh R, Kaushik S, Wang Y, Xiang Y, Novak I, Komatsu M, Tanaka K, Cuervo AM, Czaja MJ (2009) Autophagy regulates lipid metabolism. Nature 458:1131–1145

Smith JJ, Aitchison JD (2013) Peroxisomes take shape. Nat Rev Mol Cell Biol 14:803–817

Söllner T, Whiteheart SW, Brunner M, Erdjument-Bromage H, Geromanos S, Tempst P, Rothman JE (1993) SNAP receptors implicated in vesicle targeting and fusion. Nature 362:318–324

Stathopulos PB, Zheng L, Li GY, Plevin MJ, Ikura M (2008) Structural and mechanistic insights into STIM1-mediated initiation of store-operated calcium entry. Cell 135:110–122

Takeshima H, Komazaki S, Nishi M, Iino M, Kangawa K (2000) Junctophilins: a novel family of junctional membrane complex proteins. Mol Cell 6:11–22

Thomson M (2003) Does cholesterol use the mitochondrial contact site as a conduit to the steroidogenic pathway? BioEssays 25:252–258

Tubbs E, Theurey P, Vial G, Bendridi N, Bravard A, Chauvin MA, Ji-Cao J, Zoulim F, Bartosch B, Ovize M et al (2014) Mitochondria-associated endoplasmic reticulum membrane (MAM) integrity is required for insulin signaling and is implicated in hepatic insulin resistance. Diabetes 63:3279–3294

van den Bogaart G, Meyenberg K, Risselada HJ, Amin H, Willig KI, Hubrich BE, Dier M, Hell SW, Grubmuller H, Diederichsen U et al (2011a) Membrane protein sequestering by ionic protein-lipid interactions. Nature 479:552–555

van den Bogaart G, Thutupalli S, Risselada JH, Meyenberg K, Holt M, Riedel D, Diederichsen U, Herminghaus S, Grubmuller H, Jahn R (2011b) Synaptotagmin-1 may be a distance regulator acting upstream of SNARE nucleation. Nat Struct Mol Biol 18:805–812

van Zutphen T, Todde V, de Boer R, Kreim M, Hofbauer HF, Wolinski H, Veenhuis M, van der Klei IJ, Kohlwein SD (2014) Lipid droplet autophagy in the yeast *Saccharomyces cerevisiae*. Mol Biol Cell 25:290–301

Vance JE (1990) Phospholipid synthesis in a membrane fraction associated with mitochondria. J Biol Chem 265:7248–7256

Vance JE (1991) Newly made phosphatidylserine and phosphatidylethanolamine are preferentially translocated between rat liver mitochondria and endoplasmic reticulum. J Biol Chem 266:89–97

Vance JE (2015) Phospholipid synthesis and transport in mammalian cells. Traffic 16:1–18

Varnai P, Toth B, Toth DJ, Hunyady L, Balla T (2007) Visualization and manipulation of plasma membrane-endoplasmic reticulum contact sites indicates the presence of additional molecular components within the STIM1-Orai1 complex. J Biol Chem 282:29678–29690

Voeltz GK, Prinz WA, Shibata Y, Rist JM, Rapoport TA (2006) A class of membrane proteins shaping the tubular endoplasmic reticulum. Cell 124:573–586

Wang HJ, Guay G, Pogan L, Sauve R, Nabi IR (2000) Calcium regulates the association between

mitochondria and a smooth subdomain of the endoplasmic reticulum. J Cell Biol 150:1489–1498
Weaver D, Eisner V, Liu X, Varnai P, Hunyady L, Gross A, Hajnoczky G (2014) Distribution and apoptotic function of outer membrane proteins depend on mitochondrial fusion. Mol Cell 54:870–878
Westermann B (2015) The mitochondria-plasma membrane contact site. Curr Opin Cell Biol 35:1–6
Wideman JG, Gawryluk RM, Gray MW, Dacks JB (2013) The ancient and widespread nature of the ER-mitochondria encounter structure. Mol Biol Evol 30:2044–2049
Wilfling F, Haas JT, Walther TC, Farese RV Jr (2014) Lipid droplet biogenesis. Curr Opin Cell Biol 29:39–45
Wirtz KW, Zilversmit DB (1968) Exchange of phospholipids between liver mitochondria and microsomes in vitro. J Biol Chem 243:3596–3602
Wozniak MJ, Bola B, Brownhill K, Yang YC, Levakova V, Allan VJ (2009) Role of kinesin-1 and cytoplasmic dynein in endoplasmic reticulum movement in VERO cells. J Cell Sci 122:1979–1989
Wu MM, Buchanan J, Luik RM, Lewis RS (2006) Ca^{2+} store depletion causes STIM1 to accumulate in ER regions closely associated with the plasma membrane. J Cell Biol 174:803–813
Xu P, Lu J, Li Z, Yu X, Chen L, Xu T (2006) Aggregation of STIM1 underneath the plasma membrane induces clustering of Orai1. Biochem Biophys Res Commun 350:969–976
Züchner S, Mersiyanova IV, Muglia M, Bissar-Tadmouri-N, Rochelle J, Dadali EL, Zappia M, Nelis E, Patitucci A, Senderek J et al (2004) Mutations in the mitochondrial GTPase mitofusin 2 cause Charcot-Marie-Tooth neuropathy type 2A. Nat Genet 36:449–451

2 Over Six Decades of Discovery and Characterization of the Architecture at Mitochondria-Associated Membranes (MAMs)

Maria Sol Herrera-Cruz and Thomas Simmen

Abstract

The discovery of proteins regulating ER-mitochondria tethering including phosphofurin acidic cluster sorting protein 2 (PACS-2) and mitofusin-2 has pushed contact sites between the endoplasmic reticulum (ER) and mitochondria into the spotlight of cell biology. While the field is developing rapidly and controversies have come and gone multiple times during its history, it is sometimes overlooked that significant research has been done decades ago with the original discovery of these structures in the 1950s and the first characterization of their function (and coining of the term mitochondria-associated membrane, MAM) in 1990. Today, an ever-increasing array of proteins localize to the MAM fraction of the endoplasmic reticulum (ER) to regulate the interaction of this organelle with mitochondria. These mitochondria-ER contacts, sometimes referred to as MERCs, regulate a multitude of biological functions, including lipid metabolism, Ca^{2+} signaling, bioenergetics, inflammation, autophagy, mitochondrial structure, and apoptosis.

Keywords

Mitochondria-associated membrane • Lipids • Calcium • Endoplasmic reticulum • Mitochondria

2.1 Introduction

In the early 1980s, most cell biologists dismissed the idea that the endoplasmic reticulum and mitochondria make physical contacts as an artifact (Katz et al. 1983). There was little indication that these contacts exist outside of obscure model systems such as teleosts or onions (Bracker and Grove 1971; Copeland and Dalton 1959; Morre et al. 1971). This explains why Jean Vance's biochemical description of the mitochondria-associated membrane (MAM, originally called fraction X) from rat liver in 1990 only received relatively modest numbers

M.S. Herrera-Cruz • T. Simmen (✉)
Faculty of Medicine and Dentistry, Department of Cell Biology, University of Alberta, Edmonton, AB, Canada
e-mail: Thomas.Simmen@ualberta.ca

M. Tagaya, T. Simmen (eds.), *Organelle Contact Sites*, Advances in Experimental Medicine and Biology 997, DOI 10.1007/978-981-10-4567-7_2

of citations in the beginning (Vance 1990). However, today this publication is seen as the seminal work that described ER-mitochondria membrane contact sites (MCS) for the first time as essential structures needed for the activities of phospholipid biosynthetic enzymes. Vance's work demonstrated that MAMs co-fractionate with mitochondria and that they house enzymes synthesizing phosphatidylserine (PS), phosphatidylethanolamine (PE), and phosphatidylcholine (PC) (Vance 1991). Others extended these discoveries to demonstrate that PS can translocate between the ER and mitochondria directly (Voelker 1989). Today, the term MAM is often used when referring to ER subdomains, which are biochemically distinct from the remainder of the ER, but which remain attached to mitochondria as mitochondria-ER contacts (MERCs) (Sood et al. 2014).

Curiously, many cell biologists are unaware of much earlier studies of electron microscopy (EM) images from the 1950s that identified ER-mitochondria contacts. It appears safe to say that these contacts had been seen for the first time on electron micrographs of rat liver by Wilhelm Bernhard in 1952 (Bernhard et al. 1952) and 1956 (Bernhard and Rouiller 1956a, b). His studies also led to the insight that ER-mitochondria contacts depend upon feeding and fasting, a remarkable finding that was solidified only during the past 5 years (Theurey and Rieusset 2017). Following these pioneering studies, liver MERCs were also seen by others (Fawcett 1955) and were also discovered in mouse lung cells (Karrer 1956), *Tetrahymena* ciliates (Franke and Kartenbeck 1971), fungi (Bracker and Grove 1971), and some types of plants (Morre et al. 1971). Compared to this ease of detecting ER-mitochondria contacts via EM, biochemical experiments were more difficult to perform and, the co-fractionation of the ER with mitochondria was often interpreted as a contamination problem (Lever and Chappell 1958; Siekevitz 1963). A first step forward was achieved with the insight that the biochemical ER moiety can be separated into two pools, each with distinct levels of association with mitochondria (Shore and Tata 1977). Further experiments showed that the portion of the ER, which remains attached to mitochondria, is biochemically distinct from the bulk ER (Meier et al. 1981) and is mostly of the smooth subtype (Pickett et al. 1980).

Further research has shown that ER-mitochondria contacts result when the ER approaches mitochondria to a distance of about 20 nm on 20% of the mitochondrial surface (Rizzuto et al. 1998). An important recent insight is that the extent and length of these contacts depends on cellular conditions such as ER stress (Bravo et al. 2011; Csordas et al. 2006). Physical, proteinaceous tethers mediate MAM contacts, since proteinases can detach the ER from mitochondria (Achleitner et al. 1999; Csordas et al. 2006), and artificial tethers can boost ER-mitochondria ion flux (Csordas et al. 2006). The past 5 years have seen the discovery of a novel, unexpected role of proteinaceous components of the MAM in mitochondrial dynamics. From their location at ER-mitochondria contacts, human dynamin-related protein 1 (Drp1) (Friedman et al. 2011) and mitofusins determine mitochondrial fission and fusion (Klecker et al. 2014).

Today, we can look back to well over three decades of research on ER-mitochondria contacts (Fig. 2.1). These have led to the insight that the MAM accommodates the exchange of lipids, ions, and second messengers (Levine and Patel 2016), houses the machinery for mitochondrial fission (Friedman et al. 2011), serves as a point of origin for autophagosome formation (Hamasaki et al. 2013), and makes connections to signaling pathways and the cytoskeleton.

2.2 The Study of Lipid Metabolism Discovered Functional Significance of Mitochondria-ER Contacts (MERCs)

The fundamental requirement for ER-mitochondria interaction in phospholipid metabolism has been discovered around 1970 through the observation that liver-derived mitochondria themselves are unable to synthesize PS, but rather must rely on import of this lipid from the ER (McMurray and Dawson 1969; Sauner and Levy 1971; Wirtz and Zilversmit

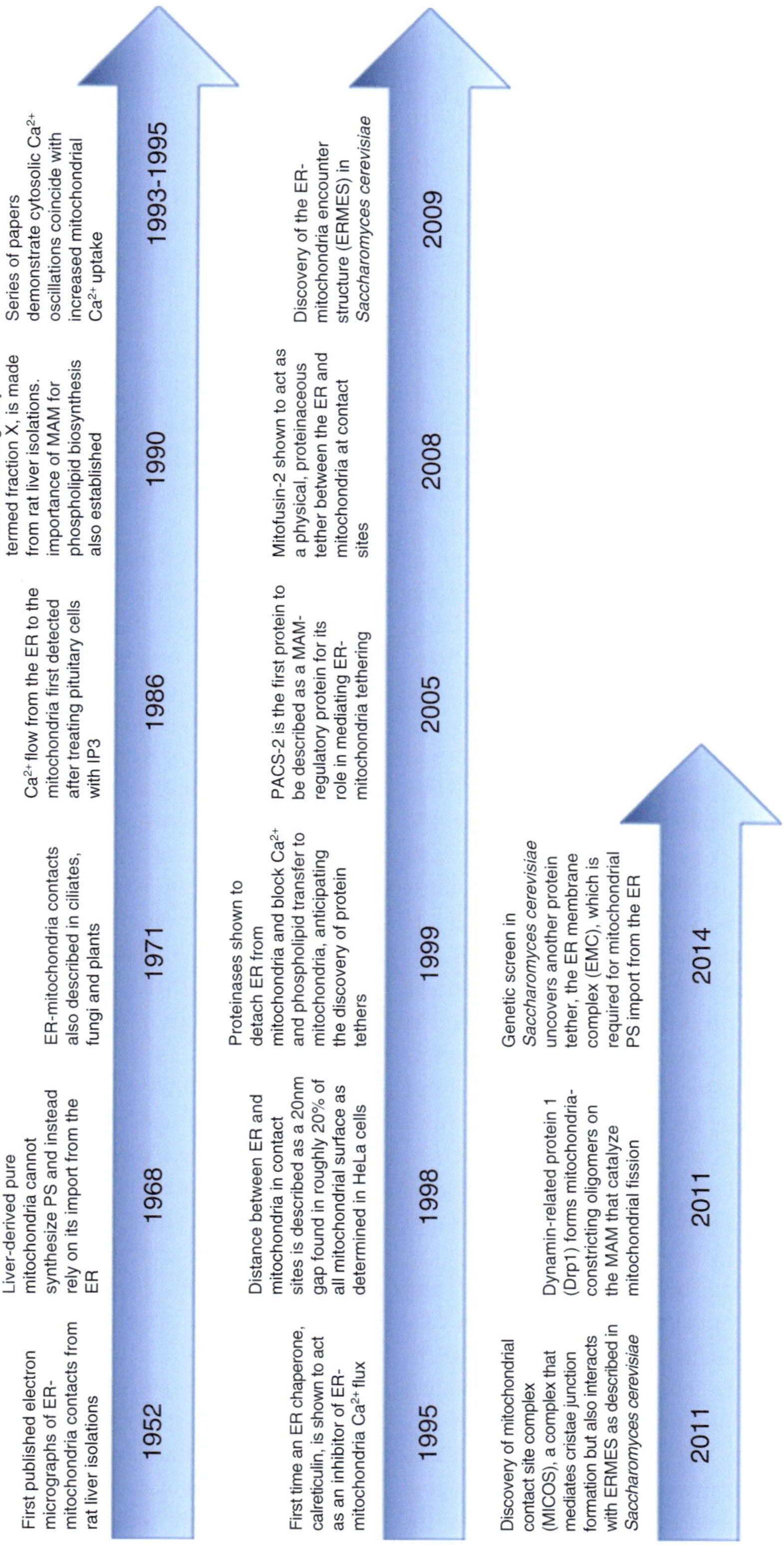

Fig. 2.1 A timeline of the key discoveries in the field of mitochondria-associated membrane (MAM) from 1952 to 2016

1968). While initial hypotheses proposed a cytoplasmic transfer mechanism based on proteinaceous shuttles between the ER and mitochondria (Dennis and Kennedy 1972), it became clear later that PS shuttles directly over to mitochondria, where it is decarboxylated to yield PE (Voelker 1989, 1990). Interestingly, this transfer does not use a vesicular transport step (Shiao and Vance 1995; Vance 1991; Voelker 1989). PS transfer likely occurs on a triple contact site where ER membranes and mitochondrial inner and outer membranes meet and house enzymes that produce PE (Ardail et al. 1991, 1993). While the mechanism for PE transfer to the ER is still unknown, the enzymes catalyzing PE methylation to yield PC localize to the MAM (Vance 1991), as determined via biochemical and EM techniques (Rusinol et al. 1994). Interestingly, MAM sorting signals in the form of cryptic mitochondrial targeting sequences have been detected in some of these enzymes, including acyl-CoA:diacylglycerol acyltransferase 2 (Stone et al. 2009).

An interesting observation is that ER-mitochondria membrane apposition (Flis and Daum 2013) and PS transfer from the ER to mitochondria require ATP in liver cells (Tatsuta et al. 2014). This finding is in contrast to what is known in yeast, where PS and PE shuttle between the two organelles (Gnamusch et al. 1992) without the need for ATP (Achleitner et al. 1999; Simbeni et al. 1993). Ubiquitination of MAM proteins could regulate this lipid shuttling, as shown by the requirement of the yeast SCF ubiquitin ligase Met30p for PS transfer from the MAM to mitochondria (Schumacher et al. 2002) and of the human ubiquitin ligase MITOL for ER-mitochondria apposition (Sugiura et al. 2013).

Lipid transfer between the ER and mitochondria is critical for mitochondria metabolism, albeit for unclear reasons (Vance 2015). For instance, PE deficiency in mammalian mitochondria impairs mitochondrial bioenergetics (Tasseva et al. 2013). Another example is cardiolipin, synthesized from phosphatidic acid (PA) within mitochondria, which is essential for mitochondrial oxidative phosphorylation, apoptosis, mitochondrial protein import, and mitochondrial membrane dynamics (Claypool and Koehler 2012). Similarly, the intracellular PC transfer protein StarD7, which controls both ER and mitochondria structure and function (Flores-Martin et al. 2016). However, further research is needed to fully explain these tight links between mitochondria metabolism and ER-mitochondria lipid transfer.

2.3 Mitochondria-ER Contacts (MERCs) Accumulate Sterols and May Orchestrate the Trafficking of Sterol Towards Mitochondria

Besides PA production, the ER also houses sterol synthesis (Porter 2015). Although not abundant within mitochondria, the sterols cholesterol and ergosterol are important for mitochondrial structure and function in mammalian cells (Lucken-Ardjomande et al. 2008) and yeast (Altmann and Westermann 2005). Research performed by several labs using human cells has identified the MAM as a cholesterol-rich membrane marked with the sterol-interacting protein caveolin (Bosch et al. 2011a, b; Sano et al. 2009). Moreover, the MAM remarkably exhibits lipid raft-like properties (Area-Gomez et al. 2012; Fujimoto et al. 2012, Hayashi and Fujimoto 2010, Williamson et al. 2011). Importantly, caveolin appears to be central to lipid-related functions of the MAM, since it helps to enrich intracellular lipid and sterol metabolism-related processes to the MAM (Sala-Vila et al. 2016). Consistent with these findings, methyl-β-cyclodextrin, an agent that depletes membrane cholesterol, disrupts the MAM platform (Ciarlo et al. 2010), leading to disrupted mitochondrial bioenergetics (Ziolkowski et al. 2010).

This suggests that sterols might use MAM contacts to enter mitochondria. Consistent with this hypothesis, mitochondria use the steroidogenic acute regulatory (StAR) protein D1 (STARD1), which binds free cholesterol in the cytoplasm and then shuttles it toward the MAM in a PKA-dependent manner (Miller 2013). Here, sterols bind to the voltage-dependent anion channel (VDAC), a component of mitochondrial Ca^{2+} import and known MAM protein (Szabadkai et al. 2006), which imports them to mitochondria (Bose et al. 2002) and distributes them within the organelle (Campbell and Chan 2007). Another example is yeast Lam6p/Ltc1p, a StAR-domain-containing protein, which partially localizes to ER-mitochondria contacts (Gatta et al. 2015). Lam6p/Ltc1p interacts with the mitochondrial Tom70/71 complex and might tether the two organelles independently from other molecular tethers (Elbaz-Alon et al. 2015; Murley et al. 2015).

Another candidate mechanism for mitochondrial sterol import is based on a family of proteins originally thought to sense cytosolic sterols, the oxysterol-binding protein (OSBP)-related proteins (ORPs) (Dawson et al. 1989; Taylor et al. 1984). Their 16 human and 7 yeast family members (called oxysterol homology or Osh proteins) frequently localize to MCS (Du et al. 2015). In order to achieve sterol translocation, ORPs translocate from the ER and exchange sterols produced at the ER for phosphatidylinositol 4-phosphate at their cognate target membrane (Drin et al. 2016). ORP5 and ORP8 are of particular interest. These two family members partially target to the MAM, but also interact with a suspected MAM tether and mitochondrial membrane protein, PTPIP51. While it is currently unknown whether these two ORPs mediate lipid or cholesterol trafficking to mitochondria, their knockdown alters biochemical activities, as well as membrane dynamics of mitochondria (Galmes et al. 2016). Together, despite a proven role of sterols in ER-mitochondria contact formation, a number of questions remain on their role for the MAM. Further research will have to investigate the relationship between plasma membrane caveolin and MAM caveolin, as well as the role of ORP proteins in E tethering.

2.4 MAM Proteins Mediate Calcium Signaling: Mitochondria-ER Contacts (MERCs) Emerge as an Intracellular Signaling Hotspot

While the functional significance of the MAM is historically best tied to lipid metabolism, this structure has broader significance for intracellular ion flux, since four mitochondrial dehydrogenases require Ca^{2+} ions to allow for respiration and the Krebs cycle (Denton 2009). Another functional connection exists between Ca^{2+} and the permeability transition pore that controls the mitochondrial protein content, as well as the mitochondrial membrane potential (Hurst et al. 2017). Early in mitochondrial research, Ca^{2+} had been identified as a factor that opens the pore (Hunter et al. 1976). In all cases, the ER acts as a point of origin for Ca^{2+} (Bravo-Sagua et al. 2013).

First evidence for this connection was seen with the observation that the ER and mitochondria cooperate to regulate cytosolic Ca^{2+} amounts (Becker et al. 1980). Moreover, an early study suggested in 1981 that the isolation of ER-mitochondria membrane contacts drastically decreases in the presence of the Ca^{2+} chelator EDTA (Meier et al. 1981). A flux of Ca^{2+} ions between the ER and mitochondria was described for the first time in 1986 using a Ca^{2+}-selective mini-electrode in digitonin-permeabilized rat pituitary GH_3 cells as a discrete pool of ER Ca^{2+} that is transferred to mitochondria upon incubation of a mixed organellar preparation with inositol 1,4,5-trisphosphate (IP_3) (Biden et al. 1986). Following these early discoveries, apoptosis research showed that mitochondria receive Ca^{2+} from the ER and that this abnormally high flux of the ion leads to apoptosis (Baffy et al. 1993).

However, the true magnitude of these discoveries and speculations became clear only through further research by the laboratories of Tullio Pozzan and Andrew Thomas. Their studies demonstrated that cytosolic Ca^{2+} oscillations coincided with mitochondrial Ca^{2+} increases in live human hepatocytes and HeLa cells, using the Rhod-2 dye or mitochondrially targeted aequorin, a Ca^{2+}-activated photoprotein (Hajnoczky et al. 1995; Rizzuto et al. 1993, 1994). Moreover, this Ca^{2+} transfer required the formation of highly concentrated microdomains used by mitochondrial Ca^{2+} uniporters (MCUs) to import Ca^{2+} (Rizzuto et al. 1993). This observation was refined by the Nabi lab, demonstrating that concentrations below 100 nM favor dissociation of contacts, while concentrations above 1 μM favor contact formation (Wang et al. 2000). Using green fluorescent protein (GFP)-tagged fusion proteins targeted to the ER and mitochondria, Rosario Rizzuto then pushed the field further by showing that mitochondria form an interconnected network that shows close apposition to the ER on 5–20% of its surface (Rizzuto et al. 1998). These points of apposition were also proposed as the location of Ca^{2+} microdomains in the same study. Here, mitochondrial ATP fuels ER Ca^{2+} uptake (Landolfi et al. 1998) upon formation of Ca^{2+} microdomains (Hajnoczky et al. 1999). The complex relationship between the two organelles regarding Ca^{2+} leads to the establishment of Ca^{2+} signal transmission between the ER and mitochondria similar to what occurs at a synapse, as formulated by the Hajnóczky lab in 1999 upon their analysis of experiments with fluorescent, mitochondria-targeted probes (Csordas et al. 1999).

ER-mitochondria Ca^{2+} crosstalk boosts mitochondrial bioenergetics (Cardenas et al. 2010), and studies by Murphy and Finkel using MCU knockout (ko) cells show that in vivo, mitochondrial Ca^{2+} import is critical for mitochondria peak performance (Pan et al. 2013). The MCU molecular discovery in 2011 jointly by the Rizzuto and Mootha labs (Baughman et al. 2011; De Stefani et al. 2011) showed that this protein forms a complex with mitochondrial calcium uptake 1 and 2 (MICU1 and MICU2) (Mallilankaraman et al. 2012; Perocchi et al. 2010; Plovanich et al. 2013). These two regulatory proteins likely perform vital functions in the control of ER-mitochondria Ca^{2+} flux as well, although this has not yet been fully investigated (Antony et al. 2016). Interestingly, *S. cerevisiae* mitochondria have lost the MCU during evolution, suggesting that *S. cerevisiae* has a drastically reduced and simplified Ca^{2+} machinery (Carafoli and Lehninger 1971). Moreover, yeast stores the majority of its Ca^{2+} inside vacuoles, not the ER (Belde et al. 1993). More detailed characterizations need to demonstrate whether this organism is able to faithfully reproduce ER-mitochondria Ca^{2+} flux as observed in metazoan cells (Plattner and Verkhratsky 2016).

On the ER side of the MAM, the investigation of cytosolic Ca^{2+} waves by Lechleiter and Camacho identified additional players. These excitatory events are triggered via the activation of IP_3Rs and depend on the presence of the Ca^{2+} pump sarco/endoplasmic reticulum Ca^{2+}-ATPase (SERCA). The properties of Ca^{2+} waves suggest they are generated from a cross-reaction with oscillating cytosolic [Ca^{2+}] and Ca^{2+}-regulated IP_3Rs (Camacho and Lechleiter 1993). Moreover, Ca^{2+} waves are tightly connected to mitochondria metabolism: while mitochondrial Ca^{2+} import and mitochondrial ROS can boost Ca^{2+} waves (Walsh et al. 2009), a block of mitochondrial respiration reduces their amplitude and frequency (Jouaville et al. 1995). The mechanistic basis for this connection has recently been elucidated in a series of elegant experiments by the Hajnóczky lab. ER-mitochondria Ca^{2+} transfer leads to the release of cristae-stored ROS that subsequently diffuse over to the ER and converge on the MAM, where they chemically modify Ca^{2+}-handling proteins (Booth et al. 2016). One group of these are inositol 1,4,5-trisphosphate receptors (IP_3Rs), which localize to special subdomains of the ER and possibly concentrate in portions that have the tendency to appose to mitochondrial membranes (Takei et al. 1994). ROS modifications result in the sensitization of IP_3Rs, thus strengthening cytosolic Ca^{2+} waves (Walsh et al. 2009). Further research needs to address the role of ER-derived ROS in this process (Appenzeller-Herzog et al. 2016). Nevertheless, these results mechanistically tie ER-mitochondria Ca^{2+} flux to ER energy demand (Bravo et al. 2011). In further research, ER chaperones, including calreticulin (Camacho and Lechleiter 1995), calnexin (Roderick et al. 2000), and ERp57 (Li and Camacho 2004) were identified as inhibitors of cytosolic Ca^{2+} waves and MAM Ca^{2+} crosstalk (Appenzeller-Herzog and Simmen 2016). Consistent with these observations, calnexin reduces the amount of Ca^{2+} that can reach mitochondria, but increases the amount of Ca^{2+} that can be taken up to the ER (Lynes et al. 2013). This function of calnexin depends on its localization to the MAM and is dependent on its phosphorylation and palmitoylation state (Lynes et al. 2012; Myhill et al. 2008). Opposing these inhibitors of ER-mitochondria Ca^{2+} flux, the ER oxidoreductase TMX1/TXNDC1 interacts in a thiol-dependent manner with SERCA2b to reduce ER Ca^{2+} import and thus augment MAM Ca^{2+} flux (Raturi et al. 2016). By doing so, TMX1 concomitantly increases mitochondrial respiration and ATP production, while mitochondria metabolism is blocked in its absence, resulting in a Warburg-style acceleration of cancer cell growth (Raturi et al. 2016).

The functional connection between ER protein folding and ER-mitochondria Ca^{2+} flux is not limited to the regulation of SERCA activity, but also extends to modulatory interactions between ER chaperones and IP_3Rs. For instance, the ER oxidoreductases ERp44 and Ero1α inhibit or activate IP_3Rs, respectively, and, by doing so, ER-mitochondria Ca^{2+} flux (Higo et al. 2005; Li et al. 2009). Again, at least in the case of Ero1α, this property is tied to its localization to the MAM (Anelli et al. 2011; Gilady et al. 2010).

Proteins of the Bcl2 family mediate further regulation of ER Ca^{2+} release and uptake (Vervliet et al. 2016). For example, Bcl-x_L interacts with IP_3R1 to activate this channel when $[IP_3]$ is low, but to inactivate it when $[IP_3]$ is high (Yang et al. 2016). The BH3-only protein Bid and pro-apoptotic Bax inhibit this regulatory interaction (White et al. 2005). The regulatory mechanism extends to the mitochondrial membrane, where Bcl-x_L blocks mitochondrial Ca^{2+} import via VDAC1 (Monaco et al. 2015). In contrast, the interaction between Bcl2 and the IP_3R largely serves to block pro-apoptotic Ca^{2+} transfer (Chen et al. 2004; Hanson et al. 2008; Monaco et al. 2012; Rong et al. 2009), similar to Bax inhibitor-1 (BI-1) (Sano et al. 2012). A similar inhibitory interaction exists between Bcl2 and the ryanodine receptor (Vervliet et al. 2014).

A prototype regulatory protein of ER-mitochondria Ca^{2+} flux is the MAM-localized sigma-1 receptor (SIGMAR1), a transmembrane protein that can act as a chaperone for IP_3Rs under conditions of low ER $[Ca^{2+}]$ (Hayashi and Su 2007). Through this activity and its localization to detergent-resistant membranes at the MAM (Hayashi and Fujimoto 2010), a property shared with mitofusin-2 and TMX1 (Lynes et al. 2012), SIGMAR1 controls ER-mitochondria Ca^{2+} flux and mitochondria lipid metabolism (Bernard-Marissal et al. 2015; Marriott et al. 2012).

The sum of these chaperone/oxidoreductase interactions with Ca^{2+}-handling proteins allows ER protein folding to increase ER-mitochondria Ca^{2+} flux under conditions of the unfolded protein response (UPR). This boosts ATP import into the ER, which can then alleviate folding problems or trigger apoptosis, should these problems persist (Bravo et al. 2011). There is therefore a tight link between ER protein folding and redox and the MAM that can telegraph the state of the ER to mitochondria (Simmen et al. 2010).

2.5 Proteins Mediating Formation of the MAM

Proteinases have been shown to untether the ER and mitochondria (Achleitner et al. 1999; Csordas et al. 2006) and thus disrupt Ca^{2+} flux as well as phospholipid import into mitochondria. This insight led to the postulate that proteinaceous tethers rather than interorganellar membrane fusion events mediate ER-mitochondria apposition (de Brito and Scorrano 2008). The past decade has seen the discovery of MAM-forming and MAM-regulating proteins that tether ER and mitochondria membranes and regulate the composition of contact sites (Fig. 2.2).

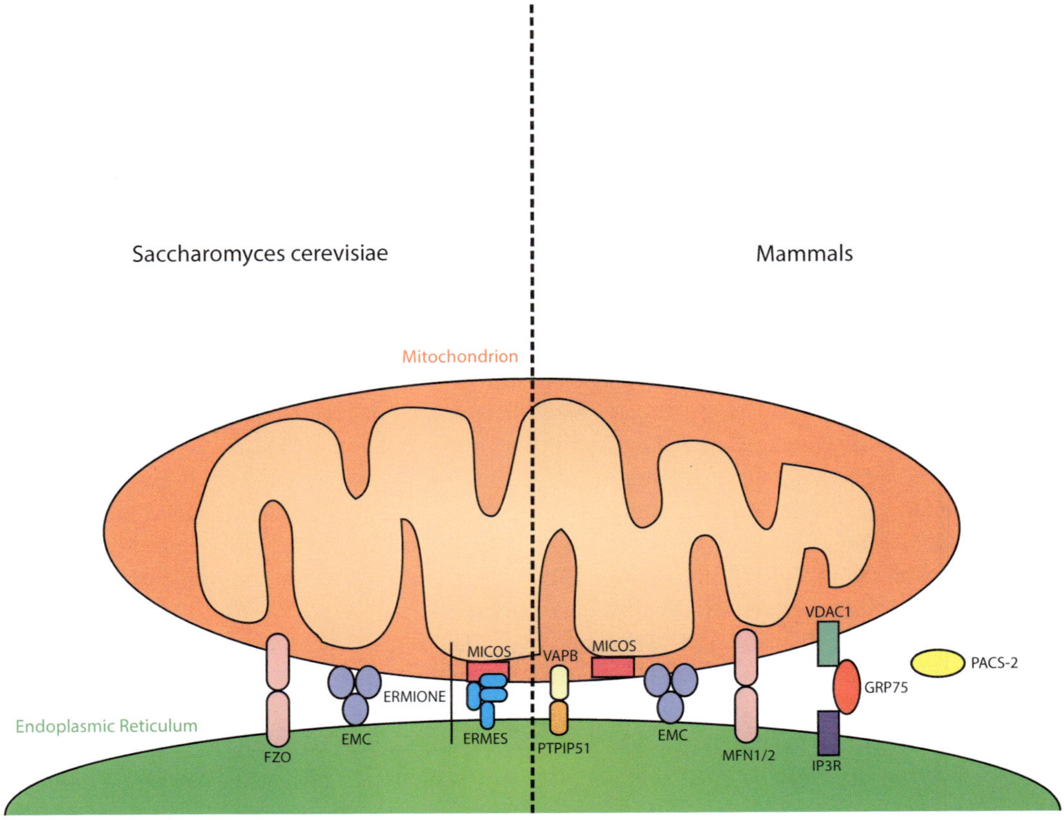

Fig. 2.2 Representation of ER-mitochondria tethers identified in yeast and mammalian models (See text for details)

A genetic screen in *S. cerevisiae* expressing an artificial ER-mitochondria tether led Benoit Kornmann to discover the ER-mitochondria encounter structure (ERMES) (Kornmann et al. 2009). This heterotetrameric protein complex is composed of Mmm1p, an ER transmembrane protein; Mdm12p, a cytosolic protein; as well as Mdm34p and Mdm10p, which target to the mitochondrial outer membrane (Lang et al. 2015). Interestingly, the ERMES subunits Mdm12p, Mdm34p, and Mmm1p all contain synaptotagmin-like lipid-binding protein domains (SMP), known modules that mediate MCS formation (Reinisch and De Camilli 2016). The Ca^{2+}-binding, rho-like GTPase Gem1p regulates the number and activity of ERMES contact points between the ER and mitochondria (Kornmann et al. 2011). Questions arose in the past whether ERMES is dispensable for PS transport from the ER to mitochondria (Nguyen et al. 2012), but this appears now firmly established (Kojima et al. 2016). Curiously, despite this strong evidence for an essential role of ERMES in yeast, this complex is not conserved in metazoa (Wideman et al. 2013). In yeast cells, the mitochondrial contact site complex (MICOS) genetically interacts with ERMES (Hoppins et al. 2011). MICOS comprises the six proteins Mic60p, Mic10p, Mic12p, Mic19p, Mic26p, and Mic27p (Harner et al. 2011; Hoppins et al. 2011; von der Malsburg et al. 2011) and is critical for the formation of cristae junctions within mitochondria (Ardail et al. 1990; Hackenbrock 1968). Since MICOS is not restricted to yeast (Kozjak-Pavlovic 2017), its role in aligning the MAM with cristae junctions could be conserved.

More recently, Prinz and Loewen identified another potential tether structure in a yeast genetic screen using *CHO2* that negatively interacts with *PSD1*. This screen identified the ER membrane protein complex (EMC), composed of Emc1p, Emc2p, Emc3p, Emc4p, Emc5p, and Emc6p, all of which are necessary for mitochondrial PS import from the ER (Lahiri et al. 2014). Subsequent work has identified the additional components Emc7, Emc8, Emc9, and Emc10 in human cells, but these proteins are absent in yeast (Wideman 2015). The EMC complex is far better conserved throughout evolution than ERMES. However, it is currently unclear how important its role in tethering is compared to its other roles in ER protein folding, since EMC catalyzes the assembly of proteins that span the membrane multiple times as well (Satoh et al. 2015).

In mammalian models, defects in MAM Ca^{2+} flux or the optical apposition between the ER and mitochondria have been used frequently to identify tethers. One example of a human protein implicated in ER-mitochondria tethering is phosphofurin acidic cluster sorting protein 2 (PACS-2), a cytosolic protein that regulates MAM formation (Simmen et al. 2005). It is unclear whether PACS-2 is a MAM tether per se, but its knockdown or knockout nevertheless uncouples the ER from mitochondria, as seen by light and electron microscopy (Simmen et al. 2005). Consistent with a role in MAM formation that is required for Ca^{2+} moving from the ER towards mitochondria, PACS-2 knockdown blocks Ca^{2+}-mediated apoptosis progression (Myhill et al. 2008). The functions of PACS-2 at the MAM are under the control of its serine 437 residue, where Akt-mediated phosphorylation enables PACS-2 to maintain the MAM (Betz et al. 2013). Interestingly, Akt is downstream of mammalian target of rapamycin complex 2 (mTORC2) that localizes to the MAM itself and activates PACS-2 via Akt (Betz et al. 2013). PACS-2 interacts with calnexin, another known regulator of ER-mitochondria Ca^{2+} flux (Lynes et al. 2013; Myhill et al. 2008), and prevents BAP31 processing mediated by caspase-8 (Simmen et al. 2005). Potentially, this interaction is the true basis of PACS-2-mediated ER-mitochondria tethering, since BAP31 directly interacts with the mitochondrial Drp1 docking protein Fis1 (Iwasawa et al. 2011). This complex becomes associated with procaspase-8 under conditions of cell stress. Under that condition, active caspase-8 triggers the formation of the BAP31 p20 fragment, an activator of mitochondrial fission (Breckenridge et al. 2003). Importantly, PACS-2 knockdown leads to this latter event by itself and thus detaches the ER from mitochondria (Simmen et al. 2005).

A second mechanism that influences ER-mitochondria tethering was discovered by the lab of Luca Scorrano in the form of mitofusin-2 (de Brito and Scorrano 2008). When examining the structures of mitofusin-2 ko cells, light and electron microscopy showed a detachment of the ER and mitochondria in mouse embryonic fibroblasts, neurons (Schneeberger et al. 2013), and cardiomyocytes (Chen et al. 2012). Mitofusin-2 ko cells also show inhibited Ca^{2+} transfer from the ER to mitochondria upon Ca^{2+} release from the ER. Consistent with this observation, ER stress does not result in efficient apoptosis onset in mitofusin-2 ko cells (Munoz et al. 2013). Further demonstrating the role for the formation of ER-mitochondria contacts, mitofusin-2 ko cells show a decrease in mitochondrial respiration capacity (Mourier et al. 2015). A characteristic shared between PACS-2 and mitofusin-2 is that their knockdown results in ER stress (Sebastian et al. 2012; Simmen et al. 2005), but silencing of the kinase PERK can reverse the mitofusin-2 ko phenotype partially (Munoz et al. 2013). Some doubts on the role of mitofusin-2 for the MAM arose with the observation that some ER tubules can be found in closer contact with mitochondria in mitofusin-2 ko cells (Cosson et al. 2012; Filadi et al. 2015). However, recent results suggest that this effect could be due to culture conditions (Naon et al. 2016), consistent with cell stress arising in mitofusin-2 ko cells quicker than in wild-type cells (Bucha et al. 2015).

Additional tethers include a protein complex between IP_3Rs, VDAC, and the outer mitochondrial membrane chaperone Grp75 (Szabadkai et al. 2006) as well as a novel tethering complex formed between the ER vesicle-associated membrane protein- associated protein B (VAPB) and PTPIP51 (De Vos et al. 2012). TDP-43 can disrupt the VAPB-PTPIP51 tethers and results in disrupted cellular Ca^{2+} homeostasis (Stoica et al. 2014). Fused in sarcoma (FUS) is another inhibitor of these tethers and similarly results in clean detachment of the ER from mitochondria (Stoica et al. 2016), suggesting that the VAPB-PTPIP51 tethers are currently among the best characterized.

2.6 The MAM Regulates Mitochondrial Dynamics

The past 5 years have seen the discovery of novel, unexpected connections between the MAM and mitochondrial dynamics. These are based on both the mitochondrial fission and fusion machinery. Connections with mitochondrial fusion could be based on the equilibrium of mitofusins mediating mitochondrial fusion and mitofusins localizing to the MAM. Here, the human ubiquitin ligase MITOL regulates MAM formation and activates mitofusin-2 by ubiquitinating its mitochondrial moiety. Apparently, this function is restricted to MAM-specific functions of mitofusin-2 and does not apply to the role of mitofusin-2 in mitochondrial fusion (Sugiura et al. 2013). A similar case is the E3 ubiquitin ligase Gp78. This ER protein normally targets mitofusin-2 for degradation, but its function appears to predominantly determine the ratio between rough ER (rER) and smooth ER (sER)-mitochondria contacts (Li et al. 2015). This finding suggests that mitofusin-2 could also serve to keep the rER apart from mitochondria while promoting the formation of contacts between mitochondria and the sER (Wang et al. 2015).

The best-studied connection between the MAM and mitochondrial dynamics involves mitochondrial fission. At mitochondria-ER contacts, human dynamin-related protein 1 (Drp1) forms mitochondria-constricting oligomers that dock to the mitochondrial proteins Fis1, Mff, and MiD49/MiD51 (Friedman et al. 2011), all localized to the MAM prior to mitochondrial fission (Murley et al. 2013). The actin-remodeling proteins WAVE-1 (Sung et al. 2008) and ER-associated inverted formin 2 (INF2) (Korobova et al. 2013) form a construction site onto which Drp1 can oligomerize (Hatch et al. 2014). A number of mitochondrial protein kinase A (PKA)-anchoring proteins (AKAPs) like the MAM trafficking GTPase Rab32 control this mechanism (Bui et al. 2010), since PKA acts as an on/off switch for Drp1 (Cribbs and Strack 2007). Another candidate is AKAP121, which can target to either ER or mitochondria (Huang et al. 1999), from where it fine-tunes the availability of Drp1 (Kim et al. 2011).

2.7 There Is More to the MAM Than MERCs the Eye

The new millennium has seen a further expansion of the palette of functions of ER-mitochondria contacts with the discoveries that important cellular signaling mechanisms use mitochondria-ER contacts as a home base for their activities. These include the formation of inflammasomes, which form here upon production of ROS within the ER or mitochondria (Zhou et al. 2011); PKA, which operates from this structure upon binding to the small GTPase Rab32 in HeLa cells (Bui et al. 2010); or the mammalian target of rapamycin complex 2 (mTORC2) that can be co-isolated with the MAM, from where it controls Akt phosphorylation of IP_3Rs and hexokinase 2 (Betz et al. 2013).

A connection between the MAM and autophagy has been made in the context of syntaxin-17, a master regulator of autophagosome formation and progression (Itakura et al. 2012). The significance of syntaxin-17 for the MAM is demonstrated by its localization to mitochondrial rafts and its requirement for Drp1 localization and activity (Arasaki et al. 2015). These activities of syntaxin-17 depend on its inhibitory action on the MAM AKAP Rab32 that uses Drp1 as an effector (Ortiz-Sandoval et al. 2014).

Autophagy is a highly conserved process that involves the formation of a double-membrane vacuole to capture cytosolic contents and deliver them to lysosomes for degradation (Ashrafi and Schwarz 2013; Lamb et al. 2013; Yang and Klionsky 2010). Autophagy is initiated with the formation of an isolation membrane or phagophore, which expands as it engulfs contents and seals to form a double-membrane vacuole known as the autophagosome (Shibutani and Yoshimori 2014). The current paradigm in the field is that the isolation membrane is derived from a core existing membrane and then formed largely de novo with the addition of vesicles and lipids (Klionsky 2007) to yield a platform for autophagosome biogenesis as an Ω-shaped membrane (Axe et al. 2008). Consistent with the importance of the MAM for autophagosome formation, multiple lines of evidence show that mitochondrial and autophagosomal membranes are transiently linked in a mitofusin-2-dependent manner (Hailey et al. 2010). These autophagosomal membranes containing the ganglioside GD3 become enriched with the MAM protein calnexin (Lynes et al. 2012) and then participate in the early autophagic process (Garofalo et al. 2016). Importantly, calnexin serves as a docking site to target the OMM protein FUNDC1 to the MAM, where it interacts with Drp1 (Wu et al. 2016). Although ER autophagic receptors Atg39p, Atg40p (yeast), and FAM134B (human) have been described recently, it is unclear whether they play a role for MAM-initiated autophagy (Khaminets et al. 2015; Mochida et al. 2015). In yeast, ERMES is critical for the progression of mitophagy (Bockler and Westermann 2014), suggesting that proteins mediating or regulating the tethering of the ER to mitochondria could play a similar role.

2.8 Conclusion

As our chapter shows, ER-mitochondria contacts have come a long way from being seen for the first time in the 1950s on EM micrographs by Bernhard and others. Following their first assignment of a biochemical function by Vance in 1990, the field has exploded (Fig. 2.1) and has served as a paradigm for the ongoing characterization of MCS between a multitude of organelles.

Although not the topic of this review, it comes therefore as no surprise that the MAM plays important roles in many diseases. Again, this was interestingly anticipated by the pioneers of the 1950s, who saw that MAM formation changes during feeding (Bernhard and Rouiller 1956b) and also during the progression of cancer (Howatson and Ham 1955). Today, researchers are unraveling the molecular mechanisms behind these early observations (Giorgi et al. 2015; Raturi et al. 2016; Sood et al. 2014; Theurey et al. 2016), showing that the MAM machinery is intricately linked to diabetes (Arruda et al. 2014; Tubbs et al. 2014) and cancer (Bononi et al. 2013). Emerging areas of research show that the MAM is likely also playing a prominent role in infection (Horner et al. 2011), as well as neurodegeneration (Area-Gomez et al. 2012; Paillusson et al. 2016). We therefore predict that the ongoing identification of MAM tethers, MAM targeting mechanisms, and novel MAM signaling mechanisms will lead to important insight into these diseases.

References

Achleitner G, Gaigg B, Krasser A, Kainersdorfer E, Kohlwein SD, Perktold A, Zellnig G, Daum G (1999) Association between the endoplasmic reticulum and mitochondria of yeast facilitates interorganelle transport of phospholipids through membrane contact. Eur J Biochem 264:545–553

Altmann K, Westermann B (2005) Role of essential genes in mitochondrial morphogenesis in *Saccharomyces cerevisiae*. Mol Biol Cell 16:5410–5417

Anelli T, Bergamelli L, Margittai E, Rimessi A, Fagioli C, Malgaroli A, Pinton P, Ripamonti M, Rizzuto R, Sitia R (2011) Ero1α regulates Ca^{2+} fluxes at the endoplasmic reticulum-mitochondria Interface (MAM). Antioxid Redox Signal 16:1077–1087

Antony AN, Paillard M, Moffat C, Juskeviciute E, Correnti J, Bolon B, Rubin E, Csordas G, Seifert EL, Hoek JB et al (2016) MICU1 regulation of mitochondrial C^{a2+} uptake dictates survival and tissue regeneration. Nat Commun 7:10955

Appenzeller-Herzog C, Simmen T (2016) ER-luminal thiol/selenol-mediated regulation of Ca^{2+} signalling. Biochem Soc Trans 44:452–459

Appenzeller-Herzog C, Banhegyi G, Bogeski I, Davies KJ, Delaunay-Moisan A, Forman HJ, Gorlach A, Kietzmann T, Laurindo F, Margittai E et al (2016) Transit of H_2O_2 across the endoplasmic reticulum membrane is not sluggish. Free Radic Biol Med 94: 157–160

Arasaki K, Shimizu H, Mogari H, Nishida N, Hirota N, Furuno A, Kudo Y, Baba M, Baba N, Cheng J et al (2015) A role for the ancient SNARE syntaxin 17 in regulating mitochondrial division. Dev Cell 32: 304–317

Ardail D, Privat JP, Egret-Charlier M, Levrat C, Lerme F, Louisot P (1990) Mitochondrial contact sites. Lipid composition and dynamics J Biol Chem 265: 18797–18802

Ardail D, Lerme F, Louisot P (1991) Involvement of contact sites in phosphatidylserine import into liver mitochondria. J Biol Chem 266:7978–7981

Ardail D, Gasnier F, Lerme F, Simonot C, Louisot P, Gateau-Roesch O (1993) Involvement of mitochondrial contact sites in the subcellular compartmentalization of phospholipid biosynthetic enzymes. J Biol Chem 268:25985–25992

Area-Gomez E, Del Carmen Lara Castillo M, Tambini MD, Guardia-Laguarta C, de Groof AJ, Madra M, Ikenouchi J, Umeda M, Bird TD, Sturley SL et al (2012) Upregulated function of mitochondria-associated ER membranes in Alzheimer disease. EMBO J 31:4106–4123

Arruda AP, Pers BM, Parlakgul G, Guney E, Inouye K, Hotamisligil GS (2014) Chronic enrichment of hepatic endoplasmic reticulum-mitochondria contact leads to mitochondrial dysfunction in obesity. Nat Med 20: 1427–1435

Ashrafi G, Schwarz TL (2013) The pathways of mitophagy for quality control and clearance of mitochondria. Cell Death Differ 20:31–42

Axe EL, Walker SA, Manifava M, Chandra P, Roderick HL, Habermann A, Griffiths G, Ktistakis NT (2008) Autophagosome formation from membrane compartments enriched in phosphatidylinositol 3-phosphate and dynamically connected to the endoplasmic reticulum. J Cell Biol 182:685–701

Baffy G, Miyashita T, Williamson JR, Reed JC (1993) Apoptosis induced by withdrawal of interleukin-3 (IL-3) from an IL-3-dependent hematopoietic cell line is associated with repartitioning of intracellular calcium and is blocked by enforced Bcl-2 oncoprotein production. J Biol Chem 268:6511–6519

Baughman JM, Perocchi F, Girgis HS, Plovanich M, Belcher-Timme CA, Sancak Y, Bao XR, Strittmatter L, Goldberger O, Bogorad RL et al (2011) Integrative genomics identifies MCU as an essential component of the mitochondrial calcium uniporter. Nature 476:341–345

Becker GL, Fiskum G, Lehninger AL (1980) Regulation of free Ca^{2+} by liver mitochondria and endoplasmic reticulum. J Biol Chem 255:9009–9012

Belde PJ, Vossen JH, Borst-Pauwels GW, Theuvenet AP (1993) Inositol 1,4,5-trisphosphate releases Ca^{2+} from vacuolar membrane vesicles of *Saccharomyces cerevisiae*. FEBS Lett 323:113–118

Bernard-Marissal N, Medard JJ, Azzedine H, Chrast R (2015) Dysfunction in endoplasmic reticulum-mitochondria crosstalk underlies SIGMAR1 loss of function mediated motor neuron degeneration. Brain 138:875–890

Bernhard W, Rouiller C (1956a) Close topographical relationship between mitochondria and ergastoplasm of liver cells in a definite phase of cellular activity. J Biophys Biochem Cytol 2:73–78

Bernhard W, Rouiller C (1956b) Microbodies and the problem of mitochondrial regeneration in liver cells. J Biophys Biochem Cytol 2:355–360

Bernhard W, Haguenau F, Gautier A, Oberling C (1952) Submicroscopical structure of cytoplasmic basophils in the liver, pancreas and salivary gland; study of ultrafine slices by electron microscope. Z Zellforsch Mikrosk Anat 37:281–300

Betz C, Stracka D, Prescianotto-Baschong C, Frieden M, Demaurex N, Hall MN (2013) Feature article: mTOR complex 2-Akt signaling at mitochondria-associated endoplasmic reticulum membranes (MAM) regulates mitochondrial physiology. Proc Natl Acad Sci U S A 110:12526–12534

Biden TJ, Wollheim CB, Schlegel W (1986) Inositol 1,4,5-trisphosphate and intracellular Ca^{2+} homeostasis in clonal pituitary cells (GH3). Translocation of Ca^{2+} into mitochondria from a functionally discrete portion of the nonmitochondrial store. J Biol Chem 261: 7223–7229

Bockler S, Westermann B (2014) ER-mitochondria contacts as sites of mitophagosome formation. Autophagy 10:1346–1347

Bononi A, Bonora M, Marchi S, Missiroli S, Poletti F, Giorgi C, Pandolfi PP, Pinton P (2013) Identification of PTEN at the ER and MAMs and its regulation of Ca^{2+} signaling and apoptosis in a protein phosphatase-dependent manner. Cell Death Differ 20:1631–1643

Booth DM, Enyedi B, Geiszt M, Varnai P, Hajnoczky G (2016) Redox nanodomains are induced by and control calcium signaling at the ER-mitochondrial interface. Mol Cell 63:240–248

Bosch M, Mari M, Gross SP, Fernandez-Checa JC, Pol A (2011a) Mitochondrial cholesterol: a connection between caveolin, metabolism, and disease. Traffic 12:1483–1489

Bosch M, Mari M, Herms A, Fernandez A, Fajardo A, Kassan A, Giralt A, Colell A, Balgoma D, Barbero E et al (2011b) Caveolin-1 deficiency causes cholesterol-dependent mitochondrial dysfunction and apoptotic susceptibility. Curr Biol 21:681–686

Bose HS, Lingappa VR, Miller WL (2002) Rapid regulation of steroidogenesis by mitochondrial protein import. Nature 417:87–91

Bracker CE, Grove SN (1971) Continuity between cytoplasmic endomembranes and outer mitochondrial membranes in fungi. Protoplasma 73:15–34

Bravo R, Vicencio JM, Parra V, Troncoso R, Munoz JP, Bui M, Quiroga C, Rodriguez AE, Verdejo HE, Ferreira J et al (2011) Increased ER-mitochondrial coupling promotes mitochondrial respiration and bioenergetics during early phases of ER stress. J Cell Sci 124:2143–2152

Bravo-Sagua R, Rodriguez AE, Kuzmicic J, Gutierrez T, Lopez-Crisosto C, Quiroga C, Diaz-Elizondo J, Chiong M, Gillette TG, Rothermel BA et al (2013) Cell death and survival through the endoplasmic reticulum-mitochondrial axis. Curr Mol Med 13:317–329

Breckenridge DG, Stojanovic M, Marcellus RC, Shore GC (2003) Caspase cleavage product of BAP31 induces mitochondrial fission through endoplasmic reticulum calcium signals, enhancing cytochrome c release to the cytosol. J Cell Biol 160:1115–1127

Bucha S, Mukhopadhyay D, Bhattacharyya NP (2015) Regulation of mitochondrial morphology and cell cycle by microRNA-214 targeting Mitofusin2. Biochem Biophys Res Commun 465:797–802

Bui M, Gilady SY, Fitzsimmons RE, Benson MD, Lynes EM, Gesson K, Alto NM, Strack S, Scott JD, Simmen T (2010) Rab32 modulates apoptosis onset and mitochondria-associated membrane (MAM) properties. J Biol Chem 285:31590–31602

Camacho P, Lechleiter JD (1993) Increased frequency of calcium waves in *Xenopus laevis* oocytes that express a calcium-ATPase. Science 260:226–229

Camacho P, Lechleiter JD (1995) Calreticulin inhibits repetitive intracellular Ca^{2+} waves. Cell 82:765–771

Campbell AM, Chan SH (2007) The voltage dependent anion channel affects mitochondrial cholesterol distribution and function. Arch Biochem Biophys 466:203–210

Carafoli E, Lehninger AL (1971) A survey of the interaction of calcium ions with mitochondria from different tissues and species. Biochem J 122:681–690

Cardenas C, Miller RA, Smith I, Bui T, Molgo J, Muller M, Vais H, Cheung KH, Yang J, Parker I et al (2010) Essential regulation of cell bioenergetics by constitutive $InsP_3$ receptor Ca^{2+} transfer to mitochondria. Cell 142:270–283

Chen R, Valencia I, Zhong F, McColl KS, Roderick HL, Bootman MD, Berridge MJ, Conway SJ, Holmes AB, Mignery GA et al (2004) Bcl-2 functionally interacts with inositol 1,4,5-trisphosphate receptors to regulate calcium release from the ER in response to inositol 1,4,5-trisphosphate. J Cell Biol 166:193–203

Chen Y, Csordas G, Jowdy C, Schneider TG, Csordas N, Wang W, Liu Y, Kohlhaas M, Meiser M, Bergem S et al (2012) Mitofusin 2-containing mitochondrial-reticular microdomains direct rapid cardiomyocyte bioenergetic responses via interorganelle Ca^{2+} crosstalk. Circ Res 111:863–875

Ciarlo L, Manganelli V, Garofalo T, Matarrese P, Tinari A, Misasi R, Malorni W, Sorice M (2010) Association of fission proteins with mitochondrial raft-like domains. Cell Death Differ 17:1047–1058

Claypool SM, Koehler CM (2012) The complexity of cardiolipin in health and disease. Trends Biochem Sci 37:32–41

Copeland DE, Dalton AJ (1959) An association between mitochondria and the endoplasmic reticulum in cells of the pseudobranch gland of a teleost. J Biophys Biochem Cytol 5:393–396

Cosson P, Marchetti A, Ravazzola M, Orci L (2012) Mitofusin-2 independent juxtaposition of endoplasmic reticulum and mitochondria: an ultrastructural study. PLoS One 7:e46293

Cribbs JT, Strack S (2007) Reversible phosphorylation of Drp1 by cyclic AMP-dependent protein kinase and calcineurin regulates mitochondrial fission and cell death. EMBO Rep 8:9399–9344

Csordas G, Thomas AP, Hajnoczky G (1999) Quasi-synaptic calcium signal transmission between endoplasmic reticulum and mitochondria. EMBO J 18:96–108

Csordas G, Renken C, Varnai P, Walter L, Weaver D, Buttle KF, Balla T, Mannella CA, Hajnoczky G (2006) Structural and functional features and significance of the physical linkage between ER and mitochondria. J Cell Biol 174:915–921

Dawson PA, Ridgway ND, Slaughter CA, Brown MS, Goldstein JL (1989) cDNA cloning and expression of oxysterol-binding protein, an oligomer with a potential leucine zipper. J Biol Chem 264:16798–16803

de Brito OM, Scorrano L (2008) Mitofusin 2 tethers endoplasmic reticulum to mitochondria. Nature 456:605–610

De Stefani D, Raffaello A, Teardo E, Szabo I, Rizzuto R (2011) A forty-kilodalton protein of the inner membrane is the mitochondrial calcium uniporter. Nature 476:336–340

De Vos KJ, Morotz GM, Stoica R, Tudor EL, Lau KF, Ackerley S, Warley A, Shaw CE, Miller CC (2012) VAPB interacts with the mitochondrial protein PTPIP51 to regulate calcium homeostasis. Hum Mol Genet 21:1299–1311

Dennis EA, Kennedy EP (1972) Intracellular sites of lipid synthesis and the biogenesis of mitochondria. J Lipid Res 13:263–267

Denton RM (2009) Regulation of mitochondrial dehydrogenases by calcium ions. Biochim Biophys Acta 1787:1309–1316

Drin G, von Filseck JM, Copic A (2016) New molecular mechanisms of inter-organelle lipid transport. Biochem Soc Trans 44:486–492

Du X, Brown AJ, Yang H (2015) Novel mechanisms of intracellular cholesterol transport: oxysterol-binding proteins and membrane contact sites. Curr Opin Cell Biol 35:37–42

Elbaz-Alon Y, Eisenberg-Bord M, Shinder V, Stiller SB, Shimoni E, Wiedemann N, Geiger T, Schuldiner M (2015) Lam6 regulates the extent of contacts between organelles. Cell Rep 12:7–14

Fawcett DW (1955) Observations on the cytology and electron microscopy of hepatic cells. J Natl Cancer Inst 15:1475–1503

Filadi R, Greotti E, Turacchio G, Luini A, Pozzan T, Pizzo P (2015) Mitofusin 2 ablation increases endoplasmic reticulum-mitochondria coupling. Proc Natl Acad Sci U S A 112:E2174–E2181

Flis VV, Daum G (2013) Lipid transport between the endoplasmic reticulum and mitochondria. Cold Spring Harb Perspect Biol 5

Flores-Martin J, Reyna L, Ridano ME, Panzetta-Dutari GM, Genti-Raimondi S (2016) Suppression of StarD7 promotes endoplasmic reticulum stress and induces ROS production. Free Radic Biol Med 99:286–295

Franke WW, Kartenbeck J (1971) Outer mitochondrial membrane continuous with endoplasmic reticulum. Protoplasma 73:35–41

Friedman JR, Lackner LL, West M, DiBenedetto JR, Nunnari J, Voeltz GK (2011) ER tubules mark sites of mitochondrial division. Science 334:358–362

Fujimoto M, Hayashi T, Su TP (2012) The role of cholesterol in the association of endoplasmic reticulum membranes with mitochondria. Biochem Biophys Res Commun 417:635–639

Galmes R, Houcine A, van Vliet AR, Agostinis P, Jackson CL, Giordano F (2016) ORP5/ORP8 localize to endoplasmic reticulum-mitochondria contacts and are involved in mitochondrial function. EMBO Rep 17: 800–810

Garofalo T, Matarrese P, Manganelli V, Marconi M, Tinari A, Gambardella L, Faggioni A, Misasi R, Sorice M, Malorni W (2016) Evidence for the involvement of lipid rafts localized at the ER-mitochondria associated membranes in autophagosome formation. Autophagy 12:917–935

Gatta AT, Wong LH, Sere YY, Calderon-Norena DM, Cockcroft S, Menon AK, Levine TP (2015) A new family of StART domain proteins at membrane contact sites has a role in ER-PM sterol transport. elife 4

Gilady SY, Bui M, Lynes EM, Benson MD, Watts R, Vance JE, Simmen T (2010) Ero1α requires oxidizing and normoxic conditions to localize to the mitochondria-associated membrane (MAM). Cell Stress Chaperones 15:619–629

Giorgi C, Bonora M, Sorrentino G, Missiroli S, Poletti F, Suski JM, Galindo Ramirez F, Rizzuto R, Di Virgilio F, Zito E et al (2015) p53 at the endoplasmic reticulum regulates apoptosis in a Ca^{2+}-dependent manner. Proc Natl Acad Sci U S A 112:1779–1784

Gnamusch E, Kalaus C, Hrastnik C, Paltauf F, Daum G (1992) Transport of phospholipids between subcellular membranes of wild-type yeast cells and of the phosphatidylinositol transfer protein-deficient strain *Saccharomyces cerevisiae* sec 14. Biochim Biophys Acta 1111:120–126

Hackenbrock CR (1968) Chemical and physical fixation of isolated mitochondria in low-energy and high-energy states. Proc Natl Acad Sci U S A 61:598–605

Hailey DW, Rambold AS, Satpute-Krishnan P, Mitra K, Sougrat R, Kim PK, Lippincott-Schwartz J (2010) Mitochondria supply membranes for autophagosome biogenesis during starvation. Cell 141:656–667

Hajnoczky G, Robb-Gaspers LD, Seitz MB, Thomas AP (1995) Decoding of cytosolic calcium oscillations in the mitochondria. Cell 82:415–424

Hajnoczky G, Hager R, Thomas AP (1999) Mitochondria suppress local feedback activation of inositol 1,4, 5-trisphosphate receptors by Ca^{2+}. J Biol Chem 274: 14157–14162

Hamasaki M, Furuta N, Matsuda A, Nezu A, Yamamoto A, Fujita N, Oomori H, Noda T, Haraguchi T, Hiraoka Y et al (2013) Autophagosomes form at ER-mitochondria contact sites. Nature 495: 389–393

Hanson CJ, Bootman MD, Distelhorst CW, Wojcikiewicz RJ, Roderick HL (2008) Bcl-2 suppresses Ca^{2+} release through inositol 1,4,5-trisphosphate receptors and inhibits Ca^{2+} uptake by mitochondria without affecting ER calcium store content. Cell Calcium 44: 324–338

Harner M, Korner C, Walther D, Mokranjac D, Kaesmacher J, Welsch U, Griffith J, Mann M, Reggiori F, Neupert W (2011) The mitochondrial contact site complex, a determinant of mitochondrial architecture. EMBO J 30:4356–4370

Hatch AL, Gurel PS, Higgs HN (2014) Novel roles for actin in mitochondrial fission. J Cell Sci 127: 4549–4560

Hayashi T, Fujimoto M (2010) Detergent-resistant microdomains determine the localization of sigma-1 receptors to the endoplasmic reticulum-mitochondria junction. Mol Pharmacol 77:517–528

Hayashi T, Su TP (2007) Sigma-1 receptor chaperones at the ER-mitochondrion interface regulate Ca^{2+} signaling and cell survival. Cell 131:596–610

Higo T, Hattori M, Nakamura T, Natsume T, Michikawa T, Mikoshiba K (2005) Subtype-specific and ER Lumenal environment-dependent regulation of Inositol 1,4,5-Trisphosphate receptor Type 1 by ERp44. Cell 120:85–98

Hoppins S, Collins SR, Cassidy-Stone A, Hummel E, Devay RM, Lackner LL, Westermann B, Schuldiner M, Weissman JS, Nunnari J (2011) A mitochondrial-focused genetic interaction map reveals a scaffold-like complex required for inner membrane organization in mitochondria. J Cell Biol 195:323–340

Horner SM, Liu HM, Park HS, Briley J, Gale M Jr (2011) Mitochondrial-associated endoplasmic reticulum membranes (MAM) form innate immune synapses and are targeted by hepatitis C virus. Proc Natl Acad Sci U S A 108:14590–14595

Howatson AF, Ham AW (1955) Electron microscope study of sections of two rat liver tumors. Cancer Res 15:62–69

Huang LJ, Wang L, Ma Y, Durick K, Perkins G, Deerinck TJ, Ellisman MH, Taylor SS (1999) NH2-terminal targeting motifs direct dual specificity A-kinase-anchoring protein 1 (D-AKAP1) to either mitochondria or endoplasmic reticulum. J Cell Biol 145: 951–959

Hunter DR, Haworth RA, Southard JH (1976) Relationship between configuration, function, and permeability in calcium-treated mitochondria. J Biol Chem 251: 5069–5077

Hurst S, Hoek J, Sheu SS (2017) Mitochondrial Ca^{2+} and regulation of the permeability transition pore. J Bioenerg Biomembr 49(1):27–47

Itakura E, Kishi-Itakura C, Mizushima N (2012) The hairpin-type tail-anchored SNARE syntaxin 17 targets to autophagosomes for fusion with endosomes/lysosomes. Cell 151:1256–1269

Iwasawa R, Mahul-Mellier AL, Datler C, Pazarentzos E, Grimm S (2011) Fis1 and Bap31 bridge the mitochondria-ER interface to establish a platform for apoptosis induction. EMBO J 30:556–568

Jouaville LS, Ichas F, Holmuhamedov EL, Camacho P, Lechleiter JD (1995) Synchronization of calcium waves by mitochondrial substrates in *Xenopus laevis* oocytes. Nature 377:438–441

Karrer HE (1956) Electron microscopic study of bronchiolar epithelium of normal mouse lung; preliminary report. Exp Cell Res 10:237–241

Katz J, Wals PA, Golden S, Raijman L (1983) Mitochondrial-reticular cytostructure in liver cells. Biochem J 214:795–813

Khaminets A, Heinrich T, Mari M, Grumati P, Huebner AK, Akutsu M, Liebmann L, Stolz A, Nietzsche S, Koch N et al (2015) Regulation of endoplasmic reticulum turnover by selective autophagy. Nature 522: 354–358

Kim H, Scimia MC, Wilkinson D, Trelles RD, Wood MR, Bowtell D, Dillin A, Mercola M, Ronai ZA (2011) Fine-tuning of Drp1/Fis1 availability by AKAP121/Siah2 regulates mitochondrial adaptation to hypoxia. Mol Cell 44:532–544

Klecker T, Bockler S, Westermann B (2014) Making connections: interorganelle contacts orchestrate mitochondrial behavior. Trends Cell Biol 24:537–545

Klionsky DJ (2007) Autophagy: from phenomenology to molecular understanding in less than a decade. Nat Rev Mol Cell Biol 8:931–937

Kojima R, Endo T, Tamura Y (2016) A phospholipid transfer function of ER-mitochondria encounter structure revealed in vitro. Sci Rep 6:30777

Kornmann B, Currie E, Collins SR, Schuldiner M, Nunnari J, Weissman JS, Walter P (2009) An ER-mitochondria tethering complex revealed by a synthetic biology screen. Science 325:477–481

Kornmann B, Osman C, Walter P (2011) The conserved GTPase Gem1 regulates endoplasmic reticulum-mitochondria connections. Proc Natl Acad Sci U S A 108:14151–14156

Korobova F, Ramabhadran V, Higgs HN (2013) An actin-dependent step in mitochondrial fission mediated by the ER-associated formin INF2. Science 339:464–467

Kozjak-Pavlovic V (2017) The MICOS complex of human mitochondria. Cell Tissue Res 367(1):83–93

Lahiri S, Chao JT, Tavassoli S, Wong AK, Choudhary V, Young BP, Loewen CJ, Prinz WA (2014) A conserved endoplasmic reticulum membrane protein complex (EMC) facilitates phospholipid transfer from the ER to mitochondria. PLoS Biol 12:e1001969

Lamb CA, Yoshimori T, Tooze SA (2013) The autophagosome: origins unknown, biogenesis complex. Nat Rev Mol Cell Biol 14:759–774

Landolfi B, Curci S, Debellis L, Pozzan T, Hofer AM (1998) Ca^{2+} homeostasis in the agonist-sensitive internal store: functional interactions between mitochondria and the ER measured in situ in intact cells. J Cell Biol 142:1235–1243

Lang A, John Peter AT, Kornmann B (2015) ER-mitochondria contact sites in yeast: beyond the myths of ERMES. Curr Opin Cell Biol 35:7–12

Lever JD, Chappell JB (1958) Mitochondria isolated from rat brown adipose tissue and liver. J Biophys Biochem Cytol 4:287–290

Levine TP, Patel S (2016) Signalling at membrane contact sites: two membranes come together to handle second messengers. Curr Opin Cell Biol 39:77–83

Li Y, Camacho P (2004) Ca^{2+}-dependent redox modulation of SERCA 2b by ERp57. J Cell Biol 164:35–46

Li G, Mongillo M, Chin KT, Harding H, Ron D, Marks AR, Tabas I (2009) Role of ERO1-α-mediated stimulation of inositol 1,4,5-triphosphate receptor activity in endoplasmic reticulum stress-induced apoptosis. J Cell Biol 186:783–792

Li L, Gao G, Shankar J, Joshi B, Foster LJ, Nabi IR (2015) p38 MAP kinase-dependent phosphorylation of the Gp78 E3 ubiquitin ligase controls ER-mitochondria association and mitochondria motility. Mol Biol Cell 26:3828–3840

Lucken-Ardjomande S, Montessuit S, Martinou JC (2008) Bax activation and stress-induced apoptosis delayed by the accumulation of cholesterol in mitochondrial membranes. Cell Death Differ 15:484–493

Lynes EM, Bui M, Yap MC, Benson MD, Schneider B, Ellgaard L, Berthiaume LG, Simmen T (2012) Palmitoylated TMX and calnexin target to the mitochondria-associated membrane. EMBO J 31: 457–470

Lynes EM, Raturi A, Shenkman M, Ortiz Sandoval C, Yap MC, Wu J, Janowicz A, Myhill N, Benson MD, Campbell RE et al (2013) Palmitoylation is the switch that assigns calnexin to quality control or ER Ca^{2+} signaling. J Cell Sci 126:3893–3903

Mallilankaraman K, Doonan P, Cardenas C, Chandramoorthy HC, Muller M, Miller R, Hoffman NE, Gandhirajan RK, Molgo J, Birnbaum MJ et al (2012) MICU1 is an essential gatekeeper for

MCU-mediated mitochondrial Ca^{2+} uptake that regulates cell survival. Cell 151:630–644
Marriott KS, Prasad M, Thapliyal V, Bose HS (2012) Sigma-1 receptor at the mitochondrial-associated endoplasmic reticulum membrane is responsible for mitochondrial metabolic regulation. J Pharmacol Exp Ther 343:578–586
McMurray WC, Dawson RM (1969) Phospholipid exchange reactions within the liver cell. Biochem J 112:91–108
Meier PJ, Spycher MA, Meyer UA (1981) Isolation and characterization of rough endoplasmic reticulum associated with mitochondria from normal rat liver. Biochim Biophys Acta 646:283–297
Miller WL (2013) Steroid hormone synthesis in mitochondria. Mol Cell Endocrinol 379:62–73
Mochida K, Oikawa Y, Kimura Y, Kirisako H, Hirano H, Ohsumi Y, Nakatogawa H (2015) Receptor-mediated selective autophagy degrades the endoplasmic reticulum and the nucleus. Nature 522:359–362
Monaco G, Decrock E, Akl H, Ponsaerts R, Vervliet T, Luyten T, De Maeyer M, Missiaen L, Distelhorst CW, De Smedt H et al (2012) Selective regulation of IP_3-receptor-mediated Ca^{2+} signaling and apoptosis by the BH4 domain of Bcl-2 versus Bcl-xl. Cell Death Differ 19:295–309
Monaco G, Decrock E, Arbel N, van Vliet AR, La Rovere RM, De Smedt H, Parys JB, Agostinis P, Leybaert L, Shoshan-Barmatz V et al (2015) The BH4 domain of anti-apoptotic Bcl-XL, but not that of the related Bcl-2, limits the voltage-dependent anion channel 1 (VDAC1)-mediated transfer of pro-apoptotic Ca^{2+} signals to mitochondria. J Biol Chem 290:9150–9161
Morre DJ, Merritt WD, Lembi CA (1971) Connections between mitochondria and endoplasmic reticulum in rat liver and onion stem. Protoplasma 73:43–49
Mourier A, Motori E, Brandt T, Lagouge M, Atanassov I, Galinier A, Rappl G, Brodesser S, Hultenby K, Dieterich C et al (2015) Mitofusin 2 is required to maintain mitochondrial coenzyme Q levels. J Cell Biol 208:429–442
Munoz JP, Ivanova S, Sanchez-Wandelmer J, Martinez-Cristobal P, Noguera E, Sancho A, Diaz-Ramos A, Hernandez-Alvarez MI, Sebastian D, Mauvezin C et al (2013) Mfn2 modulates the UPR and mitochondrial function via repression of PERK. EMBO J 32: 2348–2361
Murley A, Lackner LL, Osman C, West M, Voeltz GK, Walter P, Nunnari J (2013) ER-associated mitochondrial division links the distribution of mitochondria and mitochondrial DNA in yeast. elife 2:e00422
Murley A, Sarsam RD, Toulmay A, Yamada J, Prinz WA, Nunnari J (2015) Ltc1 is an ER-localized sterol transporter and a component of ER-mitochondria and ER-vacuole contacts. J Cell Biol 209:539–548
Myhill N, Lynes EM, Nanji JA, Blagoveshchenskaya AD, Fei H, Carmine Simmen K, Cooper TJ, Thomas G, Simmen T (2008) The subcellular distribution of calnexin is mediated by PACS-2. Mol Biol Cell 19: 2777–2788
Naon D, Zaninello M, Giacomello M, Varanita T, Grespi F, Lakshminaranayan S, Serafini A, Semenzato M, Herkenne S, Hernandez-Alvarez MI et al (2016) Critical reappraisal confirms that Mitofusin 2 is an endoplasmic reticulum-mitochondria tether. Proc Natl Acad Sci U S A 113:11249–11254
Nguyen TT, Lewandowska A, Choi JY, Markgraf DF, Junker M, Bilgin M, Ejsing CS, Voelker DR, Rapoport TA, Shaw JM (2012) Gem1 and ERMES do not directly affect phosphatidylserine transport from ER to mitochondria or mitochondrial inheritance. Traffic 13:880–890
Ortiz-Sandoval CG, Hughes SC, Dacks JB, Simmen T (2014) Interaction with the effector dynamin-related protein 1 (Drp1) is an ancient function of Rab32 subfamily proteins. Cell Logist 4:e986399
Paillusson S, Stoica R, Gomez-Suaga P, Lau DH, Mueller S, Miller T, Miller CC (2016) There's something wrong with my MAM; the ER-mitochondria axis and neurodegenerative diseases. Trends Neurosci 39: 146–157
Pan X, Liu J, Nguyen T, Liu C, Sun J, Teng Y, Fergusson MM, Rovira II, Allen M, Springer DA et al (2013) The physiological role of mitochondrial calcium revealed by mice lacking the mitochondrial calcium uniporter. Nat Cell Biol 15:1464–1472
Perocchi F, Gohil VM, Girgis HS, Bao XR, McCombs JE, Palmer AE, Mootha VK (2010) MICU1 encodes a mitochondrial EF hand protein required for Ca^{2+} uptake. Nature 467:291–296
Pickett CB, Montisano D, Eisner D, Cascarano J (1980) The physical association between rat liver mitochondria and rough endoplasmic reticulum. I. Isolation, electron microscopic examination and sedimentation equilibrium centrifugation analyses of rough endoplasmic reticulum-mitochondrial complexes. Exp Cell Res 128:343–352
Plattner H, Verkhratsky A (2016) Inseparable tandem: evolution chooses ATP and Ca^{2+} to control life, death and cellular signalling. Philos Trans R Soc Lond Ser B Biol Sci 371
Plovanich M, Bogorad RL, Sancak Y, Kamer KJ, Strittmatter L, Li AA, Girgis HS, Kuchimanchi S, De Groot J, Speciner L et al (2013) MICU2, a paralog of MICU1, resides within the mitochondrial uniporter complex to regulate calcium handling. PLoS One 8: e55785
Porter TD (2015) Electron transfer pathways in cholesterol synthesis. Lipids 50:927–936
Raturi A, Gutierrez T, Ortiz-Sandoval C, Ruangkittisakul A, Herrera-Cruz MS, Rockley JP, Gesson K, Ourdev D, Lou PH, Lucchinetti E et al (2016) TMX1 determines cancer cell metabolism as a thiol-based modulator of ER-mitochondria Ca^{2+} flux. J Cell Biol 214:433–444
Reinisch KM, De Camilli P (2016) SMP-domain proteins at membrane contact sites: structure and function. Biochim Biophys Acta 1861:924–927
Rizzuto R, Brini M, Murgia M, Pozzan T (1993) Microdomains with high Ca^{2+} close to IP_3-sensitive

channels that are sensed by neighboring mitochondria. Science 262:744–747

Rizzuto R, Bastianutto C, Brini M, Murgia M, Pozzan T (1994) Mitochondrial Ca^{2+} homeostasis in intact cells. J Cell Biol 126:1183–1194

Rizzuto R, Pinton P, Carrington W, Fay FS, Fogarty KE, Lifshitz LM, Tuft RA, Pozzan T (1998) Close contacts with the endoplasmic reticulum as determinants of mitochondrial Ca^{2+} responses. Science 280: 1763–1766

Roderick HL, Lechleiter JD, Camacho P (2000) Cytosolic phosphorylation of calnexin controls intracellular Ca $^{2+}$ oscillations via an interaction with SERCA2b. J Cell Biol 149:1235–1248

Rong YP, Bultynck G, Aromolaran AS, Zhong F, Parys JB, De Smedt H, Mignery GA, Roderick HL, Bootman MD, Distelhorst CW (2009) The BH4 domain of Bcl-2 inhibits ER calcium release and apoptosis by binding the regulatory and coupling domain of the IP3 receptor. Proc Natl Acad Sci U S A 106:14397–14402

Rusinol AE, Cui Z, Chen MH, Vance JE (1994) A unique mitochondria-associated membrane fraction from rat liver has a high capacity for lipid synthesis and contains pre-Golgi secretory proteins including nascent lipoproteins. J Biol Chem 269:27494–27502

Sala-Vila A, Navarro-Lerida I, Sanchez-Alvarez M, Bosch M, Calvo C, Lopez JA, Calvo E, Ferguson C, Giacomello M, Serafini A et al (2016) Interplay between hepatic mitochondria-associated membranes, lipid metabolism and caveolin-1 in mice. Sci Rep 6: 27351

Sano R, Annunziata I, Patterson A, Moshiach S, Gomero E, Opferman J, Forte M, d'Azzo A (2009) GM1-ganglioside accumulation at the mitochondria-associated ER membranes links ER stress to Ca^{2+}-dependent mitochondrial apoptosis. Mol Cell 36: 500–511

Sano R, Hou YC, Hedvat M, Correa RG, Shu CW, Krajewska M, Diaz PW, Tamble CM, Quarato G, Gottlieb RA et al (2012) Endoplasmic reticulum protein BI-1 regulates Ca^{2+}-mediated bioenergetics to promote autophagy. Genes Dev 26:1041–1054

Satoh T, Ohba A, Liu Z, Inagaki T, Satoh AK (2015) dPob/EMC is essential for biosynthesis of rhodopsin and other multi-pass membrane proteins in Drosophila photoreceptors.Elife 4

Sauner MT, Levy M (1971) Study of the transfer of phospholipids from the endoplasmic reticulum to the outer and inner mitochondrial membranes. J Lipid Res 12:71–75

Schneeberger M, Dietrich MO, Sebastian D, Imbernon M, Castano C, Garcia A, Esteban Y, Gonzalez-Franquesa A, Rodriguez IC, Bortolozzi A et al (2013) Mitofusin 2 in POMC neurons connects ER stress with leptin resistance and energy imbalance. Cell 155:172–187

Schumacher MM, Choi JY, Voelker DR (2002) Phosphatidylserine transport to the mitochondria is regulated by ubiquitination. J Biol Chem 277: 51033–51042

Sebastian D, Hernandez-Alvarez MI, Segales J, Sorianello E, Munoz JP, Sala D, Waget A, Liesa M, Paz JC, Gopalacharyulu P et al (2012) Mitofusin 2 (Mfn2) links mitochondrial and endoplasmic reticulum function with insulin signaling and is essential for normal glucose homeostasis. Proc Natl Acad Sci U S A 109:5523–5528

Shiao YJ, Vance JE (1995) Evidence for an ethanolamine cycle: differential recycling of the ethanolamine moiety of phosphatidylethanolamine derived from phosphatidylserine and ethanolamine. Biochem J 310: 673–679

Shibutani ST, Yoshimori T (2014) A current perspective of autophagosome biogenesis. Cell Res 24:58–68

Shore GC, Tata JR (1977) Two fractions of rough endoplasmic reticulum from rat liver. I. Recovery of rapidly sedimenting endoplasmic reticulum in association with mitochondria. J Cell Biol 72:714–725

Siekevitz P (1963) Protoplasm: endoplasmic reticulum and microsomes and their properties. Annu Rev Physiol 25:15–40

Simbeni R, Tangemann K, Schmidt M, Ceolotto C, Paltauf F, Daum G (1993) Import of phosphatidylserine into isolated yeast mitochondria. Biochim Biophys Acta 1145:1–7

Simmen T, Aslan JE, Blagoveshchenskaya AD, Thomas L, Wan L, Xiang Y, Feliciangeli SF, Hung CH, Crump CM, Thomas G (2005) PACS-2 controls endoplasmic reticulum-mitochondria communication and Bid-mediated apoptosis. EMBO J 24:717–729

Simmen T, Lynes EM, Gesson K, Thomas G (2010) Oxidative protein folding in the endoplasmic reticulum: tight links to the mitochondria-associated membrane (MAM). Biochim Biophys Acta 1798: 1465–1473

Sood A, Jeyaraju DV, Prudent J, Caron A, Lemieux P, McBride HM, Laplante M, Toth K, Pellegrini L (2014) A Mitofusin-2-dependent inactivating cleavage of Opa1 links changes in mitochondria cristae and ER contacts in the postprandial liver. Proc Natl Acad Sci U S A 111:16017–16022

Stoica R, De Vos KJ, Paillusson S, Mueller S, Sancho RM, Lau KF, Vizcay-Barrena G, Lin WL, Xu YF, Lewis J et al (2014) ER-mitochondria associations are regulated by the VAPB-PTPIP51 interaction and are disrupted by ALS/FTD-associated TDP-43. Nat Commun 5:3996

Stoica R, Paillusson S, Gomez-Suaga P, Mitchell JC, Lau DH, Gray EH, Sancho RM, Vizcay-Barrena G, De Vos KJ, Shaw CE et al (2016) ALS/FTD-associated FUS activates GSK-3β to disrupt the VAPB-PTPIP51 interaction and ER-mitochondria associations. EMBO Rep 17:1326–1342

Stone SJ, Levin MC, Zhou P, Han J, Walther TC, Farese RV Jr (2009) The endoplasmic reticulum enzyme DGAT2 is found in mitochondria-associated

membranes and has a mitochondrial targeting signal that promotes its association with mitochondria. J Biol Chem 284:5352–5361
Sugiura A, Nagashima S, Tokuyama T, Amo T, Matsuki Y, Ishido S, Kudo Y, McBride HM, Fukuda T, Matsushita N et al (2013) MITOL regulates endoplasmic reticulum-mitochondria contacts via Mitofusin2. Mol Cell 51:20–34
Sung JY, Engmann O, Teylan MA, Nairn AC, Greengard P, Kim Y (2008) WAVE1 controls neuronal activity-induced mitochondrial distribution in dendritic spines. Proc Natl Acad Sci U S A 105: 3112–3116
Szabadkai G, Bianchi K, Varnai P, De Stefani D, Wieckowski MR, Cavagna D, Nagy AI, Balla T, Rizzuto R (2006) Chaperone-mediated coupling of endoplasmic reticulum and mitochondrial Ca^{2+} channels. J Cell Biol 175:901–911
Takei K, Mignery GA, Mugnaini E, Sudhof TC, De Camilli P (1994) Inositol 1,4,5-trisphosphate receptor causes formation of ER cisternal stacks in transfected fibroblasts and in cerebellar Purkinje cells. Neuron 12:327–342
Tasseva G, Bai HD, Davidescu M, Haromy A, Michelakis E, Vance JE (2013) Phosphatidylethanolamine deficiency in Mammalian mitochondria impairs oxidative phosphorylation and alters mitochondrial morphology. J Biol Chem 288:4158–4173
Tatsuta T, Scharwey M, Langer T (2014) Mitochondrial lipid trafficking. Trends Cell Biol 24:44–52
Taylor FR, Saucier SE, Shown EP, Parish EJ, Kandutsch AA (1984) Correlation between oxysterol binding to a cytosolic binding protein and potency in the repression of hydroxymethylglutaryl coenzyme A reductase. J Biol Chem 259:12382–12387
Theurey P, Rieusset J (2017) Mitochondria-associated membranes response to nutrient availability and role in metabolic diseases. Trends Endocrinol Metab 28(1):32–45
Theurey P, Tubbs E, Vial G, Jacquemetton J, Bendridi N, Chauvin MA, Alam MR, Le Romancer M, Vidal H, Rieusset J (2016) Mitochondria-associated endoplasmic reticulum membranes allow adaptation of mitochondrial metabolism to glucose availability in the liver. J Mol Cell Biol 8:129–143
Tubbs E, Theurey P, Vial G, Bendridi N, Bravard A, Chauvin MA, Ji-Cao J, Zoulim F, Bartosch B, Ovize M et al (2014) Mitochondria-associated endoplasmic reticulum membrane (MAM) integrity is required for insulin signaling and is implicated in hepatic insulin resistance. Diabetes 63:3279–3294
Vance JE (1990) Phospholipid synthesis in a membrane fraction associated with mitochondria. J Biol Chem 265:7248–7256
Vance JE (1991) Newly made phosphatidylserine and phosphatidylethanolamine are preferentially translocated between rat liver mitochondria and endoplasmic reticulum. J Biol Chem 266:89–97
Vance JE (2015) Phospholipid synthesis and transport in mammalian cells. Traffic 16:1–18
Vervliet T, Decrock E, Molgo J, Sorrentino V, Missiaen L, Leybaert L, De Smedt H, Kasri NN, Parys JB, Bultynck G (2014) Bcl-2 binds to and inhibits ryanodine receptors. J Cell Sci 127: 2782–2792
Vervliet T, Parys JB, Bultynck G (2016) Bcl-2 proteins and calcium signaling: complexity beneath the surface. Oncogene 35:5079–5092
Voelker DR (1989) Reconstitution of phosphatidylserine import into rat liver mitochondria. J Biol Chem 264: 8019–8025
Voelker DR (1990) Characterization of phosphatidylserine synthesis and translocation in permeabilized animal cells. J Biol Chem 265:14340–14346
von der Malsburg K, Muller JM, Bohnert M, Oeljeklaus S, Kwiatkowska P, Becker T, Loniewska-Lwowska A, Wiese S, Rao S, Milenkovic D et al (2011) Dual role of mitofilin in mitochondrial membrane organization and protein biogenesis. Dev Cell 21:694–707
Walsh C, Barrow S, Voronina S, Chvanov M, Petersen OH, Tepikin A (2009) Modulation of calcium signalling by mitochondria. Biochim Biophys Acta 1787: 1374–1382
Wang HJ, Guay G, Pogan L, Sauve R, Nabi IR (2000) Calcium regulates the association between mitochondria and a smooth subdomain of the endoplasmic reticulum. J Cell Biol 150:1489–1498
Wang PT, Garcin PO, Fu M, Masoudi M, St-Pierre P, Pante N, Nabi IR (2015) Distinct mechanisms controlling rough and smooth endoplasmic reticulum contacts with mitochondria. J Cell Sci 128:2759–2765
White C, Li C, Yang J, Petrenko NB, Madesh M, Thompson CB, Foskett JK (2005) The endoplasmic reticulum gateway to apoptosis by Bcl-X_L modulation of the $InsP_3R$. Nat Cell Biol 7:1021–1028
Wideman JG (2015) The ubiquitous and ancient ER membrane protein complex (EMC): tether or not? F1000Res 4:624
Wideman JG, Gawryluk RM, Gray MW, Dacks JB (2013) The ancient and widespread nature of the ER-mitochondria encounter structure. Mol Biol Evol 30:2044–2049
Williamson CD, Zhang A, Colberg-Poley AM (2011) The human cytomegalovirus protein UL37 exon 1 associates with internal lipid rafts. J Virol 85: 2100–2111
Wirtz KW, Zilversmit DB (1968) Exchange of phospholipids between liver mitochondria and microsomes in vitro. J Biol Chem 243:3596–3602
Wu W, Lin C, Wu K, Jiang L, Wang X, Li W, Zhuang H, Zhang X, Chen H, Li S et al (2016) FUNDC1 regulates mitochondrial dynamics at the

ER-mitochondrial contact site under hypoxic conditions. EMBO J 35:1368–1384

Yang Z, Klionsky DJ (2010) Eaten alive: a history of macroautophagy. Nat Cell Biol 12:814–822

Yang J, Vais H, Gu W, Foskett JK (2016) Biphasic regulation of $InsP_3$ receptor gating by dual Ca^{2+} release channel BH3-like domains mediates Bcl-xL control of cell viability. Proc Natl Acad Sci U S A 113:E1953–E1962

Zhou R, Yazdi AS, Menu P, Tschopp J (2011) A role for mitochondria in NLRP3 inflammasome activation. Nature 469:221–225

Ziolkowski W, Szkatula M, Nurczyk A, Wakabayashi T, Kaczor JJ, Olek RA, Knap N, Antosiewicz J, Wieckowski MR, Wozniak M (2010) Methyl-beta-cyclodextrin induces mitochondrial cholesterol depletion and alters the mitochondrial structure and bioenergetics. FEBS Lett 584:4606–4610

Regulation of Mitochondrial Dynamics and Autophagy by the Mitochondria-Associated Membrane

3

Mitsuo Tagaya and Kohei Arasaki

Abstract

Mitochondria are powerhouses and central to metabolism in cells. They are highly dynamic organelles that continuously fuse, divide, and move along the cytoskeleton to form the mitochondrial network. The fusion and fission are catalyzed by four dynamin-related GTPases in mammals that are controlled by a variety of protein-protein interactions and posttranslational modifications. Mitochondrial dynamics and metabolism are linked and regulate each other. Starvation induces mitochondrial elongation, which enables the mitochondria to produce energy more efficiently and to escape from autophagic degradation. Damaged portions of mitochondria are removed from the healthy parts by division, and subsequently degraded via a specific mode of autophagy termed mitophagy. Recent studies shed light on the contribution of the endoplasmic reticulum to mitochondrial dynamics and the cooperation of the two organelles for the progression of autophagy including mitophagy. A subdomain of the endoplasmic reticulum apposed to mitochondria is called the mitochondria-associated membrane (MAM), which comprises a unique set of proteins that interact with mitochondrial proteins. Here we review our current understanding of the molecular mechanisms of mitochondrial dynamics and mitochondria-related processes in the context of the interaction with the endoplasmic reticulum.

Keywords

Autophagy • Drp1 • Endoplasmic reticulum • Mitochondria • Mitochondria-associated membrane • Syntaxin 17 • FUNDC1

M. Tagaya (✉) • K. Arasaki
School of Life Sciences, Tokyo University of Pharmacy and Life Sciences, Hachioji, Tokyo 192-0392, Japan
e-mail: tagaya@toyaku.ac.jp

3.1 Introduction

Mitochondria are powerhouses of the cell that generate ATP through oxidative phosphorylation.

M. Tagaya, T. Simmen (eds.), *Organelle Contact Sites*, Advances in Experimental Medicine and Biology 997, DOI 10.1007/978-981-10-4567-7_3

The origin of mitochondria is assumed to be an α-proteobacterium-like ancestor that invaded an Archaea-type host cell >1.5 billion years ago (Dyall et al. 2004). This invader escaped from elimination by the host cell and successfully lived together with it as a parasite. The parasite was profoundly beneficial for the host cell because it efficiently produced energy (ATP) using molecular oxygen from a limited source of food and might have allowed the host cell to develop a complex intracelllar architecture. The use of oxygen by the parasite matched up with an increase in the oxygen concentration in the atmosphere. Although being beneficial, the parasite might occasionally have caused trouble for the host cell because it produced excessive reactive oxygen species through the electron transport chain, which potentially damage membrane lipids and other architecture in cells. The Archaea-type host might have "appointed" the endoplasmic reticulum (ER) to overwatch the parasite. In cells today, 5–20% of the mitochondrial surface is in close vicinity to the ER (Rizzuto et al. 1998). The ER subdomain, now called the mitochondria-associated membrane (MAM), can regulate the activity of mitochondria using Ca^{2+} stored inside the ER. Given that Ca^{2+} is an important factor for apoptosis (Rizzuto et al. 2012), it is easily understandable that the MAM regulates apoptosis. The MAM is also the site of lipid synthesis in coordination with mitochondria.

Emerging evidence has revealed that the MAM has more functions than anticipated. It participates in the regulation of mitochondrial dynamics, ER stress signaling, autophagy, mitophagy, pathogen proliferation and responses, inflammation, type 2 diabetes, cancer, and diseases associated with neurodegeneration, most of which are topics in this book. In this review, we focus on the regulation of mitochondrial dynamics, autophagy, and mitophagy. These processes are tightly coordinated by the MAM.

3.2 Dynamics of Mitochondria

Mitochondria are dynamic organelles that continuously undergo fusion and fission and move along the cytoskeleton. Depending on the cell type and in response to cellular physiology, mitochondria exhibit a variety of structures, ranging from highly connected tubular structures distributed throughout the cells to fragmented, aggregated structures accumulated in certain areas of cells (reviewed by Labbé et al. (2014)). In general, an excess nutrient supply facilitates mitochondrial division, whereas starvation tends to cause elongation of mitochondria (reviewed by Mishra and Chan (2016)). Because of the presence of two membranes, the outer and inner membranes, two different dynamin-related GTPases mediate membrane fusion. The fusion of the outer membrane is catalyzed by mitofusin (Mfn) 1 and its close homologue Mfn2 in mammals, which correspond to yeast Fzo1. The fusion of the inner membrane is mediated by Opa1 and Mgm1 in mammals and yeast, respectively. On the other hand, mitochondrial division is mediated by another dynamin-related GTPase, Drp1 (Dnm1 in yeast). Very recently, dynamin2 was found to work in concert with Drp1 to constrict and divide mitochondria (Lee et al. 2016).

3.2.1 Mfns

Although Mfn1 and Mfn2 exhibit 60-63% amino acid identity, form homotypic and heterotypic complexes, and therefore have redundant functions, they have several nonredundant, special functions as well (reviewed by Schrepfer and Scorrano (2016)). Mfn1 displays relatively ubiquitous expression, whereas Mfn2 displays some tissue preference and is highly expressed in brain and adrenal gland (Eura et al. 2003; Santel et al. 2003). Mfn1 is crucial for mitochondrial docking and fusion, whereas Mfn2 has lower GTPase activity and has more diverse functions (Ishihara et al. 2004). Mfn1 exclusively localizes to mitochondria, whereas Mfn2 is present not only in mitochondria but also abundant in the MAM (de Brito and Scorrano 2008). Consistent with the different subcellular localization, Mfn2, but not Mfn1, regulates ER stress and insulin signaling (reviewed by Dorn et al. (2015) and Zorzano et al. (2015)). Another unique property of Mfn2 is that it interacts with proapoptotic Bcl-2 proteins, Bax and perhaps Bak (Karbowski et al. 2006; Hoppins et al. 2011). These interactions affect Mfn2-dependent processes, thereby regulating the morphology of mitochondria.

3.2.2 Opa1

The Opa1 GTPase, whose mutation causes autosomal dominant optic atrophy, undergoes processing by mitochondrial inner membrane proteases (Yme1L and Oma1) to produce short, soluble, intermembrane space-localized forms (S-Opa1) from long inner membrane-attached forms (L-Opa1) (Anand et al. 2014). In addition to the inner membrane fusion, Opa1/Mgm1 controls cristae shape (Frezza et al. 2006; Meeusen et al. 2006). This is accomplished in cooperation with the MICOS (mitochondrial contact site and cristae organizing system) (Glytsou et al. 2016). The cristae shape determines the release of cytochrome c in response to starvation and apoptotic stimuli and regulates the assembly of respiratory chain supercomplexes (Frezza et al. 2006; Patten et al. 2014; Cogliati et al. 2013). Loss of L-Opa1 through excessive processing by Oma1 facilitates mitochondrial fragmentation and cristae disruption, leading to cell death (MacVicar and Langer 2016). Opa1 also acts as a PKA-kinase anchoring protein (AKAP) on lipid droplets and mediates adrenergic control of lipolysis (Pidoux et al. 2011).

3.2.3 Drp1

In contrast to the membrane association nature of Mfns and Opa1, the fission protein Drp1 is predominantly present in the cytosol and recruited to the mitochondrial surface via Drp1 receptors/adaptors, which allows Drp1 to further oligomerize and wrap around mitochondria (reviewed by Otera et al. (2013) and Richter et al. (2015)). This and Drp1-catalyzed GTP hydrolysis constrict and finally sever the mitochondrial outer membrane concomitantly with the release of Drp1 from the mitochondria (Mears et al. 2011). The activity of Drp1 is regulated not only by its receptors/adaptors, but also by phospholipids. Quite recently, phosphatidic acid was reported to control the activity of Drp1 (Adachi et al. 2016).

3.2.3.1 Drp1 Receptors/Adaptors

Drp1 receptors/adaptors so far identified are Mff (Gandre-Babbe and van der Bliek 2008; Otera et al. 2010), MiD49/MIEF2, and MiD51/MIEF1 (Palmer et al. 2011; Zhao et al. 2011). Although yeast Fis1 functions as a receptor for Dnm1, the role of mammalian Fis1 as a Drp1 receptor/adaptor remains controversial (Labbé et al. 2014). It appears that Mff is the principal Drp1 receptor, whereas MiD proteins have regulatory roles. Mff marks Drp1-binding sites, whereas MiD proteins require Drp1 in order to associate with constriction sites (Richter et al. 2014). Drp1 GTPase activity is stimulated by Mff, but inhibited by MiD51 (Osellame et al. 2016). The inhibitory effect of MiD51 on Drp1 GTPase activity may be related to the low frequency of mitochondrial fission events even after the formation of constriction sites (Ji et al. 2015).

The presence of the multiple Drp1 receptors/adaptors makes it possible to regulate Drp1 in response to the cellular energy state and cell physiology including apoptosis and mitophagy. In this context it is interesting that MiD51 has an ADP-binding site (Losón et al. 2014; Richter et al. 2014). MiD49/MiD51, but not Mff, are required for cristae remodeling in response to apoptotic stimuli (Otera et al. 2016). During apoptosis, Fis1 and the ER membrane protein Bap31 constitute a platform for recruiting procaspase-8 to trigger the cleavage of Bap31 into a proapoptotic form (p20). The Bap31 fragment promotes the release of ER Ca^{2+}, causing the dissipation of the mitochondrial membrane potential and thereby cell death (Iwasawa et al. 2011). Deletion of Fis1 in *Caenorhabditis elegans* and mammalian cells results in a defect in mitophagy, but not autophagy, leading to the accumulation of large mitophagosome aggregates (Shen et al. 2014). Under stress conditions, Drp1 is recruited to mitochondria via Mff and then binds to Fis1 as a Ser637-phosphorylated form in a complex with the ER proteins calnexin and Bap31 (Shen et al. 2014). During mitophagy, Fis1 recruits TBC1D15 and TBC1D17, both of which are Rab-GAPs for Rab7, a small GTPase that regulates endolysosomal trafficking, and may regulate Rab7 activity for autophagosomal membrane formation at the interface between mitochondria and isolation membranes (Yamano et al. 2014).

3.2.3.2 Posttranslational Modification of Drp1

The activity of Drp1 is regulated by posttranslational modifications including phosphorylation-dephosphorylation, SUMOylation-deSUMOylation, *S*-nitrosylation, and ubiquitination (Otera et al. 2013; Richter et al. 2015). Phosphorylation of Drp1 occurs at Ser616, Ser637, and Ser693. Phosphorylation at Ser616 by protein kinases such as Cdk1 (Taguchi et al. 2007) and Cdk5 (Jahani-Asl et al. 2015; Xie et al. 2015) promotes fission activity. Ser616 phosphorylation is also mediated by MAPK (Erk) in tumor growth (Kashatus et al. 2015; Serasinghe et al. 2015) and cell reprograming (Prieto et al. 2016). Curiously, phosphorylation of Drp1 at Ser637 (Ser637 is also referred to as Ser600, 617, and 656 depending on the species and splice variant) causes inhibition (PKA and AMP kinase) or activation (Ca^{2+}/calmodulin-dependent kinase and ROCK1) of fission activity, perhaps depending on the cells, tissues, and/or cellular context (Otera et al. 2013). Phosphorylation by PKA inhibits fission activity (Chang and Blackstone 2007; Cribbs and Strack 2007). Recruitment of PKA to mitochondria is likely mediated by PKA-anchoring proteins. AKAP1 (also known as AKAP121 and AKAP149)-mediated phosphorylation of Drp1 supports neural cell survival (Merrill et al. 2011) and regulates neuronal development (Dickey and Strack 2011). The cellular content of AKAP1 is finely tuned by the ubiquitin ligase Siah2 (Carlucci et al. 2008), which degrades AKAP1 under hypoxia, leading to mitochondrial fission and apoptosis (Kim et al. 2011). AKAP1 can compensate for deficiencies in PINK1 (PTEN-induced putative kinase 1) function (Dagda et al. 2011), which is associated with a familial syndrome related to Parkinson's disease (Pickrell and Youle 2015). A recent study showed that PINK1 triggers PKA displacement from AKAP1 for mitophagy, which causes the dephosphorylation of Drp1, leading to mitochondrial fission prerequisite for mitophagy (Pryde et al. 2016). Another mitochondrial AKAP protein Rab32 localizes to the MAM and mitochondria and increases the phosphorylation of Drp1 by PKA (Bui et al. 2010; Ortiz-Sandoval et al. 2014). Dephosphorylation of Drp1 at Ser637 is catalyzed by calcineurin (Cereghetti et al. 2008), PP2A (Merrill et al. 2013), and PGAM5 (Wang et al. 2012; Kang et al. 2015; Xu et al. 2015). Similar to the phosphorylation of Drp1 at Ser637 by PKA, phosphorylation of Ser693 by GSK3β inactivates Drp1 (Chou et al. 2012).

SUMOylation-deSUMOylation of Drp1 is closely related to apoptosis (Wasiak et al. 2007; Guo et al. 2013; Fu et al. 2014). Drp1 is sumoylated by MAPL (Braschi et al. 2010), which is promoted by the apoptotic proteins Bax and Bak (Wasiak et al. 2007). A recent study showed that MAPL-mediated SUMOylation of Drp1 stabilizes the ER-mitochondria platform for mitochondrial constriction, Ca^{2+} flux, cristae remodeling, and cytochrome c release downstream of Bax and Bak (Prudent et al. 2015).

NO-induced *S*-nitrosylation of Drp1 at Cys644 highly stimulates fission activity. This is induced in neuronal cells by β-amyloid, which is thought to be a key mediator of Alzheimer disease, via *N*-methyl-D-aspartate (NMDA)-type glutamate receptor, leading to synaptic loss and neuronal damage (Cho et al. 2009). Recently, elevated glucose levels seen in the metabolic syndrome and type 2 diabetes mellitus were reported to have the same impact on synapses as β-amyloid (Akhtar et al. 2016).

Ubiquitination of Drp1 for proteasomal degradation is mediated by March5 (also known as Mitol) (Nakamura et al. 2006; Yonashiro et al. 2006) and Parkin (Wang et al. 2011). March5 is also responsible for ubiquitination and proteasomal degradation of MiD49 (Xu et al. 2016a), and its activity is negatively regulated by Drp1 and Mff (Cherok et al. 2017).

Regulation of Drp1 activity is mediated not only at the level of Drp1 modification but also at the level of Drp1 receptor/adaptor modification. Recently, Mff was found to be phosphorylated by AMP kinase in response to electron transport chain inhibition (Toyama et al. 2016). However, under starvation, a condition that activates AMP kinase, mitochondria tend to become elongated (Mishra and Chan 2016), and AMP kinase directly phosphorylates and inactivates Drp1 (Wikstrom et al. 2013), raising the question of how AMP kinase discriminates specific targets in the context of cell physiology.

3.3 Roles of the ER-Mitochondria Contact Site

In most cells, 5–20% of the mitochondrial surface is in contact with the ER (Rizzuto et al. 1998). Electron microscopic analysis revealed the distances between mitochondria and the smooth ER (~10 nm) and the rough ER (~25-40 nm) (Csordás et al. 2006). Several candidates, including Mfn2 (de Brito and Scorrano 2008), that mediate tethering between the ER and mitochondria have been reported (For details see Chap. 2). Thanks to the close apposition between the ER and mitochondria, lipids and Ca^{2+} can be efficiently transported between the two organelles. The contact of mitochondria with the smooth and rough ER membranes may be directly or indirectly regulated by Mfn1 and Mfn2, respectively (Wang et al. 2015). The MAM-localized ubiquitin ligase Gp78 participates in the regulation of the mitochondria-rough ER contact, perhaps by degrading Mfn2 (Wang et al. 2015).

3.3.1 Dynamic Nature of the ER-Mitochondria Contact

Although both the ER and mitochondria are highly dynamic organelles, they remain in close contact during continuous movement (Friedman et al. 2010 and reviewed by Friedman and Voeltz (2011)). They move bidirectionally on microtubules (MTs) in metazoan cells. The ER membranes bind to the growing tips of MTs through an ER membrane protein, STIM1, and an MT plus end-binding protein, EB1. ER tubules bind to the shafts of existing MTs and slide along the MT, in particular, on curved acetylated MTs (Friedman et al. 2010). This sliding is driven by the MT motor proteins kinesin 1 family proteins (also known as KIF5) and cytoplasmic dynein. Mitochondria use the same motor proteins as the ER (Tanaka et al. 1998). Kinesin 1 family motors become attached to mitochondria through the adaptor proteins Trak1 and Trak2 (Smith et al. 2006; MacAskill et al. 2009), which correspond to *Drosophila* Milton. The Trak receptor on mitochondria is the mitochondrial Rho-GTPase Miro (Fransson et al. 2006), whose yeast homologue is Gem1. Miro GTPase has two GTPase domains and two EF-hand motifs with a tail anchored to the mitochondrial outer membrane. Ca^{2+} binding to Miro induces kinesin 1 release from MTs and thus prevents mitochondrial movement. Alternatively, Ca^{2+} binding releases kinesin 1 from mitochondria. Thus, in areas where the Ca^{2+} concentration is high, such as at activated synaptic terminals, mitochondrial movement is arrested and mitochondria accumulate at synaptic terminals (Sheng 2017). Mfn2 participates in the axonal transport of mitochondria in association with the Miro/Milton complex (Misko et al. 2010). In this context, it is interesting that March5 polyubiquitinates and degrades *S*-nitrosylated MAP1B-LC1 (Yonashiro et al. 2012), an MT-associated protein that may regulate in concert with MAP1B-HC MT dynamics and axonal transport of mitochondria (Jiménez-Mateos et al. 2006). March5 also polyubiquitinates mitochondria-associated Mfn2 for stabilizing the ER-mitochondria contact (Sugiura et al. 2013).

3.3.2 The MAM Defines Sites of Mitochondrial Division and DNA Replication

In 2011 Voeltz and colleagues (Friedman et al. 2011) discovered a new role of the MAM. On live confocal microscopy, they found that the ER becomes in contact with mitochondria and mediates mitochondrial constriction before Drp1 and Mff recruitment. These findings inspired extensive analysis of the role of the ER in mitochondrial division. Soon after, the ER-localized INF2 (inverted formin 2) was found to mediate actin polymerization and to recruit the myosin IIa motor that pulls on the actin filaments, leading to mitochondrial constriction (Korobova et al. 2013, 2014). Binding of assembled actin to Drp1 likely stimulates Drp1 GTPase activity, leading to mitochondrial fission (Ji et al. 2015). Actin assembly and disassembly on mitochondria occur transiently, and, in addition to INF2, actin-nucleating (Spire1C and Arp2/3 complexes), -binding (cortactin), and -depolymerizing (cofilin) proteins mediate these processes (Li et al. 2015; Manor et al. 2015; Moore et al. 2016). In addition to driving constriction, actin filaments may serve as reservoirs of oligomerization-competent Drp1 that can be accessed for mitochondrial fission (Hatch et al. 2016). Other than actin, there are several proteins that are known to regulate Drp1 activity. Syntaxin 17, a SNARE (SNAP receptor) protein originally implicated in ER-Golgi intracellular trafficking, was reported to localize to the MAM and mitochondria and to define the localization and activity of Drp1 at the MAM (Arasaki et al. 2015). For this role, the SNARE domain is not required, but the C-terminal hairpin-like hydrophobic structure followed by the cytoplasmic tail is important (Arasaki et al. 2015). A recent report showed that septin 2 localizes to a subset of mitochondrial constriction sites and directly binds to Drp1 and may support Drp1-dependent fission (Pagliuso et al. 2016).

Each mitochondrion contains one or more mitochondrial DNA molecules packaged into proteinaceous complexes termed nucleoids that are distributed throughout the mitochondrial network. Mitochondrial fusion and fission are important for nucleoid inheritance (Labbé et al. 2014). In yeast, ER-linked mitochondrial division sites, which are marked by the ERMES (ER-mitochondria encounter structure) complex (Kornmann et al. 2009), are linked to nucleoids (Murley et al. 2013), and the ERMES complex interacts with actin and thereby supports nucleoid segregation (Boldogh et al. 2003). Although the ERMES is not conserved in mammals, mtDNA in mammalian cells, like in yeast, is replicated at a small subset of ER-mitochondria contact sites and segregated (Lewis et al. 2016).

3.3.3 Autophagosome Formation at the MAM

Macroautophagy (hereafter autophagy) is an intracellular degradation process in which cytoplasmic constituents are sequestered by double-membrane structures termed autophagosomes (reviewed by Mizushima et al. (2011) and Lamb et al. (2013)). The autophagosomes fuse with lysosomes to become autolysosomes, and the constituents are degraded by lysosomal enzymes. Basal autophagy occurs under normal conditions, but nutrient deprivation and other stresses stimulate autophagy. The execution of autophagy is mediated by the Atg (autophagy-related) proteins in cooperation with proteins involved in membrane trafficking. Autophagy starts with the nucleation of isolation membranes (also known as phagophores) that emerge from the omega-shaped subdomain of the ER enriched in phosphatidylinositol 3-phosphate (PI3P). The isolation membranes expand likely owing to de novo lipid synthesis, which is probably driven by cyclical activation of two complexes, ULK and PI3-kinase complexes (reviewed by Ktistakis and Tooze (2016)), as well as to membrane supply from the ER-Golgi intermediate compartment, Golgi, endosomes, and plasma membrane (Lamb et al. 2013). The nascent isolation membranes engulf cytoplasmic constituents and are finally detached from the ER concomitantly with the closure of the membranes to form autophagosomes. The fusion of autophagosomes with lysosomes is dependent on Rab7 and its effectors. A Rab7 effector, the HOPS (homotypic fusion and vacuole protein sorting) tethering complex (Jiang et al. 2014), in cooperation with another Rab7 effector, PLEKHM1 (pleckstrin

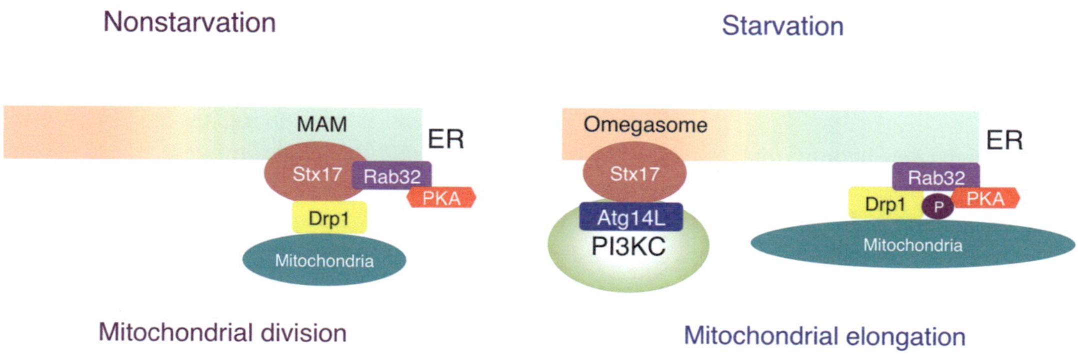

Fig. 3.1 Syntaxin 17 at the ER-mitochondria interface promotes mitochondrial fission in fed cells and autophagosome formation in starved cells by interacting with different partners. Under nonstarvation conditions, syntaxin 17 (Stx17) interacts with Drp1 and prevents its PKA-mediated phosphorylation by competing with the AKAP protein Rab32. Upon starvation, syntaxin 17 dissociates from Drp1 and recruits the PI3-kinase complex (PI3KC) to the MAM through interaction with the PI3-kinase subunit Atg14L. Drp1 is phosphorylated by PKA, leading to mitochondrial elongation

homology domain-containing protein family member) (McEwan et al. 2015), promotes fusion between autophagosomes and lysosomes, likely by tethering the two organelles. A recent study showed that Pacer mediates the function of the PI3-kinase and HOPS complexes in autophagosome maturation by engaging syntaxin 17 (Cheng et al. 2017). After tethering, autophagosomes fuse with lysosomes in a manner dependent on syntaxin 17, SNAP29, and VAMP8 (Itakura et al. 2012), with the aid of the HOPS complex (Jiang et al. 2014) and Atg14L (Diao et al. 2015).

The involvement of mitochondria in autophagy was first demonstrated by Hailey et al. (2010), who showed that mitochondria-localized cytochrome b5, which has a hairpin-like membrane anchor, and fluorescent-labeled phosphatidylserine, which is converted to fluorescent-labeled phosphatidylethanolamine in mitochondria, are transferred from mitochondria to autophagosomes upon starvation. A subsequent study revealed, at least partly, why both the ER and mitochondria contribute to the formation of autophagosomes; isolation membranes are formed at the MAM. Upon starvation, syntaxin 17 is redistributed to the MAM, where it recruits the PI3-kinase complex through interaction with the PI3-kinase subunit Atg14L (Hamasaki et al. 2013). Loss of syntaxin 17, due to short interfering RNA-mediated knockdown or cleavage by Lpg1137 encoded by *Legionella pneumophila*, abrogates PI3P formation at the early stage of autophagy (Arasaki et al. 2017). While syntaxin 17 interacts with Drp1, but not with Atg14L, under nutrient-rich conditions, starvation induces the release of syntaxin 17 from Drp1 accompanied by its binding to Atg14L (Arasaki et al. 2015). This binding partner change can elegantly explain the mitochondrial elongation observed in starved cells. The release of syntaxin 17 from Drp1 causes the inhibitory phosphorylation of Drp1 (Ser637) by PKA because starvation abrogates the inhibitory effect of syntaxin 17 on the binding of the AKAP Rab32 to Drp1, thereby allowing Rab32 to gain access to Drp1 for phosphorylation (Arasaki et al. 2015) (Fig. 3.1). Elongated mitochondria, which can produce ATP more efficiently than fragmented mitochondria, are able to escape from engulfment by isolation membranes because of their large size (Gomes et al. 2011; Rambold et al. 2011).

The recruitment and activation of PI3-kinase via Atg14L for autophagosome formation appear to be also mediated by a multiple membrane-spanning protein, PAQR3. In healthy cells PAQR3 localizes to the Golgi apparatus, but upon glucose starvation, this protein undergoes phosphorylation at Thr32 by AMP kinase and then translocates from the Golgi to the omegasomes, where it functions as a scaffold protein that facilitates the formation of PI3P (Xu et al. 2016b).

Although the finding that isolation membranes are formed at the MAM can partly explain the contribution of mitochondria to autophagosome formation, it is not clear how mitochondrial proteins and lipids are transferred from mitochondria to nascent isolation membranes. One possibility is that lipids are transferred between the MAM and mitochondria via a nonvesicular pathway, as seen in the biosynthesis of lipids. Although there is so far no direct evidence that proteins can shuttle between the MAM and mitochondria, it was reported that certain mitochondrial proteins can escape to the ER during mitophagy (Saita et al. 2013).

3.3.4 Mitophagy

Mitochondria constantly suffer from self-generated internal reactive oxygen species and in some cases from other intrinsic and extrinsic stresses, which may cause dysfunction of mitochondria. Damaged mitochondria are eliminated through a specific mode of autophagy termed mitophagy. This process may be also important for regulation of the mass of mitochondria in physiological contexts (reviewed by Hamacher-Brady and Brady (2016)).

3.3.4.1 PINK1/Parkin-Dependent Mitophagy

The best-known process is one that depends on PINK1/Parkin, which are associated with Parkinson's disease. Recent genetic and biochemical studies have revealed the basic mechanism of the PINK1/Parkin-dependent mitophagy pathway (reviewed by Pickrell and Youle 2015). In healthy mitochondria, the serine/threonine kinase PINK1 is imported into mitochondria in a membrane potential-dependent manner and then sequentially degraded by PARL (presenilin-associated rhomboid-like protein) located in the inner membrane and proteasomes in the cytosol. Upon mitochondrial damage, which dissipates the mitochondrial membrane potential, PINK1 is not imported into mitochondria, but accumulates on the depolarized mitochondrial surface, where it phosphorylates Ser65 in both ubiquitin and the ubiquitin ligase Parkin for the activation and recruitment of Parkin onto depolarized mitochondria. Then, activated Parkin ubiquitinates outer mitochondrial membrane proteins to trigger mitophagy (for references see Pickrell and Youle 2015).

A key step in mitophagy is the specific engulfment of damaged mitochondria by autophagosomes. There are a number of mitophagy receptors that interact through their LIRs (LC3-interacting regions) with autophagosome-associated LC3 (Atg8 in yeast), such as Atg32 in yeast and p62, NBR1, NDP52, TAX1BP1, OPTN (optineurin), Bnip3, Nix (also known as Bnip3L), and FUNDC1 (FUN14 domain-containing protein 1) in mammalian cells (Birgisdottir et al. 2013). Among these receptors, NDP52 and OPTN have been reported to be essential for the PINK1/Parkin pathway (Heo et al. 2015; Lazarou et al. 2015).

Similar to autophagy, mitophagy occurs at the ER-mitochondria contact site. In yeast, the ERMES complex, which tethers the ER and mitochondria (Kornmann et al. 2009), is colocalized with the sites of mitophagosome biogenesis and thereby affects the formation of the isolation membrane (Böckler and Westerman 2014). In mammalian cells, the contact regions between the ER and impaired mitochondria are the initiation sites for local LC3 recruitment and mitochondrial remodeling that support bit-by-bit Parkin-mediated mitophagy (Yang and Yang 2013). A recent study demonstrated that, following mitophagic stimuli, PINK1 relocates to the MAM where it recruits beclin1, a key component of the PI3-kinase complex, for the production of PI3P (Gelmetti et al. 2017). Gp78, a MAM-localized ubiquitin ligase, participates in Parkin-independent mitophagy (Fu et al. 2013). There seems to be no report demonstrating that Nix functions at the ER-mitochondria contact site for mitochondria clearance during blood cell development, but this protein localizes to both the ER and mitochondria and mediates the

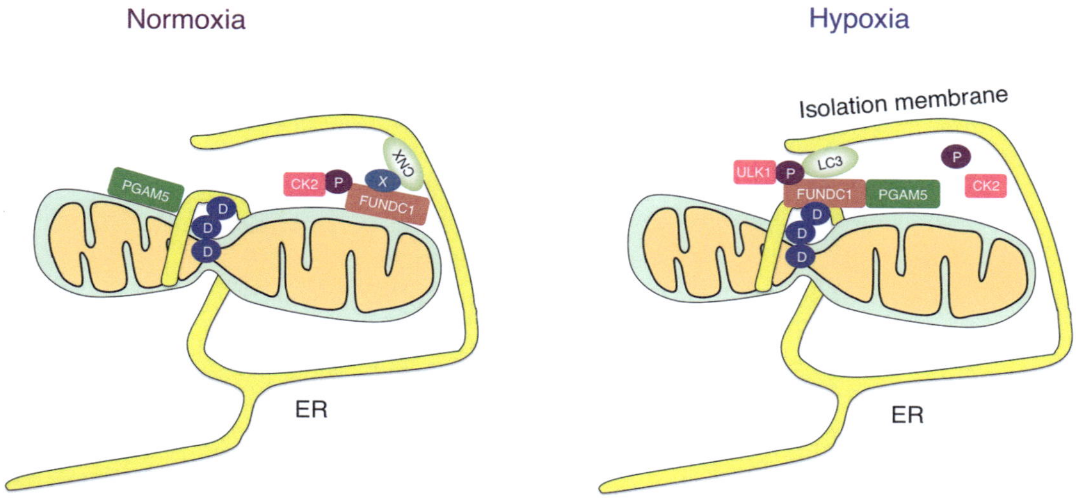

Fig. 3.2 FUNDC1 mediates mitophagy by interacting with Drp1 and LC3. Under normoxia, FUNDC1 binds to calnexin (CNX) through unknown protein(s) (X). The LIR of FUNDC1 is masked by CK2-mediated phosphorylation. In response to hypoxia, FUNDC1 at Ser13 is dephosphorylated by PGAM5, and the dephosphorylated protein promotes mitochondrial fission by interacting with Drp1 (D) and then mediates the incorporation of fragmented mitochondria into autophagosomes by acting as a receptor for LC3 attached to autophagosomal membranes. Phosphorylation of FUNDC1 at Ser17 by ULK1 also promotes the interaction of FUNDC1 with LC3

crosstalk between the two organelles in apoptosis (Diwan et al. 2009).

3.3.4.2 FUNDC1-Dependent Mitophagy

The process of mitophagy involving FUNDC1 has been extensively studied, and evidence suggests that it functions at the ER-mitochondria contact site (Fig. 3.2). FUNDC1, a mitochondrial outer membrane protein containing a LIR (LC3-interacting region), participates in hypoxia-induced mitophagy (Liu et al. 2012) and its expression is stimulated in response to hypoxia (Li et al. 2014). Under normoxia or in healthy mitochondria, Tyr18 and Ser13 in FUNDC1 are phosphorylated by Src and casein kinase 2 (CK2), respectively, which prevents the interaction between the LIR of FUNDC1 and LC3 (Liu et al. 2012; Chen et al. 2014). During hypoxia or in depolarized mitochondria, Ser13 is dephosphorylated by the mitochondrial protein phosphatase PGAM5, which allows FUNDC1 to associate with LC3, thus leading to the sequestration of dysfunctional mitochondria into autophagosomes. The phosphatase activity of PGAM5 is regulated by the anti-apoptotic protein Bcl-XL (Wu et al. 2014b). During mitophagy, the ULK1 complex, a key enzyme for the progression of autophagy, translocates to mitochondria and phosphorylates FUNDC1 at Ser17, favoring the interaction of FUNDC1 with LC3 and therefore promoting the progression of mitophagy (Wu et al. 2014a). A recent study provided compelling evidence that FUNDC1 functions at the ER-mitochondria interface. At the early stage of hypoxia, FUNDC1 increasingly accumulates at the MAM through association with MAM-localized calnexin via unknown protein(s). Then, FUNDC1 dissociates from calnexin and functions as a Drp1 receptor/adaptor, promoting the fission of mitochondria necessary for the sequestration of mitochondria into autophagosomes (Wu et al. 2016).

Mitophagy is not the only process to eliminate damaged mitochondrial components. In response to oxidative stress, oxidized components of the mitochondrial matrix and inner membrane are sorted, incorporated into MDVs (mitochondrial-derived vesicles) in a PINK1/Parkin-dependent manner, and then delivered to lysosomes (Soubannier et al. 2012; McLelland et al. 2014). The fusion of MDVs with lysosomes is mediated by syntaxin 17 associated with MDVs, SNAP29, and VAMP7 in a HOPS complex-dependent manner (McLelland et al. 2016). The membrane fusion mechanism of this system is very similar to that of autophagy except that VAMP7 is used instead of VAMP8. The involvement of the MAM in this system remains to be examined in future studies.

3.4 Concluding Remarks

Recent studies have highlighted the involvement of the ER in the regulation of mitochondrial dynamics and the interplay between the ER and mitochondria in autophagy. These processes are mediated in part by multifunctional proteins, such as syntaxin 17. Under nutrient-rich conditions, syntaxin 17 regulates Drp1 (Arasaki et al. 2015), whereas, under starvation, it mediates the recruitment of PI3-kinase complex to omegasomes at the onset of autophagy (Hamasaki et al. 2013) and catalyzes the fusion of autophagosomes with lysosomes in a late stage of autophagy (Itakura et al. 2012). Future studies should reveal the mechanism by which such proteins can change the functions and binding partners of proteins. Perhaps, posttranslational modifications such as phosphorylation-dephosphorylation play important roles in these changes. Another important question to be answered is how proteins and lipids are transferred between the ER, mitochondria, and autophagosomes. Here, it should be noted that cytochrome b5 and syntaxin 17, both of which translocate from mitochondria/MAM to autophagosomes, have a hairpin-like hydrophobic structure that is embedded only in the cytoplasmic leaflet of the membrane. Because of this feature, the energy barrier to escape from the membrane may be low, which allows these proteins to transit between organelle membranes. The raft-like structure of the MAM (Garofalo et al. 2015) may underpin this movement.

Acknowledgment We thank Dr. Ayumu Sugiura for critical reading of the manuscript. This work was supported in part by Grants-in-Aid for Scientific Research, #26111520 and #26713016 (to K.A.), and the MEXT-Supported Program for the Strategic Research Foundation at Private Universities (to M.T., and K.A.) from the Ministry of Education, Culture, Sports, Science and Technology of Japan. This paper is dedicated to the memory of Dr. Katsuko Tani, who prematurely passed away on March 27, 2016.

References

Adachi Y, Itoh K, Yamada T, Cerveny KL, Suzuki TL, Macdonald P, Frohman MA, Ramachandran R, Iijima M, Sesaki H (2016) Coincident phosphatidic acid interaction restrains Drp1 in mitochondrial division. Mol Cel 63:1034–1043

Akhtar MW, Sanz-Blasco S, Dolatabadi N, Parker J, Chon K, Lee MS, Soussou W, McKercher SR, Ambasudhan R, Nakamura T et al (2016) Elevated glucose and oligomeric β-amyloid disrupt synapses via a common pathway of aberrant protein S-nitrosylation. Nat Commun 7:10242

Anand R, Wai T, Baker MJ, Kladt N, Schauss AC, Rugarli E, Langer T (2014) The i-AAA protease YME1L and OMA1 cleave OPA1 to balance mitochondrial fusion and fission. J Cell Biol 204: 919–929

Arasaki K, Shimizu H, Mogari H, Nishida N, Hirota N, Furuno A, Kudo Y, Baba M, Baba N, Cheng J et al (2015) A role for the ancient SNARE syntaxin 17 in regulating mitochondrial division. Dev Cell 32: 304–317

Arasaki K, Mikami Y, Shames SR, Inoue H, Wakana Y, Tagaya M (2017) Legionella effector Lpg1137 shuts down ER-mitochondria communication through cleavage of syntaxin 17. Nat Commun 8:15406

Birgisdottir ÅB, Lamark T, Johansen T (2013) The LIR motif – crucial for selective autophagy. J Cell Sci 126: 3237–3247

Böckler S, Westermann B (2014) Mitochondrial ER contacts are crucial for mitophagy in yeast. Dev Cell 28:450–458

Boldogh IR, Nowakowski DW, Yang HC, Chung H, Karmon S, Royes P, Pon LA (2003) A protein complex containing Mdm10p, Mdm12p, and Mmm1p links mitochondrial membranes and DNA to the cytoskeleton-based segregation machinery. Mol Biol Cell 14:4618–4627

Braschi E, Goyon V, Zunino R, Mohanty A, Xu L, McBride HM (2010) Vps35 mediates vesicle transport between the mitochondria and peroxisomes. Curr Biol 20:1310–1315

Bui M, Gilady SY, Fitzsimmons RE, Benson MD, Lynes EM, Gesson K, Alto NM, Strack S, Scott JD, Simmen T (2010) Rab32 modulates apoptosis onset and mitochondria-associated membrane (MAM) properties. J Biol Chem 285:31590–31602

Carlucci A, Adornetto A, Scorziello A, Viggiano D, Foca M, Cuomo O, Annunziato L, Gottesman M,

Feliciello A (2008) Proteolysis of AKAP121 regulates mitochondrial activity during cellular hypoxia and brain ischaemia. EMBO J 27:1073–1084

Cereghetti GM, Stangherlin A, Martins de Brito O, Chang CR, Blackstone C, Bernardi P, Scorrano L (2008) Dephosphorylation by calcineurin regulates translocation of Drp1 to mitochondria. Proc Natl Acad Sci USA 105:15803–15808

Chang CR, Blackstone C (2007) Cyclic AMP-dependent protein kinase phosphorylation of Drp1 regulates its GTPase activity and mitochondrial morphology. J Biol Chem 282:21583–21587

Chen G, Han Z, Feng D, Chen Y, Chen L, Wu H, Huang L, Zhou C, Cai X, Fu C et al (2014) A regulatory signaling loop comprising the PGAM5 phosphatase and CK2 controls receptor-mediated mitophagy. Mol Cell 54:362–377

Cheng X, Ma X, Ding X, Li L, Jiang X, Shen Z, Chen S, Liu W, Gong W, Sun Q (2017) Pacer mediates the function of class III PI3K and HOPS complexes in autophagosome maturation by engaging Stx17. Mol Cell 65:1029–1043

Cherok E, Xu S, Li S, Das S, Meltzer WA, Zalzman M, Wang C, Karbowski M (2017) Novel regulatory roles of Mff and Drp1 in E3 ubiquitin ligase MARCH5-dependent degradation of MiD49 and Mcl1 and control of mitochondrial dynamics. Mol Biol Cell 28:396–410

Cho DH, Nakamura T, Fang J, Cieplak P, Godzik A, Gu Z, Lipton SA (2009) S-nitrosylation of Drp1 mediates β-amyloid-related mitochondrial fission and neuronal injury. Science 324:102–105

Chou CH, Lin CC, Yang MC, Wei CC, Liao HD, Lin RC, Tu WY, Kao TC, Hsu CM, Cheng JT et al (2012) GSK3β-mediated Drp1 phosphorylation induced elongated mitochondrial morphology against oxidative stress. PLoS One 7:e49112

Cogliati S, Frezza C, Soriano ME, Varanita T, Quintana-Cabrera R, Corrado M, Cipolat S, Costa V, Casarin A, Gomes LC et al (2013) Mitochondrial cristae shape determines respiratory chain supercomplexes assembly and respiratory efficiency. Cell 155:160–171

Cribbs JT, Strack S (2007) Reversible phosphorylation of Drp1 by cyclic AMP-dependent protein kinase and calcineurin regulates mitochondrial fission and cell death. EMBO Rep 8:939–944

Csordás G, Renken C, Várnai P, Walter L, Weaver D, Buttle KF, Balla T, Mannella CA, Hajnóczky G (2006) Structural and functional features and significance of the physical linkage between ER and mitochondria. J Cell Biol 174:915–921

Dagda RK, Gusdon AM, Pien I, Strack S, Green S, Li C, Van Houten B, Cherra SJ 3rd, Chu CT (2011) Mitochondrially localized PKA reverses mitochondrial pathology and dysfunction in a cellular model of Parkinson's disease. Cell Death Differ 18:1914–1923

de Brito OM, Scorrano L (2008) Mitofusin 2 tethers endoplasmic reticulum to mitochondria. Nature 456:605–610

Diao J, Liu R, Rong Y, Zhao M, Zhang J, Lai Y, Zhou Q, Wilz LM, Li J, Vivona S et al (2015) ATG14 promotes membrane tethering and fusion of autophagosomes to endolysosomes. Nature 520:563–566

Dickey AS, Strack S (2011) PKA/AKAP1 and PP2A/Bβ2 regulate neuronal morphogenesis via Drp1 phosphorylation and mitochondrial bioenergetics. J Neurosci 31:15716–15726

Diwan A, Matkovich SJ, Yuan Q, Zhao W, Yatani A, Brown JH, Molkentin JD, Kranias EG, Dorn GW 2nd (2009) Endoplasmic reticulum-mitochondria crosstalk in NIX-mediated murine cell death. J Clin Invest 119:203–212

Dorn GW II, Song M, Walsh K (2015) Functional implications of mitofusin 2-mediated mitochondrial-SR tethering. J Mol Cell Cardiol 78:123–128

Dyall SD, Brown MT, Johnson PJ (2004) Ancient invasions: from endosymbionts to organelles. Science 304:253–257

Eura Y, Ishihara N, Yokota S, Mihara K (2003) Two mitofusin proteins, mammalian homologues of FZO, with distinct functions are both required for mitochondrial fusion. J Biochem 134:333–344

Fransson S, Ruusala A, Aspenström P (2006) The atypical Rho GTPases Miro-1 and Miro-2 have essential roles in mitochondrial trafficking. Biochem Biophys Res Commun 344:500–510

Frezza C, Cipolat S, Martins de Brito O, Micaroni M, Beznoussenko GV, Rudka T, Bartoli D, Polishuck RS, Danial NN, De Strooper B et al (2006) OPA1 controls apoptotic cristae remodeling independently from mitochondrial fusion. Cell 126:177–189

Friedman JR, Voeltz GK (2011) The ER in 3D: a multifunctional dynamic membrane network. Trends Cell Biol 21:709–717

Friedman JR, Webster BM, Mastronarde DN, Verhey KJ, Voeltz GK (2010) ER sliding dynamics and ER-mitochondrial contacts occur on acetylated microtubules. J Cell Biol 190:363–375

Friedman JR, Lackner LL, West M, DiBenedetto JR, Nunnari J, Voeltz GK (2011) ER tubules mark sites of mitochondrial division. Science 334:358–362

Fu M, St-Pierre P, Shankar J, Wang PT, Joshi B, Nabi IR (2013) Regulation of mitophagy by the Gp78 E3 ubiquitin ligase. Mol Biol Cell 24:1153–1162

Fu J, Yu HM, Chiu SY, Mirando AJ, Maruyama EO, Cheng JG, Hsu W (2014) Disruption of SUMO-specific protease 2 induces mitochondria mediated neurodegeneration. PLoS Genet 10:e1004579

Gandre-Babbe S, van der Bliek AM (2008) The novel tail-anchored membrane protein Mff controls mitochondrial and peroxisomal fission in mammalian cells. Mol Biol Cell 19:2402–2412

Garofalo T, Manganelli V, Grasso M, Mattei V, Ferri A, Misasi R, Sorice M (2015) Role of mitochondrial raft-like microdomains in the regulation of cell apoptosis. Apoptosis 20:621–634

Gelmetti V, De Rosa P, Torosantucci L, Marini ES, Romagnoli A, Di Rienzo M, Arena G, Vignone D, Fimia GM, Valente EM (2017) PINK1 and BECN1 relocalize at mitochondria-associated membranes during mitophagy and promote ER-mitochondria tethering and autophagosome formation. Autophagy 13:654–669

Glytsou C, Calvo E, Cogliati S, Mehrotra A, Anastasia I, Rigoni G, Raimondi A, Shintani N, Loureiro M, Vazquez J et al (2016) Optic atrophy 1 is epistatic to the core MICOS component MIC60 in mitochondrial cristae shape control. Cell Rep 17:3024–3034

Gomes LC, Di Benedetto G, Scorrano L (2011) During autophagy mitochondria elongate, are spared from degradation and sustain cell viability. Nat Cell Biol 13:589–598

Guo C, Hildick KL, Luo J, Dearden L, Wilkinson KA, Henley JM (2013) SENP3-mediated deSUMOylation of dynamin-related protein 1 promotes cell death following ischaemia. EMBO J 32:1514–1528

Hailey DW, Rambold AS, Satpute-Krishnan P, Mitra K, Sougrat R, Kim PK, Lippincott-Schwartz J (2010) Mitochondria supply membranes for autophagosome biogenesis during starvation. Cell 141:656–667

Hamacher-Brady A, Brady NR (2016) Mitophagy programs: mechanisms and physiological implications of mitochondrial targeting by autophagy. Cell Mol Life Sci 73:775–795

Hamasaki M, Furuta N, Matsuda A, Nezu A, Yamamoto A, Fujita N, Oomori H, Noda T, Haraguchi T, Hiraoka Y et al (2013) Autophagosomes form at ER-mitochondria contact sites. Nature 495: 389–393

Hatch AL, Ji WK, Merrill RA, Strack S, Higgs HN (2016) Actin filaments as dynamic reservoirs for Drp1 recruitment. Mol Biol Cell 27:3109–3121

Heo JM, Ordureau A, Paulo JA, Rinehart J, Harper JW (2015) The PINK1-PARKIN mitochondrial ubiquitylation pathway drives a program of OPTN/NDP52 recruitment and TBK1 activation to promote mitophagy. Mol Cell 60:7–20

Hoppins S, Edlich F, Cleland MM, Banerjee S, McCaffery JM, Youle RJ, Nunnari J (2011) The soluble form of Bax regulates mitochondrial fusion via MFN2 homotypic complexes. Mol Cell 41: 150–160

Ishihara N, Eura Y, Mihara K (2004) Mitofusin 1 and 2 play distinct roles in mitochondrial fusion reactions via GTPase activity. J Cell Sci 117:6535–6546

Itakura E, Kishi-Itakura C, Mizushima N (2012) The hairpin-type tail-anchored SNARE syntaxin 17 targets to autophagosomes for fusion with endosomes/lysosomes. Cell 151:1256–1269

Iwasawa R, Mahul-Mellier AL, Datler C, Pazarentzos E, Grimm S (2011) Fis1 and Bap31 bridge the mitochondria-ER interface to establish a platform for apoptosis induction. EMBO J 30:556–568

Jahani-Asl A, Huang E, Irrcher I, Rashidian J, Ishihara N, Lagace DC, Slack RS, Park DS (2015) CDK5 phosphorylates DRP1 and drives mitochondrial defects in NMDA-induced neuronal death. Hum Mol Genet 24:4573–4583

Ji WK, Hatch AL, Merrill RA, Strack S, Higgs HN (2015) Actin filaments target the oligomeric maturation of the dynamin GTPase Drp1 to mitochondrial fission sites. Elife 4:e11553

Jiang P, Nishimura T, Sakamaki Y, Itakura E, Hatta T, Natsume T, Mizushima N (2014) The HOPS complex mediates autophagosome-lysosome fusion through interaction with syntaxin 17. Mol Biol Cell 25: 1327–1337

Jiménez-Mateos EM, González-Billault C, Dawson HN, Vitek MP, Avila J (2006) Role of MAP 1B in axonal retrograde transport of mitochondria. Biochem J 397:53–59

Kang YJ, Bang BR, Han KH, Hong L, Shim EJ, Ma J, Lerner RA, Otsuka M (2015) Regulation of NKT cell-mediated immune responses to tumours and liver inflammation by mitochondrial PGAM5-Drp1 signalling. Nat Commun 6:8371

Karbowski M, Norris KL, Cleland MM, Jeong SY, Youle RJ (2006) Role of Bax and Bak in mitochondrial morphogenesis. Nature 443:658–662

Kashatus JA, Nascimento A, Myers LJ, Sher A, Byrne FL, Hoehn KL, Counter CM, Kashatus DF (2015) Erk2 phosphorylation of Drp1 promotes mitochondrial fission and MAPK-driven tumor growth. Mol Cell 57: 537–551

Kim H, Scimia MC, Wilkinson D, Trelles RD, Wood MR, Bowtell D, Dillin A, Mercola M, Ronai ZA (2011) Fine-tuning of Drp1/Fis1 availability by AKAP121/Siah2 regulates mitochondrial adaptation to hypoxia. Mol Cell 44:532–544

Kornmann B, Currie E, Collins SR, Schuldiner M, Nunnari J, Weissman JS, Walter P (2009) An ER-mitochondria tethering complex revealed by a synthetic biology screen. Science 325:477–481

Korobova F, Ramabhadran V, Higgs HN (2013) An actin-dependent step in mitochondrial fission mediated by the ER-associated formin INF2. Science 339:464–467

Korobova F, Gauvin TJ, Higgs HN (2014) A role for myosin II in mammalian mitochondrial fission. Curr Biol 24:409–414

Ktistakis NT, Tooze SA (2016) Digesting the expanding mechanisms of autophagy. Trends Cell Biol 26:624–635

Labbé K, Murley A, Nunnari J (2014) Determinants and functions of mitochondrial behavior. Annu Rev Cell Dev Biol 30:357–391

Lamb CA, Yoshimori T, Tooze SA (2013) The autophagosome: origins unknown, biogenesis complex. Nat Rev Mol Cell Biol 14:759–774

Lazarou M, Sliter DA, Kane LA, Sarraf SA, Wang C, Burman JL, Sideris DP, Fogel AI, Youle RJ (2015) The ubiquitin kinase PINK1 recruits autophagy receptors to induce mitophagy. Nature 524:309–314

Lee JE, Westrate LM, Wu H, Page C, Voeltz GK (2016) Multiple dynamin family members collaborate to drive mitochondrial division. Nature 540:139–143

Lewis SC, Uchiyama LF, Nunnari J (2016) ER-mitochondria contacts couple mtDNA synthesis with mitochondrial division in human cells. Science 353:aaf5549

Li W, Zhang X, Zhuang H, Chen HG, Chen Y, Tian W, Wu W, Li Y, Wang S, Zhang L et al (2014) MicroRNA-137 is a novel hypoxia-responsive microRNA that inhibits mitophagy via regulation of two mitophagy receptors FUNDC1 and NIX. J Biol Chem 289:10691–10701

Li S, Xu S, Roelofs BA, Boyman L, Lederer WJ, Sesaki H, Karbowski M (2015) Transient assembly of F-actin on the outer mitochondrial membrane contributes to mitochondrial fission. J Cell Biol 208: 109–123

Liu L, Feng D, Chen G, Chen M, Zheng Q, Song P, Ma Q, Zhu C, Wang R, Qi W et al (2012) Mitochondrial outer-membrane protein FUNDC1 mediates hypoxia-induced mitophagy in mammalian cells. Nat Cell Biol 14:177–185

Losón OC, Liu R, Rome ME, Meng S, Kaiser JT, Shan SO, Chan DC (2014) The mitochondrial fission receptor MiD51 requires ADP as a cofactor. Structure 22:367–377

MacAskill AF, Brickley K, Stephenson FA, Kittler JT (2009) GTPase dependent recruitment of Grif-1 by Miro1 regulates mitochondrial trafficking in hippocampal neurons. Mol Cell Neurosci 40:301–312

MacVicar T, Langer T (2016) OPA1 processing in cell death and disease – the long and short of it. J Cell Sci 129:2297–2306

Manor U, Bartholomew S, Golani G, Christenson E, Kozlov M, Higgs H, Spudich J, Lippincott-Schwartz J (2015) A mitochondria-anchored isoform of the actin-nucleating spire protein regulates mitochondrial division. Elife 4:e08828

McEwan DG, Popovic D, Gubas A, Terawaki S, Suzuki H, Stadel D, Coxon FP, Miranda de Stegmann D, Bhogaraju S et al (2015) PLEKHM1 regulates autophagosome-lysosome fusion through HOPS complex and LC3/GABARAP proteins. Mol Cell 57:39–54

McLelland GL, Soubannier V, Chen CX, McBride HM, Fon EA (2014) Parkin and PINK1 function in a vesicular trafficking pathway regulating mitochondrial quality control. EMBO J 33:282–295

McLelland GL, Lee SA, McBride HM, Fon EA (2016) Syntaxin-17 delivers PINK1/parkin-dependent mitochondrial vesicles to the endolysosomal system. J Cell Biol 214:275–291

Mears JA, Lackner LL, Fang S, Ingerman E, Nunnari J, Hinshaw JE (2011) Conformational changes in Dnm1 support a contractile mechanism for mitochondrial fission. Nat Struct Mol Biol 18:20–26

Meeusen S, Devay R, Block J, Cassidy-Stone A, Wayson S, McCaffery JM, Nunnari J (2006) Mitochondrial inner-membrane fusion and crista maintenance requires the dynamin-related GTPase Mgm1. Cell 127:383–395

Merrill RA, Dagda RK, Dickey AS, Cribbs JT, Green SH, Usachev YM, Strack S (2011) Mechanism of neuroprotective mitochondrial remodeling by PKA/AKAP1. PLoS Biol 9:e1000612

Merrill RA, Slupe AM, Strack S (2013) N-terminal phosphorylation of protein phosphatase 2A/Bβ 2 regulates translocation to mitochondria, dynamin-related protein 1 dephosphorylation, and neuronal survival. FEBS J 280:662–673

Mishra P, Chan DC (2016) Metabolic regulation of mitochondrial dynamics. J Cell Biol 212:379–387

Misko A, Jiang S, Wegorzewska I, Milbrandt J, Baloh RH (2010) Mitofusin 2 is necessary for transport of axonal mitochondria and interacts with the Miro/Milton complex. J Neurosci 30:4232–4240

Mizushima N, Yoshimori T, Ohsumi Y (2011) The role of Atg proteins in autophagosome formation. Annu Rev Cell Dev Biol 27:107–132

Moore AS, Wong YC, Simpson CL, Holzbaur EL (2016) Dynamic actin cycling through mitochondrial subpopulations locally regulates the fissionfusion balance within mitochondrial networks. Nat Commun 7:12886

Murley A, Lackner LL, Osman C, West M, Voeltz GK, Walter P, Nunnari J (2013) ER-associated mitochondrial division links the distribution of mitochondria and mitochondrial DNA in yeast. ELife 2:e00422

Nakamura N, Kimura Y, Tokuda M, Honda S, Hirose S (2006) MARCH-V is a novel mitofusin 2- and Drp1-binding protein able to change mitochondrial morphology. EMBO Rep 7:1019–1022

Ortiz-Sandoval CG, Hughes SC, Dacks JB, Simmen T (2014) Interaction with the effector dynamin-related protein 1 (Drp1) is an ancient function of Rab32 subfamily proteins. Cell Logist 4:e986399

Osellame LD, Singh AP, Stroud DA, Palmer CS, Stojanovski D, Ramachandran R, Ryan MT (2016) Cooperative and independent roles of the Drp1 adaptors Mff, MiD49 and MiD51 in mitochondrial fission. J Cell Sci 129:2170–2181

Otera H, Wang C, Cleland MM, Setoguchi K, Yokota S, Youle RJ, Mihara K (2010) Mff is an essential factor for mitochondrial recruitment of Drp1 during mitochondrial fission in mammalian cells. J Cell Biol 191:1141–1158

Otera H, Ishihara N, Mihara K (2013) New insights into the function and regulation of mitochondrial fission. Biochim Biophys Acta 1833:1256–1268

Otera H, Miyata N, Kuge O, Mihara K (2016) Drp1-dependent mitochondrial fission via MiD49/51 is essential for apoptotic cristae remodeling. J Cell Biol 212:531–544

Pagliuso A, Tham TN, Stevens JK, Lagache T, Persson R, Salles A, Olivo-Marin JC, Oddos S, Spang A, Cossart P et al (2016) A role for septin 2 in Drp1-mediated mitochondrial fission. EMBO Rep 17:858–873

Palmer CS, Osellame LD, Laine D, Koutsopoulos OS, Frazier AE, Ryan MT (2011) MiD49 and MiD51, new components of the mitochondrial fission machinery. EMBO Rep 12:565–573

Patten DA, Wong J, Khacho M, Soubannier V, Mailloux RJ, Pilon-Larose K, MacLaurin JG, Park DS, McBride HM, Trinkle-Mulcahy L et al (2014) OPA1-dependent cristae modulation is essential for cellular adaptation to metabolic demand. EMBO J. 33:2676–2691

Pickrell AM, Youle RJ (2015) The roles of PINK1, parkin, and mitochondrial fidelity in Parkinson's disease. Neuron 85:257–273
Pidoux G, Witczak O, Jarnæss E, Myrvold L, Urlaub H, Stokka AJ, Küntziger T, Taskén K (2011) Optic atrophy 1 is an A-kinase anchoring protein on lipid droplets that mediates adrenergic control of lipolysis. EMBO J 30:4371–4386
Prieto J, León M, Ponsoda X, Sendra R, Bort R, Ferrer-Lorente R, Raya A, López-García C, Torres J (2016) Early ERK1/2 activation promotes DRP1-dependent mitochondrial fission necessary for cell reprogramming. Nat Commun 7:11124
Prudent J, Zunino R, Sugiura A, Mattie S, Shore GC, McBride HM (2015) MAPL SUMOylation of Drp1 stabilizes an ER/mitochondrial platform required for cell death. Mol Cell 59:941–955
Pryde KR, Smith HL, Chau KY, Schapira AH (2016) PINK1 disables the anti-fission machinery to segregate damaged mitochondria for mitophagy. J Cell Biol 213:163–171
Rambold AS, Kostelecky B, Elia N, Lippincott-Schwartz J (2011) Tubular network formation protects mitochondria from autophagosomal degradation during nutrient starvation. Proc Natl Acad Sci USA 108: 10190–10195
Richter V, Palmer CS, Osellame LD, Singh AP, Elgass K, Stroud DA, Sesaki H, Kvansakul M, Ryan MT (2014) Structural and functional analysis of MiD51, a dynamin receptor required for mitochondrial fission. J Cell Biol 204:477–486
Richter V, Singh AP, Kvansakul M, Ryan MT, Osellame LD (2015) Splitting up the powerhouse: structural insights into the mechanism of mitochondrial fission. Cell Mol Life Sci 72:3695–3707
Rizzuto R, Pinton P, Carrington W, Fay FS, Fogarty KE, Lifshitz LM, Tuft RA, Pozzan T (1998) Close contacts with the endoplasmic reticulum as determinants of mitochondrial Ca^{2+} responses. Science 280: 1763–1766
Rizzuto R, De Stefani D, Raffaello A, Mammucari C (2012) Mitochondria as sensors and regulators of calcium signalling. Nat Rev Mol Cell Biol 13: 566–578
Saita S, Shirane M, Nakayama KI (2013) Selective escape of proteins from the mitochondria during mitophagy. Nat Commun 4:1410
Santel A, Frank S, Gaume B, Herrler M, Youle RJ, Fuller MT (2003) Mitofusin-1 protein is a generally expressed mediator of mitochondrial fusion in mammalian cells. J Cell Sci 116:2763–2774
Schrepfer E, Scorrano L (2016) Mitofusins, from mitochondria to metabolism. Mol Cell 61:683–694
Serasinghe MN, Wieder SY, Renault TT, Elkholi R, Asciolla JJ, Yao JL, Jabado O, Hoehn K, Kageyama Y, Sesaki H et al (2015) Mitochondrial division is requisite to RAS-induced transformation and targeted by oncogenic MAPK pathway inhibitors. Mol Cell 57:521–536
Shen Q, Yamano K, Head BP, Kawajiri S, Cheung JT, Wang C, Cho JH, Hattori N, Youle RJ, van der Bliek AM (2014) Mutations in Fis1 disrupt orderly disposal of defective mitochondria. Mol Biol Cell 25:145–159
Sheng ZH (2017) The interplay of axonal energy homeostasis and mitochondrial trafficking and anchoring. Trends Cell Biol 27:403–416
Smith MJ, Pozo K, Brickley K, Stephenson FA (2006) Mapping the GRIF-1 binding domain of the kinesin, KIF5C, substantiates a role for GRIF-1 as an adaptor protein in the anterograde trafficking of cargoes. J Biol Chem 281:27216–27228
Soubannier V, McLelland GL, Zunino R, Braschi E, Rippstein P, Fon EA, McBride HM (2012) A vesicular transport pathway shuttles cargo from mitochondria to lysosomes. Curr Biol 22:135–141
Sugiura A, Nagashima S, Tokuyama T, Amo T, Matsuki Y, Ishido S, Kudo Y, McBride HM, Fukuda T, Matsushita N et al (2013) MITOL regulates endoplasmic reticulum-mitochondria contacts via Mitofusin2. Mol Cell 51:20–34
Taguchi N, Ishihara N, Jofuku A, Oka T, Mihara K (2007) Mitotic phosphorylation of dynamin-related GTPase Drp1 participates in mitochondrial fission. J Biol Chem 282:11521–11529
Tanaka Y, Kanai Y, Okada Y, Nonaka S, Takeda S, Harada A, Hirokawa N (1998) Targeted disruption of mouse conventional kinesin heavy chain, kif5B, results in abnormal perinuclear clustering of mitochondria. Cell 93:1147–1158
Toyama EQ, Herzig S, Courchet J, Lewis TL Jr, Losón OC, Hellberg K, Young NP, Chen H, Polleux F, Chan DC et al (2016) Metabolism. AMP-activated protein kinase mediates mitochondrial fission in response to energy stress. Science 351:275–281
Wang H, Song P, Du L, Tian W, Yue W, Liu M, Li D, Wang B, Zhu Y, Cao C et al (2011) Parkin ubiquitinates Drp1 for proteasome-dependent degradation: implication of dysregulated mitochondrial dynamics in Parkinson disease. J Biol Chem 286:11649–11658
Wang Z, Jiang H, Chen S, Du F, Wang X (2012) The mitochondrial phosphatase PGAM5 functions at the convergence point of multiple necrotic death pathways. Cell 148:228–243
Wang PT, Garcin PO, Fu M, Masoudi M, St-Pierre P, Panté N, Nabi IR (2015) Distinct mechanisms controlling rough and smooth endoplasmic reticulum contacts with mitochondria. J Cell Sci 128: 2759–2765
Wasiak S, Zunino R, McBride HM (2007) Bax/Bak promote sumoylation of DRP1 and its stable association with mitochondria during apoptotic cell death. J Cell Biol 177:439–450
Wikstrom JD, Israeli T, Bachar-Wikstrom E, Swisa A, Ariav Y, Waiss M, Kaganovich D, Dor Y, Cerasi E, Leibowitz G (2013) AMPK regulates ER morphology and function in stressed pancreatic β-cells via phosphorylation of DRP1. Mol Endocrinol 27:1706–1723

Wu W, Tian W, Hu Z, Chen G, Huang L, Li W, Zhang X, Xue P, Zhou C, Liu L et al (2014a) ULK1 translocates to mitochondria and phosphorylates FUNDC1 to regulate mitophagy. EMBO Rep 15:566–575

Wu H, Xue D, Chen G, Han Z, Huang L, Zhu C, Wang X, Jin H, Wang J, Zhu Y et al (2014b) The BCL2L1 and PGAM5 axis defines hypoxia-induced receptor-mediated mitophagy. Autophagy 10:1712–1725

Wu W, Lin C, Wu K, Jiang L, Wang X, Li W, Zhuang H, Zhang X, Chen H, Li S et al (2016) FUNDC1 regulates mitochondrial dynamics at the ER-mitochondrial contact site under hypoxic conditions. EMBO J 35: 1368–1384

Xie Q, Wu Q, Horbinski CM, Flavahan WA, Yang K, Zhou W, Dombrowski SM, Huang Z, Fang X, Shi Y et al (2015) Mitochondrial control by DRP1 in brain tumor initiating cells. Nat Neurosci 18:501–510

Xu W, Jing L, Wang Q, Lin CC, Chen X, Diao J, Liu Y, Sun X (2015) Bax-PGAM5L-Drp1 complex is required for intrinsic apoptosis execution. Oncotarget 6:30017–30034

Xu S, Cherok E, Das S, Li S, Roelofs BA, Ge SX, Polster BM, Boyman L, Lederer WJ, Wang C et al (2016a) Mitochondrial E3 ubiquitin ligase MARCH5 controls mitochondrial fission and cell sensitivity to stress-induced apoptosis through regulation of MiD49 protein. Mol Biol Cell 27:349–539

Xu DQ, Wang Z, Wang CY, Zhang DY, Wan HD, Zhao ZL, Gu J, Zhang YX, Li ZG, Man KY et al (2016b) PAQR3 controls autophagy by integrating AMPK signaling to enhance ATG14L-associated PI3K activity. EMBO J 35:496–514

Yamano K, Fogel AI, Wang C, van der Bliek AM, Youle RJ (2014) Mitochondrial Rab GAPs govern autophagosome biogenesis during mitophagy. eLife 3:e01612

Yang JY, Yang WY (2013) Bit-by-bit autophagic removal of parkin-labelled mitochondria. Nat Commun 4:2428

Yonashiro R, Ishido S, Kyo S, Fukuda T, Goto E, Matsuki Y, Ohmura-Hoshino M, Sada K, Hotta H, Yamamura H et al (2006) A novel mitochondrial ubiquitin ligase plays a critical role in mitochondrial dynamics. EMBO J 25:3618–3626

Yonashiro R, Kimijima Y, Shimura T, Kawaguchi K, Fukuda T, Inatome R, Yanagi S (2012) Mitochondrial ubiquitin ligase MITOL blocks S-nitrosylated MAP1B-light chain 1-mediated mitochondrial dysfunction and neuronal cell death. Proc Natl Acad Sci USA 109:2382–2387

Zhao J, Liu T, Jin S, Wang X, Qu M, Uhlén P, Tomilin N, Shupliakov O, Lendahl U, Nistér M (2011) Human MIEF1 recruits Drp1 to mitochondrial outer membranes and promotes mitochondrial fusion rather than fission. EMBO J 30:2762–2778

Zorzano A, Hernández-Alvarez MI, Sebastián D, Muñoz JP (2015) Mitofusin 2 as a driver that controls energy metabolism and insulin signaling. Antioxid Redox Signal 22:1020–1031

Endoplasmic Reticulum-Mitochondria Communication Through Ca^{2+} Signaling: The Importance of Mitochondria-Associated Membranes (MAMs)

4

Saverio Marchi, Mart Bittremieux, Sonia Missiroli, Claudia Morganti, Simone Patergnani, Luigi Sbano, Alessandro Rimessi, Martijn Kerkhofs, Jan B. Parys, Geert Bultynck, Carlotta Giorgi, and Paolo Pinton

Abstract

The execution of proper Ca^{2+} signaling requires close apposition between the endoplasmic reticulum (ER) and mitochondria. Hence, Ca^{2+} released from the ER is "quasi-synaptically" transferred to mitochondrial matrix, where Ca^{2+} stimulates mitochondrial ATP synthesis by activating the tricarboxylic acid (TCA) cycle. However, when the Ca^{2+} transfer is excessive and sustained, mitochondrial Ca^{2+} overload induces apoptosis by opening the mitochondrial permeability transition pore. A large number of regulatory proteins reside at mitochondria-associated ER membranes (MAMs) to maintain the optimal distance between the organelles and to coordinate the functionality of both ER and mitochondrial Ca^{2+} transporters or channels. In this chapter, we discuss the different pathways involved in the regulation of ER-mitochondria Ca^{2+} flux and describe the activities of the various Ca^{2+} players based on their primary intra-organelle localization.

Keywords

Endoplasmic reticulum (ER) • Mitochondria • Mitochondria associated membranes (MAMs) • Calcium • ROS • ER-mitochondria contact sites • Cell death • Apoptosis • Autophagy

S. Marchi • S. Missiroli • C. Morganti • S. Patergnani • L. Sbano • A. Rimessi • C. Giorgi (✉) • P. Pinton (✉)
Department of Morphology, Surgery and Experimental Medicine, Section of Pathology, Oncology and Experimental Biology, Laboratory for Technologies of Advanced Therapies (LTTA), University of Ferrara, Ferrara, Italy
e-mail: grgclt@unife.it; paolo.pinton@unife.it

M. Bittremieux • M. Kerkhofs • J.B. Parys • G. Bultynck
Laboratory of Molecular and Cellular Signaling, Department of Cellular and Molecular Medicine and Leuven Kanker Instituut, KU Leuven, Campus Gasthuisberg O&N-I box 802, Herestraat 49, B-3000 Leuven, Belgium

M. Tagaya, T. Simmen (eds.), *Organelle Contact Sites*, Advances in Experimental Medicine and Biology 997, DOI 10.1007/978-981-10-4567-7_4

4.1 Introduction

The endoplasmic reticulum (ER) and the mitochondrion in living cells are two essential organelles with roles that are classically quite distinct. Evidence has been accruing over the years that points to a specific interplay and cooperation between these compartments that is essential for several cellular functions, such as Ca^{2+} signaling, lipid metabolism, autophagy, inflammation, cell survival, and cell death (Decuypere et al. 2011; Lamb et al. 2013; Marchi et al. 2014; Vance 2014; Patergnani et al. 2015). These close appositions between the ER and mitochondria represent a site where microdomains with a high Ca^{2+} concentration ($[Ca^{2+}]$) are generated upon IP_3-mediated Ca^{2+} release (Rizzuto et al. 1998). Ca^{2+} accumulation in the mitochondrial matrix occurs through the mitochondrial calcium uniporter (MCU) and by the membrane potential ($\Delta\Psi_m$) that exists across the inner mitochondrial membrane (IMM) (Marchi and Pinton 2014). The MCU has a low Ca^{2+} affinity (K_d approximately 10–20 μM), and thus, the Ca^{2+} uptake rate, particularly under resting conditions, is extremely slow (Gunter and Gunter 2001). Based on this consideration, it was long assumed that mitochondrial Ca^{2+} uptake was of little importance in cell physiology. Although initial experiments suggested that mitochondrial $[Ca^{2+}]$ ($[Ca^{2+}]_m$) could increase up to ~10 μM, subsequent research revealed that $[Ca^{2+}]_m$ may in fact transiently reach near mM levels in different systems (Montero et al. 2000). It is the ER-mitochondria connection that allows for fast uptake of a large amount of Ca^{2+} inside the mitochondrial matrix, where the rapid diffusion of ions within it permits the swift tuning of mitochondrial metabolism to address the needs of the cell (Rimessi et al. 2008). ER-mitochondria Ca^{2+} transfer is fundamental to ensure the cellular energy supply by modulating key enzymes involved in mitochondrial ATP production (Bonora et al. 2012). However, excessive transfer of ER Ca^{2+} to mitochondria is a pro-apoptotic signal with important consequences for cell fate (Giorgi et al. 2012). In other words, ER-mitochondria connections are relevant for cell survival, and the maintenance of the proper spacing between the ER and mitochondria appears critical for proper cell functioning (Naon and Scorrano 2014). Tomography analysis has shown that tethers of ~10 nm or ~25 nm adjoin the two organelles, depending on whether smooth ER and/or rough ER are implicated (Csordas et al. 2006). These sites of contact are typically termed mitochondria-associated ER membranes (MAMs) (Giorgi et al. 2015c).

The development and expression of ER-mitochondria artificial linkers revealed that the length of the tethers is strategic. Using rapamycin-inducible fluorescent inter-organelle linkers, Csordas et al. (Csordas et al. 2010) elegantly demonstrated that the spatial relationship between the ER and the outer mitochondrial membrane (OMM) is a critical factor in the efficient transfer of Ca^{2+} and is likely to affect the other functions of the junction in various ways. The distance between the ER and the OMM may vary in different pathophysiological situations. In response to apoptotic agents, the ER-mitochondria gap narrows (Csordas et al. 2006). Also, during the early phases of ER stress, the number of ER-mitochondria contacts increases and their Ca^{2+} transfer is enhanced, helping the cell to overcome this emergency state through the modulation of key mitochondrial metabolic events (Bravo et al. 2011, 2012). These observations reflect the extremely dynamic nature of MAMs, which is now generally assumed to be a strategic intracellular platform that employs Ca^{2+} flux to regulate a wide range of biological processes (Naon and Scorrano 2014). For example, ER-mitochondria connections are essential for modulating mitochondrial fission. In the close appositions between the ER and mitochondria, organelle constriction occurs via a Ca^{2+}-dependent mechanism that involves recruitment of the cytosolic dynamin-related protein 1 (Drp1) and the mitochondria-associated membrane protein syntaxin17 (Friedman et al. 2011; Arasaki et al. 2015). Mitochondrial fission, in turn, is an essential event that is involved in mitochondrial network shaping, and it is required to generate small organelles to be transported or to facilitate the removal of damaged organelles by a selective form of autophagy, termed mitophagy (Haroon and Vermulst 2016). Bulk autophagy directly depends on ER-mitochondria juxtaposition. Not only does the ER membrane supply material for the formation of autophagosomes (Tooze et al. 2010), but the ER-mitochondria contacts may also be the specific regions where autophagosomes start to form, due to the MAM localization of the pre-autophagosomal ER protein Atg14 (Hamasaki et al. 2013; Lamb et al. 2013).

Therefore, the correct organization, the mutual interactions between the ER and mitochondria, and their Ca^{2+} crosstalk are linked events aimed at coordinating important functions of the two organelles, and these events determine key aspects of cell fate. For these reasons, a huge number of proteins are gathered at the ER-mitochondria interface to regulate MAM dynamics, thereby preserving the intracellular equilibrium. Indeed, alterations in both MAM architecture and composition lead to different pathological conditions, which in many cases are accompanied by a drastic dysregulation of the intracellular Ca^{2+} homeostasis and dynamics (Patergnani et al. 2011). In this chapter, we will discuss the different pathways controlling Ca^{2+} flux from the ER to the mitochondria and their impact on the physiological state of the cell, and the activities of the various Ca^{2+} players will be distinguished based on their primary intra-organelle localization.

4.2 Ca^{2+} Signaling on the ER Side

The ER is the largest store of Ca^{2+} inside the cell. In resting condition, the ER may contain hundreds of μM free Ca^{2+} (in order of magnitude, nearly three- to fourfold higher compared to the cytosol) (Hofer and Schulz 1996; Bonora et al. 2013). Specific Ca^{2+} pumps and channels operate to maintain the correct luminal $[Ca^{2+}]$ by executing the correct balance between ER Ca^{2+} uptake and release (Ashby and Tepikin 2001) (Fig. 4.1).

ER Ca^{2+} uptake is exclusively performed by the sarco-/endoplasmic reticulum Ca^{2+}-ATPase (SERCA) pumps, which actively pump Ca^{2+} into the ER in an ATP-dependent manner (Vandecaetsbeek et al. 2011). Three SERCA genes (ATP2A1, ATPA2, and ATP2A3) are present in the human genome. They generate various splice variants that differ in their C-terminal regions, and their expression is dependent on tissue type and development stage (Papp et al. 2012). Generally, SERCA1a and 1b are widely present in adult and neonatal skeletal muscle. The isoform 2a is highly expressed in cardiomyocytes, while SERCA2b is ubiquitously expressed, functioning as the housekeeping isoform. Finally, SERCA3 is the least studied and gives rise to six isoforms. Among the different SERCAs, the 2b isoform displays the highest Ca^{2+} affinity and, thus, is the main isoform involved in Ca^{2+} uptake in the ER in virtually all cells, except skeletal and cardiac muscle. Because SERCAs are the only pumps regulating ER Ca^{2+} entry, it is not surprising that their activity mediates a wide range of cellular functions controlled by proper ER Ca^{2+} homeostasis, including protein folding, lipid and steroid synthesis, and cell death and survival processes like proliferation, apoptosis, growth, and differentiation (Vandecaetsbeek et al. 2011). These functions are regulated by the luminal ER $[Ca^{2+}]$ ($[Ca^{2+}]_{ER}$) and by intracellular Ca^{2+} peaks and oscillations. SERCAs influence the amplitude, the shape, and the frequency of these modulatory events. Their Ca^{2+}-sequestering activity is regulated by several physiological actors, such as i) proteins, ii) posttranslational modifications, and iii) microRNAs (miRNAs) (Harada et al. 2014; Melo et al. 2015). Their activity can also be modulated by natural compounds and pharmacological tools that either inhibit SERCA like thapsigargin, BHQ, and CPAE (Lytton et al. 1991; Vangheluwe et al. 2009) or promote SERCA like CDN1163 (Kang et al. 2016). Important biological modulators of SERCA are phospholamban (PLN), sarcolipin (SLN), calreticulin, calnexin, TMX1, and ORMD1.

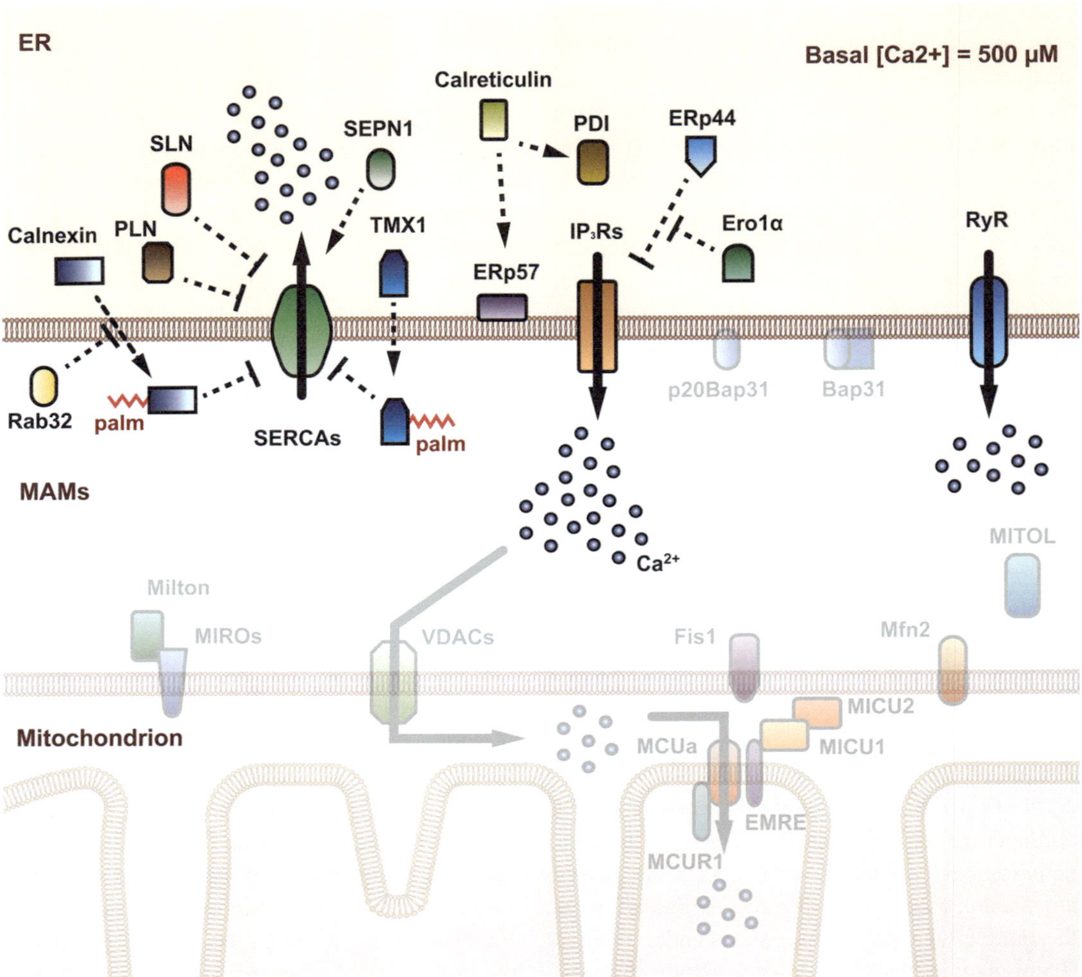

Fig. 4.1 Ca^{2+} homeostasis at the ER-mitochondria interface. Overview on the multiple molecular pathways acting at the ER side (See Sect. 4.2 for details)

Briefly, PLN regulates SERCA function through direct protein-protein interactions (Vittorini et al. 2007). Additionally, SLN directly interacts with SERCAs, and it has been demonstrated to modulate SERCA activity by lowering both the Ca^{2+} affinity and Ca^{2+} pumping rate (Asahi et al. 2003). Another protein that regulates SERCAs is selenoprotein N (SEPN1). This redox-sensitive protein is able to bind SERCAs and enhances their ER Ca^{2+} uptake activity. This feature was found sufficient to safeguard cells against reactive oxygen species (ROS) produced during oxidative protein folding (Marino et al. 2015). Also, chaperones like calreticulin and calnexin have been identified as functional SERCA interactors and modulators. Calreticulin and calnexin were proposed to inhibit SERCA based on their inhibitory impact on high-frequency Ca^{2+} waves in Xenopus oocytes (Camacho and Lechleiter 1995; Roderick et al. 2000). Further work however revealed that overexpression of calreticulin elevated steady-state $[Ca^{2+}]_{ER}$ and increased the ER refilling rates (Arnaudeau et al. 2002). Finally, it has been reported recently that the ER luminal protein disulfide isomerase TMX1 is a strong SERCA inhibitor. In fact, a lack of TMX1 led to an increased ER Ca^{2+} uptake rate and an increase in ER Ca^{2+} storage due to enhanced ER Ca^{2+} uptake activities of SERCA2b (Krols et al. 2016; Raturi et al. 2016). Another SERCA interactor and inhibitor besides TMX1 is the ER-resident, transmembrane protein ORMDL3, an asthma-associated gene product (Cantero-Recasens et al. 2010). Modulation of its expression levels implicates steady-state ER and cytosolic Ca^{2+} levels and the activation of UPR components.

As reported above, the ER also works as the main source of releasable Ca^{2+} in the cells. Many stimuli induce Ca^{2+} release, but two channel families mainly control the ER Ca^{2+}-release program: ryanodine receptors (RyRs) and inositol 1,4,5-trisphosphate (IP_3) receptors (IP_3Rs) (Foskett et al. 2007; Parys and De Smedt 2012; Amador et al. 2013; Van Petegem 2015). RyRs and IP_3Rs form large tetrameric channels (2 MDa and 1.2 MDa in size, respectively) displaying structural and functional homology. RyRs and IP_3Rs are each encoded by three different genes leading for each to the expression of three isoforms. RyR1 is predominantly expressed in skeletal muscle and RyR2 in cardiac muscle and brain, while RyR3 is expressed at low levels in various tissues. With respect to the IP_3Rs, most cell types express a combination of IP_3R1, IP_3R2, and IP_3R3 in various proportions (Vermassen et al. 2004; Ivanova et al. 2014). There is only one known biological/physiological activator of the IP_3R, i.e., IP_3 produced by phospholipase C after activation of G protein-coupled receptors or receptor tyrosine kinases by various stimuli, including growth factors and hormones (Foskett et al. 2007; Parys and De Smedt 2012). RyRs however can be activated through conformational coupling to voltage-operated Ca^{2+} channels, by direct activation by Ca^{2+} or in some cases by the second messengers cADPR and NAADP (Gerasimenko et al. 2006). The Ca^{2+} itself is for both the RyRs and the IP_3Rs, a very important regulator which acts in a biphasic way, whereby a low $[Ca^{2+}]$ activates the channels while a high $[Ca^{2+}]$ has an inhibitory action. Finally, RyRs and IP_3Rs are regulated by phosphorylation/dephosphorylation and by multiple regulatory proteins, some of them being further explicated below. These IP_3R channels are implicated in a plethora of physiological processes, including fertilization, lymphocyte activation, brain rhythms and synaptic plasticity, memory formation, endocrine, and exocrine gland function, and their dysregulation underlies pathophysiological conditions, like neurodegenerative diseases like Alzheimer's disease, Huntington's disease and amyotrophic lateral sclerosis, autism spectrum disorders, bipolar disorder, epilepsy, schizophrenia, spinocerebellar ataxia, cancer, cardiac dysfunction, and hypertrophy (Berridge 2016). Also RyR channels control different physiological functions, mainly related to skeletal and cardiac muscle contraction features (Van Petegem 2012). However, RyR dysregulation has also been implicated in neurodegenerative diseases like Alzheimer's disease (Briggs et al. 2017; Popugaeva et al. 2017) and malignancies like breast cancer (Zhang et al. 2011b).

Ca^{2+} influx into the cytosol also occurs through ORAI and TRP channels, present in the plasma membrane, in a mechanism dependent on IP_3R opening. In fact, when $[Ca^{2+}]_{ER}$ decreases during IP_3-induced Ca^{2+} release, stromal interaction molecules 1 and 2 (STIM1 and 2) are activated and, in turn, induce opening of ORAIs and, finally, lead to the so-called capacitative or store-operated Ca^{2+} influx from the extracellular space (Prakriya and Lewis 2015). As a result of this increased intracellular $[Ca^{2+}]$, key Ca^{2+}-dependent proteins (such as calpains and calmodulins) are activated, and various Ca^{2+}-dependent cellular processes are initiated (Berridge 2016). Several proteins control the activity of the Ca^{2+}-release mechanism mediated by IP_3Rs (Fig. 4.1). For example, the ER-resident oxidoreductases Ero1α and ERp44 modulate ER Ca^{2+} release by direct interaction with IP_3Rs in a redox-sensitive manner (Higo et al. 2005; Anelli et al. 2012). Specifically, the ERp44 chaperone, an ER luminal protein of the thioredoxin family, directly inhibits Ca^{2+} release (which reinforces Ca^{2+}-dependent chaperones) by inactivating the channel activity of the IP_3R in a pH-, redox-state-, and $[Ca^{2+}]_{ER}$-dependent manner (Higo et al. 2005). Furthermore, ERp44 mediates Ero1α localization through the formation of reversible mixed disulfides (Anelli et al. 2003). Ero1α is an ER-resident protein that localizes also at MAMs, and its siRNA-mediated downregulation slightly reduced the association between IP_3R and ERp44, suggesting that Ero1α might have further roles in assembling and/or maintaining MAM integrity (Anelli et al. 2012).

It is very clear that a large number of proteins regulate SERCA pumps and IP_3R channels to maintain the appropriate $[Ca^{2+}]_{ER}$. This is because Ca^{2+} signals originating from the ER are leading to Ca^{2+} oscillations, associated with several cellular processes. As mentioned in the "Introduction" section, Ca^{2+} release from the ER is one of the main determinants for mitochondrial homeostasis. In fact, basal Ca^{2+} oscillations modulate mitochondrial metabolism of ATP production, while sustained or excessive Ca^{2+} release may lead to cell death. As a demonstration of this feature, dysregulation of Ca^{2+} flux is involved in several human disorders (Patergnani et al. 2015).

The maintenance of proper ER Ca^{2+} homeostasis also controls an elaborate surveillance system called the ER quality control (ERQC) system. In fact, inside the ER lumen a series of chaperones exists, which are involved in ERQC for the correct folding of ER proteins, and their functions may vary depending on changes in Ca^{2+} concentration (Fig. 4.1). The Ca^{2+}-binding proteins calreticulin, ERp57, and protein disulfide isomerase (PDI) are classical examples of this family. In fact, PDI directly interacts with calreticulin when the $[Ca^{2+}]_{ER}$ is lower than 400 μM, whereas the complex dissociates upon higher $[Ca^{2+}]$ (Baksh et al. 1995). Conversely, ERp57-calreticulin is insensitive to variations in $[Ca^{2+}]$, but it is dependent on Ca^{2+} binding by calreticulin, showing a direct role for Ca^{2+} in the regulation and maintenance of structural and functional complexes involved in protein turnover and synthesis (Michalak et al. 2009). Interestingly, another member of the PDI family, calnexin, plays a dual role based on its localization. At resting conditions, calnexin is highly palmitoylated ("palm" in Figs.), which leads to an increase of its localization at MAMs and regulation of Ca^{2+} signaling through its interaction with SERCA2b causing its activation. During the early, adaptive phases of ER stress, calnexin becomes depalmitoylated, primarily acting at the rough ER and employing its quality control functions (Lynes et al. 2012, 2013). The MAM localization of calnexin however not only depends on a specific palmitoylation event but also on the phosphorylation state of its cytosolic domain, on its interaction with phosphofurin acidic cluster sorting protein 2 (PACS2) (Myhill et al. 2008), and on the activity of the ER Rab protein Rab32 (Bui et al. 2010). The GTPase Rab32 localizes to the ER and mitochondria, and it has been identified as a regulator of MAM properties that modulate ER Ca^{2+} handling and disrupt the specific enrichment of calnexin in MAMs. However, it does not affect the ER distribution of PDI and mitofusin-2 (Bui et al. 2010).

Ca^{2+} ions are also important for the correct maintenance of ER structure. This was first reported in starfish eggs and then confirmed in human cells, where the ER continuity was affected by elevation of intracellular Ca^{2+} levels (Terasaki et al. 1996). Interestingly, this feature seemed to be regulated by protein kinase C (Ribeiro et al. 2000), which was already found to be involved in Ca^{2+} homeostasis, regulation of cellular processes, and modulation of important proteins and kinases (Pinton et al. 2007). In addition, the authors found that the high $[Ca^{2+}]$ registered did not induce ER fragmentation, only rearrangements of the ER network (Ribeiro et al. 2000). Overall, these findings suggest that Ca^{2+} dynamics regulate ER function without promoting pathological effects.

Notably, the ER continuously exchanges protein and lipid components with the Golgi apparatus, which is also regulated by Ca^{2+}. One example may be found in thyroglobulin protein that, once synthesized and modified in the ER, is exported to the Golgi apparatus (Di Jeso et al. 2003). By treating thyroid differentiated cells with Ca^{2+} ionophores or specific inhibitors of ER channels and pumps, transport between the ER and the Golgi apparatus is blocked (Di Jeso et al. 1998). Intriguingly, it is not only the transport from the ER to the Golgi that seems to be regulated by Ca^{2+} levels but also transport from the Golgi to the ER, and intra-Golgi transport is highly dependent on intracellular Ca^{2+} variations. Moreover, the existence of a local Ca^{2+} gradient between the ER, the cytoplasm, and the Golgi apparatus has been demonstrated (Wahl et al. 1992). These findings underline the important role of intracellular Ca^{2+} in the trafficking of material between the ER and the Golgi and suggest that the mitochondrial compartment is not the only organelle likely to receive Ca^{2+} signals from the ER.

4.3 Ca^{2+} Signaling on the Mitochondrial Side

Mitochondria play a key role in many cell functions through the regulation of Ca^{2+} signaling. The increase in mitochondrial Ca^{2+} uptake activates several dehydrogenases and carriers, inducing an increase in the respiratory rate, H^+ extrusion, and ATP production necessary for the proper energy state of the cell (Rizzuto et al. 2012). As a matter of fact, overexpression of isoforms of the mitochondrial aspartate/glutamate carrier (AGC) promotes ATP production during agonist-triggered Ca^{2+} increases, revealing that AGC plays an important role in decoding Ca^{2+} signals in the activation of mitochondrial oxidative metabolism (Lasorsa et al. 2003).

However, prolonged increase in $[Ca^{2+}]_m$ leads to opening of the mitochondrial permeability transition pore (PTP) (Halestrap 2014; Jonas et al. 2015; Morciano et al. 2015), a critical event that leads to cell death by apoptosis (Rizzuto et al. 2012; Bonora et al. 2015).

As stated previously, mitochondria can rapidly achieve a high $[Ca^{2+}]_m$ due to (1) the presence of a driving force for Ca^{2+} generated by a $\Delta\Psi_m$ of −180 mV under physiological conditions, (2) the formation of a large number of Ca^{2+} microdomains at the ER-mitochondria interface, and (3) the existence of a Ca^{2+}-selective channel, termed the MCU complex, that is able to receive the Ca^{2+} signals originating from the ER (Fig. 4.2). Electrophysiological studies have shown that the uniporter is an ion channel with strikingly high conductance and selectivity (Kirichok et al. 2004). MCU is part of the uniporter holocomplex, which is also composed of two membrane proteins, MCUb and EMRE, which is regulated by MICU1 and MICU2 (Foskett and Philipson 2015; Kamer and Mootha 2015; Raffaello et al. 2016). Overexpression of MCUb reduces the amplitude of the transient mitochondrial Ca^{2+} response evoked by agonist stimulation, whereas MCUb silencing achieves the opposite effect, indicating that it acts as a dominant-negative subunit that reduces the uniporter channel activity (Raffaello et al. 2013). The role of EMRE in the regulation of MCU activity is dual: (1) it was suggested to be necessary for MCU channel activity, and indeed, its silencing abrogates Ca^{2+} entry into mitochondria (Sancak et al. 2013), and (2) it was required for the interaction of MCU with the regulatory subunits MICU1 and MICU2 (Tsai et al. 2016). Moreover, MCU complex is regulated by miRNAs (Marchi et al. 2013; Pan et al. 2015; Hong et al. 2017), underlining its role in multiple physiopathological contexts, or could be subjected to posttranslational modifications, such as phosphorylations (Joiner et al. 2012; O-Uchi et al. 2014) or methylations (Madreiter-Sokolowski et al. 2016). These recent observations confirm that mitochondrial Ca^{2+} homeostasis could be shaped by the wide molecular panel of intracellular transducers (Pinton et al. 2004).

The composition of the MCU complex has not yet been fully defined. Several proteins have been proposed to be part of it. One of the most characterized is an IMM integral protein, named MCUR1. Indeed, MCUR1 was initially shown as a regulator of the MCU complex (Mallilankaraman et al. 2012a), interacting with MCU and EMRE but not with MICU1 or MICU2, thereby functioning as a scaffolding factor (Vais et al. 2015; Tomar et al. 2016). However, MCUR1 was not identified by mass spectrometry of affinity-purified MCU complexes (Sancak et al. 2013), and it has been proposed as a cytochrome c oxidase assembly factor (Paupe et al. 2015).

The very low affinity of the MCU complex for Ca^{2+} depends on the activity of the MICU1 and MICU2 subunits, which localize at the

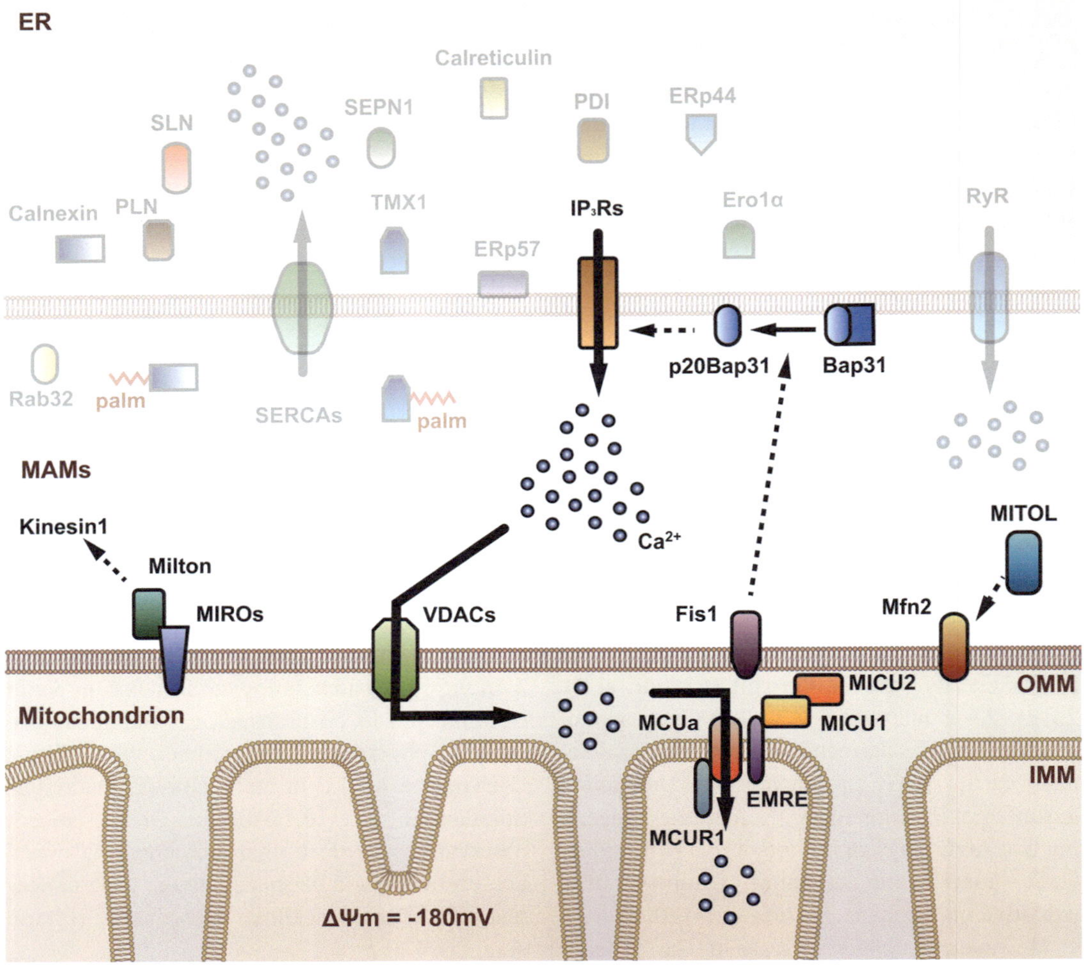

Fig. 4.2 Ca^{2+} homeostasis at the ER-mitochondria interface. Overview on the multiple molecular pathways acting at the mitochondrial side (See Sect. 4.3 for details)

mitochondrial intermembrane space and sense the cytoplasmic Ca^{2+} through their EF-hand domains, thus regulating the open/closed state of the whole uniporter complex (Patron et al. 2014). At resting conditions (cytoplasmic [Ca^{2+}]<500 nM), the MICU1-2 dimer maintains the complex in a closed state, preserving it from the continuous accumulation of Ca^{2+} inside the matrix and thus avoiding Ca^{2+}-mediated detrimental effects, such as ROS production and PTP opening (Mallilankaraman et al. 2012b; Csordas et al. 2013). However, at high cytoplasmic Ca^{2+} levels ([Ca^{2+}]>1–2 μM) or during agonist stimulation, the MICU1–2 dimer undergoes conformational changes that allow opening of the channel, ensuring a prompt and complete mitochondrial Ca^{2+} response. The "high [Ca^{2+}] microdomains theory" implies that the MCU complexes along the IMM should distribute at ER-mitochondria associations to promote effective Ca^{2+} transfer. Indeed, De La Fuente et al. recently showed that in cardiac mitochondria the MCU complexes were enriched in the IMM-OMM contact sites, positioned more to the mitochondrial periphery than inside the cristae, indicating high accessibility to cytoplasm-derived Ca^{2+} inputs (De La Fuente et al. 2016). In other words, mitochondrial Ca^{2+} channels are close to IP_3Rs and RYRs on the ER or sarcoplasmic reticulum. Therefore, high [Ca^{2+}] "hotspots" ([Ca^{2+}]>10 μM) can be formed transiently in these regions of close apposition between the two organelles.

The conformational coupling between the Ca^{2+} channels of the two organelles highlights the importance of the macromolecular complexes located in the MAMs for their functional interaction (Fig. 4.2). In this respect, the OMM, although traditionally considered freely permeable, is a critical determinant of mitochondrial Ca^{2+} accumulation (Rapizzi et al. 2002). Ca^{2+} import across the OMM occurs through the voltage-dependent anion channel (VDAC), the most abundant protein of the OMM. The VDAC protein family consists of three isoforms (VDAC1–3), sharing a 75% sequence similarity. VDACs are expressed in almost all mammalian tissues, but recent studies indicate their nonredundant role in a plethora of cell functions (Naghdi and Hajnoczky 2016). Silencing of either of the VDAC isoforms limits mitochondrial Ca^{2+} uptake, but only VDAC1 was found to mediate pro-apoptotic Ca^{2+} transfer to mitochondria with consequent cell death (De Stefani et al. 2012; Ben-Hail and Shoshan-Barmatz 2016). In contrast, VDAC2 appears to be the pivotal isoform to locally couple mitochondrial Ca^{2+} uptake with RyR-mediated Ca^{2+} release in cardiac cells (Shimizu et al. 2015). Further evidence of the role of VDACs was supported by the demonstration of the physical link between VDACs and the IP_3R. Indeed, the molecular chaperone glucose-regulated protein 75 (GRP75) was demonstrated to mediate the molecular interaction between VDACs and IP_3Rs, allowing a positive regulation of mitochondrial Ca^{2+} uptake (Szabadkai et al. 2006). Small interfering RNA (siRNA) silencing of GRP75 abolishes the functional coupling between IP_3Rs and VDACs, thereby reducing mitochondrial Ca^{2+} uptake in response to agonist stimulation (Szabadkai et al. 2006). Interestingly, thymocyte-expressed, positive selection-associated gene 1 (Tespa1) has been demonstrated to mediate Ca^{2+} transfer from mitochondria-associated ER to mitochondria interacting with GRP75 (Matsuzaki et al. 2013).

A crucial implication of a microdomain-based signaling mechanism is that the positioning and the shape of mitochondria within the cell become critical determinants of their responsiveness to Ca^{2+} inputs. Because of this, several mitochondrial proteins involved in the regulation of mitochondrial movement and morphology have also been considered key regulators of MAM integrity and functionality.

Mitochondrial trafficking is regulated by a subfamily of the Ras GTPases, the proteins Miro 1 and 2, which are located at the OMM through a short C-terminal anchor domain and have two EF-hand Ca^{2+}-binding domains through which they are able to sense high levels of Ca^{2+} (Liu and Hajnoczky 2009). Miro proteins have an important role in tethering the mitochondria to the cytoskeleton by binding a cytoplasmic factor, Milton, which binds the kinesin 1 heavy chain on microtubules (Glater et al. 2006). Miro is proposed to be a Ca^{2+} sensor that stops mitochondrial movement in response to increasing Ca^{2+} levels. In fact, increased cytoplasmic Ca^{2+} levels stop mitochondrial movement, and this effect is suppressed when Miro is depleted or a Miro EF-hand is mutated (Fransson et al. 2003; Saotome et al. 2008).

In addition to the positioning of mitochondria, fusion and fission events regulating the shape of the organelles drastically influenced the mitochondrial Ca^{2+} responses (Patron et al. 2013). Recent studies suggest a link between components of mitochondrial dynamics and Ca^{2+} signaling. Mitochondrial fission is primarily driven by Drp1, a cytoplasmic protein that is recruited to the mitochondrial membrane, where it circumscribes the OMM as a helical oligomer (Smirnova et al. 2001; Rowland and Voeltz 2012). It is also interesting that high Ca^{2+} levels lead to activation of Drp1, which increases mitochondrial fission, cooperating with Miro (Saotome et al. 2008).

It has also been shown that the mitochondrial fission protein fission 1 homolog (Fis1) conveys an apoptotic signal from the mitochondria to the

ER by interacting with Bap31 at the ER. Therefore, Fis1 facilitates the cleavage of Bap31 to its pro-apoptotic form, p20Bap31, promoting the recruitment of procaspase-8. Moreover, this signaling pathway establishes a feedback loop by releasing Ca^{2+} from the ER and, consequently, results in Ca^{2+} accumulation in mitochondria, amplifying cell death by activating the apoptotic pathway in many mitochondria that are in close proximity to the ER (Iwasawa et al. 2011).

Other mitochondrial dynamin-related GTPases involved in mitochondrial Ca^{2+} regulation include mitofusin 1 and 2. In particular, mitofusin 2 (MFN2) is a critical component of the mitochondrial fusion/fission machinery. This OMM profusion protein is also observed in the MAMs where it couples to MFN1 or MFN2 on the mitochondria to physically tether the organelles. Indeed, in 2008 de Brito and Scorrano showed that MFN2 is enriched at contact sites between the ER and mitochondria, regulating ER morphology and directly tethering the two organelles (de Brito and Scorrano 2008). Moreover, the distance between the ER and mitochondria increases in cells lacking MFN2, and this leads to impaired mitochondrial Ca^{2+} uptake, further verifying the validity of the Ca^{2+} microdomains theory.

The tethering role of MFN2 was confirmed by other laboratories (Chen et al. 2012; Sebastian et al. 2012; Schneeberger et al. 2013), but its function was recently challenged by different experimental approaches. Contrary to previous studies, electron microscopy analyses suggested that loss of MFN2 increased, rather than reduced, ER-mitochondria juxtaposition (Cosson et al. 2012; Filadi et al. 2015). Moreover, it was demonstrated that reduced Ca^{2+} transfer in MFN2-knockout cells is the result of a lower expression of MCU and is independent of ER-mitochondria juxtapositions (Filadi et al. 2015). Therefore, they proposed a different role for MFN2 in ER-mitochondria coupling, in which the protein, rather than being a component of the tethering complex, acts as a negative regulator of organelle apposition. However, very recently a critical reappraisal of MFN2's role in the ER-mitochondria connection was published, supporting previous results and identifying MFN2 as a physical tether between the two organelles in multiple tissues (Naon et al. 2016).

The activity of MFN2 at the ER-mitochondria interface is regulated by a mitochondrial ubiquitin ligase called MITOL (Sugiura et al. 2013). MITOL interacts with mitochondrial MFN2, but not with ER MFN2, and mediates the addition of lysine 63-linked polyubiquitin chains to MFN2 but not its proteasomal degradation. This polyubiquitination event induces MFN2 oligomerization, allowing ER-mitochondria tethering and Ca^{2+} uptake in the mitochondria upon stimulation with histamine. The reduction in mitochondrial Ca^{2+} uptake that occurs in MITOL-deficient cells highlights its key role as a MAM regulator and confirms the idea that the distance between the ER and mitochondria is crucial for proper Ca^{2+} transfer. However, the role of MFN2 at MAM is still highly debated, and new experimental evidence is required for definitively establish its anti- or pro-tethering functions.

4.4 Ca^{2+} Signaling at the MAM Interface

MAMs represent the physical association between the ER and mitochondria, an entity with a defined structure and architecture with distinct biochemical properties and a characteristic set of proteins. The MAM fraction was first separated and characterized by J. E. Vance (Vance 1990), who described the isolation from rat liver of a unique membrane, initially termed "fraction X," that was associated with mitochondria and had a high specific activity for several proteins attributed to the ER. After this seminal observation, several biochemical protocols have been described to isolate the MAMs fraction, both from organs and cells, and these studies confirmed that MAMs are composed of membrane fragments from both the ER and the OMM (Wieckowski et al. 2009).

In recent years, different proteomics studies identified the molecular components of the MAMs fraction, starting from human fibroblasts (Zhang et al. 2011a) and mouse brain (Poston et al. 2013), demonstrating that more than 1000

"MAM proteins" reside in this fraction. More recently, Sala-Vila et al. performed a very rigorous high-throughput mass spectrometry-based proteomics characterization of MAMs from mouse liver, identifying 1052 MAM-enriched proteins, which included several Ca^{2+} players, such as SERCA2, IP_3R, and the Ca^{2+}-binding mitochondrial carrier SLC25A12 (Sala-Vila et al. 2016). Interestingly, they observed the MAM localization of caveolin 1 (CAV1), and CAV1-deficient cells displayed ER and mitochondrial aberrations, as well as reduced contact sites between the two organelles (Sala-Vila et al. 2016). In line with this evidence, our group showed that in transformed cells, H-RAS_{12V} expression was associated with CAV1 downregulation and a drastic alteration in Ca^{2+} homeostasis (Rimessi et al. 2014).

Other central players in the ER-mitochondria Ca^{2+} flux include a series of chaperones and oxidoreductases, which also localize to the ER/MAMs compartment (Fig. 4.3). In addition to the previously cited Ca^{2+}-binding chaperone calnexin (see Sect. 4.2), sigma-1 receptor (Sig-1R) is one of the pivotal Ca^{2+} regulators residing at MAMs (Hayashi and Su 2007). Under normal conditions, Sig-1R resides specifically at MAMs and forms a complex with BiP (also named grp78) when the ER Ca^{2+} level is 0.5–1 mM. However, when the IP_3R is activated, the subsequent drop of the $[Ca^{2+}]_{ER}$ causes the dissociation of Sig-1R from BiP, unleashing the chaperone activity of the receptor. Interestingly, IP_3R3 seems to be enriched at MAMs (Mendes et al. 2005), and its stabilization by Sig-1R ensures proper Ca^{2+} influx into mitochondria.

A link between MAMs and Ca^{2+} signaling also appears in the context of ER stress-mediated apoptosis. The RNA-dependent protein kinase (PKR)-like ER kinase (PERK), a key ER stress sensor of the unfolded protein response, is uniquely enriched at MAMs (Verfaillie et al. 2012). PERK-knockout cells display an aberrant ER morphology, disturbed Ca^{2+} signaling, and weaker contact sites between ER and mitochondria; consequently, PERK likely serves as a structural tether at the ER-mitochondria interface. Collectively, these data highlight that a conserved MAM structure is indispensable for transmitting Ca^{2+} signaling as well as ROS-mediated signals to the mitochondria after ROS-based ER stress.

Ca^{2+} homeostasis is fundamental for numerous cellular mechanisms, including cell death; thus, it is not surprising that several oncogenes and tumor suppressors localize at the MAMs, where they play a crucial role in the control of ER-mitochondria Ca^{2+} flux, favoring either survival or cell death (Bittremieux et al. 2016) (Fig. 4.3).

A few years ago, our group showed that the promyelocytic leukemia protein (PML), a tumor suppressor known as an essential component of nuclear structures termed PML nuclear bodies, was localized to the ER and MAMs. PML regulates apoptosis by modulating ER Ca^{2+} release (Giorgi et al. 2010). In MAMs, PML was found to coordinate a complex that includes IP_3R3, Akt kinase, and phosphatase PP2A. In the absence of PML, Akt phosphorylation and activity was increased at the ER due to impaired PP2A activity, which resulted in impaired Ca^{2+} flux through the IP_3R because of its Akt-mediated hyperphosphorylated state. Indeed, IP_3R is a target of Akt kinase activity (Khan et al. 2006; Szado et al. 2008), and Akt activation drastically reduces IP_3R-dependent ER Ca^{2+} release (Marchi et al. 2008; Szado et al. 2008), especially through IP_3R3 phosphorylation (Marchi et al. 2012). These data have been confirmed by an independent study, which reported the localization of mechanistic target of rapamycin (mTOR) complex 2 (mTORC2) at the MAMs (Betz et al. 2013). The serine/threonine kinase (mTOR) is a pivotal regulator of autophagy and exists in two protein complexes, mTORC1 and mTORC2. The latter complex phosphorylates and activates Akt, which phosphorylates MAMs-resident proteins PACS2, IP_3R, and hexokinase 2 (HK2) to regulate MAMs integrity, Ca^{2+} flux, and energy metabolism, respectively (Betz et al. 2013).

More recently, we demonstrated that MAMs-enriched PML also exerts an important Ca^{2+}-dependent role in the autophagic process, through the AMPK/mTOR/Ulk1 pathway (Missiroli et al. 2016). We overexpressed MCU

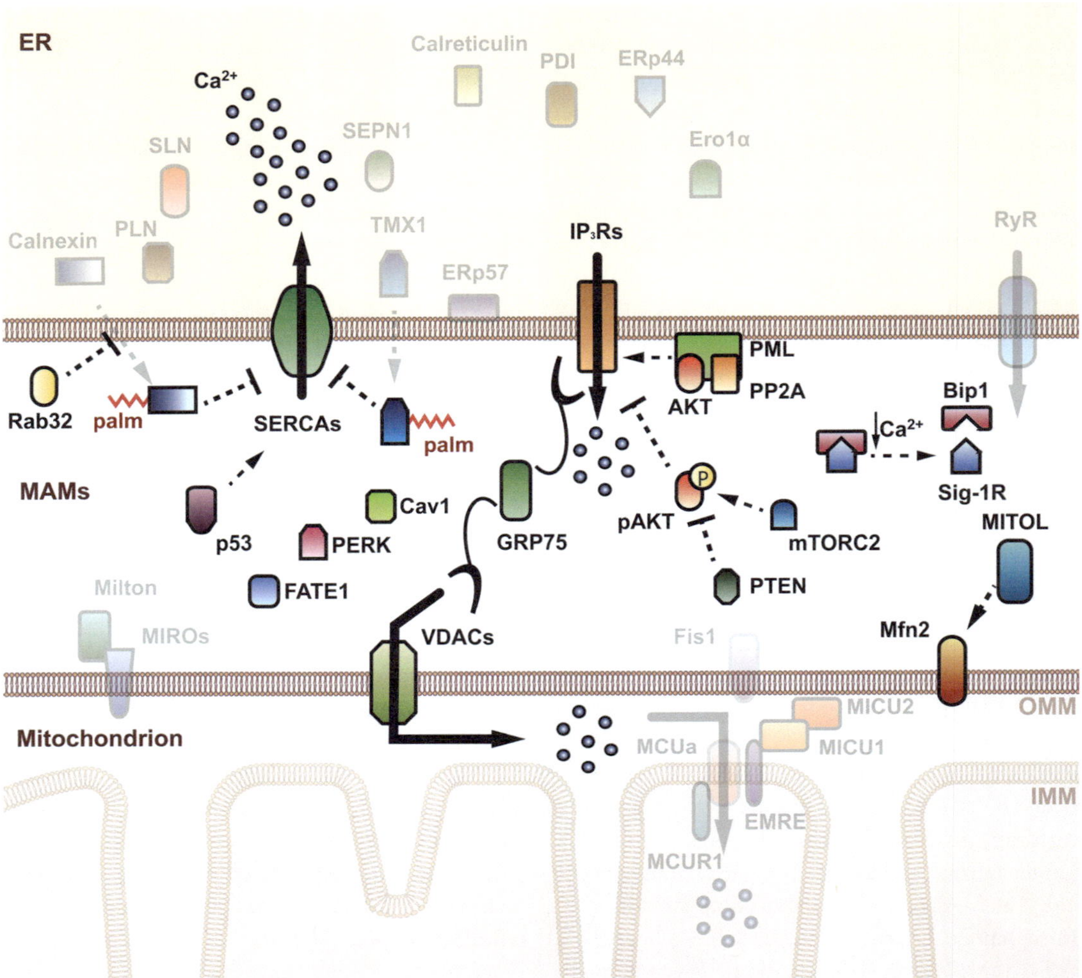

Fig. 4.3 Ca^{2+} homeostasis at the ER-mitochondria interface. Overview on the multiple molecular pathways acting at the MAMs side (See Sect. 4.4 for further details)

in PML-KO cells to verify whether downregulated ER-mitochondria Ca^{2+} transfer is important for the induction of autophagy. We demonstrated that increasing the ability of mitochondria to accumulate Ca^{2+} in PML-KO cells suppressed AMPK activity, thereby repressing autophagic flux. These data suggest that PML controls autophagy at MAMs through its effects on Ca^{2+} homeostasis and that the loss of PML from MAMs results in autophagy activation, a feature that promotes cell survival under stress conditions and thus facilitates malignant cell growth.

Among tumor suppressors, phosphatase and tensin homolog deleted on chromosome 10 (PTEN) was found to be localized at the ER and MAMs, where it modulates Ca^{2+} transfer from ER to mitochondria in a protein phosphatase-dependent manner that counteracts the Akt-mediated reduction in Ca^{2+} release via IP_3Rs (Bononi et al. 2013). Moreover, the tumor suppressor p53 also resides at ER and MAMs, modulating ER Ca^{2+} efflux to the mitochondria through modulation of the oxidative state of SERCA pumps (Giorgi et al. 2015b; Giorgi et al. 2015a).

Very recently FATE1, a cancer-testis antigen, has been implicated in the regulation of ER-mitochondria distance and Ca^{2+} uptake by mitochondria (Doghman-Bouguerra et al. 2016). FATE1 is localized at the interface between the ER and mitochondria and decreases sensitivity to mitochondrial Ca^{2+}-dependent pro-apoptotic stimuli and to chemotherapeutic drugs. This study emphasized how the ER-mitochondria uncoupling activity of FATE1 is harnessed by cancer cells to escape apoptotic death and resist the action of chemotherapeutic drugs.

Taken together, these observations highlight the role of ER-dependent Ca^{2+} release as a general mediator in many cell deaths or cell survival scenarios and reinforce the importance of MAMs in Ca^{2+} handling.

4.5 Conclusions

The importance of MAMs in the control of various cellular processes and its relevance for human health is underpinned by the disease that has linked to dysregulation and dysfunction of the ER-mitochondria interface and architecture. These diseases include obesity (Arruda et al. 2014) and type II diabetes (Tubbs et al. 2014), as well as Parkinson's and Alzheimer's diseases (Paillusson et al. 2016). MAM disorganization results in abnormal ER-mitochondria Ca^{2+} flux, which contributes to the formation of aberrant mitochondrial structures and deep metabolic alterations that are typical features of these pathological conditions. Based on the recent observations suggesting an optimal distance of 30–85 nm between IP_3R and the MCU complex to achieve effective Ca^{2+} transfer and generation of Ca^{2+} inputs (Qi et al. 2015), we need novel technologies that enable an accurate and highly precise measurement of the functional changes occurring at MAMs to elucidate the local Ca^{2+} transport and signaling mechanisms.

Acknowledgments P.P. is grateful to Camilla degli Scrovegni for continuous support. P.P. is supported by: Telethon (GGP15219/B), the Italian Association for Cancer Research (AIRC) (IG-18624), the Italian Cystic Fibrosis Research Foundation (19/2014), the Italian Ministry of Education, University and Research (COFIN no. 20129JLHSY_002, FIRB no. RBAP11FXBC_002, and Futuro in Ricerca no. RBFR10EGVP_001), local funds from the University of Ferrara and the Italian Ministry of Health. C.G is supported by AIRC (MFAG-13521), the Italian Ministry of Health, Cariplo and local funds from the University of Ferrara. S.P. is supported by a research fellowship FISM—Fondazione Italiana Sclerosi Multipla—cod. 2014/B/3. GB and JBP are supported by grants from the Research Foundation - Flanders (FWO) (grants G.0571.12N, G.0819.13N, G.0C91.14N, G.0927.15N and G.0A34.16N) and the Research Council - KU Leuven (OT14/101). MB and MK are holders of a Ph.D. Fellowship from the FWO.

References

Amador FJ, Stathopulos PB, Enomoto M, Ikura M (2013) Ryanodine receptor calcium release channels: lessons from structure-function studies. FEBS J 280:5456–5470

Anelli T, Alessio M, Bachi A, Bergamelli L, Bertoli G, Camerini S, Mezghrani A, Ruffato E, Simmen T, Sitia R (2003) Thiol-mediated protein retention in the endoplasmic reticulum: the role of ERp44. EMBO J 22:5015–5022

Anelli T, Bergamelli L, Margittai E, Rimessi A, Fagioli C, Malgaroli A, Pinton P, Ripamonti M, Rizzuto R, Sitia R (2012) Ero1α regulates Ca^{2+} fluxes at the endoplasmic reticulum-mitochondria interface (MAM). Antioxid Redox Signal 16:1077–1087

Arasaki K, Shimizu H, Mogari H, Nishida N, Hirota N, Furuno A, Kudo Y, Baba M, Baba N, Cheng J et al (2015) A role for the ancient SNARE syntaxin 17 in regulating mitochondrial division. Dev Cell 32:304–317

Arnaudeau S, Frieden M, Nakamura K, Castelbou C, Michalak M, Demaurex N (2002) Calreticulin differentially modulates calcium uptake and release in the endoplasmic reticulum and mitochondria. J Biol Chem 277:46696–46705

Arruda AP, Pers BM, Parlakgul G, Guney E, Inouye K, Hotamisligil GS (2014) Chronic enrichment of hepatic endoplasmic reticulum-mitochondria contact leads to mitochondrial dysfunction in obesity. Nat Med 20:1427–1435

Asahi M, Nakayama H, Tada M, Otsu K (2003) Regulation of sarco(endo)plasmic reticulum Ca^{2+} adenosine triphosphatase by phospholamban and sarcolipin: implication for cardiac hypertrophy and failure. Trends Cardiovasc Med 13:152–157

Ashby MC, Tepikin AV (2001) ER calcium and the functions of intracellular organelles. Semin Cell Dev Biol 12:11–17

Baksh S, Burns K, Andrin C, Michalak M (1995) Interaction of calreticulin with protein disulfide isomerase. J Biol Chem 270:31338–31344

Ben-Hail D, Shoshan-Barmatz V (2016) VDAC1-interacting anion transport inhibitors inhibit VDAC1 oligomerization and apoptosis. Biochim Biophys Acta 1863:1612–1623

Berridge MJ (2016) The Inositol trisphosphate/calcium signaling pathway in health and disease. Physiol Rev 96:1261–1296

Betz C, Stracka D, Prescianotto-Baschong C, Frieden M, Demaurex N, Hall MN (2013) Feature Article: mTOR complex 2-Akt signaling at mitochondria-associated endoplasmic reticulum membranes (MAM) regulates mitochondrial physiology. Proc Natl Acad Sci USA 110:12526–12534

Bittremieux M, Parys JB, Pinton P, Bultynck G (2016) ER functions of oncogenes and tumor suppressors: Modulators of intracellular Ca^{2+} signaling. Biochim Biophys Acta 1863:1364–1378

Bononi A, Bonora M, Marchi S, Missiroli S, Poletti F, Giorgi C, Pandolfi PP, Pinton P (2013) Identification of PTEN at the ER and MAMs and its regulation of Ca $^{2+}$ signaling and apoptosis in a protein phosphatase-dependent manner. Cell Death Differ 20:1631–1643

Bonora M, Patergnani S, Rimessi A, De Marchi E, Suski JM, Bononi A, Giorgi C, Marchi S, Missiroli S, Poletti F et al (2012) ATP synthesis and storage. Purinergic Signal 8:343–357

Bonora M, Giorgi C, Bononi A, Marchi S, Patergnani S, Rimessi A, Rizzuto R, Pinton P (2013) Subcellular calcium measurements in mammalian cells using jellyfish photoprotein aequorin-based probes. Nat Protoc 8:2105–2118

Bonora M, Wieckowski MR, Chinopoulos C, Kepp O, Kroemer G, Galluzzi L, Pinton P (2015) Molecular mechanisms of cell death: central implication of ATP synthase in mitochondrial permeability transition. Oncogene 34:1475–1486

Bravo R, Vicencio JM, Parra V, Troncoso R, Munoz JP, Bui M, Quiroga C, Rodriguez AE, Verdejo HE, Ferreira J et al (2011) Increased ER-mitochondrial coupling promotes mitochondrial respiration and bioenergetics during early phases of ER stress. J Cell Sci 124:2143–2152

Bravo R, Gutierrez T, Paredes F, Gatica D, Rodriguez AE, Pedrozo Z, Chiong M, Parra V, Quest AF, Rothermel BA et al (2012) Endoplasmic reticulum: ER stress regulates mitochondrial bioenergetics. Int J Biochem Cell Biol 44:16–20

Briggs CA, Chakroborty S, Stutzmann GE (2017) Emerging pathways driving early synaptic pathology in Alzheimer's disease. Biochem Biophys Res Commun 483(4):988–997

Bui M, Gilady SY, Fitzsimmons RE, Benson MD, Lynes EM, Gesson K, Alto NM, Strack S, Scott JD, Simmen T (2010) Rab32 modulates apoptosis onset and mitochondria-associated membrane (MAM) properties. J Biol Chem 285:31590–31602

Camacho P, Lechleiter JD (1995) Calreticulin inhibits repetitive intracellular Ca^{2+} waves. Cell 82:765–771

Cantero-Recasens G, Fandos C, Rubio-Moscardo F, Valverde MA, Vicente R (2010) The asthma-associated ORMDL3 gene product regulates endoplasmic reticulum-mediated calcium signaling and cellular stress. Hum Mol Genet 19:111–121

Chen Y, Csordas G, Jowdy C, Schneider TG, Csordas N, Wang W, Liu Y, Kohlhaas M, Meiser M, Bergem S et al (2012) Mitofusin 2-containing mitochondrial-reticular microdomains direct rapid cardiomyocyte bioenergetic responses via interorganelle Ca^{2+} crosstalk. Circ Res 111:863–875

Cosson P, Marchetti A, Ravazzola M, Orci L (2012) Mitofusin-2 independent juxtaposition of endoplasmic reticulum and mitochondria: an ultrastructural study. PLoS One 7:e46293

Csordas G, Renken C, Varnai P, Walter L, Weaver D, Buttle KF, Balla T, Mannella CA, Hajnoczky G (2006) Structural and functional features and significance of the physical linkage between ER and mitochondria. J Cell Biol 174:915–921

Csordas G, Varnai P, Golenar T, Roy S, Purkins G, Schneider TG, Balla T, Hajnoczky G (2010) Imaging interorganelle contacts and local calcium dynamics at the ER-mitochondrial interface. Mol Cell 39:121–132

Csordas G, Golenar T, Seifert EL, Kamer KJ, Sancak Y, Perocchi F, Moffat C, Weaver D, de la Fuente PS, Bogorad R et al (2013) MICU1 controls both the threshold and cooperative activation of the mitochondrial Ca^{2+} uniporter. Cell Metab 17:976–987

de Brito OM, Scorrano L (2008) Mitofusin 2 tethers endoplasmic reticulum to mitochondria. Nature 456:605–610

De La Fuente S, Fernandez-Sanz C, Vail C, Agra EJ, Holmstrom K, Sun J, Mishra J, Williams D, Finkel T, Murphy E et al (2016) Strategic positioning and biased activity of the mitochondrial calcium uniporter in cardiac muscle. J Biol Chem 291:23343–23362

De Stefani D, Bononi A, Romagnoli A, Messina A, De Pinto V, Pinton P, Rizzuto R (2012) VDAC1 selectively transfers apoptotic Ca^{2+} signals to mitochondria. Cell Death Differ 19:267–273

Decuypere JP, Monaco G, Bultynck G, Missiaen L, De Smedt H, Parys JB (2011) The IP_3 receptor-mitochondria connection in apoptosis and autophagy. Biochim Biophys Acta 1813:1003–1013

Di Jeso B, Pereira R, Consiglio E, Formisano S, Satrustegui J, Sandoval IV (1998) Demonstration of a Ca^{2+} requirement for thyroglobulin dimerization and export to the golgi complex. Eur J Biochem 252:583–590

Di Jeso B, Ulianich L, Pacifico F, Leonardi A, Vito P, Consiglio E, Formisano S, Arvan P (2003) Folding of thyroglobulin in the calnexin/calreticulin pathway and its alteration by loss of Ca^{2+} from the endoplasmic reticulum. Biochem J 370:449–458

Doghman-Bouguerra M, Granatiero V, Sbiera S, Sbiera I, Lacas-Gervais S, Brau F, Fassnacht M, Rizzuto R, Lalli E (2016) FATE1 antagonizes calcium- and drug-induced apoptosis by uncoupling ER and mitochondria. EMBO Rep 17:1264–1280

Filadi R, Greotti E, Turacchio G, Luini A, Pozzan T, Pizzo P (2015) Mitofusin 2 ablation increases endoplasmic reticulum-mitochondria coupling. Proc Natl Acad Sci USA 112:E2174–E2181

Foskett JK, Philipson B (2015) The mitochondrial Ca^{2+} uniporter complex. J Mol Cell Cardiol 78:3–8

Foskett JK, White C, Cheung KH, Mak DO (2007) Inositol trisphosphate receptor Ca^{2+} release channels. Physiol Rev 87:593–658

Fransson A, Ruusala A, Aspenstrom P (2003) Atypical Rho GTPases have roles in mitochondrial homeostasis and apoptosis. J Biol Chem 278:6495–6502

Friedman JR, Lackner LL, West M, DiBenedetto JR, Nunnari J, Voeltz GK (2011) ER tubules mark sites of mitochondrial division. Science 334:358–362

Gerasimenko JV, Sherwood M, Tepikin AV, Petersen OH, Gerasimenko OV (2006) NAADP, cADPR and IP_3 all release Ca^{2+} from the endoplasmic reticulum and an acidic store in the secretory granule area. J Cell Sci 119:226–238

Giorgi C, Ito K, Lin HK, Santangelo C, Wieckowski MR, Lebiedzinska M, Bononi A, Bonora M, Duszynski J, Bernardi R et al (2010) PML regulates apoptosis at endoplasmic reticulum by modulating calcium release. Science 330:1247–1251

Giorgi C, Baldassari F, Bononi A, Bonora M, De Marchi E, Marchi S, Missiroli S, Patergnani S, Rimessi A, Suski JM et al (2012) Mitochondrial Ca^{2+} and apoptosis. Cell Calcium 52:36–43

Giorgi C, Bonora M, Missiroli S, Poletti F, Ramirez FG, Morciano G, Morganti C, Pandolfi PP, Mammano F, Pinton P (2015a) Intravital imaging reveals p53-dependent cancer cell death induced by phototherapy via calcium signaling. Oncotarget 6:1435–1445

Giorgi C, Bonora M, Sorrentino G, Missiroli S, Poletti F, Suski JM, Galindo Ramirez F, Rizzuto R, Di Virgilio F, Zito E et al (2015b) p53 at the endoplasmic reticulum regulates apoptosis in a Ca^{2+}-dependent manner. Proc Natl Acad Sci USA 112:1779–1784

Giorgi C, Missiroli S, Patergnani S, Duszynski J, Wieckowski MR, Pinton P (2015c) Mitochondria-associated membranes: composition, molecular mechanisms, and physiopathological implications. Antioxid Redox Signal 22:995–1019

Glater EE, Megeath LJ, Stowers RS, Schwarz TL (2006) Axonal transport of mitochondria requires milton to recruit kinesin heavy chain and is light chain independent. J Cell Biol 173:545–557

Gunter TE, Gunter KK (2001) Uptake of calcium by mitochondria: transport and possible function. IUBMB Life 52:197–204

Halestrap AP (2014) The C ring of the F_1F_o ATP synthase forms the mitochondrial permeability transition pore: a critical appraisal. Front Oncol 4:234

Hamasaki M, Furuta N, Matsuda A, Nezu A, Yamamoto A, Fujita N, Oomori H, Noda T, Haraguchi T, Hiraoka Y et al (2013) Autophagosomes form at ER-mitochondria contact sites. Nature 495:389–393

Harada M, Luo X, Murohara T, Yang B, Dobrev D, Nattel S (2014) MicroRNA regulation and cardiac calcium signaling: role in cardiac disease and therapeutic potential. Circ Res 114:689–705

Haroon S, Vermulst M (2016) Linking mitochondrial dynamics to mitochondrial protein quality control. Curr Opin Genet Dev 38:68–74

Hayashi T, Su TP (2007) Sigma-1 receptor chaperones at the ER-mitochondrion interface regulate Ca^{2+} signaling and cell survival. Cell 131:596–610

Higo T, Hattori M, Nakamura T, Natsume T, Michikawa T, Mikoshiba K (2005) Subtype-specific and ER lumenal environment-dependent regulation of inositol 1,4,5-trisphosphate receptor type 1 by ERp44. Cell 120:85–98

Hofer AM, Schulz I (1996) Quantification of intraluminal free [Ca] in the agonist-sensitive internal calcium store using compartmentalized fluorescent indicators: some considerations. Cell Calcium 20:235–242

Hong Z, Chen KH, Dasgupta A, Potus F, Dunham-Snary-K, Bonnet S, Tian L, Fu J, Breuils-Bonnet S, Provencher S et al (2017) miR-138 and miR-25 Downregulate MCU, causing pulmonary arterial Hypertension's cancer phenotype. Am J Respir Crit Care Med 195(4):515–529

Ivanova H, Vervliet T, Missiaen L, Parys JB, De Smedt H, Bultynck G (2014) Inositol 1,4,5-trisphosphate receptor-isoform diversity in cell death and survival. Biochim Biophys Acta 1843:2164–2183

Iwasawa R, Mahul-Mellier AL, Datler C, Pazarentzos E, Grimm S (2011) Fis1 and Bap31 bridge the mitochondria-ER interface to establish a platform for apoptosis induction. EMBO J 30:556–568

Joiner ML, Koval OM, Li J, He BJ, Allamargot C, Gao Z, Luczak ED, Hall DD, Fink BD, Chen B et al (2012) CaMKII determines mitochondrial stress responses in heart. Nature 491:269–273

Jonas EA, Porter GA Jr, Beutner G, Mnatsakanyan N, Alavian KN (2015) Cell death disguised: The mitochondrial permeability transition pore as the c-subunit of the F_1F_o ATP synthase. Pharmacol Res 99:382–392

Kamer KJ, Mootha VK (2015) The molecular era of the mitochondrial calcium uniporter. Nat Rev Mol Cell Biol 16:545–553

Kang S, Dahl R, Hsieh W, Shin A, Zsebo KM, Buettner C, Hajjar RJ, Lebeche D (2016) Small molecular allosteric activator of the sarco/endoplasmic reticulum Ca^{2+}-ATPase (SERCA) attenuates diabetes and metabolic disorders. J Biol Chem 291:5185–5198

Khan MT, Wagner L 2nd, Yule DI, Bhanumathy C, Joseph SK (2006) Akt kinase phosphorylation of inositol 1,4,5-trisphosphate receptors. J Biol Chem 281:3731–3737

Kirichok Y, Krapivinsky G, Clapham DE (2004) The mitochondrial calcium uniporter is a highly selective ion channel. Nature 427:360–364

Krols M, Bultynck G, Janssens S (2016) ER-Mitochondria contact sites: A new regulator of cellular calcium flux comes into play. J Cell Biol 214:367–370

Lamb CA, Yoshimori T, Tooze SA (2013) The autophagosome: origins unknown, biogenesis complex. Nat Rev Mol Cell Biol 14:759–774

Lasorsa FM, Pinton P, Palmieri L, Fiermonte G, Rizzuto R, Palmieri F (2003) Recombinant expression of the Ca^{2+}-sensitive aspartate/glutamate carrier increases mitochondrial ATP production in agonist-stimulated Chinese hamster ovary cells. J Biol Chem 278:38686–38692

Liu X, Hajnoczky G (2009) Ca^{2+}-dependent regulation of mitochondrial dynamics by the Miro-Milton complex. Int J Biochem Cell Biol 41:1972–1976

Lynes EM, Bui M, Yap MC, Benson MD, Schneider B, Ellgaard L, Berthiaume LG, Simmen T (2012) Palmitoylated TMX and calnexin target to the mitochondria-associated membrane. EMBO J 31:457–470

Lynes EM, Raturi A, Shenkman M, Ortiz Sandoval C, Yap MC, Wu J, Janowicz A, Myhill N, Benson MD, Campbell RE et al (2013) Palmitoylation is the switch that assigns calnexin to quality control or ER Ca^{2+} signaling. J Cell Sci 126:3893–3903

Lytton J, Westlin M, Hanley MR (1991) Thapsigargin inhibits the sarcoplasmic or endoplasmic reticulum Ca-ATPase family of calcium pumps. J Biol Chem 266:17067–17071

Madreiter-Sokolowski CT, Klec C, Parichatikanond W, Stryeck S, Gottschalk B, Pulido S, Rost R, Eroglu E, Hofmann NA, Bondarenko AI et al (2016) PRMT1-mediated methylation of MICU1 determines the UCP2/3 dependency of mitochondrial Ca^{2+} uptake in immortalized cells. Nat Commun 7:12897

Mallilankaraman K, Cardenas C, Doonan PJ, Chandramoorthy HC, Irrinki KM, Golenar T, Csordas G, Madireddi P, Yang J, Muller M et al (2012a) MCUR1 is an essential component of mitochondrial Ca^{2+} uptake that regulates cellular metabolism. Nat Cell Biol 14:1336–1343

Mallilankaraman K, Doonan P, Cardenas C, Chandramoorthy HC, Muller M, Miller R, Hoffman NE, Gandhirajan RK, Molgo J, Birnbaum MJ et al (2012b) MICU1 is an essential gatekeeper for MCU-mediated mitochondrial Ca^{2+} uptake that regulates cell survival. Cell 151:630–644

Marchi S, Pinton P (2014) The mitochondrial calcium uniporter complex: molecular components, structure and physiopathological implications. J Physiol 592:829–839

Marchi S, Rimessi A, Giorgi C, Baldini C, Ferroni L, Rizzuto R, Pinton P (2008) Akt kinase reducing endoplasmic reticulum Ca^{2+} release protects cells from Ca $^{2+}$-dependent apoptotic stimuli. Biochem Biophys Res Commun 375:501–505

Marchi S, Marinello M, Bononi A, Bonora M, Giorgi C, Rimessi A, Pinton P (2012) Selective modulation of subtype III IP_3R by Akt regulates ER Ca^{2+} release and apoptosis. Cell Death Dis 3:e304

Marchi S, Lupini L, Patergnani S, Rimessi A, Missiroli S, Bonora M, Bononi A, Corra F, Giorgi C, De Marchi E et al (2013) Downregulation of the mitochondrial calcium uniporter by cancer-related miR-25. Curr Biol 23:58–63

Marchi S, Patergnani S, Pinton P (2014) The endoplasmic reticulum-mitochondria connection: one touch, multiple functions. Biochimica et biophysica acta 1837:461–469

Marino M, Stoilova T, Giorgi C, Bachi A, Cattaneo A, Auricchio A, Pinton P, Zito E (2015) SEPN1, an endoplasmic reticulum-localized selenoprotein linked to skeletal muscle pathology, counteracts hyperoxidation by means of redox-regulating SERCA2 pump activity. Hum Mol Genet 24:1843–1855

Matsuzaki H, Fujimoto T, Tanaka M, Shirasawa S (2013) Tespa1 is a novel component of mitochondria-associated endoplasmic reticulum membranes and affects mitochondrial calcium flux. Biochem Biophys Res Commun 433:322–326

Melo SF, Barauna VG, Neves VJ, Fernandes T, Lara Lda S, Mazzotti DR, Oliveira EM (2015) Exercise training restores the cardiac microRNA-1 and -214 levels regulating Ca^{2+} handling after myocardial infarction. BMC Cardiovasc Disord 15:166

Mendes CC, Gomes DA, Thompson M, Souto NC, Goes TS, Goes AM, Rodrigues MA, Gomez MV, Nathanson MH, Leite MF (2005) The type III inositol 1,4,5-trisphosphate receptor preferentially transmits apoptotic Ca^{2+} signals into mitochondria. J Biol Chem 280:40892–40900

Michalak M, Groenendyk J, Szabo E, Gold LI, Opas M (2009) Calreticulin, a multi-process calcium-buffering chaperone of the endoplasmic reticulum. Biochem J 417:651–666

Missiroli S, Bonora M, Patergnani S, Poletti F, Perrone M, Gafa R, Magri E, Raimondi A, Lanza G, Tacchetti C et al (2016) PML at mitochondria-associated membranes is critical for the repression of autophagy and cancer development. Cell Rep 16:2415–2427

Montero M, Alonso MT, Carnicero E, Cuchillo-Ibanez I, Albillos A, Garcia AG, Garcia-Sancho J, Alvarez J (2000) Chromaffin-cell stimulation triggers fast millimolar mitochondrial Ca^{2+} transients that modulate secretion. Nat Cell Biol 2:57–61

Morciano G, Giorgi C, Bonora M, Punzetti S, Pavasini R, Wieckowski MR, Campo G, Pinton P (2015) Molecular identity of the mitochondrial permeability transition pore and its role in ischemia-reperfusion injury. J Mol Cell Cardiol 78:142–153

Myhill N, Lynes EM, Nanji JA, Blagoveshchenskaya AD, Fei H, Carmine Simmen K, Cooper TJ, Thomas G, Simmen T (2008) The subcellular distribution of calnexin is mediated by PACS-2. Mol Biol Cell 19:2777–2788

Naghdi S, Hajnoczky G (2016) VDAC2-specific cellular functions and the underlying structure. Biochim Biophys Acta 1863:2503–2514

Naon D, Scorrano L (2014) At the right distance: ER-mitochondria juxtaposition in cell life and death. Biochim Biophys Acta 1843:2184–2194

Naon D, Zaninello M, Giacomello M, Varanita T, Grespi F, Lakshminaranayan S, Serafini A, Semenzato M, Herkenne S, Hernandez-Alvarez MI et al (2016) Critical reappraisal confirms that Mitofusin 2 is an endoplasmic reticulum-mitochondria tether. Proc Natl Acad Sci USA 113:11249–11254

O-Uchi J, Jhun BS, Xu S, Hurst S, Raffaello A, Liu X, Yi B, Zhang H, Gross P, Mishra J et al (2014) Adrenergic signaling regulates mitochondrial Ca^{2+} uptake through Pyk2-dependent tyrosine phosphorylation of the mitochondrial Ca^{2+} uniporter. Antioxid Redox Signal 21:863–879

Paillusson S, Stoica R, Gomez-Suaga P, Lau DH, Mueller S, Miller T, Miller CC (2016) There's something wrong with my MAM; the ER-mitochondria axis and neurodegenerative diseases. Trends Neurosci 39:146–157

Pan L, Huang BJ, Ma XE, Wang SY, Feng J, Lv F, Liu Y, Li CM, Liang DD, Li J et al (2015) MiR-25 protects cardiomyocytes against oxidative damage by targeting the mitochondrial calcium uniporter. Int J Mol Sci 16:5420–5433

Papp B, Brouland JP, Arbabian A, Gelebart P, Kovacs T, Bobe R, Enouf J, Varin-Blank N, Apati A (2012) Endoplasmic reticulum calcium pumps and cancer cell differentiation. Biomolecules 2:165–186

Parys JB, De Smedt H (2012) Inositol 1,4,5-trisphosphate and its receptors. Adv Exp Med Biol 740:255–279

Patergnani S, Suski JM, Agnoletto C, Bononi A, Bonora M, De Marchi E, Giorgi C, Marchi S, Missiroli S, Poletti F et al (2011) Calcium signaling around Mitochondria Associated Membranes (MAMs). Cell Commun Signal 9:19

Patergnani S, Missiroli S, Marchi S, Giorgi C (2015) Mitochondria-associated endoplasmic reticulum membranes microenvironment: targeting autophagic and apoptotic pathways in cancer therapy. Frontiers in oncology 5:173

Patron M, Raffaello A, Granatiero V, Tosatto A, Merli G, De Stefani D, Wright L, Pallafacchina G, Terrin A, Mammucari C et al (2013) The mitochondrial calcium uniporter (MCU): molecular identity and physiological roles. J Biol Chem 288:10750–10758

Patron M, Checchetto V, Raffaello A, Teardo E, Vecellio Reane D, Mantoan M, Granatiero V, Szabo I, De Stefani D, Rizzuto R (2014) MICU1 and MICU2 finely tune the mitochondrial Ca^{2+} uniporter by exerting opposite effects on MCU activity. Mol Cell 53:726–737

Paupe V, Prudent J, Dassa EP, Rendon OZ, Shoubridge EA (2015) CCDC90A (MCUR1) is a cytochrome c oxidase assembly factor and not a regulator of the mitochondrial calcium uniporter. Cell Metab 21:109–116

Pinton P, Leo S, Wieckowski MR, Di Benedetto G, Rizzuto R (2004) Long-term modulation of mitochondrial Ca^{2+} signals by protein kinase C isozymes. J Cell Biol 165:223–232

Pinton P, Rimessi A, Marchi S, Orsini F, Migliaccio E, Giorgio M, Contursi C, Minucci S, Mantovani F, Wieckowski MR et al (2007) Protein kinase C β and prolyl isomerase 1 regulate mitochondrial effects of the life-span determinant p66Shc. Science 315:659–663

Popugaeva E, Pchitskaya E, Bezprozvanny I (2017) Dysregulation of neuronal calcium homeostasis in Alzheimer's disease – A therapeutic opportunity? Biochem Biophys Res Commun 483(4):998–1004

Poston CN, Krishnan SC, Bazemore-Walker CR (2013) In-depth proteomic analysis of mammalian mitochondria-associated membranes (MAM). J Proteomics 79:219–230

Prakriya M, Lewis RS (2015) Store-Operated Calcium Channels. Physiol Rev 95:1383–1436

Qi H, Li L, Shuai J (2015) Optimal microdomain crosstalk between endoplasmic reticulum and mitochondria for Ca^{2+} oscillations. Sci Rep 5:7984

Raffaello A, De Stefani D, Sabbadin D, Teardo E, Merli G, Picard A, Checchetto V, Moro S, Szabo I, Rizzuto R (2013) The mitochondrial calcium uniporter is a multimer that can include a dominant-negative pore-forming subunit. EMBO J 32:2362–2376

Raffaello A, Mammucari C, Gherardi G, Rizzuto R (2016) Calcium at the center of cell signaling: interplay between endoplasmic reticulum, mitochondria, and lysosomes. Trends Biochem Sci 41 (12):1035–1049

Rapizzi E, Pinton P, Szabadkai G, Wieckowski MR, Vandecasteele G, Baird G, Tuft RA, Fogarty KE, Rizzuto R (2002) Recombinant expression of the voltage-dependent anion channel enhances the transfer of Ca^{2+} microdomains to mitochondria. J Cell Biol 159:613–624

Raturi A, Gutierrez T, Ortiz-Sandoval C, Ruangkittisakul A, Herrera-Cruz MS, Rockley JP, Gesson K, Ourdev D, Lou PH, Lucchinetti E et al

(2016) TMX1 determines cancer cell metabolism as a thiol-based modulator of ER-mitochondria Ca^{2+} flux. J Cell Biol 214:433–444

Ribeiro CM, McKay RR, Hosoki E, Bird GS, Putney JW Jr (2000) Effects of elevated cytoplasmic calcium and protein kinase C on endoplasmic reticulum structure and function in HEK293 cells. Cell Calcium 27:175–185

Rimessi A, Giorgi C, Pinton P, Rizzuto R (2008) The versatility of mitochondrial calcium signals: from stimulation of cell metabolism to induction of cell death. Biochim Biophys Acta 1777:808–816

Rimessi A, Marchi S, Patergnani S, Pinton P (2014) H-Ras-driven tumoral maintenance is sustained through caveolin-1-dependent alterations in calcium signaling. Oncogene 33:2329–2340

Rizzuto R, Pinton P, Carrington W, Fay FS, Fogarty KE, Lifshitz LM, Tuft RA, Pozzan T (1998) Close contacts with the endoplasmic reticulum as determinants of mitochondrial Ca^{2+} responses. Science 280:1763–1766

Rizzuto R, De Stefani D, Raffaello A, Mammucari C (2012) Mitochondria as sensors and regulators of calcium signalling. Nat Rev Mol Cell Biol 13:566–578

Roderick HL, Lechleiter JD, Camacho P (2000) Cytosolic phosphorylation of calnexin controls intracellular Ca^{2+} oscillations via an interaction with SERCA2b. J Cell Biol 149:1235–1248

Rowland AA, Voeltz GK (2012) Endoplasmic reticulum-mitochondria contacts: function of the junction. Nat Rev Mol Cell Biol 13:607–625

Sala-Vila A, Navarro-Lerida I, Sanchez-Alvarez M, Bosch M, Calvo C, Lopez JA, Calvo E, Ferguson C, Giacomello M, Serafini A et al (2016) Interplay between hepatic mitochondria-associated membranes, lipid metabolism and caveolin-1 in mice. Sci Rep 6:27351

Sancak Y, Markhard AL, Kitami T, Kovacs-Bogdan E, Kamer KJ, Udeshi ND, Carr SA, Chaudhuri D, Clapham DE, Li AA et al (2013) EMRE is an essential component of the mitochondrial calcium uniporter complex. Science 342:1379–1382

Saotome M, Safiulina D, Szabadkai G, Das S, Fransson A, Aspenstrom P, Rizzuto R, Hajnoczky G (2008) Bidirectional Ca^{2+}-dependent control of mitochondrial dynamics by the Miro GTPase. Proc Natl Acad Sci USA 105:20728–20733

Schneeberger M, Dietrich MO, Sebastian D, Imbernon M, Castano C, Garcia A, Esteban Y, Gonzalez-Franquesa A, Rodriguez IC, Bortolozzi A et al (2013) Mitofusin 2 in POMC neurons connects ER stress with leptin resistance and energy imbalance. Cell 155:172–187

Sebastian D, Hernandez-Alvarez MI, Segales J, Sorianello E, Munoz JP, Sala D, Waget A, Liesa M, Paz JC, Gopalacharyulu P et al (2012) Mitofusin 2 (Mfn2) links mitochondrial and endoplasmic reticulum function with insulin signaling and is essential for normal glucose homeostasis. Proc Natl Acad Sci USA 109:5523–5528

Shimizu H, Schredelseker J, Huang J, Lu K, Naghdi S, Lu F, Franklin S, Fiji HD, Wang K, Zhu H et al (2015) Mitochondrial Ca^{2+} uptake by the voltage-dependent anion channel 2 regulates cardiac rhythmicity. Elife 4:4e0481

Smirnova E, Griparic L, Shurland DL, van der Bliek AM (2001) Dynamin-related protein Drp1 is required for mitochondrial division in mammalian cells. Mol Biol Cell 12:2245–2256

Sugiura A, Nagashima S, Tokuyama T, Amo T, Matsuki Y, Ishido S, Kudo Y, McBride HM, Fukuda T, Matsushita N et al (2013) MITOL regulates endoplasmic reticulum-mitochondria contacts via Mitofusin2. Mol Cell 51:20–34

Szabadkai G, Bianchi K, Varnai P, De Stefani D, Wieckowski MR, Cavagna D, Nagy AI, Balla T, Rizzuto R (2006) Chaperone-mediated coupling of endoplasmic reticulum and mitochondrial Ca^{2+} channels. J Cell Biol 175:901–911

Szado T, Vanderheyden V, Parys JB, De Smedt H, Rietdorf K, Kotelevets L, Chastre E, Khan F, Landegren U, Soderberg O et al (2008) Phosphorylation of inositol 1,4,5-trisphosphate receptors by protein kinase B/Akt inhibits Ca^{2+} release and apoptosis. Proc Natl Acad Sci USA 105:2427–2432

Terasaki M, Jaffe LA, Hunnicutt GR, Hammer JA 3rd (1996) Structural change of the endoplasmic reticulum during fertilization: evidence for loss of membrane continuity using the green fluorescent protein. Dev Biol 179:320–328

Tomar D, Dong Z, Shanmughapriya S, Koch DA, Thomas T, Hoffman NE, Timbalia SA, Goldman SJ, Breves SL, Corbally DP et al (2016) MCUR1 Is a scaffold factor for the MCU complex function and promotes mitochondrial bioenergetics. Cell Rep 15:1673–1685

Tooze SA, Jefferies HB, Kalie E, Longatti A, McAlpine FE, McKnight NC, Orsi A, Polson HE, Razi M, Robinson DJ et al (2010) Trafficking and signaling in mammalian autophagy. IUBMB Life 62:503–508

Tsai MF, Phillips CB, Ranaghan M, Tsai CW, Wu Y, Willliams C, Miller C (2016) Dual functions of a small regulatory subunit in the mitochondrial calcium uniporter complex. Elife 5:e15545

Tubbs E, Theurey P, Vial G, Bendridi N, Bravard A, Chauvin MA, Ji-Cao J, Zoulim F, Bartosch B, Ovize M et al (2014) Mitochondria-associated endoplasmic reticulum membrane (MAM) integrity is required for insulin signaling and is implicated in hepatic insulin resistance. Diabetes 63:3279–3294

Vais H, Tanis JE, Muller M, Payne R, Mallilankaraman K, Foskett JK (2015) MCUR1, CCDC90A, is a regulator of the mitochondrial calcium uniporter. Cell Metab 22:533–535

Van Petegem F (2012) Ryanodine receptors: structure and function. J Biol Chem 287:31624–31632
Van Petegem F (2015) Ryanodine receptors: allosteric ion channel giants. J Mol Biol 427:31–53
Vance JE (1990) Phospholipid synthesis in a membrane fraction associated with mitochondria. J Biol Chem 265:7248–7256
Vance JE (2014) MAM (mitochondria-associated membranes) in mammalian cells: lipids and beyond. Biochimica et biophysica acta 1841:595–609
Vandecaetsbeek I, Vangheluwe P, Raeymaekers L, Wuytack F, Vanoevelen J (2011) The Ca^{2+} pumps of the endoplasmic reticulum and Golgi apparatus. Cold Spring Harb Perspect Biol 3
Vangheluwe P, Sepulveda MR, Missiaen L, Raeymaekers L, Wuytack F, Vanoevelen J (2009) Intracellular Ca^{2+}- and Mn^{2+}-transport ATPases. Chem Rev 109:4733–4759
Verfaillie T, Rubio N, Garg AD, Bultynck G, Rizzuto R, Decuypere JP, Piette J, Linehan C, Gupta S, Samali A et al (2012) PERK is required at the ER-mitochondrial contact sites to convey apoptosis after ROS-based ER stress. Cell Death Differ 19:1880–1891
Vermassen E, Parys JB, Mauger JP (2004) Subcellular distribution of the inositol 1,4,5-trisphosphate receptors: functional relevance and molecular determinants. Biol Cell 96:3–17
Vittorini S, Storti S, Parri MS, Cerillo AG, Clerico A (2007) SERCA2a, phospholamban, sarcolipin, and ryanodine receptors gene expression in children with congenital heart defects. Mol Med 13:105–111
Wahl M, Sleight RG, Gruenstein E (1992) Association of cytoplasmic free Ca^{2+} gradients with subcellular organelles. J Cell Physiol 150:593–609
Wieckowski MR, Giorgi C, Lebiedzinska M, Duszynski J, Pinton P (2009) Isolation of mitochondria-associated membranes and mitochondria from animal tissues and cells. Nat Protoc 4:1582–1590
Zhang A, Williamson CD, Wong DS, Bullough MD, Brown KJ, Hathout Y, Colberg-Poley AM (2011a) Quantitative proteomic analyses of human cytomegalovirus-induced restructuring of endoplasmic reticulum-mitochondrial contacts at late times of infection. Mol Cell Proteomics 10(M111):009936
Zhang L, Liu Y, Song F, Zheng H, Hu L, Lu H, Liu P, Hao X, Zhang W, Chen K (2011b) Functional SNP in the microRNA-367 binding site in the 3'UTR of the calcium channel ryanodine receptor gene 3 (RYR3) affects breast cancer risk and calcification. Proc Natl Acad Sci USA 108:13653–13658

Ceramide Transport from the Endoplasmic Reticulum to the *Trans* Golgi Region at Organelle Membrane Contact Sites

5

Kentaro Hanada

Abstract

Lipids are the major constituents of all cell membranes and play dynamic roles in organelle structure and function. Although the spontaneous transfer of lipids between different membranes rarely occurs, lipids are appropriately transported between different organelles for their metabolism and to exert their functions in living cells. Proteins that have the biochemical capability to catalyze the intermembrane transfer of lipids are called lipid transfer proteins (LTPs). All organisms possess many types of LTPs. Recent studies revealed that LTPs are key players in the interorganelle transport of lipids at organelle membrane contact sites (MCSs). This chapter depicts how LTPs rationally operate at MCSs by using the ceramide transport protein CERT as a typical model for the LTP-mediated interorganelle transport of lipids.

Keywords

Ceramide • Sphingomyelin • CERT • Lipid transfer proteins

5.1 Introduction

Lipids are the major constituents of all cell membranes and play dynamic roles in organelle structure and function. The intracellular transport of lipids from the sites of their synthesis to their appropriate destination is a critical step in lipid metabolism because various steps in lipid biosynthesis occur in different intracellular compartments (Hanada and Voelker 2014; Holthuis and Menon 2014; Vance 2015). Historically, several mechanisms have been proposed for the transport of lipids between different organelles.

1. The spontaneous transfer of a lipid as a monomer or micelle: although this mechanism is thermodynamically possible for water-miscible lipids, it remains unclear whether it is biologically meaningful in cells.

K. Hanada (✉)
Department of Biochemistry & Cell Biology, National Institute of Infectious Diseases, 1-23-1 Toyama, Shinjuku-ku, Tokyo 162-8640, Japan
e-mail: hanak@nih.go.jp

M. Tagaya, T. Simmen (eds.), *Organelle Contact Sites*, Advances in Experimental Medicine and Biology 997, DOI 10.1007/978-981-10-4567-7_5

2. Proteinaceous carrier-mediated transfer: this mechanism is predominant, as described below (also see Table 5.1).
3. Transport vesicle-mediated delivery provides the main intracellular transport system for lipids which are oriented to the luminal side in the lipid bilayers of organelle membranes and are predominantly delivered within transport vesicles, as are secretory or transmembrane proteins. However, its contribution to the interorganelle trafficking of lipids oriented to the cytosol appears to be very limited.
4. Lateral diffusion at a transiently fused site between different organelles: although this is mechanistically possible, it remains hypothetical as a physiological system for the interorganelle trafficking of lipids.

The efficiency of the transport of an object is generally dependent on the distance between the start and end points. This distance dependence has been clearly demonstrated in simple diffusion and random walking processes. The diffusion formula (L^2/D, where L is length and D is the diffusion rate) estimated the time periods needed by an oxygen molecule (D in water = 2.4×10^{-9} m^2/s) to traverse a typical human lymphocyte (~10 m in diameter) and human tissue (~10 cm in diameter) as ~0.04 s and ~50 days, respectively. Hence, the diffusion process is not sufficient to deliver ample oxygen throughout a tissue, and, thus, a network system of capillary vessels exists in all tissues. When the distance between the start and end points is only 10 nm, the delivery of oxygen may be accomplished in ~40 ns. However, the rapid simple diffusion of hydrophobic molecules between different membranes through the aqueous cytosol rarely occurs even when the distance between the two membranes is only 10 nm because the transfer of a hydrophobic molecule from a hydrophobic environment to an aqueous phase encounters a high thermodynamic energy barrier. A catalyst that masks the hydrophobic regions of lipid molecules in the aqueous phase is needed in order to overcome this energy barrier. Proteinaceous carriers possessing these biochemical properties are called lipid transfer proteins (LTPs).

The site at which different organelle membranes come into close apposition within *meso*-scale ranges (~10 nm) is now commonly referred to as membrane contact sites (MCSs) (Helle et al. 2013; Phillips and Voeltz 2016). The close proximity of different organelle membranes was observed in the early era of cell biology by classical electron microscopy (Palade and Porter 1954), and more detailed images of contact structures were later obtained by 3D image reconstructing cryo-electron tomography (Mogelsvang et al. 2004). Nevertheless, the functions of organelle MCSs had been elusive or speculative.

In 2000 and the early 2000s, two types of key molecules were discovered: junctophilins that physically bridge the plasma membrane and endoplasmic reticulum (ER) in myocytes and play a crucial role in cytoplasmic calcium homeostasis in the muscle (Takeshima et al. 2000) and the ceramide transport protein CERT, an LTP that mediates the transport of lipid ceramides from the ER to the Golgi apparatus presumably at ER-Golgi MCSs (Hanada et al. 2003). These findings generated the evidence-based concept that organelle MCSs are functional sites for the interorganelle transport of small molecules such as ions and lipids in cells. This concept has been widely accepted since other biological systems functioning at MCSs were identified, as described in the other chapters of this book.

5.2 Biosynthesis and Translocation of Sphingolipids in Mammalian Cells

The biosynthetic pathway of sphingolipids in mammalian cells is depicted in Fig. 5.1 (Hanada 2014; Yamaji and Hanada 2015). After the condensation of l-serine and palmitoyl CoA catalyzed by serine palmitoyltransferase, its product 3-ketodihydrosphingosine is rapidly converted to dihydrosphingosine by 3-ketodihydrosphingosine reductase, and dihydrosphingosine is then *N*-acylated to

Table 5.1 LTPs that mediate the inter- or intraorganelle transport of lipids at MCSs in mammalian cells

Human LTP (yeast ortholog)	Lipid	From-to	References
CERT	Ceramide	ER-Golgi	Hanada et al. (2003)
OSBP	Sterol/PI4P	ER-Golgi	Mesmin et al. (2013)
Nir2	PI/PC and/or PI/PA	ER-Golgi and/or ER-PM	Chang and Liou (2015), Kim et al. (2015), and Yadav et al. (2015)
FAPP2	GlcCer	Golgi-Golgi/TGN and/or ER-Golgi	D'Angelo et al. (2007) and Halter et al. (2007)
Ceramide-1-phosphate transfer protein	Ceramide-1-phosphate	TGN-PM (?)	Simanshu et al. (2013)
ORP5 and 8 (Osh6 and 7)	PS/PI4P	ER-PM	Chung et al. (2015), Maeda et al. (2013), and Moser von Filseck et al. (2015)
α-tocopherol transfer protein	α-tocopherol (vitamin E)	Lysosome-PM (?)	Kono et al. (2013)
GramD1(?) (Lam4/Ysp2/Ltc2)	Sterol	ER-PM	Gatta et al. (2015)
GramD1(?) (Lam6/Ltc1)	Sterol	ER-Mito, and ER-vacuole	Murley et al. (2015)
StAR	Sterol	OM-IM of Mito	Kallen et al. (1998)
PRELID1 (Ups1)+ TRIAP1(Mdm35)	PA	OM-IM of Mito	Connerth et al. (2012), Potting et al. (2013), and Tamura et al. (2012)
PRELID2 (Ups2)+ TRIAP1 (Mdm35)	PS	OM-IM of Mito	Aaltonen et al. (2016), Miyata et al. (2016), and Tamura et al. (2012)
NPC2	Sterol	MVB-NPC1 in LE (?)	Cheruku et al. (2006) and Gong et al. (2016)
Extended synaptotagmin 2[a]	Various glycerophospholipids	ER-PM	Giordano et al. (2013) and Schauder et al. (2014)

[a]Extended synaptotagmins may have to be regarded as a lipid tunnel, not lipid transfer proteins as discussed in (Schauder et al. 2014)

PI phosphatidylinositol, *PC* phosphatidylcholine, *PA* phosphatidic acid, *PS* phosphatidylserine, *TG* triacylglycerol, *TGN trans* Golgi network, *OM-IM of Mito* from the outer membrane to the inner membrane of mitochondria, *NPC* Niemann-Pick type C, *MVB-NPC1 in LE* from multivesicular bodies to the NPC1 protein in late endosomes

generate dihydroceramide by enzymes of the CerS family (also referred to as dihydrosphingosine-*N*-acyltransferases). Dihydroceramide is desaturated to generate ceramide. These reactions to produce ceramide, which is the common intermediate for the biosynthesis of sphingomyelin (SM) and glycosphingolipids (GSLs), occur in the cytosolic leaflet of the ER bilayer. Ceramide is then delivered to the luminal side of the Golgi apparatus and converted to SM by SM synthase, which catalyzes the transfer of phosphocholine from phosphatidylcholine to ceramide. The human genome contains genes for two different SM synthases, SMS1 and SMS2. SMS1, which is responsible for the de novo synthesis of SM, is localized to the *trans* Golgi region, while SMS2 mainly resides at the plasma membrane and plays a key role in the resynthesis of SM from ceramide generated at the plasma membrane. Ceramide is also converted to glucosylceramide (GlcCer) by GlcCer synthase (catalyzing the transfer of glucose from UDP-glucose to ceramide), the catalytic site of which is oriented to the cytosolic side. The mammalian genome possesses the sole gene (*UGCG*) of GlcCer synthase. GlcCer synthase is mainly distributed to the *cis*/*medial* Golgi region, whereas a small amount of the enzyme may also be distributed to a subregion of the ER. After being transported to the luminal side of the Golgi apparatus, GlcCer is converted to more complex glycosphingolipids.

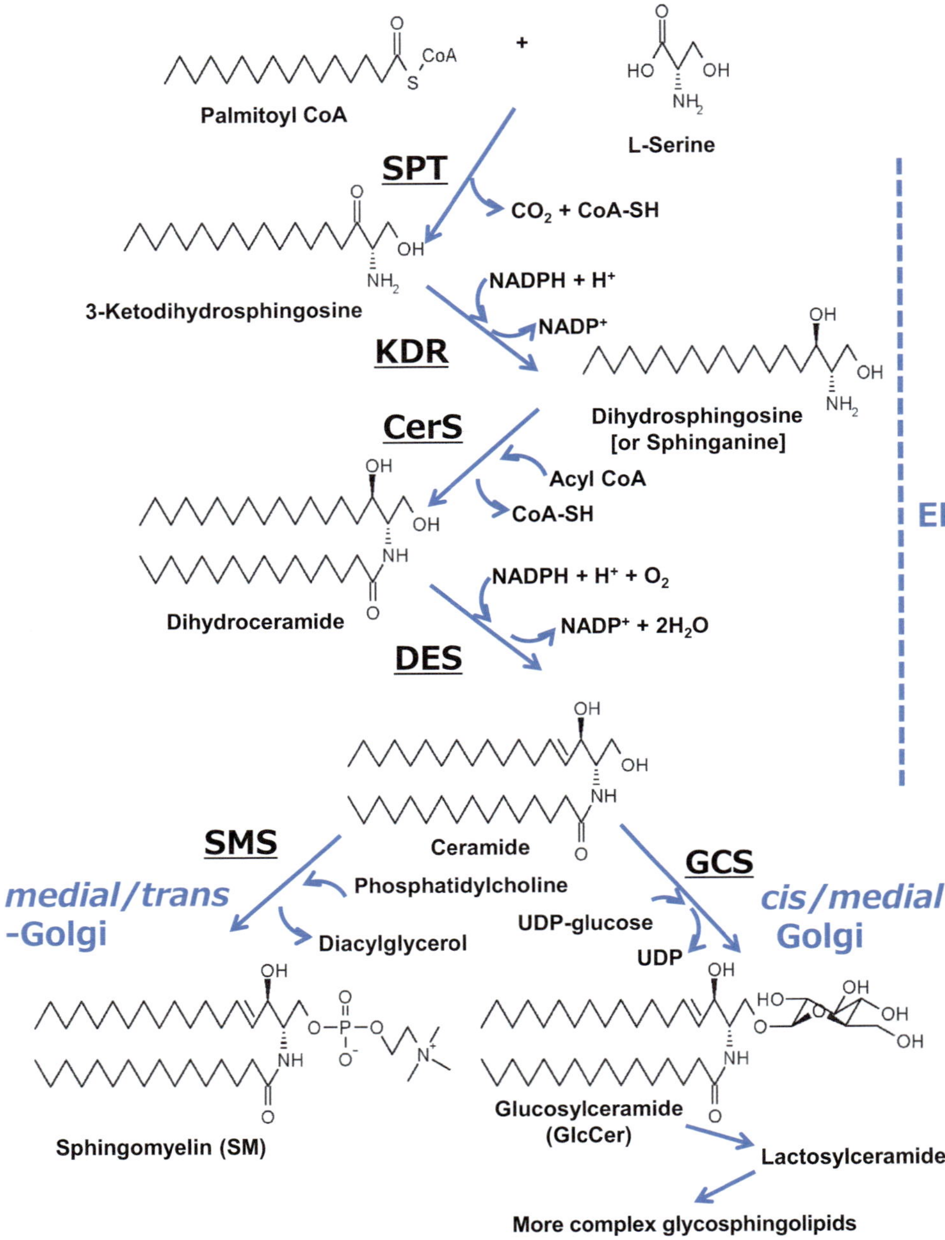

Fig. 5.1 Biosynthetic pathway of sphingolipids in mammalian cells. The biosynthetic pathway of typical sphingolipids in mammalian cells is shown. Underlined names represent enzymes: SPT, serine palmitoyltransferase; KDR, 3-ketodihydrosphingosine reductase; CerS, ceramide synthase (or dihydrosphingosine-*N*-acyltransferase); DES, dihydroceramide desaturase; SMS, sphingomyelin synthase; GCS, glucosylceramide synthase. Blue letters indicate organelles in which the enzymes are localized

5.3 Brief Remark on the Discovery of CERT

Lysenin is an earthworm-derived cytolysin, the membrane receptor of which is SM (Yamaji-Hasegawa et al. 2003). In our attempt to isolate Chinese hamster ovary (CHO) cell mutants defective in SM metabolism, several cell mutants were established as lysenin-resistant variants (Hanada et al. 1998). One variant named the LY-A cell line was found to be defective in the de novo synthesis of SM, but not glycosphingolipids, without any discernible defect in the enzymatic activity of SM synthase (Hanada et al. 1998). LY-A cells were subsequently shown to be defective in the transport of ceramide from the ER to the Golgi apparatus, whereas the ER to Golgi transport of membrane-bound secretory proteins was normal, strongly suggesting that the ER to Golgi transport mechanism for ceramide is distinct from that for proteins (Fukasawa et al. 1999). Cytosol exchange experiments within plasma membrane-permeabilized cells revealed that a cytosolic factor(s) required for ceramide transport is impaired in LY-A cells (Funakoshi et al. 2000). After screening of cDNA that may confer phenotypic recovery to LY-A cells, a human cDNA encoding a cytosolic ~70 kD-protein (named CERT) was identified (Hanada et al. 2003). CERT was found to have several functional modules, which appeared to be suitable for the molecular machinery to catalyze the transfer of ceramide from the ER to the Golgi apparatus (Hanada et al. 2003) (see below for more details). The *CERT* gene of LY-A cells was shown to have a missense mutation, which causes a dysfunction in CERT (Hanada et al. 2003). CERT is genetically identical to Goodpasture antigen-binding protein (GPBP) or its splicing variants, which were initially annotated as a serine/threonine kinase of type 4 collagen (Raya et al. 2000; Raya et al. 1999). This kinase annotation may have to be revisited, because there is no discernible kinase domain or ATP-binding motif in GPBP or CERT (Hanada 2014; Raya et al. 1999).

5.4 Functional Domains and Motifs in CERT

CERT is composed of several functional domains and motifs (Fig. 5.2).

5.4.1 PH Domain for Associating with the Golgi Apparatus

The N-terminal region consisting of ~100 amino acids forms a pleckstrin homology (PH) domain. The PH domain of CERT preferentially binds to phosphatidylinositol 4-monophosphate (PI4P) among the various phosphoinositides (Hanada et al. 2003; Levine and Munro 2002; Sugiki et al. 2012). PI4P is mainly distributed to the plasma membrane, Golgi apparatus, *trans* Golgi network, and late endosomes/lysosomes and serves not only as the main precursor for phosphatidylinositol (4,5)-bisphosphate but also as an organelle-selective functional module by itself (Hammond et al. 2014; Ji et al. 2015; Santiago-Tirado and Bretscher 2011; Venditti et al. 2016). The PH domain of CERT selectively associates with the Golgi complex. However, the precise mechanisms by which the CERT PH domain distinguishes between the Golgi and other PI4P-containing organelles currently remain unknown. Several LTPs such as oxysterol-binding protein (OSBP), OSBP-related proteins (ORPs), and four-phosphate-adaptor proteins (FAPPs) have PI4P-preferential PH domains, the amino acid sequences of which are similar to that of the CERT PH domain. The PH domains of OSBP and FAPP1 interact with the ARF1 protein in addition to PI4P in order to target the Golgi membrane, which may explain the Golgi preference of these PH domains (He et al. 2011; Levine and Munro 2002). Nevertheless, the PH domain of CERT does not appear to interact with ARF1 (Roy and Levine 2004), and no alternative accessory proteins for the CERT PH domain have been identified to date. Coexisting lipids with PI4P in membranes may play a role in the Golgi membrane preference of the CERT PH domain as described below.

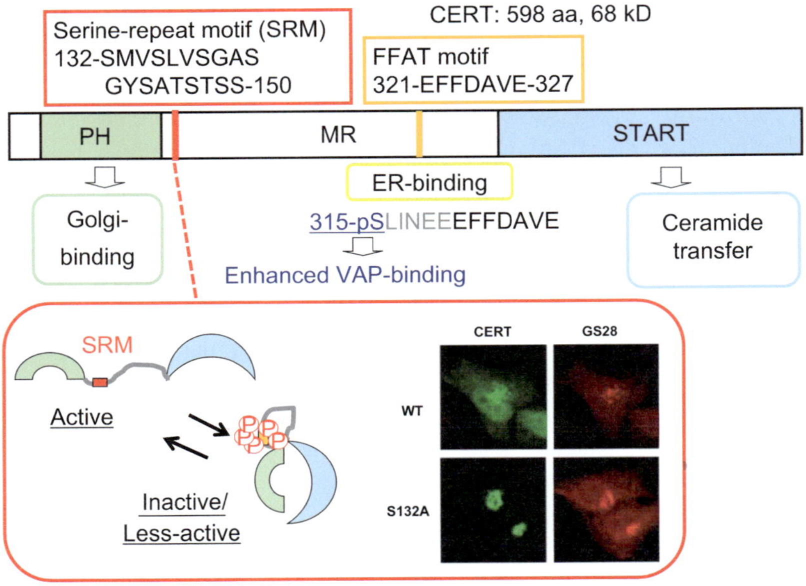

Fig. 5.2 CERT consists of various functional modules. The elucidated functions of domains and motifs in human CERT are illustrated. Regarding the detailed functions of the respective domains and motifs, refer to the text. *Inset*, downregulation of CERT by the hyper-phosphorylation of the SRM. *Left panel*, schematic model; *right panel*, intracellular distribution of wild-type CERT and its S132A mutant in HeLa cells (the panel is cited from our previous study (Kumagai et al. 2007)). GS28, a Golgi marker

The 3D structure of the CERT PH domain has been solved by solution NMR (Sugiki et al. 2012) and X-ray crystallography (Prashek et al. 2013). The findings of an NMR study showed that the CERT PH domain has two types of binding regions: a conventional phosphoinositide-binding pocket and a previously unrecognized region named a basic groove with a loop having a tryptophan pair (Sugiki et al. 2012). The basic groove appears to mildly interact with phospholipid bilayers containing acidic lipids such as phosphatidylserine even without PI4P (Sugiki et al. 2012). The synergetic effects of the two regions result in high affinity of the CERT PH for PI4P-embedded phospholipid membranes such as the Golgi membrane (Sugiki et al. 2012).

5.4.2 START Domain for the Intermembrane Transfer of Ceramide

START domains are a family of lipid transfer domains, and the human genome has 15 START domains. Although START domains have mutual similarities not only in their amino acid sequences but also their 3D structures, different START domains have different lipid substrate specificities (Clark 2012; Thorsell et al. 2011). The C-terminal ~210 amino acid region of CERT forms a START domain (Fig. 5.2). Natural long-chain ceramide produced in cells is firmly embedded in membranes because of its high hydrophobicity. The START domain of CERT efficiently extracts ceramide from membranes and facilitates the intermembrane transfer of ceramide (Hanada et al. 2003).

The lipid transfer activity of the CERT START domain is highly specific to ceramides, but flexible in various ceramide species (Hanada et al. 2003; Kumagai et al. 2005); CERT has the ability to transfer ceramide subspecies with different *N*-acyl chain lengths and natural ceramide isoforms (dihydroceramide and phytoceramide). Its activity to transfer diacylglycerol, which structurally resembles ceramide, is only 5–10% of the activity toward ceramide in vitro, and no discernible activity was detected for sphingosine, SM, or cholesterol. In co-crystals of the START domain of CERT in complex with ceramide, one ceramide molecule is buried in a long amphiphilic cavity, and at the far end of the cavity, the amide and hydroxyl groups of ceramide form a hydrogen bond network with specific amino acid residues (Kudo et al. 2008). In the amphipathic cavity, there is limited space around the C1 hydroxyl group of the bound ceramide molecule in the co-crystals, which indicates that it is spatially impossible to accommodate complex sphingolipids (such as SM and GSLs) with a bulky head group in the cavity, while different sets of amino acid residues in the cavity participate in hydrophobic interactions with the different types of hydrocarbon chains of ceramides (Kudo et al. 2008). These structural features underlie the strict substrate specificity to lipid classes and flexibility to ceramide species of the CERT START domain.

5.4.3 FFAT Motif for Associating with the ER

The middle region (amino acid residues 118–370; hereafter referred to as MR) between the PH and START domains in CERT is predicted to be largely unstructured (Hanada 2014). Although the biological meaning of the apparently unstructured regions of proteins had long been neglected, their advantageous aspects have manifested in this decade under the concept of "intrinsically disordered proteins" (Uversky 2016). In spite of its unstructured nature, the MR has several modules that are indispensable for the function and regulation of CERT.

One of these modules is a short peptidic motif named "two phenylalanines in an acidic tract" (FFAT), which associates with VAMP-associated protein (VAP), an ER-resident membrane protein (Loewen et al. 2003). Several LTPs that have been demonstrated or predicted to transfer lipids from the ER to other organelles have FFAT or FFAT-like motifs (Loewen et al. 2003; Mikitova and Levine 2012). The CERT FFAT motif (321-EFFDAVE-327), which matches the canonical consensus sequence of FFAT motifs (EFFDAxE), has been shown to interact with VAP and be crucial for the function of CERT in the ER to Golgi trafficking of ceramide (Fig. 5.2) (Kawano et al. 2006). X-ray crystallography and solution NMR analyses of complexes of a truncated soluble VAP with an FFAT motif peptide showed the 3D structures of symmetrical VAP dimers with two FFAT peptides bound at the interface of VAP by electrostatic interactions (Furuita et al. 2010; Kaiser et al. 2005).

5.4.4 Phosphorylation Sites in the MR

5.4.4.1 Downregulation by Phosphorylation

When CERT is expressed in HeLa cells, multiple (up to 10) amino acid residues in a serine-repeated motif (SRM; 132-SMVSLVSGAS-GYSATSTSS-150, serine/threonine residues in the motif are underlined) of the protein are phosphorylated (Fig. 5.2) (Kumagai et al. 2007). When these ten serine/threonine residues of the SRM are replaced with glutamic acid residues (the so-called 10E mutation), the resultant CERT 10E mutant exhibits reduced PI4P-binding and ceramide transfer activities (Kumagai et al. 2007). Casein kinase 1γ2 acts as a kinase to sequentially phosphorylate the serine/threonine residues in the SRM (Tomishige et al. 2009). When this kinase is overexpressed, the SRM of CERT is hyper-phosphorylated, and its function is repressed (Tomishige et al. 2009). Therefore, the CERT 10E mutant appears to functionally mimic endogenous hyper-phosphorylated CERT in the SRM. The repression of PI4P-binding activity by the 10E mutation may be rescued by removing the START domain, while the repression of ceramide transfer activity may also be rescued by removing the PH domain (Kumagai et al. 2007). These results suggest an inhibition mechanism by conformational changes to mask the PH and START domains with each other (Fig. 5.2).

Under normal culture conditions, most of the population of wild-type CERT is highly phosphorylated in the SRM in HeLa cells and is distributed to the cytosol, whereas the remaining unphosphorylated or hypophosphorylated population is distributed to the Golgi region at the fluorescent microscopic level (Kumagai et al.

2007). In contrast, constitutively active CERT S132A is predominantly localized to the Golgi region (Fig. 5.2, *inset*) (Kumagai et al. 2007). Thus, the hyper-phosphorylated SRM form of CERT appears to serve as an inactive pool sequestered into the cytosol, and upon its de-phosphorylation, CERT is recruited to ER-Golgi membrane contact sites at which it operates its physiological function (for this model, see below) (Kumagai et al. 2007).

A priming phosphorylation event at Ser132 is a prerequisite for the casein kinase 1-catalyzed sequential phosphorylation of other serine/threonine residues in the SRM of CERT (Kumagai et al. 2007; Tomishige et al. 2009). Therefore, the alanine replacement of S132 causes the loss of SRM phosphorylation, thereby rendering the CERT mutant constitutively active (Fugmann et al. 2007; Kumagai et al. 2007; Tomishige et al. 2009). A large-scale human genetic study has recently revealed that a mental retardation disorder with dominant inheritance is caused by a missense mutation in the human *CERT* (or *COL4A3BP*) gene, which replaces S132 with leucine (Deciphering Developmental Disorders 2015). The S132L substitution in CERT presumably destroys the repressive regulation mechanism mediated by the hyper-phosphorylated SRM on CERT, and the resultant constitutively active CERT may be detrimental to the development of the central neuronal network system. The SRM is conserved well among the CERT orthologs of various animals (Hanada 2014), suggesting that the regulatory role of the SRM is physiologically important for various organisms.

5.4.4.2 Upregulation by Phosphorylation

Protein phosphorylation preferentially occurs within intrinsically disordered protein regions (Bah and Forman-Kay 2016; Iakoucheva et al. 2004). In addition to the SRM, CERT has another phosphorylation site in the MR (Fig. 5.2). This site is Ser315, which is adjacent to the FFAT motif for binding to VAP (Olsen et al. 2006). The phosphorylation of Ser315 strongly enhances the FFAT motif-dependent binding of CERT to VAP, thereby upregulating the function of CERT (Kumagai et al. 2014). The interplay between these phosphorylation sites (i.e., the SRM and Ser315) attains a fine regulatory system for the ceramide trafficking function of CERT (Kumagai et al. 2014).

5.5 CERT Mediates the Interorganelle Trafficking of Ceramide at ER-Golgi Membrane Contact Sites

ER-Golgi membrane contact sites are subregions of the ER that are located very close (~10 nm) to *trans* Golgi stacks in mammalian cells (Mogelsvang et al. 2004). VAP, an ER resident membrane protein, displays a reticular distribution pattern, which represents the ER, in cells (Kawano et al. 2006; Kumagai et al. 2014). When wild-type CERT is co-overproduced with VAP, a small population of VAP is relocated to the perinuclear Golgi region, in which CERT is co-localized together (Kawano et al. 2006). This relocation of VAP is not induced by the co-overproduction of FFAT motif-deficient or PH domain-deficient CERT (Kawano et al. 2006). These findings together with the structural features of CERT have suggested that CERT has the potential to simultaneously bind to Golgi-localizing PI4P and ER-localizing VAP at ER-Golgi membrane contact sites, which led us to a model in which CERT mediates the interorganelle trafficking of ceramide efficiently at the contact sites in a non-vesicular manner (Fig. 5.3) (Hanada 2006; Hanada et al. 2007; Yamaji and Hanada 2015).

If the PH domain and the FFAT motif of CERT simultaneously associate with the Golgi apparatus and ER, respectively, the extraction and transfer of ceramide from the ER to the Golgi apparatus may be attained by the "neck swing" movement of the START domain. Alternatively, CERT may quickly shuttle between the two organelle membranes at the sites of contact. It is also possible that these two mechanisms are operating at ER-Golgi contact sites in cells.

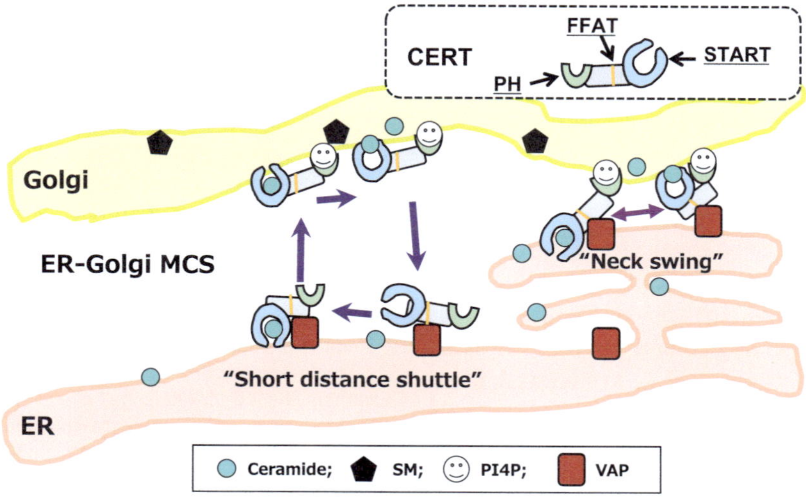

Fig. 5.3 A model of the CERT-mediated trafficking of ceramide at ER-Golgi MCSs. In this model, CERT extracts newly synthesized ceramide from the ER, depending on its START domain, carries the ceramide molecule in a non-vesicular manner, and targets the Golgi apparatus, depending on its PI4P-recognizing PH domain. After the release of ceramide at the Golgi apparatus, CERT returns to the ER, depending on its VAP-interacting FFAT motif. The rapid trafficking of ceramide through "short-distance shuttling" by CERT or through the "neck swing" movement of the START domain may occur at sites of contact between the ER and *trans* Golgi cisternae. In the neck swing model, CERT simultaneously associates with the ER (via the FFAT motif) and Golgi apparatus (via the PH domain) and catalyzes the interorganelle transfer of ceramide without the step involving the cytosolic diffusion of the protein

5.6 Consideration on the Evolutionary Aspect of CERT

Gene disruption experiments confirmed that CERT is important for the de novo synthesis of SM in human cervical carcinoma-derived HeLa cells (Yamaji and Hanada 2014) and in mice (Rao et al. 2014; Wang et al. 2009) as well as ceramide phosphoethanolamine, a close relative of SM, in the fruit fly (Rao et al. 2007). Clear orthologs of human CERT exist in various multicellular animals, but not in single-cellular eukaryotes, plants, or prokaryotes, and this phylogenetic distribution pattern of CERT is similar to that of SM (Hanada 2014). SM is chemically more robust and inert than many other phospholipid types, but is non-covalently interactive due to its acyl amide group and the 3-hyrdoxyl group of the sphingoid base moiety, both of which are absent in glycerophospholipids (Hanada 2014). SM is preferentially localized to the exoplasmic leaflet of the plasma membrane in animal cells. CERT may have evolved to produce SM because multicellular animals may somehow have necessitated the creation of a robust and inert phospholipid as a major constituent in the cell surface and, to meet this necessity, may have produced the new machinery CERT to deliver ceramide rapidly from the ER to the *trans* Golgi compartment, in which the major SM synthase SMS1 is localized, while bypassing earlier Golgi compartments, in which another anabolic enzyme GlcCer synthase is localized (Hanada 2014).

5.7 Future Directions

In this decade, it has been widely accepted that the LTP-mediated transfer of lipids at MCSs is the predominant mechanism underlying the interorganelle transport of lipids (Table 5.1). The discovery of CERT paved the way for the establishment of the prevailing concept of the interorganelle transport of lipids. There are, however, various unsolved questions in the intracellular trafficking of lipids. The aforementioned model depicted by CERT does not explain the uphill movement of lipids. This may not be problematic for the transport of ceramide from the ER to the Golgi apparatus because ceramide is changed to SM upon its arrival at the Golgi apparatus, which means that the movement direction may be energetically downhill. In contrast, several lipid types obviously move against their concentration gradients in cells. For example, although cholesterol is synthesized in the ER, its level is low in this organelle and the highest in the PM at a steady state. Hence, an uphill movement step must be included in the transport of cholesterol from the ER to the PM. One mechanism underlying this uphill movement is LTP-mediated lipid exchange between different organelles with metabolic coupling of the dephosphorylation and resynthesis of PI4P, through which one ATP is consumed in one complete reaction cycle, generating a driving force for the uphill movement of the lipid (Mesmin et al. 2013). Although this mechanism appears to operate in the ORP/Osh protein-mediated transport of sterol(s) from the ER to the Golgi apparatus (Mesmin et al. 2013) and in that of phosphatidylserine from the ER to the PM (Chung et al. 2015; Moser von Filseck et al. 2015), it currently remains unclear whether other LTP-mediated uphill movements of lipids may be accounted for by analogous mechanisms. It is important to note that since cytosolic LTPs are topologically inaccessible to lipids oriented to the luminal side of organelles, intramembrane transport systems and interorganelle vesicular transport systems are also important for the network of lipid trafficking events. Although the interplay of these systems with LTP-mediated systems presumably operates in cells, this interplay remains poorly understood. Therefore, further studies are needed in order to elucidate these unresolved issues.

Acknowledgments I deeply thank my current and previous co-workers for their invaluable contributions to studies concerning CERT biology. This work was supported by AMED-CREST.

References

Aaltonen MJ, Friedman JR, Osman C, Salin B, di Rago JP, Nunnari J, Langer T, Tatsuta T (2016) MICOS and phospholipid transfer by Ups2-Mdm35 organize membrane lipid synthesis in mitochondria. J Cell Biol 213:525–534

Bah A, Forman-Kay JD (2016) Modulation of intrinsically disordered protein function by post-translational modifications. J Biol Chem 291:6696–6705

Chang CL, Liou J (2015) Phosphatidylinositol 4,5-Bisphosphate Homeostasis Regulated by Nir2 and Nir3 Proteins at Endoplasmic Reticulum-Plasma Membrane Junctions. J Biol Chem 290:14289–14301

Cheruku SR, Xu Z, Dutia R, Lobel P, Storch J (2006) Mechanism of cholesterol transfer from the Niemann-Pick type C2 protein to model membranes supports a role in lysosomal cholesterol transport. J Biol Chem 281:31594–31604

Chung J, Torta F, Masai K, Lucast L, Czapla H, Tanner LB, Narayanaswamy P, Wenk MR, Nakatsu F et al (2015) PI4P/phosphatidylserine countertransport at ORP5- and ORP8-mediated ER-plasma membrane contacts. Science 349:428–432

Clark BJ (2012) The mammalian START domain protein family in lipid transport in health and disease. J Endocrinol 212:257–275

Connerth M, Tatsuta T, Haag M, Klecker T, Westermann B, Langer T (2012) Intramitochondrial transport of phosphatidic acid in yeast by a lipid transfer protein. Science 338:815–818

D'Angelo G, Polishchuk E, Di Tullio G, Santoro M, Di Campli A, Godi A, West G, Bielawski J, Chuang CC et al (2007) Glycosphingolipid synthesis requires FAPP2 transfer of glucosylceramide. Nature 449: 62–67

Deciphering Developmental Disorders S (2015) Large-scale discovery of novel genetic causes of developmental disorders. Nature 519:223–228

Fugmann T, Hausser A, Schoffler P, Schmid S, Pfizenmaier K, Olayioye MA (2007) Regulation of secretory transport by protein kinase D-mediated phosphorylation of the ceramide transfer protein. J Cell Biol 178:15–22

Fukasawa M, Nishijima M, Hanada K (1999) Genetic evidence for ATP-dependent endoplasmic reticulum-to-Golgi apparatus trafficking of ceramide for sphingomyelin synthesis in Chinese hamster ovary cells. J Cell Biol 144:673–685

Funakoshi T, Yasuda S, Fukasawa M, Nishijima M, Hanada K (2000) Reconstitution of ATP- and cytosol-dependent transport of de novo synthesized ceramide to the site of sphingomyelin synthesis in semi-intact cells. J Biol Chem 275:29938–29945

Furuita K, Jee J, Fukada H, Mishima M, Kojima C (2010) Electrostatic interaction between oxysterol-binding protein and VAMP-associated protein A revealed by NMR and mutagenesis studies. J Biol Chem 285: 12961–12970

Gatta AT, Wong LH, Sere YY, Calderon-Norena DM, Cockcroft S, Menon AK, Levine TP (2015) A new family of StART domain proteins at membrane contact sites has a role in ER-PM sterol transport. Elife 4

Giordano F, Saheki Y, Idevall-Hagren O, Colombo SF, Pirruccello M, Milosevic I, Gracheva EO, Bagriantsev SN, Borgese N et al (2013) PI(4,5)P(2)-dependent and Ca(2+)-regulated ER-PM interactions mediated by the extended synaptotagmins. Cell 153:1494–1509

Gong X, Qian H, Zhou X, Wu J, Wan T, Cao P, Huang W, Zhao X, Wang X et al (2016) Structural Insights into the Niemann-Pick C1 (NPC1)-Mediated Cholesterol Transfer and Ebola Infection. Cell 165:1467–1478

Halter D, Neumann S, van Dijk SM, Wolthoorn J, de Maziere AM, Vieira OV, Mattjus P, Klumperman J, van Meer G et al (2007) Pre- and post-Golgi translocation of glucosylceramide in glycosphingolipid synthesis. J Cell Biol 179:101–115

Hammond GR, Machner MP, Balla T (2014) A novel probe for phosphatidylinositol 4-phosphate reveals multiple pools beyond the Golgi. J Cell Biol 205: 113–126

Hanada K (2006) Discovery of the molecular machinery CERT for endoplasmic reticulum-to-Golgi trafficking of ceramide. Mol Cell Biochem 286:23–31

Hanada K (2014) Co-evolution of sphingomyelin and the ceramide transport protein CERT. Biochim Biophys Acta 1841:704–719

Hanada K, Voelker D (2014) Interorganelle trafficking of lipids: preface for the thematic review series. Traffic 15:889–894

Hanada K, Hara T, Fukasawa M, Yamaji A, Umeda M, Nishijima M (1998) Mammalian cell mutants resistant to a sphingomyelin-directed cytolysin. Genetic and biochemical evidence for complex formation of the LCB1 protein with the LCB2 protein for serine palmitoyltransferase. J Biol Chem 273:33787–33794

Hanada K, Kumagai K, Yasuda S, Miura Y, Kawano M, Fukasawa M, Nishijima M (2003) Molecular machinery for non-vesicular trafficking of ceramide. Nature 426:803–809

Hanada K, Kumagai K, Tomishige N, Kawano M (2007) CERT and intracellular trafficking of ceramide. Biochim Biophys Acta 1771:644–653

He J, Scott JL, Heroux A, Roy S, Lenoir M, Overduin M, Stahelin RV, Kutateladze TG (2011) Molecular basis of phosphatidylinositol 4-phosphate and ARF1 GTPase recognition by the FAPP1 pleckstrin homology (PH) domain. J Biol Chem 286:18650–18657

Helle SC, Kanfer G, Kolar K, Lang A, Michel AH, Kornmann B (2013) Organization and function of membrane contact sites. Biochim Biophys Acta 1833:2526–2541

Holthuis JC, Menon AK (2014) Lipid landscapes and pipelines in membrane homeostasis. Nature 510: 48–57

Iakoucheva LM, Radivojac P, Brown CJ, O'Connor TR, Sikes JG, Obradovic Z, Dunker AK (2004) The importance of intrinsic disorder for protein phosphorylation. Nucleic Acids Res 32:1037–1049

Ji C, Zhang Y, Xu P, Xu T, Lou X (2015) Nanoscale landscape of phosphoinositides revealed by specific pleckstrin homology (PH) domains using single-molecule superresolution imaging in the plasma membrane. J Biol Chem 290:26978–26993

Kaiser SE, Brickner JH, Reilein AR, Fenn TD, Walter P, Brunger AT (2005) Structural basis of FFAT motif-mediated ER targeting. Structure 13:1035–1045

Kallen CB, Billheimer JT, Summers SA, Stayrook SE, Lewis M, Strauss JF 3rd (1998) Steroidogenic acute regulatory protein (StAR) is a sterol transfer protein. J Biol Chem 273:26285–26288

Kawano M, Kumagai K, Nishijima M, Hanada K (2006) Efficient trafficking of ceramide from the endoplasmic reticulum to the Golgi apparatus requires a VAMP-associated protein-interacting FFAT motif of CERT. J Biol Chem 281:30279–30288

Kim YJ, Guzman-Hernandez ML, Wisniewski E, Balla T (2015) Phosphatidylinositol-Phosphatidic Acid Exchange by Nir2 at ER-PM Contact Sites Maintains Phosphoinositide Signaling Competence. Dev Cell 33: 549–561

Kono N, Ohto U, Hiramatsu T, Urabe M, Uchida Y, Satow Y, Arai H (2013) Impaired alpha-TTP-PIPs interaction underlies familial vitamin E deficiency. Science 340:1106–1110

Kudo N, Kumagai K, Tomishige N, Yamaji T, Wakatsuki S, Nishijima M, Hanada K, Kato R (2008) Structural basis for specific lipid recognition by CERT responsible for nonvesicular trafficking of ceramide. Proc Natl Acad Sci U S A 105:488–493

Kumagai K, Yasuda S, Okemoto K, Nishijima M, Kobayashi S, Hanada K (2005) CERT mediates intermembrane transfer of various molecular species of ceramides. J Biol Chem 280:6488–6495

Kumagai K, Kawano M, Shinkai-Ouchi F, Nishijima M, Hanada K (2007) Interorganelle trafficking of ceramide is regulated by phosphorylation-dependent cooperativity between the PH and START domains of CERT. J Biol Chem 282:17758–17766

Kumagai K, Kawano-Kawada M, Hanada K (2014) Phosphoregulation of the ceramide transport protein CERT at serine 315 in the interaction with VAMP-associated protein (VAP) for inter-organelle trafficking of ceramide in mammalian cells. J Biol Chem 289:10748–10760

Levine TP, Munro S (2002) Targeting of Golgi-specific pleckstrin homology domains involves both PtdIns 4-kinase-dependent and -independent components. Curr Biol 12:695–704

Loewen CJ, Roy A, Levine TP (2003) A conserved ER targeting motif in three families of lipid binding

proteins and in Opi1p binds VAP. EMBO J 22: 2025–2035
Maeda K, Anand K, Chiapparino A, Kumar A, Poletto M, Kaksonen M, Gavin AC (2013) Interactome map uncovers phosphatidylserine transport by oxysterol-binding proteins. Nature 501:257–261
Mesmin B, Bigay J, Moser von Filseck J, Lacas-Gervais S, Drin G, Antonny B (2013) A four-step cycle driven by PI(4)P hydrolysis directs sterol/PI(4)P exchange by the ER-Golgi tether OSBP. Cell 155:830–843
Mikitova V, Levine TP (2012) Analysis of the key elements of FFAT-like motifs identifies new proteins that potentially bind VAP on the ER, including two AKAPs and FAPP2. PLoS One 7:e30455
Miyata N, Watanabe Y, Tamura Y, Endo T, Kuge O (2016) Phosphatidylserine transport by Ups2-Mdm35 in respiration-active mitochondria. J Cell Biol 214: 77–88
Mogelsvang S, Marsh BJ, Ladinsky MS, Howell KE (2004) Predicting function from structure: 3D structure studies of the mammalian Golgi complex. Traffic 5:338–345
Moser von Filseck J, Copic A, Delfosse V, Vanni S, Jackson CL, Bourguet W, Drin G (2015) Phosphatidylserine transport by ORP/Osh proteins is driven by phosphatidylinositol 4-phosphate. Science 349: 432–436
Murley A, Sarsam RD, Toulmay A, Yamada J, Prinz WA, Nunnari J (2015) Ltc1 is an ER-localized sterol transporter and a component of ER-mitochondria and ER-vacuole contacts. J Cell Biol 209:539–548
Olsen JV, Blagoev B, Gnad F, Macek B, Kumar C, Mortensen P, Mann M (2006) Global, in vivo, and site-specific phosphorylation dynamics in signaling networks. Cell 127:635–648
Palade GE, Porter KR (1954) Studies on the endoplasmic reticulum. I. Its identification in cells in situ. J Exp Med 100:641–656
Phillips MJ, Voeltz GK (2016) Structure and function of ER membrane contact sites with other organelles. Nat Rev Mol Cell Biol 17:69–82
Potting C, Tatsuta T, Konig T, Haag M, Wai T, Aaltonen MJ, Langer T (2013) TRIAP1/PRELI complexes prevent apoptosis by mediating intramitochondrial transport of phosphatidic acid. Cell Metab 18:287–295
Prashek J, Truong T, Yao X (2013) Crystal structure of the pleckstrin homology domain from the ceramide transfer protein: implications for conformational change upon ligand binding. PLoS One 8:e79590
Rao RP, Yuan C, Allegood JC, Rawat SS, Edwards MB, Wang X, Merrill AH Jr, Acharya U, Acharya JK (2007) Ceramide transfer protein function is essential for normal oxidative stress response and lifespan. Proc Natl Acad Sci U S A 104:11364–11369
Rao RP, Scheffer L, Srideshikan SM, Parthibane V, Kosakowska-Cholody T, Masood MA, Nagashima K, Gudla P, Lockett S et al (2014) Ceramide transfer protein deficiency compromises organelle function and leads to senescence in primary cells. PLoS One 9:e92142
Raya A, Revert F, Navarro S, Saus J (1999) Characterization of a novel type of serine/threonine kinase that specifically phosphorylates the human goodpasture antigen. J Biol Chem 274:12642–12649
Raya A, Revert-Ros F, Martinez-Martinez P, Navarro S, Rosello E, Vieites B, Granero F, Forteza J, Saus J (2000) Goodpasture antigen-binding protein, the kinase that phosphorylates the goodpasture antigen, is an alternatively spliced variant implicated in autoimmune pathogenesis. J Biol Chem 275:40392–40399
Roy A, Levine TP (2004) Multiple pools of phosphatidylinositol 4-phosphate detected using the pleckstrin homology domain of Osh2p. J Biol Chem 279: 44683–44689
Santiago-Tirado FH, Bretscher A (2011) Membrane-trafficking sorting hubs: cooperation between PI4P and small GTPases at the trans-Golgi network. Trends Cell Biol 21:515–525
Schauder CM, Wu X, Saheki Y, Narayanaswamy P, Torta F, Wenk MR, De Camilli P, Reinisch KM (2014) Structure of a lipid-bound extended synaptotagmin indicates a role in lipid transfer. Nature 510: 552–555
Simanshu DK, Kamlekar RK, Wijesinghe DS, Zou X, Zhai X, Mishra SK, Molotkovsky JG, Malinina L, Hinchcliffe EH et al (2013) Non-vesicular trafficking by a ceramide-1-phosphate transfer protein regulates eicosanoids. Nature 500:463–467
Sugiki T, Takeuchi K, Yamaji T, Takano T, Tokunaga Y, Kumagai K, Hanada K, Takahashi H, Shimada I (2012) Structural basis for the Golgi association by the pleckstrin homology domain of the ceramide trafficking protein (CERT). J Biol Chem 287: 33706–33718
Takeshima H, Komazaki S, Nishi M, Iino M, Kangawa K (2000) Junctophilins: a novel family of junctional membrane complex proteins. Mol Cell 6:11–22
Tamura Y, Onguka O, Hobbs AE, Jensen RE, Iijima M, Claypool SM, Sesaki H (2012) Role for two conserved intermembrane space proteins, Ups1p and Ups2p, [corrected] in intra-mitochondrial phospholipid trafficking. J Biol Chem 287:15205–15218
Thorsell AG, Lee WH, Persson C, Siponen MI, Nilsson M, Busam RD, Kotenyova T, Schuler H, Lehtio L (2011) Comparative structural analysis of lipid binding START domains. PLoS One 6:e19521
Tomishige N, Kumagai K, Kusuda J, Nishijima M, Hanada K (2009) Casein kinase I{gamma}2 down-regulates trafficking of ceramide in the synthesis of sphingomyelin. Mol Biol Cell 20:348–357
Uversky VN (2016) Dancing Protein Clouds: The Strange Biology and Chaotic Physics of Intrinsically Disordered Proteins. J Biol Chem 291:6681–6688
Vance JE (2015) Phospholipid synthesis and transport in mammalian cells. Traffic 16:1–18

Venditti R, Masone MC, Wilson C, De Matteis MA (2016) PI(4)P homeostasis: Who controls the controllers? Adv Biol Regul 60:105–114

Wang X, Rao RP, Kosakowska-Cholody T, Masood MA, Southon E, Zhang H, Berthet C, Nagashim K, Veenstra TK et al (2009) Mitochondrial degeneration and not apoptosis is the primary cause of embryonic lethality in ceramide transfer protein mutant mice. J Cell Biol 184:143–158

Yadav S, Garner K, Georgiev P, Li M, Gomez-Espinosa E, Panda A, Mathre S, Okkenhaug H, Cockcroft S et al (2015) RDGBalpha, a PtdIns-PtdOH transfer protein, regulates G-protein-coupled PtdIns(4,5)P2 signalling during Drosophila phototransduction. J Cell Sci 128: 3330–3344

Yamaji T, Hanada K (2014) Establishment of HeLa cell mutants deficient in sphingolipid-related genes using TALENs. PLoS One 9:e88124

Yamaji T, Hanada K (2015) Sphingolipid metabolism and interorganellar transport: localization of sphingolipid enzymes and lipid transfer proteins. Traffic 16:101–122

Yamaji-Hasegawa A, Makino A, Baba T, Senoh Y, Kimura-Suda H, Sato SB, Terada N, Ohno S, Kiyokawa E et al (2003) Oligomerization and pore formation of a sphingomyelin-specific toxin, lysenin. J Biol Chem 278:22762–22770

6 Endoplasmic Reticulum – Plasma Membrane Crosstalk Mediated by the Extended Synaptotagmins

Yasunori Saheki

Abstract

The endoplasmic reticulum (ER) possesses multiplicity of functions including protein synthesis, membrane lipid biogenesis, and Ca^{2+} storage and has broad localization throughout the cell. While the ER and most other membranous organelles are highly interconnected via vesicular traffic that relies on membrane budding and fusion reactions, the ER forms direct contacts with virtually all other membranous organelles, including the plasma membrane (PM), without membrane fusion. Growing evidence suggests that these contacts play major roles in cellular physiology, including the regulation of Ca^{2+} homeostasis and signaling and control of cellular lipid homeostasis. Extended synaptotagmins (E-Syts) are evolutionarily conserved family of ER-anchored proteins that tether the ER to the PM in PM $PI(4,5)P_2$-dependent and cytosolic Ca^{2+}-regulated manner. In addition, E-Syts possess a cytosolically exposed lipid-harboring module that confers the ability to transfer/exchange glycerolipids between the ER and the PM at E-Syts-mediated ER-PM contacts. In this chapter, the functions of ER-PM contacts and their role in non-vesicular lipid transport with special emphasis on the crosstalk between the two bilayers mediated by E-Syts will be discussed.

Keywords

E-Syts • SMP domain • Ca^{2+} • Non-vesicular lipid transport • Lipid transfer proteins

6.1 Introduction

The ER forms a complex network of tubules and sheets that extends throughout the interior of cells. Although ER membranes are functionally

Y. Saheki (✉)
Lee Kong Chian School of Medicine, Nanyang Technological University, Singapore 308232, Singapore
e-mail: yasunori.saheki@ntu.edu.sg

M. Tagaya, T. Simmen (eds.), *Organelle Contact Sites*, Advances in Experimental Medicine and Biology 997, DOI 10.1007/978-981-10-4567-7_6

connected to all membranes of secretory and endocytic pathways via vesicular transport, they only fuse with each other and with vesicles involved in retrograde transport from the Golgi complex. However, accumulating evidence suggests that the close appositions between the ER and the membranes of all other membranous organelles, including the PM, regulates cellular physiology without membrane fusion. These sites are involved in numerous functions, including the regulation of Ca^{2+} homeostasis, control of ER-localized enzymes that act in trans on the apposed membrane, organelle dynamics, and exchange of lipids between membranes (Holthuis and Levine 2005; Elbaz and Schuldiner 2011; Friedman and Voeltz 2011; Toulmay and Prinz 2011; Stefan et al. 2011; Mesmin et al. 2013; Helle et al. 2013; Prinz 2014; Gallo et al. 2016; Phillips and Voeltz 2016).

ER-PM contacts are universal structures that are present in all eukaryotic cells. In cells of higher eukaryotes, ER-PM contacts were first described by electron microscopy in the 1950s in muscle (Porter and Palade 1957). They were subsequently also observed in neurons (Rosenbluth 1962; Metuzals et al. 1997; Spacek and Harris 1997). Since then, it has become clear that they represent a general feature of all eukaryotic cells, from unicellular organisms to specialized mammalian cells (Friedman and Voeltz 2011; Phillips and Voeltz 2016; Perez-Sancho et al. 2016).

Although ER-PM contacts had been observed for decades, the molecular mechanisms that govern the formation of these contacts were largely elusive. The distance between the ER and the PM at contact sites is estimated to be 10–30 nm based on electron micrographs, suggesting the presence of molecular tethers that define the distance and functional property of these contacts (Orci et al. 2009; West et al. 2011; Fernandez-Busnadiego et al. 2015; Perni et al. 2015; Varnai et al. 2007).

Identification of molecules that are responsible for ER-PM tethering has contributed significantly to our understanding of the function of ER-PM contacts in cellular physiology. The properties of these molecules are directly linked to numerous functions of ER-PM contacts. For example, physical coupling of ER-localized STIM1 and PM-localized Orai Ca^{2+} channel is responsible for store-operated Ca^{2+} entry (SOCE) (Liou et al. 2005; Zhang et al. 2005; Feske et al. 2006; Lewis 2007; Orci et al. 2009). A number of ER-localized enzymes including protein and lipid phosphatases have been reported to regulate the activity of PM substrates in trans (Braithwaite et al. 2006; Stuible and Tremblay 2010; Stefan et al. 2011; Dickson et al. 2016).

Moreover, growing evidence suggests the critical roles of lipid transfer proteins in the regulation of cellular lipid homeostasis without inducing membrane fusion via their property to transfer lipids from one compartment to another at membrane contact sites (Prinz 2014; Holthuis and Menon 2014; Gallo et al. 2016). Some of these lipid transfer proteins localize and function at ER-PM contacts and also act as tethering molecules of these two membranes. The analyses of these tethers have revealed fundamental roles of the membrane contact sites in lipid regulation. One of these tethers is the E-Syts (tricalbins in yeast). In this chapter, the crosstalk between the ER and the PM mediated by the E-Syts and its implication in lipid regulation will be discussed.

6.2 ER-PM Crosstalk Mediated by the E-Syts

6.2.1 Identification of E-Syts/Tricalbins as ER-PM Tether

While the STIM1-Orai coupling enhances ER-PM contacts, ER-PM contacts exist even before recruitment of STIM1 to the sites. Furthermore, the molecules responsible for this coupling are not conserved in yeast, while ER-PM contacts are evolutionarily conserved in all eukaryotic cells. In yeast, the bulk of the ER (greater than 50%) is cortically localized (West et al. 2011). Supporting the role of ER-PM contacts in lipid regulation, lipid synthesizing enzymes and lipid transport machinery are enriched at these contacts (Pichler et al. 2001; Tavassoli et al. 2013; Maeda et al. 2013; Chung et al. 2015; Moser von Filseck et al. 2015; Chang and Liou 2015; Kim et al. 2015; Gatta et al.

2015). Thus, proteins that mediate cortical ER formation in yeast might regulate fundamental properties of ER-PM contacts, including lipid regulation. Given the importance of lipid homeostasis in all eukaryotic cells, the mechanisms that regulate ER-PM tethering are likely to be highly conserved.

In 2012, the Prinz and Emr labs reported that three yeast ER proteins, the tricalbins (Tcb1p, Tcb2p, and Tcb3p), are selectively enriched in cortical ER and involved in ER-PM tethering (Toulmay and Prinz 2012; Manford et al. 2012). Mammalian homologs of the tricalbins are E-Syts; both tricalbins and E-Syts share similar domain organization with an N-terminal hydrophobic region, followed by a cytosolic synaptotagmin-like mitochondrial-lipid-binding protein (SMP) domain and several C2 domains.

As their name indicates, E-Syts/tricalbins also share a similar domain organization with classical synaptotagmins (Min et al. 2007). However, E-Syts contain cytosolic SMP domain, which is absent in synaptotagmins, and possess strikingly distinct molecular functions. While synaptotagmins are anchored to secretory or synaptic vesicles as Ca^{2+} sensors for SNAP (soluble NSF attachment protein) Receptor (SNARE)-mediated membrane fusion (Chapman 2008; Sudhof 2002, 2012), E-Syts are anchored to ER membranes and involved in ER-PM tethering and in harboring and transferring lipids via their SMP domain (see below). In the following sections, the role of E-Syts in ER-PM tethering as well as in control of PM lipid homeostasis will be discussed in details.

6.2.2 E-Syts Mediate PI(4,5)P_2-Dependent and Ca^{2+}-Regulated ER-PM Contacts

There are three E-Syts (E-Syt1, E-Syt2, and E-Syt3) in mammalian cells, which form homo- and heterodimers (Giordano et al. 2013; Saheki et al. 2016). They all localize at the ER through their N-terminal hydrophobic hairpin sequence that anchors them to ER membrane. The N-terminal hydrophobic region is followed by an SMP domain and multiple C2 domains [i.e., five in the case of E-Syt1 (C2A, C2B, C2C, C2D, C2E) and three in the case of E-Syt2 and E-Syt3 (C2A, C2B, C2C)] (Fig. 6.1a). E-Syts tether the ER to the PM via their C2 domain-dependent interactions with the PM that critically depends on the presence of PI(4,5)P_2 in the PM (Giordano et al. 2013).

Overexpression of E-Syt2 and E-Syt3 induces massive formation of ER-PM contacts through their C2C domain-dependent interaction with PM PI(4,5)P_2. E-Syt1, on the other hand, recruits the ER to the PM in the presence of high levels of cytosolic Ca^{2+} by Ca^{2+}-sensing property of its C2C domain (Giordano et al. 2013; Fernandez-Busnadiego et al. 2015; Idevall-Hagren et al. 2015). As they form heteromeric complexes, the E-Syts confer cytosolic Ca^{2+} regulation to PI(4,5)P_2-regulated ER-PM contact formation. Supporting the role of E-Syt heteromers in Ca^{2+}-dependent ER-PM contact formation, the depletion of E-Syts leads to the loss of Ca^{2+} sensitivity in ER-PM tethering, which can be rescued by re-expression of E-Syt1 (Giordano et al. 2013; Saheki et al. 2016).

Measurement of Ca^{2+} concentration required for E-Syt1-recruitment to the PM in semi-intact cells revealed the EC50 of such recruitment in low micromolar Ca^{2+} range, similar to synaptotagmins. Accordingly, E-Syt1 accumulation at ER-PM contacts occurred only upon experimental manipulations known to achieve these levels of Ca^{2+} via its influx from outside of the cells, such as membrane depolarization in excitable cells and SOCE (Idevall-Hagren et al. 2015). However, E-Syts-dependent contacts are not required for SOCE (Giordano et al. 2013). Thus, ER-PM contacts mediated by the E-Syts are functionally distinct from those mediated by STIM1 and Orai (Giordano et al. 2013; Quintana et al. 2015; Jing et al. 2015).

6.2.3 E-Syts Possess the SMP Domain, a Lipid-Harboring Module That Belongs to TULIP Superfamily

The presence of the SMP domain, a member of a family of lipid-binding modules that is often present in proteins localized at membrane contact sites (Kornmann et al. 2009; Kopec et al. 2010;

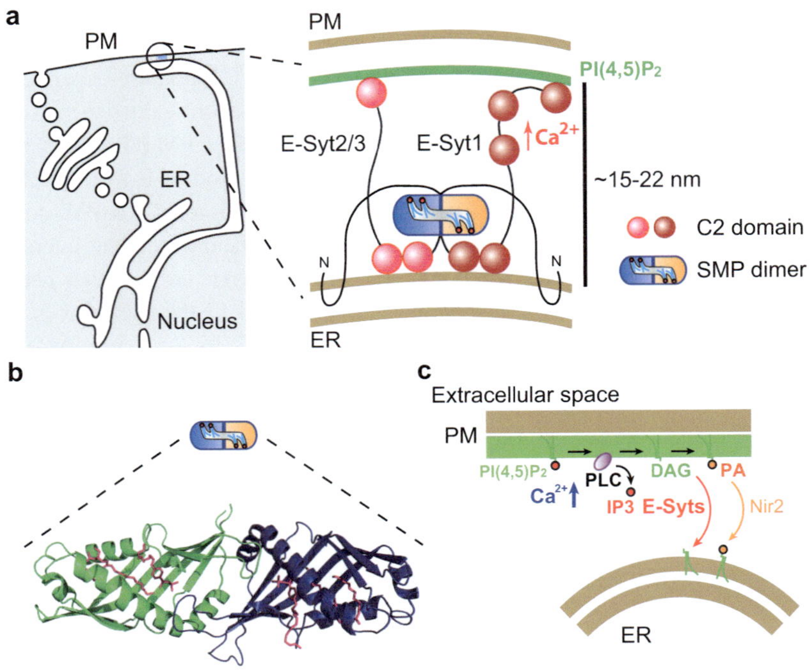

Fig. 6.1 E-Syts mediate ER-PM tethering and regulate lipid transfer. (**a**) E-Syts, anchored to ER membrane, form homo- and heterodimer and control ER-PM tethering via their C2 domain-dependent interaction with the PM that is additionally regulated by cytosolic Ca^{2+}. The SMP dimer likely mediates lipid transfer by shuttling the two membranes. (**b**) The crystal structure of the SMP domain. The SMP domain forms a dimer; hydrophobic residues line the channel. Lipid fatty acids are shown in red. Adapted from Schauder et al. (2014). (**c**) Schematics of the potential role of E-Syts in the control of PM lipid homeostasis in response to PLC activation. E-Syts may cooperate with Nir2 in this process. *IP3* inositol 1,4,5-trisphosphate, *DAG* diacylglycerol, *PA* phosphatidic acid, *PLC* phospholipase C

Toulmay and Prinz 2012; Kopec et al. 2011; Alva and Lupas 2016; Reinisch and De Camilli 2016), suggested that E-Syts may transfer lipids between membranes of the ER and the PM. Structural study of the SMP domain of E-Syt2 revealed that the SMP domain has a β-barrel structure and dimerizes to form a ~9 nm long cylinder traversed by a channel lined entirely with hydrophobic residues (Schauder et al. 2014) (Fig. 6.1b). A mass spectrometry analysis of the lipids harbored by the SMP domain revealed the presence of a wide variety of glycerolipids without a particular specificity against their head groups (Schauder et al. 2014). This is in good agreement with the solved structure of the SMP domain in which acyl chains are inserted to its deep hydrophobic groove while the head groups are exposed to the outside the groove (Schauder et al. 2014). The similar lipid-harboring property was also reported for the SMP domains of the ERMES [endoplasmic reticulum (ER)-mitochondria encounter structure] complex, which localizes at ER-mitochondria contact sites (AhYoung et al. 2015).

The structure of the SMP domain of E-Syts confirmed its homology to the tubular lipid-binding (TULIP) superfamily (Kopec et al. 2010, 2011; Schauder et al. 2014; Reinisch and De Camilli 2016). Other TULIP domain-containing proteins are also involved in harboring and, at least in some cases, transporting lipids via their TULIP domains. Although many of these proteins act in extracellular environment rather than intracellular environment (like proteins with the SMP domain), the similar structure and the shared lipid-harboring property of the SMP domain and other TULIP domains indicate their common functions in lipid harboring, sensing, and transport (Alva and Lupas 2016).

For example, cholesterol ester transfer protein (CETP) is a well-characterized serum protein that transports lipids between high-density and low-density lipoproteins. Several different lipids, including cholesterol esters, can be harbored by its TULIP domain (Qiu et al. 2007). Other serum proteins, lipopolysaccharide-binding protein (LBP) and bacterial/permeability-increasing protein (BPI), play an important role in the innate immune response against invading Gram-negative bacteria by binding to lipopolysaccharides/endotoxins present in the membrane of the bacteria (Beamer et al. 1997; Weiss 2003). A BPI-family protein NRF-5 has been reported to extract phosphatidylserine (PS) from the surface of apoptotic cells and transport it to engulfing cells during their clearance in *C. elegans* (Zhang et al. 2012b). Interestingly, NRF-5 is also involved in the process of regenerative axonal fusion during injury, which requires PS recognition (Neumann et al. 2015).

These studies suggested that the SMP domain might be involved in direct lipid transport and supported the role of membrane contact sites, which are populated by SMP domain-containing proteins, in lipid exchange reactions.

6.2.4 E-Syts Transfer Glycerolipids In Vitro via the SMP Domain

Given the presence of the SMP domain that can harbor glycerolipids promiscuously, E-Syts were thought to transport multiple lipids between membranes. In vitro analysis of lipid transport ability of E-Syt1, the Ca^{2+}-sensing E-Syt, using artificial membranes in cell-free system, revealed that E-Syt1 could indeed transport glycerolipids between membranes regardless of their head groups (Saheki et al. 2016; Yu et al. 2016); E-Syt1 even transports diacylglycerol (DAG), which is most hydrophobic due to the lack of phosphorylated head group (Saheki et al. 2016). In contrast to the ability of SNAREs to induce membrane fusion of two opposing membranes, E-Syts mediate lipid transport between tethered membranes without membrane fusion. The lipid transport ability of E-Syt1 was completely abolished when its lipid-harboring SMP domain was deleted and significantly impaired when the acyl chain insertion sites of the SMP domain were mutated (i.e., lipids can no longer be accommodated in the lipid-insertion defective mutant protein), while these mutations did not inhibit tethering of the membranes (Saheki et al. 2016; Yu et al. 2016). Synaptotagmin1 (Syt1), on the other hand, only tethers membranes without lipid transport. However, the SMP domain of E-Syt1 can confer lipid transport ability to Syt1 when this domain was artificially conjugated to the cytosolic region of Syt1 (Yu et al. 2016). Therefore, the SMP domain is necessary and sufficient for the lipid transport ability of E-Syt1.

E-Syt1-dependent lipid transport is significantly enhanced by the presence of Ca^{2+} in the micromolar range, and this effect requires PI(4,5) P_2 in one of the two opposing membranes (Saheki et al. 2016). To further support the role of Ca^{2+} in this process, E-Syt1 carrying mutations in the Ca^{2+}-binding pockets of C2C domain, the domain responsible for Ca^{2+}-dependent interaction with the PM, show much reduced lipid transfer activity (Yu et al. 2016). Given the PI(4,5)P_2-dependent and Ca^{2+}-regulated ER-PM tethering mediated by E-Syts in cells, the major action of the Ca^{2+} is likely to enhance tethering of the opposing bilayers rather than acting directly on the SMP dimer as the SMP domain does not possess Ca^{2+}-binding sites (nor predicted changes in conformation via Ca^{2+}-binding). However, it is also possible that Ca^{2+}-dependent binding to lipid bilayers of its C2A domain (Giordano et al. 2013; Chang et al. 2013), which also contains a Ca^{2+}-binding site, may assist lipid transfer by a direct and local effect on the bilayer. As a Ca^{2+}-sensitive C2A domain is present in E-Syt2 and E-Syt3 (Xu et al. 2014), even these two proteins, whose ER-PM tethering function is not regulated by cytosolic Ca^{2+}, may have a Ca^{2+}-dependent role in lipid transfer.

Collectively, these studies show that the lipid-harboring SMP domain indeed confers E-Syt1, and possibly all the E-Syts, the ability to transport lipids (Saheki et al. 2016; Yu et al. 2016).

6.2.5 Two Models of SMP Domain-Dependent Lipid Transfer: Shuttle Model and Tunnel Model

How the E-Syt SMP dimer associates with membranes to extract and deliver lipids remains elusive. In one model, termed as "tunnel model," the SMP dimer would act as bridges between the ER and the PM. Lipids could then get transferred along the hydrophobic channel as if they "snorkel" through aqueous cytosol with their hydrophobic fatty acid moieties protected in the channel, while their hydrophilic lipid head groups protrude into the cytosol (Reinisch and De Camilli 2016). The similar model was proposed for lipid transfer between lipoproteins by CETP (Qiu et al. 2007; Zhang et al. 2012a). In the tunnel model, the SMP dimer needs to directly contact with the two membranes at its ends, and lipids enter and exit there. For this model to work, the distance between the ER and the PM at ER-PM contacts has to be as close as ~9 nm. However, the electron micrographs of E-Syt-induced contacts do not support the major occurrence of such short distance (Fernandez-Busnadiego et al. 2015). Therefore, the second model, termed as "shuttle" model, is more plausible. In this model, SMP dimer is tethered between the ER and the PM by its franking PM interacting modules (C2 domains) and ER membrane anchor, shuttling freely to transfer harbored-lipids between the two bilayers. This is also consistent with the modes of lipid transfer mediated by other lipid transfer proteins (Reinisch and De Camilli 2016; Chiapparino et al. 2016). The elucidation of the precise mechanisms of lipid extraction and delivery by the SMP domain requires further investigation.

6.2.6 Control of PM Lipid Homeostasis by E-Syts

E-Syt1-dependent lipid transport is bidirectional and driven by the concentration gradient of the lipid (Saheki et al. 2016). Microsomal vesicles, representing crude ER fraction, isolated from preadipocytes were able to acquire lipids from artificial membranes; this activity relies on the expression of E-Syt1 (Yu et al. 2016). As E-Syt1-dependent lipid transfer between artificial membranes is facilitated by an increase in Ca^{2+}, E-Syts may participate in lipid transport/exchange between the membranes of the ER and the PM particularly when the levels of cytosolic Ca^{2+} increase in cells. In fact, cells lacking E-Syts do not exhibit abnormalities in the major glycerolipids at a resting condition (Saheki et al. 2016), arguing against their primary roles in the steady-state regulation of PM lipid compositions.

Because E-Syt1 can transfer lipids down their concentration gradient, E-Syts may counteract changes in membrane lipid compositions that occur during the increase in cytosolic Ca^{2+} and facilitate the transport of membrane lipids from one membrane to another in order to reset their concentration to the normal levels. For example, the formation of E-Syt-dependent ER-PM tethers in response to stimuli that elevate Ca^{2+} may help reverse accumulation of DAG in the PM by transferring it from the PM to the ER for metabolic recycling. Supporting such function, cells lacking E-Syts exhibit enhanced and sustained accumulation of DAG in the PM following $PI(4,5)P_2$ hydrolysis in response to phospholipase C (PLC) activation (Saheki et al. 2016). This effect can be rescued by re-expression of E-Syt1, but not by mutant E-Syt1 lacking the SMP domain, supporting the role of the SMP domain in lipid transfer in cells (Saheki et al. 2016).

ER-PM contacts are populated by various ER-associated lipid transfer proteins (Maeda et al. 2013; Chung et al. 2015; Moser von Filseck et al. 2015; Mesmin et al. 2013; Chang et al. 2013; Chang and Liou 2015; Kim et al. 2015; Gatta et al. 2015). Therefore, the function of E-Syts may be coupled with other lipid transfer proteins such as Nir2, which interacts with ubiquitous ER protein VAP (Lev et al. 2008; Murphy and Levine 2016) and binds to PM lipids such as DAG and phosphatidic acid (PA) (Chang and Liou 2015; Kim et al. 2015). In fact, a strong functional coupling between E-Syts and Nir2 was observed at ER-PM contacts (Saheki et al. 2016). In the absence of E-Syts, more Nir2 is recruited to the PM upon PLC activation during an increase in cytosolic Ca^{2+}. Nir2 transfers PA, a phosphorylated product of DAG, from the PM to the ER during PLC-dependent $PI(4,5)P_2$ hydrolysis for the recycling of PA for phosphatidylinositol (PI) resynthesis in the ER (Cockcroft et al. 2016; Kim et al. 2015; Yadav et al. 2015). Therefore, one of the functions of E-Syts may be to control the recycling pathway of $PI(4,5)P_2$ metabolites for lipid resynthesis in the ER (Fig. 6.1c).

6.2.7 Elusive Functions of E-Syts

Other functions of E-Syts, including the regulation of signaling, have been also proposed (Herdman and Moss 2016). In plant, all proteins previously considered as synaptotagmin homologs contain the SMP domains (Craxton 2001, 2007; Levy et al. 2015). Interestingly, one of these proteins in *Arabidopsis thaliana* is required for plant freezing tolerance as well as for resistance to mechanical stresses (Schapire et al. 2008; Yamazaki et al. 2008; Perez-Sancho et al. 2015). These studies indicate the additional role of this family of proteins in the regulation of lipid integrity in the PM. This is also consistent with the reported function of yeast tricalbins in PM integrity (Aguilar et al. 2007; Omnus et al. 2016).

Mutant mice lacking all the three E-Syts are viable and fertile, and they do not show major abnormality in development (Sclip et al. 2016; Tremblay and Moss 2016), and yeast mutants lacking tricalbins (homologs of E-Syts in yeast) grow normally (Manford et al. 2012). Therefore, these proteins are not essential for life. Given the lack of major phenotypes in E-Syt triple knockout mice, the function of E-Syts must be at least partially overlapping and redundant with the function of other proteins such as Nir2.

Strong evolutionary conservation of E-Syts from yeast to human indicates important function of these proteins in eukaryotes. In mammals, E-Syts are heterogeneously expressed in different tissues with highest expression in the lung, testis, and the immune system (Tremblay and Moss 2016). Therefore, functional studies of E-Syts in these tissues may help reveal the still elusive function of these proteins. Elucidating the physiological roles of E-Syts as well as other SMP domain-containing proteins localized at membrane contact sites will be important to gain further insights into how non-vesicular lipid transport contributes to lipid compartmentalization in cells.

6.3 Future Prospective: Neuronal Function of Non-vesicular Lipid Transport at ER-PM Contacts

Brain activity and neuronal survival rely on PM lipids that play a major role in neurotransmission (Saheki and De Camilli 2012). Thus, abnormal lipid homeostasis is linked to numerous neurological disorders (Blackstone et al. 2011; Lev et al. 2008; Di Paolo and Kim 2011).

Neurons are specialized cells with extensions of neuronal processes for efficient neurotransmission; synaptic membranes are highly dynamic and can be separated from the cell body by a significant distance. At distant nerve terminals, vesicular transport is not sufficiently rapid to replenish the loss of PM lipids. Thus, non-vesicular lipid transport is likely to have significant roles in maintenance of the neuronal PM. In fact, neuronal processes, including axons and dendrites, are highly decorated with a continuous network of the ER (Spacek and Harris 1997; Fernandez-Busnadiego et al. 2015) (Fig. 6.2a, b). Significantly, mutations in ER morphogenetic proteins as well as regulators of cellular lipid homeostasis have been identified in neurodegenerative disorders, including hereditary spastic paraplegia (HSP) and amyotrophic lateral sclerosis (ALS) (Blackstone et al. 2011; Lev et al. 2008; Dong et al. 2016). In both HSP and ALS, motor neurons with long axons are selectively lost during disease progression. The communication of neuronal PM, including synaptic membranes, with the ER occurs directly at ER-PM contacts. Therefore, ER-PM contacts may play an essential role in the maintenance of PM lipids via regulation of non-vesicular lipid transport, whose loss might contribute to the reduced viability of neurons with particularly long axons.

Importantly, genetic studies suggest a common genetic basis between these rare disorders and more common neurodegenerative conditions such as

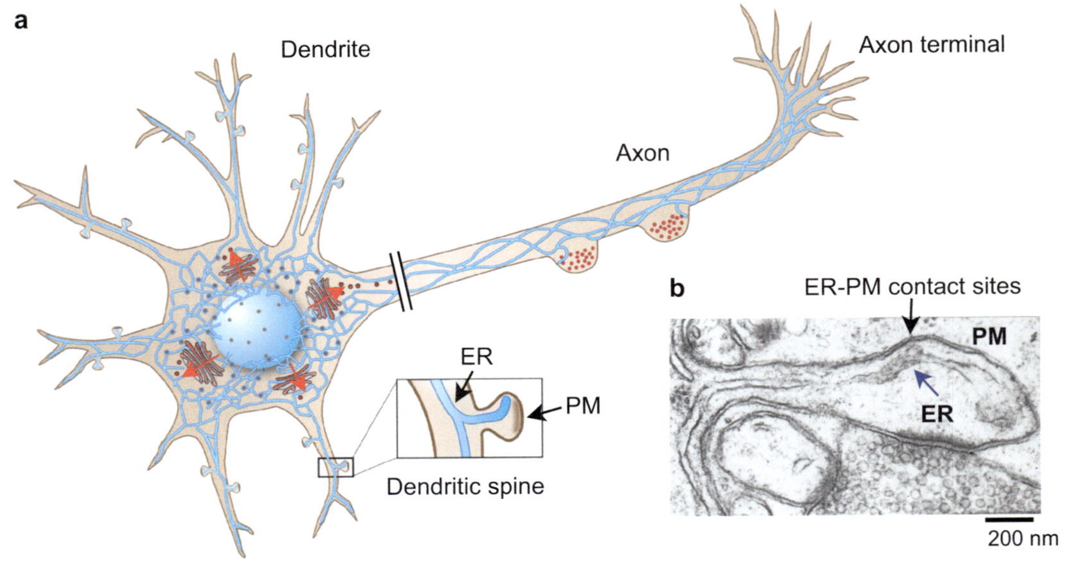

Fig. 6.2 **Maintenance and regulation of the neuronal plasma membrane.** (**a**) The ER extends throughout neuronal processes and forms physical contacts with the PM. These ER-PM contacts are implicated in lipid transfer, signaling, Ca^{2+} homeostasis, and synaptic plasticity. Red arrows indicate classical vesicular transport via the Golgi apparatus, which is far from many cell compartments and neuronal processes in neurons. (**b**) An electron micrograph showing ER-PM contacts at a dendritic spine (Adapted from Synapse Web, Kristen M. Harris, PI, http://synapseweb.clm.utexas.edu/)

Alzheimer's disease and Parkinson's disease (Novarino et al. 2014; Singleton and Hardy 2016; Small and Petsko 2015). Therefore, elucidating the mechanisms of non-vesicular lipid transport and cellular lipid homeostasis will not only advance our knowledge in fundamental cell biology but also our understanding of neurodegeneration in general.

Acknowledgements We apologize to all the investigators whose work could not be cited due to space limitations. We thank Pietro De Camilli for helpful suggestions. Work from the authors related to this review has been supported in part by a Nanyang Assistant Professorship grant (L0403050) and a Lee Kong Chian School of Medicine Start-up grant (L0412410), Nanyang Technological University, to Y.S.

References

Aguilar PS, Engel A, Walter P (2007) The plasma membrane proteins Prm1 and Fig1 ascertain fidelity of membrane fusion during yeast mating. Mol Biol Cell 18:547–556

AhYoung AP, Jiang J, Zhang J, Khoi Dang X, Loo JA, Zhou ZH, Egea PF (2015) Conserved SMP domains of the ERMES complex bind phospholipids and mediate tether assembly. Proc Natl Acad Sci USA 112:E3179–E3188

Alva V, Lupas AN (2016) The TULIP superfamily of eukaryotic lipid-binding proteins as a mediator of lipid sensing and transport. Biochim Biophys Acta 1861:913–923

Beamer LJ, Carroll SF, Eisenberg D (1997) Crystal structure of human BPI and two bound phospholipids at 2.4 angstrom resolution. Science 276:1861–1864

Blackstone C, O'Kane CJ, Reid E (2011) Hereditary spastic paraplegias: membrane traffic and the motor pathway. Nat Rev Neurosci 12:31–42

Braithwaite SP, Paul S, Nairn AC, Lombroso PJ (2006) Synaptic plasticity: one STEP at a time. Trends Neurosci 29:452–458

Chang CL, Liou J (2015) Phosphatidylinositol 4,5-Bisphosphate homeostasis regulated by Nir2 and Nir3 proteins at endoplasmic reticulum-plasma membrane junctions. J Biol Chem 290:14289–14301

Chang CL, Hsieh TS, Yang TT, Rothberg KG, Azizoglu DB, Volk E, Liao JC, Liou J (2013) Feedback regulation of receptor-induced Ca^{2+} signaling mediated by E-Syt1 and Nir2 at endoplasmic reticulum-plasma membrane junctions. Cell Rep 5:813–825

Chapman ER (2008) How does synaptotagmin trigger neurotransmitter release? Annu Rev Biochem 77:615–641

Chiapparino A, Maeda K, Turei D, Saez-Rodriguez J, Gavin AC (2016) The orchestra of lipid-transfer proteins at the crossroads between metabolism and signaling. Prog Lipid Res 61:30–39

Chung J, Torta F, Masai K, Lucast L, Czapla H, Tanner LB, Narayanaswamy P, Wenk MR, Nakatsu F, De

Camilli P (2015) PI4P/phosphatidylserine countertransport at ORP5- and ORP8-mediated ER-plasma membrane contacts. Science 349:428–432

Cockcroft S, Garner K, Yadav S, Gomez-Espinoza E, Raghu P (2016) RdgBα reciprocally transfers PA and PI at ER-PM contact sites to maintain $PI(4,5)P_2$ homoeostasis during phospholipase C signalling in Drosophila photoreceptors. Biochem Soc Trans 44:286–292

Craxton M (2001) Genomic analysis of synaptotagmin genes. Genomics 77:43–49

Craxton M (2007) Evolutionary genomics of plant genes encoding N-terminal-TM-C2 domain proteins and the similar FAM62 genes and synaptotagmin genes of metazoans. BMC Genom 8:259

Di Paolo G, Kim TW (2011) Linking lipids to Alzheimer's disease: cholesterol and beyond. Nat Rev Neurosci 12:284–296

Dickson EJ, Jensen JB, Vivas O, Kruse M, Traynor-Kaplan AE, Hille B (2016) Dynamic formation of ER-PM junctions presents a lipid phosphatase to regulate phosphoinositides. J Cell Biol 213:33–48

Dong R, Saheki Y, Swarup S, Lucast L, Harper JW, De Camilli P (2016) Endosome-ER contacts control actin nucleation and retromer function through VAP-dependent regulation of PI4P. Cell 166:408–423

Elbaz Y, Schuldiner M (2011) Staying in touch: the molecular era of organelle contact sites. Trends Biochem Sci 36:616–623

Fernandez-Busnadiego R, Saheki Y, De Camilli P (2015) Three-dimensional architecture of extended synaptotagmin-mediated endoplasmic reticulum-plasma membrane contact sites. Proc Natl Acad Sci USA 112(16):E2004-13

Feske S, Gwack Y, Prakriya M, Srikanth S, Puppel SH, Tanasa B, Hogan PG, Lewis RS, Daly M, Rao A (2006) A mutation in Orai1 causes immune deficiency by abrogating CRAC channel function. Nature 441:179–185

Friedman JR, Voeltz GK (2011) The ER in 3D: a multifunctional dynamic membrane network. Trends Cell Biol 21:709–717

Gallo A, Vannier C, Galli T (2016) Endoplasmic reticulum-plasma membrane associations:structures and functions. Annu Rev Cell Dev Biol 32:279–301

Gatta AT, Wong LH, Sere YY, Calderon-Norena DM, Cockcroft S, Menon AK, Levine TP (2015) A new family of StART domain proteins at membrane contact sites has a role in ER-PM sterol transport. Elife 4

Giordano F, Saheki Y, Idevall-Hagren O, Colombo SF, Pirruccello M, Milosevic I, Gracheva EO, Bagriantsev SN, Borgese N, De Camilli P (2013) PI(4,5)P2-dependent and Ca^{2+}-regulated ER-PM interactions mediated by the extended synaptotagmins. Cell 153:1494–1509

Helle SC, Kanfer G, Kolar K, Lang A, Michel AH, Kornmann B (2013) Organization and function of membrane contact sites. Biochim Biophys Acta 1833:2526–2541

Herdman C, Moss T (2016) Extended-Synaptotagmins (E-Syts); the extended story. Pharmacol Res 107:48–56

Holthuis JC, Levine TP (2005) Lipid traffic: floppy drives and a superhighway. Nat Rev Mol Cell Biol 6:209–220

Holthuis JC, Menon AK (2014) Lipid landscapes and pipelines in membrane homeostasis. Nature 510:48–57

Idevall-Hagren O, Lu A, Xie B, De Camilli P (2015) Triggered Ca^{2+} influx is required for extended synaptotagmin 1-induced ER-plasma membrane tethering. EMBO J 34:2291–2305

Jing J, He L, Sun A, Quintana A, Ding Y, Ma G, Tan P, Liang X, Zheng X, Chen L et al (2015) Proteomic mapping of ER-PM junctions identifies STIMATE as a regulator of Ca^{2+} influx. Nat Cell Biol 17:1339–1347

Kim YJ, Guzman-Hernandez ML, Wisniewski E, Balla T (2015) Phosphatidylinositol-phosphatidic acid exchange by Nir2 at ER-PM contact sites maintains phosphoinositide signaling competence. Dev Cell 33:549–561

Kopec KO, Alva V, Lupas AN (2010) Homology of SMP domains to the TULIP superfamily of lipid-binding proteins provides a structural basis for lipid exchange between ER and mitochondria. Bioinformatics 26:1927–1931

Kopec KO, Alva V, Lupas AN (2011) Bioinformatics of the TULIP domain superfamily. Biochem Soc Trans 39:1033–1038

Kornmann B, Currie E, Collins SR, Schuldiner M, Nunnari J, Weissman JS, Walter P (2009) An ER-mitochondria tethering complex revealed by a synthetic biology screen. Science 325:477–481

Lev S, Ben Halevy D, Peretti D, Dahan N (2008) The VAP protein family: from cellular functions to motor neuron disease. Trends Cell Biol 18:282–290

Levy A, Zheng JY, Lazarowitz SG (2015) Synaptotagmin SYTA forms ER-plasma membrane junctions that are recruited to plasmodesmata for plant virus movement. Curr Biol 25:2018–2025

Lewis RS (2007) The molecular choreography of a store-operated calcium channel. Nature 446:284–287

Liou J, Kim ML, Heo WD, Jones JT, Myers JW, Ferrell JE Jr, Meyer T (2005) STIM is a Ca^{2+} sensor essential for Ca^{2+}-store-depletion-triggered Ca^{2+} influx. Curr Biol 15:1235–1241

Maeda K, Anand K, Chiapparino A, Kumar A, Poletto M, Kaksonen M, Gavin AC (2013) Interactome map uncovers phosphatidylserine transport by oxysterol-binding proteins. Nature 501:257–261

Manford AG, Stefan CJ, Yuan HL, Macgurn JA, Emr SD (2012) ER-to-plasma membrane tethering proteins regulate cell signaling and ER morphology. Dev Cell 23:1129–1140

Mesmin B, Bigay J, Moser von Filseck J, Lacas-Gervais S, Drin G, Antonny B (2013) A four-step cycle driven by PI(4)P hydrolysis directs sterol/PI(4)P exchange by the ER-Golgi tether OSBP. Cell 155:830–843

Metuzals J, Chang D, Hammar K, Reese TS (1997) Organization of the cortical endoplasmic reticulum in the squid giant axon. J Neurocytol 26:529–539

Min SW, Chang WP, Sudhof TC (2007) E-Syts, a family of membranous Ca^{2+}-sensor proteins with multiple C2 domains. Proc Natl Acad Sci USA 104:3823–3828

Moser von Filseck J, Copic A, Delfosse V, Vanni S, Jackson CL, Bourguet W, Drin G (2015) Phosphatidylserine transport by ORP/Osh proteins is driven by phosphatidylinositol 4-phosphate. Science 349:432–436

Murphy SE, Levine TP (2016) VAP, a versatile access point for the endoplasmic reticulum: review and analysis of FFAT-like motifs in the VAPome. Biochim Biophys Acta 1861:952–961

Neumann B, Coakley S, Giordano-Santini R, Linton C, Lee ES, Nakagawa A, Xue D, Hilliard MA (2015) EFF-1-mediated regenerative axonal fusion requires components of the apoptotic pathway. Nature 517:219–222

Novarino G, Fenstermaker AG, Zaki MS, Hofree M, Silhavy JL, Heiberg AD, Abdellateef M, Rosti B, Scott E, Mansour L et al (2014) Exome sequencing links corticospinal motor neuron disease to common neurodegenerative disorders. Science 343:506–511

Omnus DJ, Manford AG, Bader JM, Emr SD, Stefan CJ (2016) Phosphoinositide kinase signaling controls ER-PM cross-talk. Mol Biol Cell 27:1170–1180

Orci L, Ravazzola M, Le Coadic M, Shen WW, Demaurex N, Cosson P (2009) From the Cover: STIM1-induced precortical and cortical subdomains of the endoplasmic reticulum. Proc Natl Acad Sci USA 106:19358–19362

Perez-Sancho J, Vanneste S, Lee E, McFarlane HE, Esteban Del Valle A, Valpuesta V, Friml J, Botella MA, Rosado A (2015) The Arabidopsis synaptotagmin1 is enriched in endoplasmic reticulum-plasma membrane contact sites and confers cellular resistance to mechanical stresses. Plant Physiol 168:132–143

Perez-Sancho J, Tilsner J, Samuels AL, Botella MA, Bayer EM, Rosado A (2016) Stitching organelles: organization and function of specialized membrane contact sites in plants. Trends Cell Biol 26:705–717

Perni S, Dynes JL, Yeromin AV, Cahalan MD, Franzini-Armstrong C (2015) Nanoscale patterning of STIM1 and Orai1 during store-operated Ca^{2+} entry. Proc Natl Acad Sci USA 112:E5533–E5542

Phillips MJ, Voeltz GK (2016) Structure and function of ER membrane contact sites with other organelles. Nat Rev Mol Cell Biol 17:69–82

Pichler H, Gaigg B, Hrastnik C, Achleitner G, Kohlwein SD, Zellnig G, Perktold A, Daum G (2001) A subfraction of the yeast endoplasmic reticulum associates with the plasma membrane and has a high capacity to synthesize lipids. Eur J Biochem 268:2351–2361

Porter KR, Palade GE (1957) Studies on the endoplasmic reticulum. III. Its form and distribution in striated muscle cells. J Biophys Biochem Cytol 3:269–300

Prinz WA (2014) Bridging the gap: Membrane contact sites in signaling, metabolism, and organelle dynamics. J Cell Biol 205:759–769

Qiu X, Mistry A, Ammirati MJ, Chrunyk BA, Clark RW, Cong Y, Culp JS, Danley DE, Freeman TB, Geoghegan KF et al (2007) Crystal structure of cholesteryl ester transfer protein reveals a long tunnel and four bound lipid molecules. Nat Struct Mol Biol 14:106–113

Quintana A, Rajanikanth V, Farber-Katz S, Gudlur A, Zhang C, Jing J, Zhou Y, Rao A, Hogan PG (2015) TMEM110 regulates the maintenance and remodeling of mammalian ER-plasma membrane junctions competent for STIM-ORAI signaling. Proc Natl Acad Sci USA 112:E7083–E7092

Reinisch KM, De Camilli P (2016) SMP-domain proteins at membrane contact sites: Structure and function. Biochim Biophys Acta 1861:924–927

Rosenbluth J (1962) Subsurface cisterns and their relationship to the neuronal plasma membrane. J Cell Biol 13:405–421

Saheki Y, De Camilli P (2012) Synaptic vesicle endocytosis. Cold Spring Harb Perspect Biol 4:a005645

Saheki Y, Bian X, Schauder CM, Sawaki Y, Surma MA, Klose C, Pincet F, Reinisch KM, De Camilli P (2016) Control of plasma membrane lipid homeostasis by the extended synaptotagmins. Nat Cell Biol 18:504–515

Schapire AL, Voigt B, Jasik J, Rosado A, Lopez-Cobollo-R, Menzel D, Salinas J, Mancuso S, Valpuesta V, Baluska F et al (2008) Arabidopsis synaptotagmin 1 is required for the maintenance of plasma membrane integrity and cell viability. Plant Cell 20:3374–3388

Schauder CM, Wu X, Saheki Y, Narayanaswamy P, Torta F, Wenk MR, De Camilli P, Reinisch KM (2014) Structure of a lipid-bound extended synaptotagmin indicates a role in lipid transfer. Nature 510:552–555

Sclip A, Bacaj T, Giam LR, Sudhof TC (2016) Extended synaptotagmin (ESyt) triple knock-out mice are viable and fertile without obvious endoplasmic reticulum dysfunction. PLoS One 11:e0158295

Singleton A, Hardy J (2016) The evolution of genetics: Alzheimer's and Parkinson's diseases. Neuron 90:1154–1163

Small SA, Petsko GA (2015) Retromer in Alzheimer disease, Parkinson disease and other neurological disorders. Nat Rev Neurosci 16:126–132

Spacek J, Harris KM (1997) Three-dimensional organization of smooth endoplasmic reticulum in hippocampal CA1 dendrites and dendritic spines of the immature and mature rat. J Neurosci 17:190–203

Stefan CJ, Manford AG, Baird D, Yamada-Hanff J, Mao Y, Emr SD (2011) Osh proteins regulate phosphoinositide metabolism at ER-plasma membrane contact sites. Cell 144:389–401

Stuible M, Tremblay ML (2010) In control at the ER: PTP1B and the down-regulation of RTKs by dephosphorylation and endocytosis. Trends Cell Biol 20:672–679

Sudhof TC (2002) Synaptotagmins: why so many? J Biol Chem 277:7629–7632

Sudhof TC (2012) Calcium control of neurotransmitter release. Cold Spring Harb Perspect Biol 4:a011353

Tavassoli S, Chao JT, Young BP, Cox RC, Prinz WA, de Kroon AI, Loewen CJ (2013) Plasma membrane--endoplasmic reticulum contact sites regulate phosphatidylcholine synthesis. EMBO Rep 14:434–440

Toulmay A, Prinz WA (2011) Lipid transfer and signaling at organelle contact sites: the tip of the iceberg. Curr Opin Cell Biol 23:458–463

Toulmay A, Prinz WA (2012) A conserved membrane-binding domain targets proteins to organelle contact sites. J Cell Sci 125:49–58

Tremblay MG, Moss T (2016) Loss of all 3 extended synaptotagmins does not affect normal mouse development, viability or fertility. Cell Cycle 15:2360–2366

Varnai P, Toth B, Toth DJ, Hunyady L, Balla T (2007) Visualization and manipulation of plasma membrane-endoplasmic reticulum contact sites indicates the presence of additional molecular components within the STIM1-Orai1 Complex. J Biol Chem 282:29678–29690

Weiss J (2003) Bactericidal/permeability-increasing protein (BPI) and lipopolysaccharide-binding protein (LBP): structure, function and regulation in host defence against Gram-negative bacteria. Biochem Soc Trans 31:785–790

West M, Zurek N, Hoenger A, Voeltz GK (2011) A 3D analysis of yeast ER structure reveals how ER domains are organized by membrane curvature. J Cell Biol 193:333–346

Xu J, Bacaj T, Zhou A, Tomchick DR, Sudhof TC, Rizo J (2014) Structure and Ca^{2+}-binding properties of the tandem C(2) domains of E-Syt2. Structure 22:269–280

Yadav S, Garner K, Georgiev P, Li M, Gomez-Espinosa E, Panda A, Mathre S, Okkenhaug H, Cockcroft S, Raghu P (2015) RDGBα, a PtdIns-PtdOH transfer protein, regulates G-protein-coupled PtdIns(4,5)P_2 signalling during Drosophila phototransduction. J Cell Sci 128:3330–3344

Yamazaki T, Kawamura Y, Minami A, Uemura M (2008) Calcium-dependent freezing tolerance in Arabidopsis involves membrane resealing via synaptotagmin SYT1. Plant Cell 20:3389–3404

Yu H, Liu Y, Gulbranson DR, Paine A, Rathore SS, Shen J (2016) Extended synaptotagmins are Ca^{2+}-dependent lipid transfer proteins at membrane contact sites. Proc Natl Acad Sci USA 113:4362–4367

Zhang SL, Yu Y, Roos J, Kozak JA, Deerinck TJ, Ellisman MH, Stauderman KA, Cahalan MD (2005) STIM1 is a Ca^{2+} sensor that activates CRAC channels and migrates from the Ca^{2+} store to the plasma membrane. Nature 437:902–905

Zhang L, Yan F, Zhang S, Lei D, Charles MA, Cavigiolio G, Oda M, Krauss RM, Weisgraber KH, Rye KA et al (2012a) Structural basis of transfer between lipoproteins by cholesteryl ester transfer protein. Nat Chem Biol 8:342–349

Zhang Y, Wang H, Kage-Nakadai E, Mitani S, Wang X (2012b) C. elegans secreted lipid-binding protein NRF-5 mediates PS appearance on phagocytes for cell corpse engulfment. Curr Biol 22:1276–1284

Endoplasmic Reticulum-Plasma Membrane Contacts Regulate Cellular Excitability

7

Eamonn J. Dickson

Abstract

Cells that have intrinsic electrical excitability utilize changes in membrane potential to communicate with neighboring cells and initiate cellular cascades. Excitable cells like neurons and myocytes have evolved highly specialized subcellular architectures to translate these electrical signals into cellular events. One such structural specialization is sarco-/endoplasmic reticulum-plasma membrane contact sites. These membrane contact sites are positioned by specific membrane-membrane tethering proteins and contain an ever-expanding list of additional proteins that organize information transfer across the junctional space (~ 15–25 nm distance) to shape membrane identity and control cellular excitability. In this chapter we discuss how contacts between the sarco-/endoplasmic reticulum and plasma membrane are essential for regulated excitation-contraction coupling in striated muscle and control of lipid-dependent ion channels.

Keywords

Membrane-membrane contact sites • ER-PM contact sites • Phosphoinositides • PI(4,5)P_2 • Ion channels • Endoplasmic reticulum • Tethering proteins • Excitation-contraction coupling

This chapter concerns the regulation of cellular excitability at endoplasmic reticulum-plasma membrane contacts (ER-PM). When referring to "excitability," we primarily mean electrical excitability. Cells that have intrinsic electrical excitability, such as neurons, muscle cells, and endocrine cells, respond to stimulation by altering their electrical properties to initiate intracellular processes, including synaptic transmitter release, muscle contraction, or hormone secretion. The basis for cellular excitability is the ability to generate a labile resting membrane potential that is dependent on electrochemical gradients and energy consumption. The

E.J. Dickson (✉)
Department of Physiology and Membrane Biology, University of California, Davis, CA 95616, USA
e-mail: ejdickson@ucdavs.edu

M. Tagaya, T. Simmen (eds.), *Organelle Contact Sites*, Advances in Experimental Medicine and Biology 997, DOI 10.1007/978-981-10-4567-7_7

distinctive feature of electrically excitable cells is the generation of electrical current in response to membrane depolarization. Whereas a non-excitable cell that has been slightly depolarized will return to its original resting membrane potential, an electrically excitable cell that is depolarized to the same extent will respond with an action potential and/or contraction. These responses occur due to the movement of small charged particles (ions) such as Na^+, K^+, Ca^{2+}, and Cl^- across the plasma membrane through ion channels, pumps, and transporters. Electrically excitable cells produce an action potential because of the presence of voltage-gated channels in the plasma membrane. Thus, ion channels (pumps and transporters) are the foundation of electrical excitability. These ion channels can have their properties tuned by post-translational modification, accessory subunits, phosphorylation, and also lipids.

In addition to ion channels, eukaryotic cells have evolved compartmentalization to orchestrate and separate a myriad of cellular processes. One organizing tool is the partitioning of the cytoplasm into membrane-enclosed compartments called organelles. These specialized subcellular structures have distinct functions and communicate to organize specific cellular tasks via vesicular transport and via membrane-membrane junctions or contact sites. These intimate sites (~ 10–30 nm) of information transfer can be stable or dynamic and are mediated via specific proteins (Figs. 7.1, 7.2, and 7.4). Two important functions of ER-PM contact sites are (1) the transfer of calcium ions to generate and coordinate cytoplasmic calcium signaling events, and (2) the transfer and regulation of lipids. This chapter briefly highlights the major principles and molecular identity of the proteins involved in regulating cellular excitability at ER-PM contact sites in mammalian cells.

7.1 Role of Calcium in Regulating Cellular Excitability at Membrane Contact Sites

The most widely studied and physiologically important example of how ER-PM contact sites regulate cellular excitability takes place during excitation-contraction (EC) coupling in striated muscle cells. EC coupling is the connection between the electrical action potential and mechanical muscle contraction in cardiac, skeletal, and smooth muscle myocytes. This process of converting an electrical stimulus into a mechanical response is the essential event for many fundamental physiological processes including walking (skeletal muscle), pumping blood around the body (cardiac muscle), and peristalsis (smooth muscle) and relies on the tight physical coupling between the plasma membrane (sarcolemma) and sarcoplasmic reticulum. In the next few sections, we focus our attentions on cardiac and skeletal muscle EC coupling. For cell-specific review of smooth muscle EC coupling mechanisms, see (Berridge 2008; Kotlikoff 2003; Navedo and Santana 2013; Sanders et al. 2012).

7.1.1 Cardiac Muscle Excitation-Contraction Coupling

In cardiac muscle (Fig. 7.1a), deep sarcolemma invaginations called transverse tubules (T-tubules) penetrate into the myocyte and are functionally coupled to sarcoplasmic reticulum (SR) cisterna in a 1:1 ratio (1 T-tubule/1 SR terminal cisterna (Moore and Ruska 1957)). These morphological units are often referred to as cardiac dyads (or diads). T-tubule biogenesis has been extensively characterized, and the molecular identity of specific signaling proteins responsible for initiating and regulating this event is constantly evolving. Several proteins thought to have fundamental roles in T-tubule biogenesis include caveolin 3 (CAV3), amphiphysin 2 (BIN1), and dysferlin (DYSF). Among these, amphiphysin 2 (BIN1) appears to have a particularly interesting role through its dual involvement in T-tubule formation (Lee et al. 2002) and shuttling of L-type Ca^{2+} channels to the T-tubule sarcolemma (Hong et al. 2010).

The membrane-membrane tethering protein primarily responsible for structural coupling between the sarcolemma and SR (PM-SR) appears to be junctophilin-2 (JPH2, Fig. 7.1a).

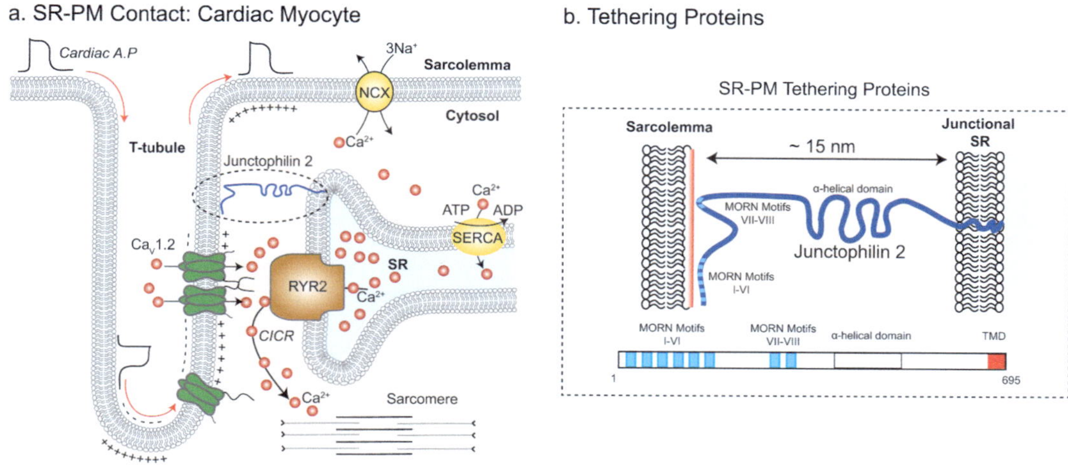

Fig. 7.1 Excitation-contraction coupling in cardiac myocytes. (**a**) Cartoon of a cardiac myocyte sarcoplasmic reticulum-plasmalemma (SR-PM) contact site. A single T-tubule is paired with a single terminal cisterna of the sarcoplasmic reticulum (dyad). Cardiac action potentials (A.P) propagating along cardiac myocyte T-tubules open clustered voltage-gated calcium channels ($Ca_V1.2$) to facilitate calcium-induced calcium release (CICR) from ryanodine receptors (RYR2) on the SR. Elevations in cytoplasmic calcium lead to myocyte contractions. (**b**) Top: organization and interactions of junctophilin-2 and the plasmalemma. Note the hypothesized interaction between the repeating MORN motifs of junctophilin-2 and the negative charges (red line) of the plasmalemma. Bottom: linear representation of junctophilin-2 protein. NCX, sodium calcium exchanger; SERCA, sarco-/endoplasmic reticulum Ca^{2+}-ATPase; CC, coiled coil; PBD, polybasic domain; TMD, transmembrane domain

JPH2 promotes coupling through its C-terminal hydrophobic segment spanning the SR membrane and its eight, N-terminus, MORN motifs (membrane occupation and recognition motifs; Fig. 7.1b) potentially interacting with sarcolemma phospholipids (Bennett et al. 2013; Lee et al. 2002) or proteins (Minamisawa et al. 2004). Further emphasizing the importance of JPH2 in mediating PM-SR contact sites, JPH2 knockout myocytes have deficient functional membrane complexes, impaired Ca^{2+} signaling, and are embryonic lethal (Takeshima et al. 2000; van Oort et al. 2011). Thus, JPH2 is a tether for the junctional membrane-membrane complex in cardiac EC coupling.

The functional coupling between PM-SR contact sites in cardiac muscle arises from the close apposition between dihydropyridine-sensitive voltage-gated calcium channels ($Ca_V1.2$) on the sarcolemma and ryanodine receptors (RYR2) on the SR membrane (Fig. 7.1a). Cardiac EC coupling is initiated when an action potential (AP) depolarizes the cell membrane of a cardiac myocyte, stimulating the brief opening of $Ca_V1.2$ channels. Influx of Ca^{2+} across a single $Ca_V1.2$ channel is known as a "Ca^{2+} sparklet" and can be visualized with Ca^{2+}-sensitive fluorescent dyes (Wang et al. 2001). These elemental calcium events allow the Ca^{2+} concentration within the junctional cleft to increase from ~100 nM to ~10 μM (Bers and Guo 2005). Recent data suggests that in order to achieve such elevations in calcium, $Ca_V1.2$ channels form clusters (average six channels per cluster), with each channel simultaneously opening (cooperative gating) during a cardiac AP (Dixon et al. 2015; Dixon et al. 2012; Inoue and Bridge 2003; Navedo et al. 2010; Sobie and Ramay 2009). This local Ca^{2+} microdomain is by itself alone not sufficient to initiate substantial contraction; rather it triggers further "Ca^{2+}-induced Ca^{2+} release" (CICR: (Fabiato 1983)) from clustered RYR2 on the SR, producing a "Ca^{2+} spark" (Cheng and Lederer 2008; Cheng et al. 1993). It is estimated that approximately 25 L-type Ca^{2+} channels and 100 RyRs closely appose one another within the

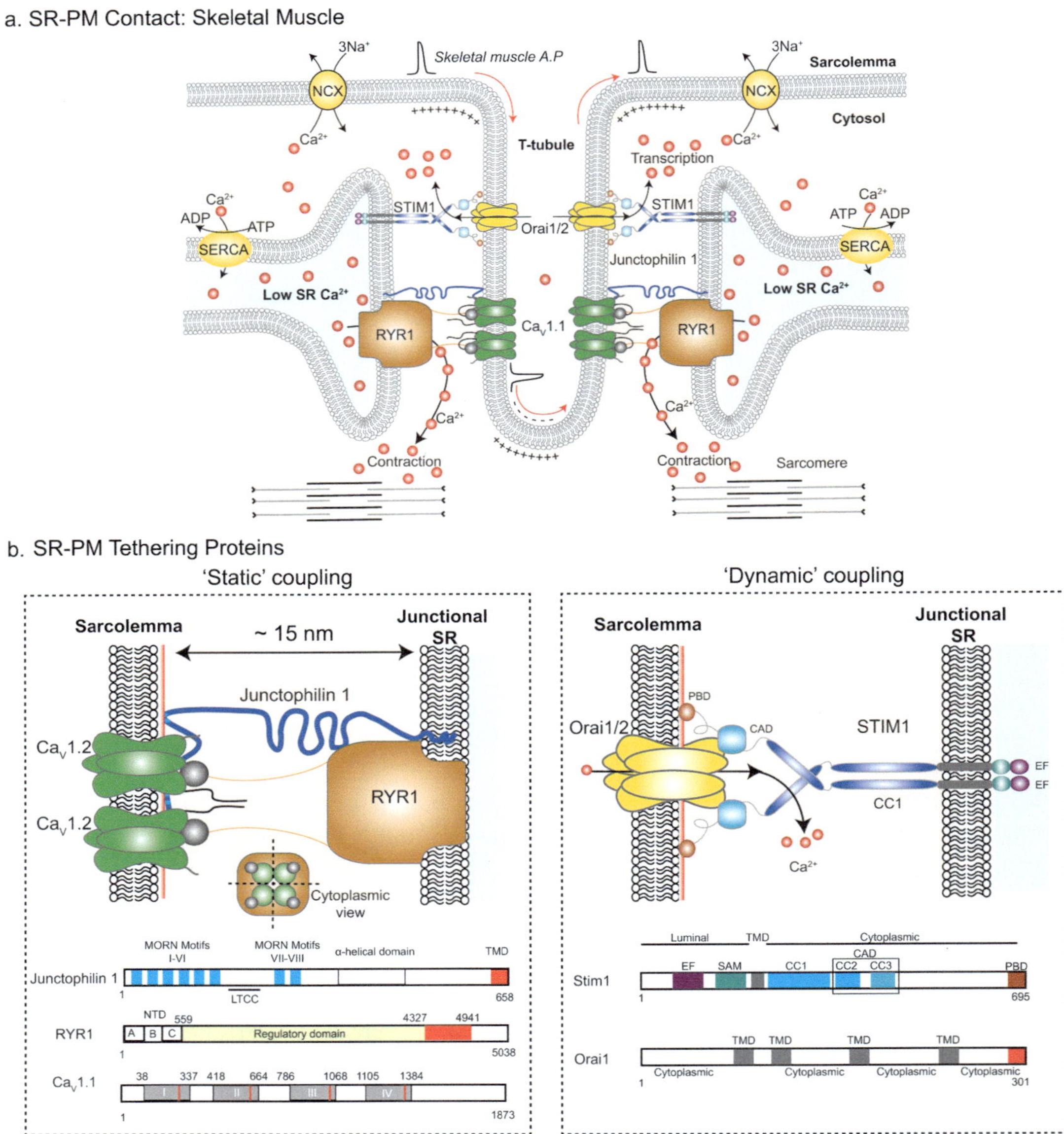

Fig. 7.2 Excitation-contraction coupling in skeletal muscle. (**a**) Schematic representation of skeletal muscle SR-PM contacts. A single T-tubule paired with two terminal cisternae of the SR (triad). Skeletal muscle action potentials (A.P) invade skeletal myocyte T-tubules to activate voltage-gated calcium channels ($Ca_V1.1$). Movement of the $Ca_V1.1$ voltage sensor physically induces release of calcium from ryanodine receptors (RYR1) to elicit muscle contraction. Note, unlike Fig. 7.1 in cardiac myocytes, this is a calcium-independent process. In a separate, voltage-independent, calcium release event, store-operated calcium entry (SOCE) is initiated following reduction of SR calcium concentration. Diagram represents the dynamic reorganization of STIM1 dimers in sarcoplasmic reticulum (SR) membranes and Orai1 hexamers in the plasmalemma. Orai1-STIM1 binding initiates calcium influx to the cytosol, the refilling of SR stores via the SERCA pump, and stimulation of transcription factors. (**b**) *Left*: "static" SR-PM contact site. *Top*: schematic representation of the organization and interactions of $Ca_V1.1$ and RYR1. *Bottom*: linear representation of junctophilin-1, RYR1, and $Ca_V1.1$. *Right*: "dynamic" SR-PM contact site. *Top*: molecular elements involved in store-operated calcium entry. *Bottom*: linear representation of STIM1 and Orai1 proteins

junctional cleft to form a "couplon" (Bers and Guo 2005). Synchronized activation of multiple sparks leads to a global elevation in intracellular Ca^{2+} sufficient to engage the contractile machinery, thereby promoting myocyte contraction and providing the force for pumping blood around the body.

$Ca_V1.2$ and RYR2 channels are both subject to regulation by multiple signaling pathways. Such signaling pathways have the ability to rapidly alter the amplitude and/or spatial properties of myocyte Ca^{2+} signaling to shape cardiac myocyte contraction. For example, $Ca_V1.2$ are subject to regulation by protein kinases, including PKA (β-adrenergic receptor stimulation, e.g., during the fight or flight response) and PKC (activated by stimulation of α_1-adrenergic, angiotensin II, and endothelin G_q-coupled receptors). RyRs are part of larger macromolecular signaling complexes and as such can have their activity modulated by one or more accessory proteins including calsequestrin, calmodulin, CaMKII-dependent phosphorylation, FK506-binding proteins, and PKA (for review, see (Lanner et al. 2010)).

Following elevations in cardiac myocyte cytosolic Ca^{2+}, and consequent activation of the cardiac contractile apparatus, Ca^{2+} must be rapidly extruded from the cytosol in preparation for the subsequent action potential. The main mechanisms for Ca^{2+} efflux in cardiac myocytes are the sarcolemma sodium/calcium exchanger and the sarco/endoplasmic reticulum Ca^{2+} ATPase (SERCA) pump on the SR (Bers 2002).

It is clear that cellular architecture, specifically the maintenance of PM-SR contacts, is critically important to the fidelity of EC coupling. Work from several groups has established that T-tubules are lost or atrophied as cardiac hypertrophy develops, resulting in the progressive orphaning of RYR2 (Gomez et al. 2001; Song et al. 2005). Deficiencies in EC coupling because of T-tubule remodeling have also been observed during the earlier stages of hypertrophy; here the width of the junctional cleft is marginally increased, altering the functional coupling between membrane depolarization and Ca^{2+} release from the SR (Xu et al. 2007). Thus, cardiac EC coupling is dependent on extracellular Ca^{2+} and functional coupling between $Ca_V1.2$ and RYR2, which relies on the close spatial proximity between the PM and SR membranes.

7.1.2 Skeletal Muscle Excitation-Contraction Coupling

Skeletal muscle cells also require precise SR-PM contact site arrangement for functional EC coupling. The highly specialized structure, responsible for translating a skeletal muscle AP into intracellular calcium release and muscle contraction, is known as the triad and consists of a single T-tubule closely (~ 15 nm) apposed to two adjacent SR terminal cisternae (1 T-tubule/2 SR terminal cisternae; Fig. 7.2a; (Block et al. 1988; Franzini-Armstrong and Protasi 1997)).

Functional coupling between these adjacent membranes arises from the direct physical contact between $Ca_V1.1$-containing heteromultimers arranged into groups of four ("tetrads") on the sarcolemma, aligned with the four subunits of every other RyR1 homotetramer on the SR membrane (Fig. 7.1b; (Block et al. 1988; Protasi et al. 2002; Protasi et al. 2000)). Accordingly, during propagation of a skeletal muscle A.P along the T-tubule, $Ca_V1.1$ channels act as voltage sensors to physically translate membrane depolarizations directly to RYR1, independently of Ca^{2+} flux (Armstrong et al. 1972), to trigger release from SR stores. Critical structural determinants required for efficient EC coupling appear to be the $Ca_V1.1\alpha_1$ II–III loop (Kugler et al. 2004; Nakai et al. 1998b; Tanabe et al. 1990), $Ca_V1.1\beta_{1a}$ carboxyl terminus (Beurg et al. 1999; Sheridan et al. 2003), and multiple regions of RyR1 (Nakai et al. 1998a; Protasi et al. 2002; Sheridan et al. 2006). A unique property of the conformational coupling between $Ca_V1.1$ and RyR1 in skeletal muscle is that it occurs bi-directionally (Nakai et al. 1996; Nakai et al. 1998a), with both an orthograde signal passing from the $Ca_V1.1$ to RyR1 and also a retrograde signal from RyR1 to the $Ca_V1.1\alpha_1$ subunit that enhances Ca^{2+} flow across the T-tubule membrane (Nakai et al. 1996).

In skeletal muscle, junctophilin-1 (JPH1) is involved in physically linking the T- tubules to the SR membrane (Fig. 7.1b; Nishi et al. 2000). Consistent with its role as a PM-SR tethering protein, ablation of JPH1 results in distinct morphological abnormalities of the junctional cleft, including deficiencies in SR-PM complexes, swollen terminal cisternae, and decreased triad number. These deficiencies result in impaired contractile force, abnormal sensitivity to extracellular Ca^{2+}, and premature death (Ito et al. 2001; Komazaki et al. 2002).

7.1.3 Store-Operated Calcium Entry

Another example of how ER/SR-PM communication can shape cellular excitability is store-operated calcium entry (SOCE). The process of store-operated SOCE, whereby Ca^{2+} influx across the PM is activated in response to depletion of intracellular ER/SR Ca^{2+} stores, plays a critical role in the regulation of gene expression, motility, immune responses, and organ development. The key molecular determinants orchestrating this calcium-signaling cascade (for a comprehensive review, see (Prakriya and Lewis 2015)) are STIM1 and STIM2 (stromal interaction molecules) in the ER/SR membrane and Orai channels (Orai1, Orai2, Orai3) in the plasma membrane (Fig. 7.2a, b). At rest, STIM molecules (typically dimers) and Orai channels (hexamers (Hou et al. 2012)) freely diffuse along the ER/SR and PM, respectively. Physiological stimuli that initiate net depletion of ER/SR Ca^{2+} stores promote loss of calcium from the luminal EF-hand domains on STIM1 dimers and induce a conformational change in their structure that favors oligomerization and translocation to ER/SR-PM contacts (Fig. 7.2b). The clustering of multiple STIM1 dimers at ER-PM junctions then traps the freely diffusing Orai hexamers and promotes formation of higher-order oligomers. This intimate molecular choreography results in the influx of calcium into the ER-PM junction (Fig. 7.2b) which (1) helps refill depleted ER/SR calcium stores and (2) activates calcineurin, which dephosphorylates the transcription factor NFAT (nuclear factor of activated T-cells), exposing a nuclear localization sequence that enables its translocation into the nucleus to help control a wide variety of genes (Feske et al. 2001). ER-PM STIM1-Orai1 puncta deoligomerize, and SOCE is terminated when ER/SR Ca^{2+} luminal calcium levels are replenished through the ATP-dependent action of the SERCA pump on ER/SR membranes.

Historically, a role for SOCE in skeletal muscle was considered unlikely since sarcolemmal Ca^{2+} entry is not directly required for contraction (see section 1.2). However recent investigations have demonstrated a key role for SOCE in the development and function of skeletal muscle fibers. For example, skeletal myocytes express high levels of both STIM1 and Orai1, with STIM1 localized to the muscle SR at triadic junctions appearing essential for myotube development (Stiber et al. 2008). Finally, loss of STIM1 or Orai1 expression in humans is associated with a congenital nonprogressive myopathy (McCarl et al. 2009). Thus, store-operated and voltage-dependent Ca^{2+} pathways may reflect two distinct molecular channel complexes within the triad junction that may enable trans-sarcolemmal Ca^{2+} entry.

7.2 Lipid Regulation of Cellular Excitability

We now consider how minority lipids, regulated at ER/SR-PM junctions, can alter cellular excitability. The original hypothesis presented following the work of Hilgemann and Ball (Hilgemann and Ball 1996) was that ion channels in the PM of mammalian cells require interactions with phosphoinositides for proper function. It is clear by now that a large and varied range of ion channels and transporters are regulated by these low-abundance negatively charged phospholipids.

7.2.1 Phosphoinositides

Phosphoinositides are a family of eight low-abundance phospholipids found on the cytoplasmic leaflet of all cellular membranes (Balla

2013; Di Paolo and De Camilli 2006). They have an inositol ring that can be phosphorylated or dephosphorylated by a variety of lipid kinases or lipid phosphatases on either the three, four, or five positions, resulting in the synthesis of seven combinatorial PIs (PI(3)P, PI(4)P, PI(5)P, PI(3,4)P_2, PI(3,5)P_2, PI(4,5)P_2, and PI(3,4,5)P_3). The consequent addition of each phosphate group to the inositol ring confers a -2 charge at the neutral pH of the cytosol. This means that the PM phosphoinositide PI(4,5)P_2, discussed at length in this chapter, has a -4 charge at pH 7.0. These negative charges enable phosphoinositides to act as negatively charged molecular beacons recruiting cytosolic proteins to specific lipid membranes or binding to the cytosolic domain (s) of membrane proteins. Through the ability of proteins to recognize the distinctive high negative charge and head group geometries of phosphoinositides, PIs play a major role in regulating a wide range of processes including membrane budding and fusion, actin dynamics, and ion channel activity.

Why have mammalian cells evolved lipid-dependent regulation of ion channel function? Two major hypotheses have been presented in recent years. The first is regulation in time by receptors. While integral membrane proteins like ion channels are residing in the plasma membrane, receptors coupled to PLC may become activated and transiently change key membrane lipids sufficiently to modulate channel function (Fig. 7.3 (Brown and Adams 1980)). Thus, electrical and ion-transport properties of the cell may be regulated dynamically by physiological events, such as neurotransmitter or hormone release. A second hypothesis is that phosphoinositide dependence allows control of ion channel activity as they are trafficked to targeted membranes, such as the plasma membrane. During trafficking, these important integral membrane proteins would be quiescent until they enter their favored phosphoinositide environment where they would be subsequently turned on (Hilgemann et al. 2001). Such location-specific control of ion channel and transport activity would ensure that these proteins do not perturb organelle function during trafficking to the plasma membrane.

7.2.2 Phosphoinositide-Regulated Ion Channel Function

All integral membrane proteins sit in intimate contact with a bed of diverse lipids, so it is not surprising that many proteins have evolved lipid dependence. The regulation of PM ion channel activity by phosphoinositides is now widely recognized as an integral component of electrical signaling in cells. To date around 100 heterogeneous and functionally diverse ion channels and transporters have been shown to be dependent on PM PI(4,5)P_2 (for comprehensive list, see (Hille et al. 2015)). In the next three sections, we describe highlights of what is known about the phosphoinositide dependence of several classes of ion channels who play important roles in controlling the electrical excitability of cells. For detailed review of PI(4,5)P_2-sensitive ion channels, please refer to (Hille et al. 2015; Logothetis et al. 2010).

7.2.2.1 Inwardly Rectifying Potassium (Kir) Channels

How exactly does plasma membrane PI(4,5)P_2 regulate the activity of so many ion channels? Two recent crystal structures of PI(4,5)P_2 in complex with members of the inwardly rectifying potassium (Kir) channel family, Kir2.2 (Hansen et al. 2011), and G protein-coupled inwardly rectifying potassium channel GIRK2 (Kir3.2; (Whorton and MacKinnon 2011)) show how specific protein-lipid interactions facilitate conformational changes in ion channel structure. The x-ray crystallographic structure of Kir2.2 in complex with a short-chain analog of PI(4,5)P_2 reveals a homotetrameric channel composed of a transmembrane domain that forms the potassium-selective pore and a large cytoplasmic domain (Fig. 7.3a). Each of the four channel subunits is in complex with one PI(4,5)P_2 molecule (1:1 ratio) at the interface between the transmembrane and cytoplasmic domains. The positioning of the negatively charged 4- and 5-phosphates of the inositol head group of PI(4,5)P_2 promotes an electrostatic interaction with positively charged residues of the cytoplasmic domain (Fig. 7.3b) and induces

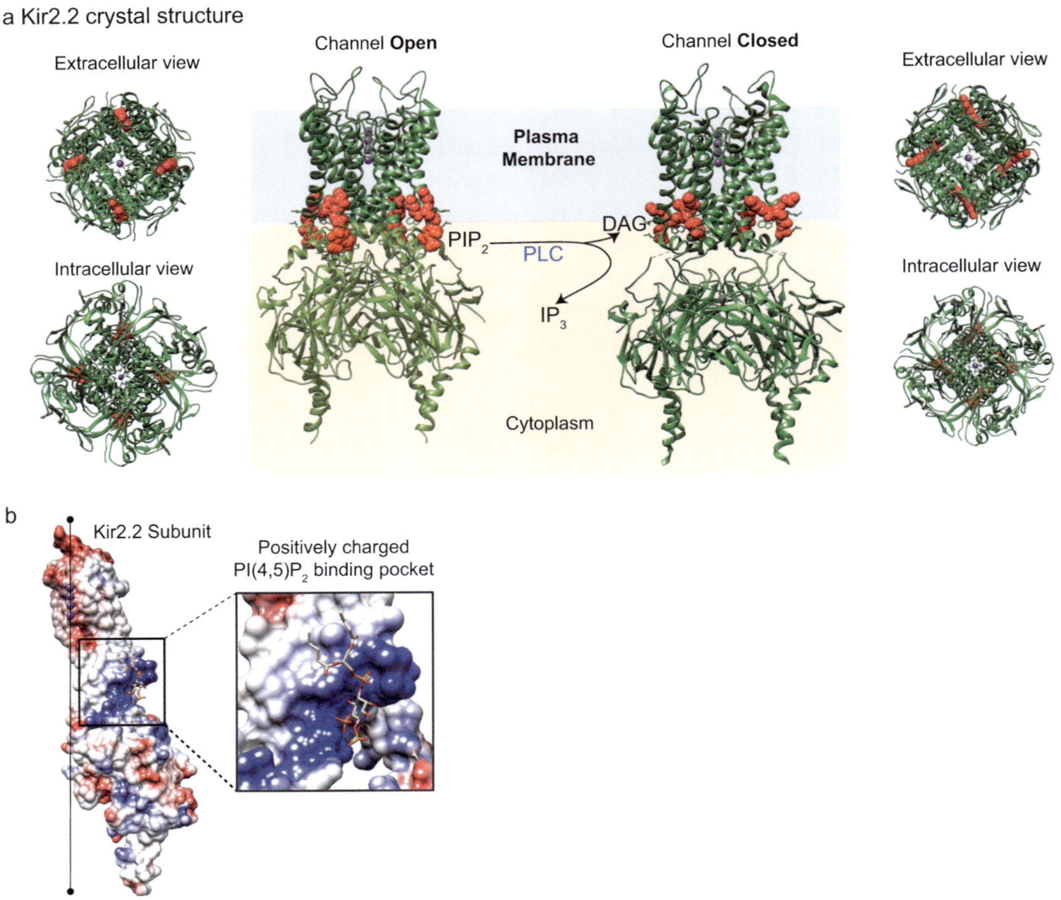

Fig. 7.3 Regulation of phosphoinositide-dependent ion channels at ER/SR-PM junctions. (**a**) Hypothetical model of Kir2.2 channel closure following hydrolysis of PM PI(4,5)P_2 by activated phospholipase C (PLC). Left: open configuration. Crystal structure of four Kir2.2 channel subunits in complex with one PI(4,5)P_2 molecule (1:1 ratio; pdb 3SPI). Closed configuration. Hydrolysis of PM PI(4,5)P_2 (pdb 3SPC) induces a translation (~ 6 Å) of the cytoplasmic domain away from the plasma membrane and closure of the inner helices. Note this is a hypothetical scenario. The channel structure in the "closed" configuration is Kir2.2 in complex with pyrophosphatidic acid (PPA), which is similar in structure to DAG. (**b**) Electrostatic map of a single Kir2.2 subunit in complex with a single PI(4,5)P_2 molecule (*blue* is positive; *red* is negative; *white* is neutral). The *vertical line* is the pore axis with four K^+ ions (*purple*) visible. The PI(4,5)P_2-binding site comprises numerous basic residues (*blue*) that interact electrostatically with the negatively charged phosphates of PI(4,5)P_2

a translation (~ 6 Å) of the cytoplasmic domain upward toward the membrane domain and promotes rotation of the inner helices (Fig. 7.3a). This lipid-mediated rearrangement of channel structure favors the active conformation of the channel (Fig. 7.2a) and provides evidence for lipid binding being a condition for channel activation. In the structure of a constitutively active GIRK2 channel (Kir3.2 R201A) with PI(4,5)P_2, there are similar striking changes in channel conformation, with a comparable rotation of the cytoplasmic domains and splaying apart of the inner helices, promoting gate opening (Whorton and MacKinnon 2011). Given that all Kir channels appear to be PI(4,5)P_2 dependent (Rohács et al. 2003; Zhang et al. 1999), and have conserved residues within their PI(4,5)P_2 binding pockets (Hansen et al. 2011), it is reasonable to suggest that such lipid-dependent, electrostatically-mediated conformational changes in protein structure may present a conserved gating mechanism for Kir's and possibly other ion channels.

7.2.2.2 Voltage-Gated Potassium Channels

Voltage-gated potassium channels are critical determinants in setting (resting membrane potential) and shaping (duration/repolarization of the action potential) the membrane potential in excitable cells. Despite the large number of voltage-gated potassium channels (K_v1-K_v11), the clearest and most prominent examples of regulation by PI(4,5)P_2 are members of the slowly activating, non-inactivating K_v7 (K_v7.1–7.5) potassium channel family. All members of the K_v7 channel family are PI(4,5)P_2 dependent (Kruse et al. 2012; Suh et al. 2004; Zaydman and Cui 2014; Zaydman et al. 2013). In particular it is accepted for the neuronal K_v7.2/K_v7.3 heterotetrameric potassium channels that PI (4,5)P_2 is essential for their voltage-gated activity and that depletion of PI(4,5)P_2 renders them inactive so the neuron becomes more excitable until the lipid is resynthesized (Hille et al. 2014; Vivas et al. 2014; Zhang et al. 2003).

For the other tested members of the K_V channel family, PI(4,5)P_2 is apparently not essential, and whether it has modulatory effects under physiological conditions is controversial in several examples (Kruse and Hille 2013). Why then, in contrast to all Kir channels, do only a subpopulation of K_V channels appear to be regulated by PM PI(4,5)P_2? One hypothesis is that despite the observed increase in excitability during PI(4,5)P_2 depletion, a reserve of potassium channel activity is needed to repolarize the membrane potential (Kruse and Hille 2013). For that reason, the PI (4,5)P_2 insensitivity of many other important voltage-gated potassium channels (such as K_V1.1/$K_V\beta$1.1, K_V1.5/$K_V\beta$1.3, K_V3, and K_V4) might be a necessity to initiate action potential repolarization during periods of receptor-mediated PI(4,5)P_2 depletion.

7.2.2.3 Voltage-Gated Calcium Channels

Voltage-gated calcium (Ca_V) channels mediate calcium influxes upon membrane depolarization that initiate the effector actions of electrically excitable cells (see sections 1.2, 1.3). Over the past few years, several voltage-activated Ca_V channels have been shown to have partial (30–50%) PI(4,5)P_2 dependence (Suh et al. 2012; Wu et al. 2002). Intriguingly, this phosphoinositide dependence of Ca_V channels changes with coexpression of different Ca_V β-subunits (Suh et al. 2012). Ca_V channels that interact with β_{2a} (lipidated) have little PI(4,5)P_2 sensitivity, whereas Ca_V β_3 subunits (non-lipidated) show greater PI(4,5)P_2 sensitivity (Suh et al. 2012). These findings raise the interesting concept that excitable cells may be able to "tune" the lipid dependence of specific ion channels by regulating the expression of not only the lipid or pore-forming α-subunit but also auxiliary β-subunits.

7.3 Regulation of Phosphoinositide Metabolism at Membrane Contact Sites

In this section we discuss how membrane contact sites represent platforms for regulating PM phosphoinositides and thus ion channel function and cellular excitability. As partially discussed in section 1, and in other chapters in this book, membrane contact sites are regions of close proximity (< 20 nm) between adjacent membranes in mammalian cells (Fig. 7.4a). They should be considered portals of information transfer, distinct from classic vesicular transport, and are essential for regulating cellular calcium dynamics and membrane lipid identity. Membrane contact sites are established by organelle-specific membrane-membrane tethering proteins and contain a growing number of additional proteins that organize information transfer to shape membrane phosphoinositide composition.

7.3.1 Receptor-Mediated PI(4,5)P_2 Depletion

In excitable cells a major mechanism by which PM PI(4,5)P_2 levels are reduced is through activation of the G protein-coupled receptor

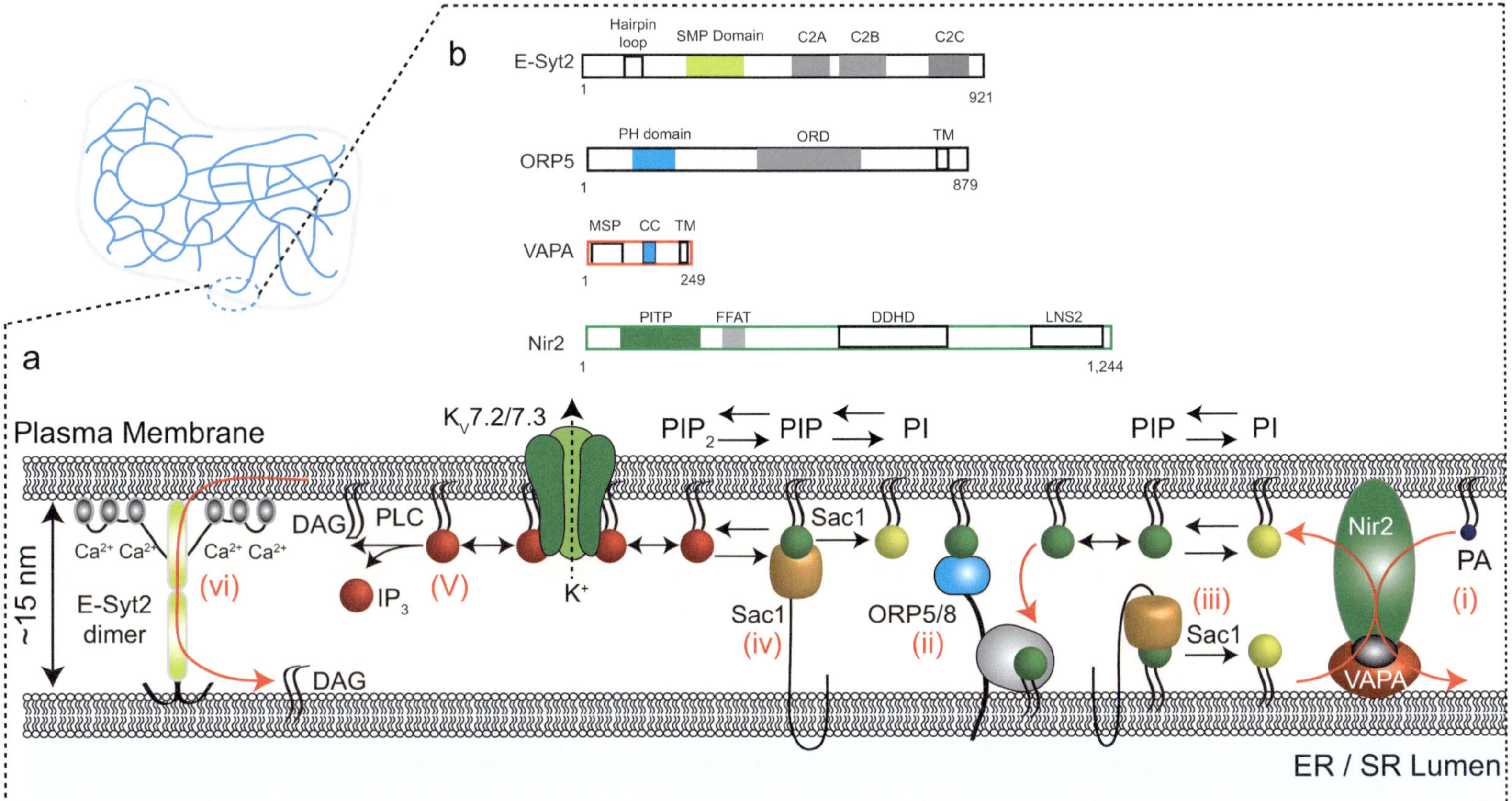

Fig. 7.4 **Phosphoinositide metabolism at membrane contact sites.** (**a**) Endoplasmic reticulum-plasma membrane junction (ER-PM). *(i)* During periods of net PI(4,5)P_2 depletion, Nir2 translocates and binds to VAPA/B on ER membranes to mediate the transfer of PI. *(ii)* ORP5/8 proteins tether ER-PM junctions via interactions between their PH domain and facilitate the countertransport of PM PI(4)P for ER PS. *(iii)* ER PI(4)P is dephosphorylated to PI by the ER-resident lipid phosphatase Sac1 acting in cis. *(iv)* PM PI(4)P is dephosphorylated to PI via Sac1 acting in trans. *(v)* During periods of PM PI(4,5)P_2 depletion, PI(4,5)P_2-dependent ion channels like K_V7.2/7.3 will have their activities suppressed to modulate cellular electrical excitability. *(vi)* Extended synaptotagmin (E-Syt) proteins tether ER-PM junctions via interactions with PM PI(4,5)P_2 via their C2 domains. The SMP domain of E-Syt2 is also responsible for clearing PM DAG. The distance between ER and PM membrane junctions is approximately 15 nm. (**b**) Linear representation of the tethering proteins present at ER-PM contacts

signaling cascade. Activation of G_q subsequently activates PLC, which cleaves PM $PI(4,5)P_2$ into the second messengers, cytosolic messenger inositol 1,4,5-trisphosphate (IP_3), and membrane-bound diacylglycerol (DAG; see Fig. 7.4). IP_3 binding to IP_3 receptors (IP_3Rs) at the ER triggers the release of Ca^{2+} into the cytosol. Calcium and DAG activate PKC. Net depletion of PI $(4,5)P_2$ will also result in the closure of PI(4,5) P_2-dependent ion channels, like $K_v7.2/7.3$ (Fig. 7.4). Following G_q-receptor stimulation and subsequent PM $PI(4,5)P_2$ hydrolysis, there is an immediate need for cells to replenish their depleted PM $PI(4,5)P_2$ pools. In mammalian cells, $PI(4,5)P_2$ is generated in two steps: PI is phosphorylated by a PI 4-kinase to generate PI(4) P, which is subsequently phosphorylated by PI(4) P 5-kinase to generate $PI(4,5)P_2$. One role for ER-PM contact sites is to help regulate the resynthesis of PM $PI(4,5)P_2$ following its depletion.

7.3.2 Transfer of PI to the Plasma Membrane

Following G_q-receptor-mediated PM $PI(4,5)P_2$ depletion, phosphoinositide transfer domains, called Nir2 and Nir3, are recruited to ER-PM junctions (Fig. 7.4 a(i)) (Chang et al. 2013; Kim et al. 2013; Kim et al. 2015). This recruitment is triggered through the combined actions of DAG and phosphatidic acid (Kim et al. 2015) and mediated through an interaction with the FFAT motif on the ER vesicle-associated membrane protein (VAMP)-associated proteins A and B (VAP-A and VAP-B) (Amarilio et al. 2005) (Fig. 7.4a, b). The stimulated recruitment of Nir2 and Nir3 to contact sites serves to deliver PI to the PM to replenish PM PI(4)P. The same process also delivers phosphatidic acid to the ER for future PI synthesis (Fig. 7.4a) (Kim et al. 2015).

7.3.3 Transfer and Regulation of PI(4)P at ER-PM Junctions

The majority of PM PI(4)P, the precursor to PM $PI(4,5)P_2$, seems to be generated by the PM-targeted lipid kinase PI4KIIIα (Chung et al. 2015a; Nakatsu et al. 2012) and delivery of PI(4) P from the Golgi (Dickson et al. 2014; Szentpetery et al. 2010). Recently, several integral ER proteins have been identified as key regulators of PM PI(4)P. The first of these are the oxysterol-binding protein (OSBP)-related protein 5 (ORP5) and ORP8 (Chung et al. 2015b). Through the action of a hydrophobic tail sequence that anchors it in the ER membrane and a pleckstrin homology (PH) domain that interacts with PM PI(4)P, ORP5/8 serves dual functions at ER-PM contact sites (Fig. 7.4a (ii)). Not only do these proteins facilitate contact between the two membranes, i.e., acts as a tether, but they also facilitate the exchange of phosphatidylserine (PS) for PI(4)P at these junctions (Chung et al. 2015b; Maeda et al. 2013). In addition to the ORP proteins, an ER PI(4)P phosphatase, Sac1, is present at ER-PM junctions (Fig. 7.4 (iii and iv)) and can regulate PM PI(4)P (Dickson et al. 2016; Stefan et al. 2011) directly and PM $PI(4,5)P_2$ indirectly (Dickson et al. 2016). The precise mode of action of Sac1, i.e., acting in trans (Fig. 7.4 (iv)), across the junctional cleft to directly dephosphorylate PM PI(4)P, or in cis (Fig. 7.4 (iii)) to dephosphorylate transferred ER PI(4)P, remains to be fully determined.

7.3.4 $PI(4,5)P_2$ at ER-PM Contact Sites

Many ER-PM tethering proteins share a unifying architecture with an ER-anchoring domain and lipid-interacting domain (Bennett et al. 2013; Giordano et al. 2013). For example, interactions between the ER and PM can be enhanced following increases in intracellular calcium (E-Syt1) (Idevall-Hagren et al. 2015) or dissolved following receptor-mediated $PI(4,5)P_2$ depletion (Chung et al. 2015b; Dickson et al. 2016; Giordano et al. 2013). Such dynamic lipid-protein interactions mean that $PI(4,5)P_2$-dependent contact sites are inherently dynamic and also self-regulating (Fig. 7.4). At steady-state net depletion of PM $PI(4,5)P_2$ reduces the number of ER-PM contacts and reduces the population of Sac1 proximate to the PM. Loss of both

Sac1 and ORP5 would release an inhibitory brake on the PM PI(4)P pool, allowing PM PI (4)P and PI(4,5)P_2 levels to accumulate. Increases in PM PI(4,5)P_2 help reestablish ER-PM contacts and, after PI(4)P accumulation, bring ORP5 and also Sac1 back into close proximity of the PM, reapplying the brake on PI(4)P and PI(4,5)P_2 accumulation. Physiologically, the dynamic correlation between phosphoinositides and ER-PM contacts would be beneficial for the recovery of each of the PM phosphoinositide species following periods of net depletion and provide a mechanism for the temporal and spatial positioning of ER-PM contacts depending on the requirements of the cell.

7.4 Perspectives

The concept that ER and PM contacts represent "intracellular synapses" for the transfer of information to shape membrane identity and excitability has been discussed for 60 years. With the advancement of new technologies to image at higher resolution in living cells, we are beginning to understand the molecular identity and activities of many more proteins and effectors that exist within these intimate regions of membrane-membrane contact. There is still much information to determine. Future advances must address not only the nanoscopic structural organization and function of ER-PM contacts in excitable cells but also their role(s) in coordinating macroscopic cellular events. It remains to be seen if single endogenous tethering proteins work independently to create their own distinct signaling nano-domains or if multiple tethers work synergistically/cooperatively in primary cells, like myocytes or neurons.

7.5 Conclusion

Endoplasmic reticulum-plasma membrane contact sites orchestrate, segregate, and organize specific cellular tasks in a spatially defined, compartmentalized manner to shape cellular excitability. These contact sites serve important functions in regulating membrane proteins, lipids, and ion composition. We anticipate that future studies will uncover additional proteins involved in information transfer at membrane-membrane junctions and elucidate their role in cellular pathways and physiological events.

Acknowledgments Special thanks to Dr. Rose Ellen Dixon for critical comments on the chapter.

Competing Interests The author declares that no competing interests exist.

References

Amarilio R, Ramachandran S, Sabanay H, Lev S (2005) Differential regulation of endoplasmic reticulum structure through VAP-Nir protein interaction. J Biol Chem 280:5934–5944

Armstrong CM, Bezanilla FM, Horowicz P (1972) Twitches in the presence of ethylene glycol bis(β-aminoethyl ether)-N,N′-tetracetic acid. Biochim Biophys Acta 267:605-608

Balla T (2013) Phosphoinositides: tiny lipids with giant impact on cell regulation. Physiol Rev 93:1019–1137

Bennett HJ, Davenport JB, Collins RF, Trafford AW, Pinali C, Kitmitto A (2013) Human junctophilin-2- undergoes a structural rearrangement upon binding PtdIns(3,4,5)P_3 and the S101R mutation identified in hypertrophic cardiomyopathy obviates this response. Biochem J 456:205–217

Berridge MJ (2008) Smooth muscle cell calcium activation mechanisms. J Physiol 586:5047–5061

Bers DM (2002) Cardiac excitation-contraction coupling. Nature 415:198–205

Bers DM, Guo T (2005) Calcium signaling in cardiac ventricular myocytes. Ann N Y Acad Sci 1047:86–98

Beurg M, Ahern CA, Vallejo P, Conklin MW, Powers PA, Gregg RG, Coronado R (1999) Involvement of the carboxy-terminus region of the dihydropyridine receptor β1a subunit in excitation-contraction coupling of skeletal muscle. Biophys J 77:2953–2967

Block BA, Imagawa T, Campbell KP, Franzini-Armstrong C (1988) Structural evidence for direct interaction between the molecular components of the transverse tubule/sarcoplasmic reticulum junction in skeletal muscle. J Cell Biol 107:2587–2600

Brown DA, Adams PR (1980) Muscarinic suppression of a novel voltage-sensitive K^+ current in a vertebrate neurone. Nature 283:673–676

Chang CL, Hsieh TS, Yang TT, Rothberg KG, Azizoglu DB, Volk E, Liao JC, Liou J (2013) Feedback regulation of receptor-induced Ca^{2+} signaling mediated by E-Syt1 and Nir2 at endoplasmic reticulum-plasma membrane junctions. Cell Rep 5:813–825

Cheng H, Lederer WJ (2008) Calcium sparks. Physiol Rev 88:1491–1545

Cheng H, Lederer WJ, Cannell MB (1993) Calcium sparks: elementary events underlying excitation-contraction coupling in heart muscle. Science 262:740–744

Chung J, Nakatsu F, Baskin JM, De Camilli P (2015a) Plasticity of PI4KIIIα interactions at the plasma membrane. EMBO Rep 16:312–320

Chung J, Torta F, Masai K, Lucast L, Czapla H, Tanner LB, Narayanaswamy P, Wenk MR, Nakatsu F, De Camilli P (2015b) PI4P/phosphatidylserine countertransport at ORP5- and ORP8-mediated ER-plasma membrane contacts. Science 349:428–432

Di Paolo G, De Camilli P (2006) Phosphoinositides in cell regulation and membrane dynamics. Nature 443:651–657

Dickson EJ, Jensen JB, Hille B (2014) Golgi and plasma membrane pools of PI(4)P contribute to plasma membrane PI(4,5)P_2 and maintenance of KCNQ2/3 ion channel current. Proc Natl Acad Sci USA 111: E2281–E2290

Dickson EJ, Jensen JB, Vivas O, Kruse M, Traynor-Kaplan AE, Hille B (2016) Dynamic formation of ER-PM junctions presents a lipid phosphatase to regulate phosphoinositides. J Cell Biol 213:33–48

Dixon RE, Yuan C, Cheng EP, Navedo MF, Santana LF (2012) Ca^{2+} signaling amplification by oligomerization of L-type $Ca_v1.2$ channels. Proc Natl Acad Sci USA 109:1749–1754

Dixon RE, Moreno CM, Yuan C, Opitz-Araya X, Binder MD, Navedo MF, Santana LF (2015) Graded Ca^{2+}/calmodulin-dependent coupling of voltage-gated $Ca_V1.2$ channels. Elife 4

Fabiato A (1983) Calcium-induced release of calcium from the cardiac sarcoplasmic reticulum. Am J Physiol 245:C1–14

Feske S, Giltnane J, Dolmetsch R, Staudt LM, Rao A (2001) Gene regulation mediated by calcium signals in T lymphocytes. Nat Immunol 2:316–324

Franzini-Armstrong C, Protasi F (1997) Ryanodine receptors of striated muscles: a complex channel capable of multiple interactions. Physiol Rev 77:699–729

Giordano F, Saheki Y, Idevall-Hagren O, Colombo SF, Pirruccello M, Milosevic I, Gracheva EO, Bagriantsev SN, Borgese N, De Camilli P (2013) PI(4,5)P_2-dependent and Ca^{2+}-regulated ER-PM interactions mediated by the extended synaptotagmins. Cell 153:1494–1509

Gomez AM, Guatimosim S, Dilly KW, Vassort G, Lederer WJ (2001) Heart failure after myocardial infarction: altered excitation-contraction coupling. Circulation 104:688–693

Hansen SB, Tao X, MacKinnon R (2011) Structural basis of PIP_2 activation of the classical inward rectifier K^+ channel Kir2.2. Nature 477:495–498

Hilgemann DW, Ball R (1996) Regulation of cardiac Na^+, Ca^{2+} exchange and K_{ATP} potassium channels by PIP_2. Science 273:956–959

Hilgemann DW, Feng S, Nasuhoglu C (2001) The complex and intriguing lives of PIP_2 with ion channels and transporters. Sci STKE 2001:re19

Hille B, Dickson E, Kruse M, Falkenburger B (2014) Dynamic metabolic control of an ion channel. Prog iMol Biol Transl Sci 123:219–247

Hille B, Dickson EJ, Kruse M, Vivas O, Suh BC (2015) Phosphoinositides regulate ion channels. Biochim Biophys Acta 1851:844–856

Hong TT, Smyth JW, Gao D, Chu KY, Vogan JM, Fong TS, Jensen BC, Colecraft HM, Shaw RM (2010) BIN1 localizes the L-type calcium channel to cardiac T-tubules. PLoS Biol 8:e1000312

Hou XW, Pedi L, Diver MM, Long SB (2012) Crystal structure of the calcium release-activated calcium channel Orai. Science 338:1308–1313

Idevall-Hagren O, Lu A, Xie B, De Camilli P (2015) Triggered Ca^{2+} influx is required for extended synaptotagmin 1-induced ER-plasma membrane tethering. EMBO J 34:2291–2305

Inoue M, Bridge JH (2003) Ca^{2+} sparks in rabbit ventricular myocytes evoked by action potentials: involvement of clusters of L-type Ca^{2+} channels. Circ Res 92:532–538

Ito K, Komazaki S, Sasamoto K, Yoshida M, Nishi M, Kitamura K, Takeshima H (2001) Deficiency of triad junction and contraction in mutant skeletal muscle lacking junctophilin type 1. J Cell Biol 154:1059–1067

Kim S, Kedan A, Marom M, Gavert N, Keinan O, Selitrennik M, Laufman O, Lev S (2013) The phosphatidylinositol-transfer protein Nir2 binds phosphatidic acid and positively regulates phosphoinositide signalling. EMBO Rep 14:891–899

Kim YJ, Guzman-Hernandez ML, Wisniewski E, Balla T (2015) Phosphatidylinositol-phosphatidic acid exchange by Nir2 at ER-PM contact sites maintains phosphoinositide signaling competence. Dev Cell 33:549–561

Komazaki S, Ito K, Takeshima H, Nakamura H (2002) Deficiency of triad formation in developing skeletal muscle cells lacking junctophilin type 1. FEBS Lett 524:225–229

Kotlikoff MI (2003) Calcium-induced calcium release in smooth muscle: the case for loose coupling. Prog iBiophys Mol Biol 83:171–191

Kruse M, Hille B (2013) The phosphoinositide sensitivity of the K_V channel family. Channels (Austin, Tex) 7:530–536

Kruse M, Hammond GR, Hille B (2012) Regulation of voltage-gated potassium channels by PI(4,5)P_2. J Gen Physiol 140:189–205

Kugler G, Weiss RG, Flucher BE, Grabner M (2004) Structural requirements of the dihydropyridine receptor α_{1S} II–III loop for skeletal-type excitation-contraction coupling. J Biol Chem 279:4721–4728

Lanner JT, Georgiou DK, Joshi AD, Hamilton SL (2010) Ryanodine receptors: structure, expression, molecular details, and function in calcium release. Cold Spring Harb Perspect Biol 2:a003996

Lee E, Marcucci M, Daniell L, Pypaert M, Weisz OA, Ochoa GC, Farsad K, Wenk MR, De Camilli P (2002) Amphiphysin 2 (Bin1) and T-tubule biogenesis in muscle. Science 297:1193–1196

Logothetis DE, Petrou VI, Adney SK, Mahajan R (2010) Channelopathies linked to plasma membrane phosphoinositides. Pflugers Arch 460:321–341

Maeda K, Anand K, Chiapparino A, Kumar A, Poletto M, Kaksonen M, Gavin AC (2013) Interactome map uncovers phosphatidylserine transport by oxysterol-binding proteins. Nature 501:257–261

McCarl CA, Picard C, Khalil S, Kawasaki T, Rother J, Papolos A, Kutok J, Hivroz C, Ledeist F, Plogmann K et al (2009) ORAI1 deficiency and lack of store-operated Ca^{2+} entry cause immunodeficiency, myopathy, and ectodermal dysplasia. J Allergy Clin immunol 124:1311–1318.e1317

Minamisawa S, Oshikawa J, Takeshima H, Hoshijima M, Wang Y, Chien KR, Ishikawa Y, Matsuoka R (2004) Junctophilin type 2 is associated with caveolin-3 and is down-regulated in the hypertrophic and dilated cardiomyopathies. Biochem Biophys Res Commun 325:852–856

Moore DH, Ruska H (1957) Electron microscope study of mammalian cardiac muscle cells. J Biophys Biochem Cytol 3:261–268

Nakai J, Dirksen RT, Nguyen HT, Pessah IN, Beam KG, Allen PD (1996) Enhanced dihydropyridine receptor channel activity in the presence of ryanodine receptor. Nature 380:72–75

Nakai J, Sekiguchi N, Rando TA, Allen PD, Beam KG (1998a) Two regions of the ryanodine receptor involved in coupling with L-type Ca^{2+} channels. J Biol Chem 273:13403–13406

Nakai J, Tanabe T, Konno T, Adams B, Beam KG (1998b) Localization in the II-III loop of the dihydropyridine receptor of a sequence critical for excitation-contraction coupling. J Biol Chem 273:24983–24986

Nakatsu F, Baskin JM, Chung J, Tanner LB, Shui G, Lee SY, Pirruccello M, Hao M, Ingolia NT, Wenk MR et al (2012) PtdIns4P synthesis by PI4KIIIα at the plasma membrane and its impact on plasma membrane identity. J Cell Biol 199:1003–1016

Navedo MF, Santana LF (2013) $Ca_V1.2$ sparklets in heart and vascular smooth muscle. J Mol Cell Cardiol 58:67–76

Navedo MF, Cheng EP, Yuan C, Votaw S, Molkentin JD, Scott JD, Santana LF (2010) Increased coupled gating of L-type Ca^{2+} channels during hypertension and Timothy syndrome. Circ Res 106:748–756

Nishi M, Mizushima A, Nakagawara K, Takeshima H (2000) Characterization of human junctophilin subtype genes. Biochem Biophys Res Commun 273:920–927

Prakriya M, Lewis RS (2015) Store-operated calcium channels. Physiol Rev 95:1383–1436

Protasi F, Takekura H, Wang Y, Chen SR, Meissner G, Allen PD, Franzini-Armstrong C (2000) RYR1 and RYR3 have different roles in the assembly of calcium release units of skeletal muscle. Biophys J 79:2494–2508

Protasi F, Paolini C, Nakai J, Beam KG, Franzini-Armstrong C, Allen PD (2002) Multiple regions of RyR1 mediate functional and structural interactions with α_{1S}-dihydropyridine receptors in skeletal muscle. Biophys J 83:3230–3244

Rohács T, Lopes CM, Jin T, Ramdya PP, Molnar Z, Logothetis DE (2003) Specificity of activation by phosphoinositides determines lipid regulation of Kir channels. Proc Natl Acad Sci USA 100:745–750

Sanders KM, Koh SD, Ro S, Ward SM (2012) Regulation of gastrointestinal motility – insights from smooth muscle biology. Nat Rev Gastroenterol Hepatol 9:633–645

Sheridan DC, Cheng W, Ahern CA, Mortenson L, Alsammarae D, Vallejo P, Coronado R (2003) Truncation of the carboxyl terminus of the dihydropyridine receptor β1a subunit promotes Ca^{2+} dependent excitation-contraction coupling in skeletal myotubes. Biophys J 84:220–237

Sheridan DC, Takekura H, Franzini-Armstrong C, Beam KG, Allen PD, Perez CF (2006) Bidirectional signaling between calcium channels of skeletal muscle requires multiple direct and indirect interactions. Proc Natl Acad Sci USA 103:19760–19765

Sobie EA, Ramay HR (2009) Excitation-contraction coupling gain in ventricular myocytes: insights from a parsimonious model. J Physiol 587:1293–1299

Song LS, Guatimosim S, Gomez-Viquez L, Sobie EA, Ziman A, Hartmann H, Lederer WJ (2005) Calcium biology of the transverse tubules in heart. Ann N Y Acad Sci 1047:99–111

Stefan CJ, Manford AG, Baird D, Yamada-Hanff J, Mao Y, Emr SD (2011) Osh proteins regulate phosphoinositide metabolism at ER-plasma membrane contact sites. Cell 144:389–401

Stiber J, Hawkins A, Zhang ZS, Wang S, Burch J, Graham V, Ward CC, Seth M, Finch E, Malouf N et al (2008) STIM1 signalling controls store-operated calcium entry required for development and contractile function in skeletal muscle. Nat Cell Biol 10:688–697

Suh BC, Horowitz LF, Hirdes W, Mackie K, Hille B (2004) Regulation of KCNQ2/KCNQ3 current by G protein cycling: the kinetics of receptor-mediated signaling by G_q. J Gen Physiol 123:663–683

Suh BC, Kim DI, Falkenburger BH, Hille B (2012) Membrane-localized β-subunits alter the PIP_2 regulation of high-voltage activated Ca^{2+} channels. Proc Natl Acad Sci USA 109:3161–3166

Szentpetery Z, Varnai P, Balla T (2010) Acute manipulation of Golgi phosphoinositides to assess their importance in cellular trafficking and signaling. Proc Natl Acad Sci USA 107:8225–8230

Takeshima H, Komazaki S, Nishi M, Iino M, Kangawa K (2000) Junctophilins: a novel family of junctional membrane complex proteins. Mol Cell 6:11–22

Tanabe T, Beam KG, Adams BA, Niidome T, Numa S (1990) Regions of the skeletal muscle dihydropyridine receptor critical for excitation-contraction coupling. Nature 346:567–569 van Oort RJ, Garbino A, Wang W, Dixit SS, Landstrom AP, Gaur N, De Almeida AC, Skapura DG, Rudy Y, Burns AR et al (2011) Disrupted junctional membrane complexes and hyperactive ryanodine receptors after acute junctophilin knockdown in mice. Circulation 123:979–988

Vivas O, Kruse M, Hille B (2014) Nerve growth factor sensitizes adult sympathetic neurons to the proinflammatory peptide bradykinin. J Neurosci 34:11959–11971

Wang SQ, Song LS, Lakatta EG, Cheng H (2001) Ca^{2+} signalling between single L-type Ca^{2+} channels and ryanodine receptors in heart cells. Nature 410:592–596

Whorton MR, MacKinnon R (2011) Crystal structure of the mammalian GIRK2 K^+ channel and gating regulation by G proteins, PIP_2, and sodium. Cell 147:199–208

Wu L, Bauer CS, X-g Z, Xie C, Yang J (2002) Dual regulation of voltage-gated calcium channels by PtdIns(4,5)P_2. Nature 419:947–952

Xu M, Zhou P, Xu SM, Liu Y, Feng X, Bai SH, Bai Y, Hao XM, Han Q, Zhang Y et al (2007) Intermolecular failure of L-type Ca^{2+} channel and ryanodine receptor signaling in hypertrophy. PLoS Biol 5:e21

Zaydman MA, Cui J (2014) PIP_2 regulation of KCNQ channels: biophysical and molecular mechanisms for lipid modulation of voltage-dependent gating. Front Physiol 5:195

Zaydman MA, Silva JR, Delaloye K, Li Y, Liang H, Larsson HP, Shi J, Cui J (2013) Kv7.1 ion channels require a lipid to couple voltage sensing to pore opening. Proc Natl Acad Sci USA 110:13180–13185

Zhang H, He C, Yan X, Mirshahi T, Logothetis DE (1999) Activation of inwardly rectifying K^+ channels by distinct PtdIns(4,5)P_2 interactions. Nat Cell Biol 1:183–188

Zhang H, Craciun LC, Mirshahi T, Rohács T, Lopes CM, Jin T, Logothetis DE (2003) PIP_2 activates KCNQ channels, and its hydrolysis underlies receptor-mediated inhibition of M currents. Neuron 37:963–975

The Lipid Droplet and the Endoplasmic Reticulum

8

Yuki Ohsaki, Kamil Sołtysik, and Toyoshi Fujimoto

Abstract

Lipid droplets (LDs) are often found adjacent to the endoplasmic reticulum (ER). The ER–LD association may appear morphologically similar to the prototypical membrane contact sites found between the ER and other organelles, but the functional relationship between the ER and LDs is unique in that highly hydrophobic lipid esters are transported between them. This transportation is thought to occur through some form of membrane continuity, but its details are yet to be defined. Lipin, seipin, and FIT proteins, which are located at the ER–LD interface, may be involved in the lipid ester transport and probably play important roles for functional connectivity of the two organelles. More recently, LDs in the nucleus were found to be closely adhered to the inner nuclear membrane, representing a specialized form of the ER–LD association. In this article, we will give an overview of the ER–LD association, which is still filled with many unanswered questions.

Keywords

Lipid droplet • Endoplasmic reticulum • Lipid ester • Phospholipid monolayer • Triglyceride • Cholesterol ester • Nuclear membrane • Lipin • Seipin • FIT protein

8.1 Introduction

Lipid droplets (LDs) are composed of a surface phospholipid monolayer and a lipid ester core, which in most cells is a mixture of triacylglycerols (TAG) and cholesterol esters (CE). Historically, LD was largely regarded as an inert repository for lipids, but this view was changed by the discovery of a number of functional proteins in and around LDs and of LD functions that extend beyond simple lipid deposition. Currently, LDs are recognized as independent organelles with a dynamic nature (Walther

Y. Ohsaki • K. Sołtysik • T. Fujimoto (✉)
Department of Anatomy and Molecular Cell Biology, Nagoya University Graduate School of Medicine, 65 Tsurumai, Showa, Nagoya 466-8550, Japan
e-mail: tfujimot@med.nagoya-u.ac.jp

M. Tagaya, T. Simmen (eds.), *Organelle Contact Sites*, Advances in Experimental Medicine and Biology 997, DOI 10.1007/978-981-10-4567-7_8

and Farese 2012; Ohsaki et al. 2014; Gao and Goodman 2015; Welte 2015).

Conventional electron microscopy often shows LDs adjacent to the ER, and the ER–LD relationship may look similar to the relationships between the ER and other organelles at membrane contact sites (MCS) (Phillips and Voeltz 2016). Yet the ER–LD association is probably different from those at authentic MCS, such as the mitochondria-associated membrane (MAM), because the ER is where LDs form (Note: LDs may also form through the division of existing LDs, but this mechanism probably plays a relatively minor role in most cells). Phospholipids and sterols might be transported at the ER–LD interface as in other MCS, but only at the ER–LD interface can lipid esters synthesized in the ER be transferred to LDs. Given the hydrophobic nature of lipid esters, the transport may require direct continuity between the ER membrane and the LD surface, even though transiently. There has been much speculation on how this transport is performed, but definite answer based on solid experimental data has not been obtained.

8.2 Putative Membrane Continuity Between the ER and LDs

The enzymes required in the final steps of the TG and CE synthetic pathways, diacylglycerol O-acyltransferase 1 (DGAT1) and DGAT2, and acyl-CoA acyltransferase 1 (ACAT1) and ACAT2, respectively, are present in the ER membrane (Yen et al. 2008; Chang et al. 2009). The activity center of these enzymes faces either the cytoplasmic or the luminal side of the membrane, but generated lipid esters are thought to be assimilated to the ER membrane rather than deposited to the aqueous phase. Small amounts of lipid esters can exist within the phospholipid layers of the membrane (Gorrissen et al. 1980, 1982), but when their quantities exceed a solubility limit, they are likely to "oil out," i.e., to precipitate forming a lens-shaped mass between the two phospholipid layers (Khandelia et al. 2010). As this lens-shaped intramembrane mass becomes larger, it may bud toward the cytoplasm, being covered by the cytoplasmic leaflet of the ER membrane, and eventually become an independent LD (please see Ohsaki et al. 2014 for other hypotheses).

Many aspects of the LD formation process have not been established experimentally. No matter how the process of nascent LD formation proceeds, however, it is plausible that hydrophobic lipid esters are always shielded from the aqueous phase by a phospholipid monolayer as in mature LDs (Tauchi-Sato et al. 2002), which would entail the presence of membrane continuity between the ER and newly forming LDs. This idea is consistent with the observation that some membrane proteins move between the ER and LDs (Zehmer et al. 2009; Jacquier et al. 2011). In particular, yeast Dga1p was observed to move in both directions in an energy-independent manner (Jacquier et al. 2011), indicating that its translocation is not mediated by vesicular trafficking but rather, probably, by lateral diffusion through membrane continuity.

Morphological observation of the ER–LD continuity has been reported in only a few instances, however. When the conventional chemical fixation and embedding method have been used, continuity has been observed only under special conditions (Wanner et al. 1981; Ohsaki et al. 2008). Seemingly direct ER–LD contact was observed through electron microscopy of quick-frozen and freeze-substituted specimens (Wilfling et al. 2013), but it is not clear yet whether the cytoplasmic leaflet of the ER membrane and the phospholipid monolayer of the LD surface are directly continuous. The difficulty of observing structural ER–LD continuity may be due to the inability of chemical fixatives to stabilize the structure. In light of the relatively long time that is required for Dga1p to move from the ER to an LD (Jacquier et al. 2011), and of the facilitation of its retrograde movement from an LD to the ER by Ice2p (Markgraf et al. 2014), it is likely that the connection between the ER and LDs is not a stable structure, but one that forms only transiently in a regulated manner.

In this context, it is noteworthy that the activation of the Arf1/COPI machinery induces the budding of small-sized LDs from larger LDs (Thiam et al. 2013a; Wilfling et al. 2014). Because smaller LDs have a higher surface-to-volume ratio than larger ones, the budding of small-sized LDs is thought to decrease the phospholipid density in the larger LD surface, causing packing defects and stimulating fusion between the large LD and the adjacent ER membrane (Wilfling et al. 2014). The presence of a putative Arf1 regulator ELMOD2 in LDs is consistent with the action of Arf1/COPI (Suzuki et al. 2015). Nevertheless, the Arf1/COPI-dependent process may not be the sole mechanism by which the ER–LD connection is generated, because the translocation of Dga1p in yeast was observed even in the absence of COPI (Jacquier et al. 2011).

8.3 Proteins at the ER–LD Interface

The putative membrane continuity between the ER and LDs is likely to be regulated by some protein machineries. In this section, we will discuss several proteins that have been reported to exist at the ER–LD interface and have been implicated in the functional linkage between the ER and LDs. It must be noted, however, that our knowledge of whether a protein is located at the ER–LD interface is largely limited to what we can see through fluorescence microscopy. Thus it is not always clear whether a protein is actually present at a structure connecting the ER and LDs as we tend to imagine from schematic diagrams like Fig. 8.1a.

8.3.1 Lipin

Pah1p, a yeast homolog of mammalian lipin, is a phosphatidic acid (PA) phosphatase that generates diacylglycerol (DAG) from PA. In yeast lacking Pah1p, though the total amount of lipid esters is not changed, LD formation is suppressed and aberrant accumulation of lipid esters is observed around the ER (Adeyo et al. 2011). The defect observed in *pah1Δ* is not due to a decrease in TAG synthesis caused by the decrease in DAG, because *dga1Δlro1Δ* that lacks TAG but contains sterol esters has LDs in the presence of Pah1p but not in *pah1Δ* (Adeyo et al. 2011).

Pah1p is inactive when it exists as a soluble cytosolic protein in its phosphorylated form. It is activated by being dephosphorylated by the phosphatase complex Nem1p–Spo7p and recruited to the ER membrane via an amphipathic helix (Karanasios et al. 2010). Nem1p, the catalytic subunit of the phosphatase complex, was observed as a punctum adjacent to an LD through fluorescence microscopy (Adeyo et al. 2011). The overexpression of Nem1p and Spo7p recruited Pah1p to LDs (Karanasios et al. 2013). More recently, a catalytically inactive form of Pah1p was shown to be targeted to two distinct puncta flanking the nucleus-vacuolar junction where LDs form in the diauxic phase (Barbosa et al. 2015). These results suggest that Pah1p is involved in LD formation at its interface with the ER.

Lipins in mammalian cells are expressed in five isoforms (i.e., lipin1α, lipin1β, lipin1γ, lipin2, and lipin3) encoded by three different genes (Csaki et al. 2013). Lipin1α and lipin1β in macrophages (Valdearcos et al. 2011) and lipin1γ in Cos7 cells (Wang et al. 2011) were found to be associated with LDs, and adipocytes depleted of lipin1 and lipin2 showed defects in LDs (Sembongi et al. 2013). These results suggest that mammalian lipins also play some role in LD formation. The mammalian orthologs of Nem1p and Spo7p are C-terminal domain nuclear envelope phosphatase 1 (CTDNEP1) (Kim et al. 2007) and nuclear envelope phosphatase 1-regulatory subunit 1 (NEP1-R1) (Han et al. 2012), respectively, but whether these proteins are involved in LD formation is yet to be studied.

8.3.2 Seipin

Mammalian seipin and yeast Sei1p (formerly Fld1p) are integral ER membrane proteins that are thought to exist at the ER–LD junction (Lundin et al. 2006; Szymanski et al. 2007).

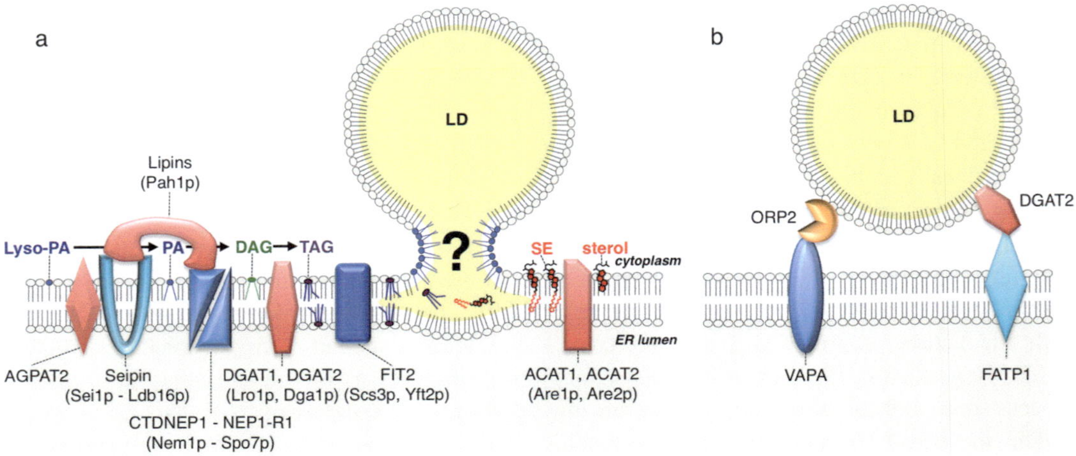

Fig. 8.1 The ER–LD interface. (**a**) Molecules related to LD formation. Lipin (Pah1p) is activated by the CTDENP1-NEP1-R1 (Nem1p-Spo7p) complex and catalyzes the conversion from PA to DAG. Seipin (Sei1p-Ldb16p complex in yeast) may be related to PA metabolism through interaction with AGPAT2 and lipin, which are ER enzymes that produce PA from lyso-PA and DAG from PA, respectively. FIT1 and FIT2 (Scs3p, Yft2p) facilitate incorporation of TAG into LDs. DGAT2 in mammals and Dga1p in yeast are depicted as proteins in the ER, but they also exist in LDs. Whether the ER and LDs are connected by a membrane bridge as depicted in this figure is not clear. (**b**) Molecules in the putative ER–LD MCS. FATP1 in the ER binds with DGAT2 in LDs. VAPA in the ER interacts with the soluble cytosolic protein ORP2, which may be bound to LDs

Seipin was originally discovered as a protein that was mutated in Berardinelli-Seip congenital lipodystrophy patients, who lack adipose tissue and fat deposition in the liver and muscles (Magre et al. 2001). In cells defective in seipin, LDs are small and aggregated. This phenotype is shared by yeast lacking Sei1p, which shows small irregular droplets in some cells, but harbors super-sized LDs in other cells (Szymanski et al. 2007; Fei et al. 2008). Recently, another transmembrane ER protein, Ldb16p, was found to make a complex with Sei1p, and *ldb16Δ* recapitulated the LD phenotype of *sei1Δ*. Interestingly, overexpression of seipin can rescue abnormalities of both *sei1Δ* and *ldb16Δ*, indicating that mammalian seipin is the functional homolog of the yeast Sei1p-Ldb16p complex (Wang et al. 2014; Cartwright et al. 2015; Grippa et al. 2015).

Seipin and Sei1p have a similar secondary structure, i.e., two transmembrane regions, a central loop in the ER lumen and N- and C-termini in the cytoplasm, and show homo-oligomerization (Binns et al. 2010; Fei et al. 2011a). Neither seipin nor the Sei1p-Ldb16p complex seems to have enzymatic activity, suggesting that their functions may be structural or carried out by interacting partners. In this context, it is notable that seipin was shown to interact with 14-3-3β protein (Yang et al. 2014), sarco/endoplasmic reticulum Ca^{2+}-ATPase (SERCA) (Bi et al. 2014), and perilipin2 (Mori et al. 2016) as well as lipin1 (Sim et al. 2013) and 1-acylglycerol-3-phosphate O-acyltransferase2 (AGPAT2) (Talukder et al. 2015). Some of these interactions, however, are not likely to be directly related to the LD phenotype shown by seipin mutants; for example, the SERCA ortholog is not present in yeast, and the deletion of putative Ca^{2+}-ATPase in yeast organelles does not cause LD abnormality (Cartwright et al. 2015).

On the other hand, it is notable that a seipin oligomer can simultaneously bind with both AGPAT2 and lipin1 (Talukder et al. 2015), which catalyzes the conversion of lysoPA to PA and of PA to DAG, respectively. Although it is not known whether the Sei1p-Ldb16p complex interacts with proteins related to PA metabolism, deletion of Sei1p was shown to increase the PA level in microsomes (Fei et al. 2011b). Curiously, puncta marked by PA-binding protein probes were observed at the ER–LD interface in *sei1Δ* , even when the cellular PA level was thought to be decreased (Grippa et al. 2015; Han et al. 2015). This result led to the proposal that a function of Sei1p may be preventing PA from being trapped by influencing the membrane curvature at the ER–LD juncture (Han et al. 2015). Here it is of note that mutation of AGPAT2 is also a cause of Berardinelli-Seip congenital lipodystrophy (Agarwal et al. 2002).

On the other hand, in *sei1Δ*, defects in the phospholipid monolayer and a decrease in LD-specific proteins were observed, and thus Sei1p was hypothesized to be a diffusion barrier involved in establishing the LD identity (Grippa et al. 2015). In support of this idea, in the absence of seipin, nascent LDs failed to become mature LDs, indicating that transport of newly synthesized lipid esters at the ER–LD junction is defective (Wang et al. 2016). The latter study failed to observe an increase in cellular PA or PA accumulation at any specific site in *Drosophila* S2 cells, suggesting that abnormality in LDs may not be caused by aberrant PA metabolism, at least in this cell type (Wang et al. 2016).

Another interesting phenotype associated with *sei1Δ* and *ldb16Δ* is the aberrant formation of LDs in the nucleus (Cartwright et al. 2015; Wolinski et al. 2015). The result suggests that the Sei1p-Ldb16p complex is also involved in determining the proper direction of LD formation, but the molecular mechanism behind this function is not known.

In addition to abnormalities in LD formation at the cellular level, seipin deficiency causes defects in adipocyte differentiation (Payne et al. 2008; Yang et al. 2013). Regulation of protein kinase A-dependent lipolysis (Chen et al. 2012) and actin cytoskeleton remodeling via 14-3-3β protein (Yang et al. 2014) were shown as seipin functionalities required for adipocyte differentiation. The C-terminal domain appears to be critical for seipin's adipogenic function (Yang et al. 2013), but the molecular mechanism regulating adipogenesis is still poorly understood.

Seipin is glycosylated in the central luminal loop domain, and mutations in its glycosylation sites cause neurodegeneration, leading to motor neuron diseases called seipinopathies that include distal hereditary motor neuropathy type V and Silver syndrome. Seipinopathies are thought to be caused downstream of ER stress and unfolded protein response due to a toxic gain-of-function and are probably not caused by a defect in the LD-related functionality of seipin (Windpassinger et al. 2004; Ito and Suzuki 2007).

8.3.3 FIT Proteins

The mammalian fat storage-inducing transmembrane (FIT) proteins, FIT1 and FIT2, are six-pass transmembrane proteins in the ER with both termini facing the cytosol. FIT1 is mainly found in the heart and skeletal muscle, while FIT2 is ubiquitously expressed with its highest expression in white and brown adipose tissue. Both FIT1 and FIT2 bind to TAG and DAG in vitro (Gross et al. 2011).

Overexpression of FIT proteins in cultured cells increased TAG incorporation into LDs (Kadereit et al. 2008). On the other hand, knockdown of FIT2 in 3T3-L1 adipocytes decreased the number and size of LDs (Kadereit et al. 2008). Consistent with these in vitro results, FIT2-knockout mice showed a decrease in LDs in preadipocytes without changes in TAG synthesis and degradation (Miranda et al. 2014), and postnatal FIT2 depletion caused LD arrest in the ER lumen of enterocytes, leading to lethal enteropathy and malabsorption (Goh et al. 2015). Furthermore, a more recent study on yeast lacking FIT2 homologs, i.e., Scs3p and Yft2p, as well as 3T3-L1 cells after FIT2 knockdown, showed aberrant accumulation of LDs, which were enclosed by the ER membrane (Choudhary et al. 2015). These findings are consistent with the hypothesis that FIT proteins may be related to TAG packaging and budding of LDs (Fig. 8.1b).

Nevertheless, FIT protein functionality may not be simple. Another study on *scs3Δyft2Δ* yeast did not detect any change in the number and size of LDs, but found a defective response to ER stress, which may have been caused by modulation of the ER membrane property (Moir et al. 2012). It is not known whether FIT proteins are concentrated at the ER–LD interface.

8.4 Membrane Contact Sites Between the ER and LDs

MCS is generally defined as a site where two adjacent membranes are apposed with a narrow gap but do not fuse each other. Through conventional electron microscopy using chemical fixation and plastic embedding, we can frequently observe a close ER–LD association which fits the morphological criteria for MCS. However, as discussed above, the structure between the ER and LDs may not be stable and may be difficult to preserve using conventional sample preparation methods. Therefore we must be aware of the possibility that a seemingly MCS-like structure at the ER–LD interface may not represent the real structure in vivo.

Though its structural characteristics remain ambiguous, certain proteins may interact at the possible MCS between the ER and LDs. The first protein combination is fatty acid transport protein (FATP1)/acyl-CoA synthase and DGAT2; their binding has been observed in *C. elegans* and mammalian cells (Xu et al. 2012). The FATP1 is an ER protein while DGAT2 is present in LDs, and the interaction between them is thought to be linked to TAG synthesis and its incorporation into LDs. The second protein combination is VAMP-associated protein (VAP) A, a transmembrane protein in the ER, and ORP2, an oxysterol-binding protein (OSBP)/OSBP-related proteins (ORPs) family protein in the cytosol. The latter interaction was found between overexpressed proteins and adjacent clustered LDs, which induced increases in LDs and TAG synthesis (Weber-Boyvat et al. 2015). VAP proteins and OSBP/ORPs are engaged in the transport of small molecules at MCS between the ER and late endosome (Raiborg et al. 2015), suggesting that a similar transport may be occurring between the ER and LDs. These results suggest that the ER and LDs may be linked functionally even without direct membrane continuity.

It is noteworthy that forced reduction of perilipin2, a protein constitutively existing in the LD surface, was found to increase the ER–LD association. Perilipin2 in LDs was decreased either by RNAi of perilipin2, by overexpression of Rab18, or by treatment with brefeldin A; in all these conditions, apposition between the ER and LDs was significantly enhanced (Ozeki et al. 2005). The phospholipid monolayer in the LD surface has a propensity to adhere to other membranes due to its innate high surface tension, and this is eased by the presence of proteins like perilipin2 (Thiam et al. 2013b). It is thus plausible that the ER–LD association is facilitated by the downregulation of perilipin2. It is tempting to think that the ER–LD interaction is regulated passively through such a shielding function of LD-associated proteins as well as actively by interaction of specific protein partners.

8.5 The Nuclear Membrane and Nuclear LDs

LDs are often present in the nucleus, but the abundance of nuclear LDs varies between cell types and is not correlated with that of cytoplasmic LDs. Recently we found that nuclear LDs are closely associated with the nucleoplasmic reticulum (NR) (Ohsaki et al. 2016), which is the extension of the nuclear membrane into the nucleoplasm (Malhas et al. 2011) (Fig. 8.2). The NR harbors DGAT2, and nuclear LDs can become larger within the interphase nucleus by incorporating newly synthesized lipid esters (Ohsaki et al. 2016). These findings seem to suggest that the relationship between nuclear LDs and the NR is essentially the same as that between cytoplasmic LDs and the ER.

There are, however, a number of notable differences between nuclear and cytoplasmic LDs (Ohsaki et al. 2016). First, nuclear LDs are associated with the promyelocytic leukemia (PML) nuclear body (Bernardi and Pandolfi 2007). This association suggests that nuclear LDs might be related to some of the diverse

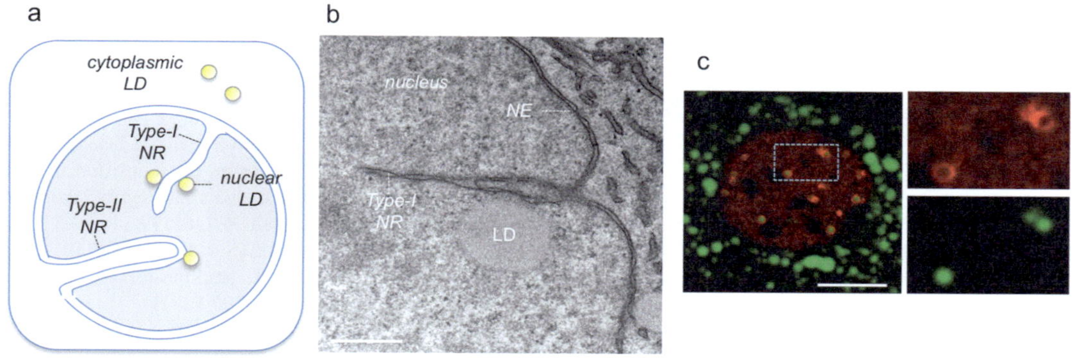

Fig. 8.2 Nuclear LDs. (**a**) Nuclear LDs are associated with the NR. The type I NR is the extension of the inner nuclear membrane alone, whereas the type II NR is the invagination of the whole nuclear envelope, made of both the outer and the inner nuclear membranes. (**b**) The association between a nuclear LD and the type I NR observed in Huh7 cells (hepatocellular carcinoma cell line) that were cultured with 0.4 mM oleic acids for 12 hr. Bar, 0.5 μm. (**c**) Accumulation of CCTα (*red*) in nuclear LDs (*green*) in Huh7 cells treated as in (**b**). *Insets* show magnified images of the rectangular area (*dotted line*). Bar, 10 μm

functions ascribed to the PML nuclear body, but this possibility is yet to be addressed. Second, overexpression of PML-II, a PML protein isoform, increased both NR and nuclear LDs, but intriguingly, the effect of PML-II overexpression was observed only in limited cell types, in which PML-II showed binding to the nuclear membrane. This cell type specificity is apparently correlated with the properties of the nuclear membrane, which is known to be highly variable in the protein composition (de Las Heras et al. 2013). Third, nuclear LDs harbored proteins resident in cytoplasmic LDs, such as perilipin3, Rab18, and adipocyte triglyceride lipase, but lacked perilipin2. This difference may affect the functionality of LDs. Fourth, CTP: phosphocholine cytidylyltransferase alpha (CCTα) was recruited to nuclear LDs. CCTα is the rate-limiting enzyme of the Kennedy pathway for phosphatidylcholine synthesis and is activated upon binding to a phospholipid layer. The nuclear membrane and cytoplasmic LDs were shown to be the platform for CCTα activation in other cell types (Gehrig et al. 2008; Krahmer et al. 2011), but nuclear LDs may also play this role, at least in some cell types (Fig. 8.2c). These characteristics suggest that nuclear LDs are not generated by accidental entry of cytoplasmic LDs into the nucleoplasm, but that they have unique properties and functions. Accordingly, we speculate that their association with the NR may also have a functional significance distinct from that of the association between the ER and cytoplasmic LDs.

8.6 Concluding Remarks

MAM, the prototypical MCS (Vance 2014), was originally described through electron microscopy as a site where an ER cistern and a mitochondrion are located closely side by side, usually within a distance of 10–20 nm. Later, MAM was obtained as a membrane fraction containing not just the ER membrane directly facing mitochondria but also the ER membrane outside of that region. By treating this isolated preparation as MAM, the definition of MAM was essentially broadened. Nevertheless this redefinition was deemed reasonable because the MAM fraction was enriched with a particular set of molecules in comparison to the canonical ER fraction.

As a consequence of this redefinition, however, the morphological criteria by which MAM and other MCS are identified have become loose; adjacent organelles separated by distances much greater than 10–20 nm are often classified as MCS. Here we think it is important to reconfirm that the molecular basis of MCS is adhesion mediated by distinct protein machineries. In other words, organelles intercalated with a gap too wide to be spanned by protein(s) must not be thought to form MCS in the absence of further information.

The ER–LD relationship is not a canonical MCS as discussed in this article. The ER is where LDs are generated, and in hepatocytes and small intestinal epithelial cells that synthesize lipoproteins, the ER is also where lipid esters in LDs are hydrolyzed and then re-esterified to make lipoproteins (Gibbons et al. 2000; Quiroga and Lehner 2012). The molecules involved in these functions are unique, and thus the structural requirements are most likely different from those of authentic MCS found in other locations.

Now that plentiful molecular information is becoming available, it is all the more important to observe the morphology of the ER–LD interface in detail and to define the location of each molecule in an ultrastructural context. Only by this approach will the functional relationship between the ER and LDs be understood more clearly.

Acknowledgments We are grateful to our lab members for their helpful discussions.

Funding This work was supported by Grants-in-Aid for Scientific Research from the Ministry of Education, Culture, Sports, Science, and Technology of the Government of Japan to YO (15K08152) and TF (15H02500, 15H05902).

References

Adeyo O, Horn PJ, Lee S, Binns DD, Chandrahas A, Chapman KD, Goodman JM (2011) The yeast lipin orthologue Pah1p is important for biogenesis of lipid droplets. J Cell Biol 192:1043–1055

Agarwal AK, Arioglu E, De Almeida S, Akkoc N, Taylor SI, Bowcock AM, Barnes RI, Garg A (2002) AGPAT2 is mutated in congenital generalized lipodystrophy linked to chromosome 9q34. Nat Genet 31:21–23

Barbosa AD, Savage DB, Siniossoglou S (2015) Lipid droplet-organelle interactions: emerging roles in lipid metabolism. Curr Opin Cell Biol 35:91–97

Bernardi R, Pandolfi PP (2007) Structure, dynamics and functions of promyelocytic leukaemia nuclear bodies. Nat Rev Mol Cell Biol 8:1006–1016

Bi J, Wang W, Liu Z, Huang X, Jiang Q, Liu G, Wang Y, Huang X (2014) Seipin promotes adipose tissue fat storage through the ER Ca^{2+}-ATPase SERCA. Cell Metab 19:861–871

Binns D, Lee S, Hilton CL, Jiang QX, Goodman JM (2010) Seipin is a discrete homooligomer. Biochemistry 49:10747–10755

Cartwright BR, Binns DD, Hilton CL, Han S, Gao Q, Goodman JM (2015) Seipin performs dissectible functions in promoting lipid droplet biogenesis and regulating droplet morphology. Mol Biol Cell 26:726–739

Chang TY, Li BL, Chang CC, Urano Y (2009) Acyl-coenzyme A:cholesterol acyltransferases. Am J Physiol Endocrinol Metab 297:E1–E9

Chen W, Chang B, Saha P, Hartig SM, Li L, Reddy VT, Yang Y, Yechoor V, Mancini MA, Chan L (2012) Berardinelli-seip congenital lipodystrophy 2/seipin is a cell-autonomous regulator of lipolysis essential for adipocyte differentiation. Mol Cell Biol 32:1099–1111

Choudhary V, Ojha N, Golden A, Prinz WA (2015) A conserved family of proteins facilitates nascent lipid droplet budding from the ER. J Cell Biol 211:261–271

Csaki LS, Dwyer JR, Fong LG, Tontonoz P, Young SG, Reue K (2013) Lipins, lipinopathies, and the modulation of cellular lipid storage and signaling. Prog Lipid Res 52:305–316

de Las Heras JI, Meinke P, Batrakou DG, Srsen V, Zuleger N, Kerr AR, Schirmer EC (2013) Tissue specificity in the nuclear envelope supports its functional complexity. Nucleus 4:460–477

Fei W, Shui G, Gaeta B, Du X, Kuerschner L, Li P, Brown AJ, Wenk MR, Parton RG, Yang H (2008) Fld1p, a functional homologue of human seipin, regulates the size of lipid droplets in yeast. J Cell Biol 180:473–482

Fei W, Li H, Shui G, Kapterian TS, Bielby C, Du X, Brown AJ, Li P, Wenk MR, Liu P et al (2011a) Molecular characterization of seipin and its mutants: implications for seipin in triacylglycerol synthesis. J Lipid Res 52:2136–2147

Fei W, Shui G, Zhang Y, Krahmer N, Ferguson C, Kapterian TS, Lin RC, Dawes IW, Brown AJ, Li P et al (2011b) A role for phosphatidic acid in the formation of "supersized" lipid droplets. PLoS Genet 7:e1002201

Gao Q, Goodman JM (2015) The lipid droplet-a well-connected organelle. Front Cell Dev Biol 3:49

Gehrig K, Cornell RB, Ridgway ND (2008) Expansion of the nucleoplasmic reticulum requires the coordinated activity of lamins and CTP:phosphocholine cytidylyltransferase α. Mol Biol Cell 19:237–247

Gibbons GF, Islam K, Pease RJ (2000) Mobilisation of triacylglycerol stores. Biochim Biophys Acta 1483:37–57

Goh VJ, Tan JS, Tan BC, Seow C, Ong WY, Lim YC, Sun L, Ghosh S, Silver DL (2015) Postnatal Deletion of Fat Storage-inducing Transmembrane Protein 2 (FIT2/FITM2) Causes Lethal Enteropathy. J Biol Chem 290:25686–25699

Gorrissen H, Tulloch AP, Cushley RJ (1980) Deuterium magnetic resonance of selectively deuterated cholesteryl esters in phosphatidylcholine vesicles. Biochemistry 19:3422–3429

Gorrissen H, Tulloch AP, Cushley RJ (1982) Deuterium magnetic resonance of triacylglycerols in phospholipid bilayers. Chem Phys Lipids 31:245–255
Grippa A, Buxo L, Mora G, Funaya C, Idrissi FZ, Mancuso F, Gomez R, Muntanya J, Sabido E, Carvalho P (2015) The seipin complex Fld1/Ldb16 stabilizes ER-lipid droplet contact sites. J Cell Biol 211:829–844
Gross DA, Zhan C, Silver DL (2011) Direct binding of triglyceride to fat storage-inducing transmembrane proteins 1 and 2 is important for lipid droplet formation. Proc Natl Acad Sci USA 108:19581–19586
Han S, Bahmanyar S, Zhang P, Grishin N, Oegema K, Crooke R, Graham M, Reue K, Dixon JE, Goodman JM (2012) Nuclear envelope phosphatase 1-regulatory subunit 1 (formerly TMEM188) is the metazoan Spo7p ortholog and functions in the lipin activation pathway. J Biol Chem 287:3123–3137
Han S, Binns DD, Chang YF, Goodman JM (2015) Dissecting seipin function: the localized accumulation of phosphatidic acid at ER/LD junctions in the absence of seipin is suppressed by Sei1p(DeltaNterm) only in combination with Ldb16p. BMC Cell Biol 16:29
Ito D, Suzuki N (2007) Molecular pathogenesis of seipin/BSCL2-related motor neuron diseases. Ann Neurol 61:237–250
Jacquier N, Choudhary V, Mari M, Toulmay A, Reggiori F, Schneiter R (2011) Lipid droplets are functionally connected to the endoplasmic reticulum in Saccharomyces cerevisiae. J Cell Sci 124:2424–2437
Kadereit B, Kumar P, Wang WJ, Miranda D, Snapp EL, Severina N, Torregroza I, Evans T, Silver DL (2008) Evolutionarily conserved gene family important for fat storage. Proc Natl Acad Sci USA 105:94–99
Karanasios E, Han GS, Xu Z, Carman GM, Siniossoglou S (2010) A phosphorylation-regulated amphipathic helix controls the membrane translocation and function of the yeast phosphatidate phosphatase. Proc Natl Acad Sci USA 107:17539–17544
Karanasios E, Barbosa AD, Sembongi H, Mari M, Han GS, Reggiori F, Carman GM, Siniossoglou S (2013) Regulation of lipid droplet and membrane biogenesis by the acidic tail of the phosphatidate phosphatase Pah1p. Mol Biol Cell 24:2124–2133
Khandelia H, Duelund L, Pakkanen KI, Ipsen JH (2010) Triglyceride blisters in lipid bilayers: implications for lipid droplet biogenesis and the mobile lipid signal in cancer cell membranes. PLoS One 5:e12811
Kim Y, Gentry MS, Harris TE, Wiley SE, Lawrence JC Jr, Dixon JE (2007) A conserved phosphatase cascade that regulates nuclear membrane biogenesis. Proc Natl Acad Sci USA 104:6596–6601
Krahmer N, Guo Y, Wilfling F, Hilger M, Lingrell S, Heger K, Newman HW, Schmidt-Supprian M, Vance DE, Mann M et al (2011) Phosphatidylcholine synthesis for lipid droplet expansion is mediated by localized activation of CTP:phosphocholine cytidylyltransferase. Cell Metab 14:504–515
Lundin C, Nordstrom R, Wagner K, Windpassinger C, Andersson H, von Heijne G, Nilsson I (2006) Membrane topology of the human seipin protein. FEBS Lett 580:2281–2284
Magre J, Delepine M, Khallouf E, Gedde-Dahl T Jr, Van Maldergem L, Sobel E, Papp J, Meier M, Megarbane A, Bachy A et al (2001) Identification of the gene altered in Berardinelli-Seip congenital lipodystrophy on chromosome 11q13. Nat Genet 28:365–370
Malhas A, Goulbourne C, Vaux DJ (2011) The nucleoplasmic reticulum: form and function. Trends Cell Biol 21:362–373
Markgraf DF, Klemm RW, Junker M, Hannibal-Bach HK, Ejsing CS, Rapoport TA (2014) An ER protein functionally couples neutral lipid metabolism on lipid droplets to membrane lipid synthesis in the ER. Cell Rep 6:44–55
Miranda DA, Kim JH, Nguyen LN, Cheng W, Tan BC, Goh VJ, Tan JS, Yaligar J, Kn BP, Velan SS et al (2014) Fat storage-inducing transmembrane protein 2 is required for normal fat storage in adipose tissue. J Biol Chem 289:9560–9572
Moir RD, Gross DA, Silver DL, Willis IM (2012) SCS3 and YFT2 link transcription of phospholipid biosynthetic genes to ER stress and the UPR. PLoS Genet 8: e1002890
Mori E, Fujikura J, Noguchi M, Nakao K, Matsubara M, Sone M, Taura D, Kusakabe T, Ebihara K, Tanaka T et al (2016) Impaired adipogenic capacity in induced pluripotent stem cells from lipodystrophic patients with BSCL2 mutations. Metabolism 65:543–556
Ohsaki Y, Cheng J, Suzuki M, Fujita A, Fujimoto T (2008) Lipid droplets are arrested in the ER membrane by tight binding of lipidated apolipoprotein B-100. J Cell Sci 121:2415–2422
Ohsaki Y, Suzuki M, Fujimoto T (2014) Open questions in lipid droplet biology. Chem Biol 21:86–96
Ohsaki Y, Kawai T, Yoshikawa Y, Cheng J, Jokitalo E, Fujimoto T (2016) PML isoform II plays a critical role in nuclear lipid droplet formation. J Cell Biol 212:29–38
Ozeki S, Cheng J, Tauchi-Sato K, Hatano N, Taniguchi H, Fujimoto T (2005) Rab18 localizes to lipid droplets and induces their close apposition to the endoplasmic reticulum-derived membrane. J Cell Sci 118:2601–2611
Payne VA, Grimsey N, Tuthill A, Virtue S, Gray SL, Dalla Nora E, Semple RK, O'Rahilly S, Rochford JJ (2008) The human lipodystrophy gene BSCL2/seipin may be essential for normal adipocyte differentiation. Diabetes 57:2055–2060
Phillips MJ, Voeltz GK (2016) Structure and function of ER membrane contact sites with other organelles. Nat Rev Mol Cell Biol 17:69–82
Quiroga AD, Lehner R (2012) Liver triacylglycerol lipases. Biochim Biophys Acta 1821:762–769
Raiborg C, Wenzel EM, Stenmark H (2015) ER-endosome contact sites: molecular compositions and functions. EMBO J 34:1848–1858

Sembongi H, Miranda M, Han GS, Fakas S, Grimsey N, Vendrell J, Carman GM, Siniossoglou S (2013) Distinct roles of the phosphatidate phosphatases lipin 1 and 2 during adipogenesis and lipid droplet biogenesis in 3T3-L1 cells. J Biol Chem 288:34502–34513

Sim MF, Talukder MM, Dennis RJ, O'Rahilly S, Edwardson JM, Rochford JJ (2013) Analysis of naturally occurring mutations in the human lipodystrophy protein seipin reveals multiple potential pathogenic mechanisms. Diabetologia 56:2498–2506

Suzuki M, Murakami T, Cheng J, Kano H, Fukata M, Fujimoto T (2015) ELMOD2 is anchored to lipid droplets by palmitoylation and regulates adipocyte triglyceride lipase recruitment. Mol Biol Cell 26:2333–2342

Szymanski KM, Binns D, Bartz R, Grishin NV, Li WP, Agarwal AK, Garg A, Anderson RG, Goodman JM (2007) The lipodystrophy protein seipin is found at endoplasmic reticulum lipid droplet junctions and is important for droplet morphology. Proc Natl Acad Sci USA 104:20890–20895

Talukder MM, Sim MF, O'Rahilly S, Edwardson JM, Rochford JJ (2015) Seipin oligomers can interact directly with AGPAT2 and lipin 1, physically scaffolding critical regulators of adipogenesis. Mol Metab 4:199–209

Tauchi-Sato K, Ozeki S, Houjou T, Taguchi R, Fujimoto T (2002) The surface of lipid droplets is a phospholipid monolayer with a unique fatty acid composition. J Biol Chem 277:44507–44512

Thiam AR, Antonny B, Wang J, Delacotte J, Wilfling F, Walther TC, Beck R, Rothman JE, Pincet F (2013a) COPI buds 60-nm lipid droplets from reconstituted water-phospholipid-triacylglyceride interfaces, suggesting a tension clamp function. Proc Natl Acad Sci USA 110:13244–13249

Thiam AR, Farese RV Jr, Walther TC (2013b) The biophysics and cell biology of lipid droplets. Nat Rev Mol Cell Biol 14:775–786

Valdearcos M, Esquinas E, Meana C, Gil-de-Gomez L, Guijas C, Balsinde J, Balboa MA (2011) Subcellular localization and role of lipin-1 in human macrophages. J Immunol 186:6004–6013

Vance JE (2014) MAM (mitochondria-associated membranes) in mammalian cells: lipids and beyond. Biochim Biophys Acta 1841:595–609

Walther TC, Farese RV Jr (2012) Lipid droplets and cellular lipid metabolism. Annu Rev Biochem 81:687–714

Wang H, Zhang J, Qiu W, Han GS, Carman GM, Adeli K (2011) Lipin-1gamma isoform is a novel lipid droplet-associated protein highly expressed in the brain. FEBS Lett 585:1979–1984

Wang CW, Miao YH, Chang YS (2014) Control of lipid droplet size in budding yeast requires the collaboration between Fld1 and Ldb16. J Cell Sci 127:1214–1228

Wang H, Becuwe M, Housden BE, Chitraju C, Porras AJ, Graham MM, Liu XN, Thiam AR, Savage DB, Agarwal AK et al (2016) Seipin is required for converting nascent to mature lipid droplets. Elife 5: e16582

Wanner G, Formanek H, Theimer RR (1981) The ontogeny of lipid bodies (spherosomes) in plant cells : Ultrastructural evidence. Planta 151:109–123

Weber-Boyvat M, Kentala H, Peranen J, Olkkonen VM (2015) Ligand-dependent localization and function of ORP-VAP complexes at membrane contact sites. Cell Mol Life Sci 72:1967–1987

Welte MA (2015) Expanding roles for lipid droplets. Curr Biol 25:R470–R481

Wilfling F, Wang H, Haas JT, Krahmer N, Gould TJ, Uchida A, Cheng JX, Graham M, Christiano R, Frohlich F et al (2013) Triacylglycerol synthesis enzymes mediate lipid droplet growth by relocalizing from the ER to lipid droplets. Dev Cell 24:384–399

Wilfling F, Thiam AR, Olarte MJ, Wang J, Beck R, Gould TJ, Allgeyer ES, Pincet F, Bewersdorf J, Farese RV Jr et al (2014) Arf1/COPI machinery acts directly on lipid droplets and enables their connection to the ER for protein targeting. Elife 3:e01607

Windpassinger C, Auer-Grumbach M, Irobi J, Patel H, Petek E, Horl G, Malli R, Reed JA, Dierick I, Verpoorten N et al (2004) Heterozygous missense mutations in BSCL2 are associated with distal hereditary motor neuropathy and Silver syndrome. Nat Genet 36:271–276

Wolinski H, Hofbauer HF, Hellauer K, Cristobal-Sarramian A, Kolb D, Radulovic M, Knittelfelder OL, Rechberger GN, Kohlwein SD (2015) Seipin is involved in the regulation of phosphatidic acid metabolism at a subdomain of the nuclear envelope in yeast. Biochim Biophys Acta 1851:1450–1464

Xu N, Zhang SO, Cole RA, McKinney SA, Guo F, Haas JT, Bobba S, Farese RV Jr, Mak HY (2012) The FATP1-DGAT2 complex facilitates lipid droplet expansion at the ER-lipid droplet interface. J Cell Biol 198:895–911

Yang W, Thein S, Guo X, Xu F, Venkatesh B, Sugii S, Radda GK, Han W (2013) Seipin differentially regulates lipogenesis and adipogenesis through a conserved core sequence and an evolutionarily acquired C-terminus. Biochem J 452:37–44

Yang W, Thein S, Wang X, Bi X, Ericksen RE, Xu F, Han W (2014) BSCL2/seipin regulates adipogenesis through actin cytoskeleton remodelling. Hum Mol Genet 23:502–513

Yen CL, Stone SJ, Koliwad S, Harris C, Farese RV Jr (2008) Thematic review series: glycerolipids. DGAT enzymes and triacylglycerol biosynthesis. J Lipid Res 49:2283–2301

Zehmer JK, Bartz R, Bisel B, Liu P, Seemann J, Anderson RG (2009) Targeting sequences of UBXD8 and AAM-B reveal that the ER has a direct role in the emergence and regression of lipid droplets. J Cell Sci 122:3694–3702

Role of Intra- and Inter-mitochondrial Membrane Contact Sites in Yeast Phospholipid Biogenesis

9

Yasushi Tamura and Toshiya Endo

Abstract

Eukaryotic cells exhibit intracellular compartments called organelles wherein various specialized enzymatic reactions occur. Despite the specificity of the characteristic functions of organelles, recent studies have shown that distinct organelles physically connect and communicate with each other to maintain the integrity of their functions. In yeast, multiple inter- and intramitochondrial membrane contact sites (MCSs) were identified to date and were proposed to be involved in phospholipid biogenesis. In the present article, we focus on inter- and intra-organellar MCSs involving mitochondria and their tethering factors, such as the ERMES (endoplasmic reticulum (ER)–mitochondria encounter structure) complex and EMC (conserved ER membrane protein complex) between mitochondria and the ER, vCLAMP (vacuole and mitochondria patch) between mitochondria and vacuoles, and the MICOS (mitochondrial contact site) complex between the mitochondrial outer and inner membranes (MOM and MIM). All of these membrane-tethering factors were proposed to be involved in phospholipid biogenesis. Furthermore, the existence of functional interconnections among multiple organelle contact sites is suggested. In the present article, we summarize the latest discoveries in regard to MCSs and MCS-forming factors involving mitochondria and discuss their molecular functions, with particular focus on phospholipid metabolism in yeast.

Keywords

Mitochondria • Phospholipid • ERMES • vCLAMP • EMC • MICOS • Yeast

Y. Tamura (✉)
Department of Material and Biological Chemistry, Faculty of Science, Yamagata University, Yamagata 990-8560, Japan
e-mail: tamura@sci.kj.yamagata-u.ac.jp

T. Endo
Faculty of Life Sciences, Kyoto Sangyo University, Kamigamo-motoyama, Kita-ku, Kyoto 603-8555, Japan

M. Tagaya, T. Simmen (eds.), *Organelle Contact Sites*, Advances in Experimental Medicine and Biology 997, DOI 10.1007/978-981-10-4567-7_9

9.1 Identification of Inter-mitochondrial Membrane Tethering Factors

9.1.1 Mitochondria–ER Tethering 1: The ERMES Complex

In yeast, the ERMES (ER–mitochondria encounter structure) complex is demonstrated to directly tether the ER and the mitochondrial outer membrane (MOM). The ERMES consists of four core components: Mmm1, Mdm10, Mdm12, and Mdm34 and three peripheral subunits: Gem1, Lam6/Ltc1, and Tom7. Mmm1, Mdm10, and Mdm12 were identified during the 1990s as temperature-sensitive (ts) alleles that affect mitochondrial distribution and morphology (Burgess et al. 1994; Sogo and Yaffe 1994; Berger et al. 1997). Upon cultivation at nonpermissive temperature (37 °C), *mmm1*, *mdm10*, and *mdm12* ts-mutant cells show aberrant mitochondrial morphology, i.e., large spherical mitochondria (Fig. 9.1a). Then, genome-wide screening using a yeast deletion mutant library allowed identification of an additional ERMES-core subunit, Mdm34 (Dimmer et al. 2002). Deletion of Mdm34 leads to an abnormal mitochondrial shape similar to those observed in *mmm1*, *mdm10*, and *mdm12* mutant cells. Subsequently, *MDM34* was also identified as a gene that is synthetically lethal in combination with the *mmm1* ts mutation (Youngman et al. 2004). Importantly, when expressed as a fusion protein with a fluorescent protein or when analyzed by indirect immunofluorescence microscopy, Mmm1, Mdm10, Mdm12, and Mdm34 were observed as punctate structures localized on the mitochondria tubule (Hobbs et al. 2001; Boldogh et al. 2003; Youngman et al. 2004) (Fig. 9.1b). Moreover, co-immunoprecipitation studies revealed interactions among Mmm1, Mdm10, Mdm12, Mdm34, and Tom7 (Boldogh et al. 2003; Meisinger et al. 2004, 2006, 2007; Yamano et al. 2010; Stroud et al. 2011). A fraction of Mdm10 also interacts with the sorting and assembly machinery (SAM) complex, which is a translocator complex in the MOM that facilitates the import and assembly of β-barrel proteins (Meisinger et al. 2004, 2006, 2007; Yamano et al. 2010). Although all the ERMES core subunits were initially thought to be mitochondrial proteins, Mmm1 was later found to be N-glycosylated, indicating that Mmm1 is an integral ER protein (Kornmann et al. 2009). Therefore, Mmm1 connects the ER membrane and the MOM through the interactions with other ERMES subunits localized in the MOM, such as Mdm10 and Mdm34 (Kornmann et al. 2009). This physical connection between the ER and mitochondria was shown to be important for efficient phospholipid transport between these organelles (the lipid-transfer function of ERMES is discussed in Section 2). Interestingly, expression of ChiMERA, an artificial ER–mitochondria tethering protein, was found to at least partially rescue defects in growth, mitochondrial morphology, and phospholipid transport in cells lacking an ERMES subunit, which supports the proposed ER–mitochondria tethering function by ERMES (Kornmann et al. 2009).

Subsequent studies further identified the additional ERMES components Gem1 and Lam6/Ltc1 as factors copurified with human influenza hemagglutinin (HA)-tagged or green fluorescent protein (GFP)-tagged ERMES subunits (Kornmann et al. 2011; Stroud et al. 2011; Elbaz-Alon et al. 2015; Murley et al. 2015). Gem1 belongs to the Miro (mitochondrial Rho) family and contains two GTPase domains and calcium-binding EF-hand motifs, respectively. Two mammalian Gem1 orthologs, Miro-1 and Miro-2, were first shown to anchor to the MOM and to be involved in mitochondrial biogenesis (Fransson et al. 2003). Subsequently, a study revealed that Gem1, a yeast Miro protein, is also tail-anchored in the MOM and that its GTPase and EF-hand domains are exposed to the cytosol (Frederick et al. 2004). Lam6/Ltc1 is also highly conserved from yeast to human and possesses glucosyltransferases, Rab-like GTPase activators and myotubularins (GRAM) domain, and a VAD1 analog of StAR-related lipid transfer (VASt) domain, which are structurally similar to the pleckstrin homology (PH) and

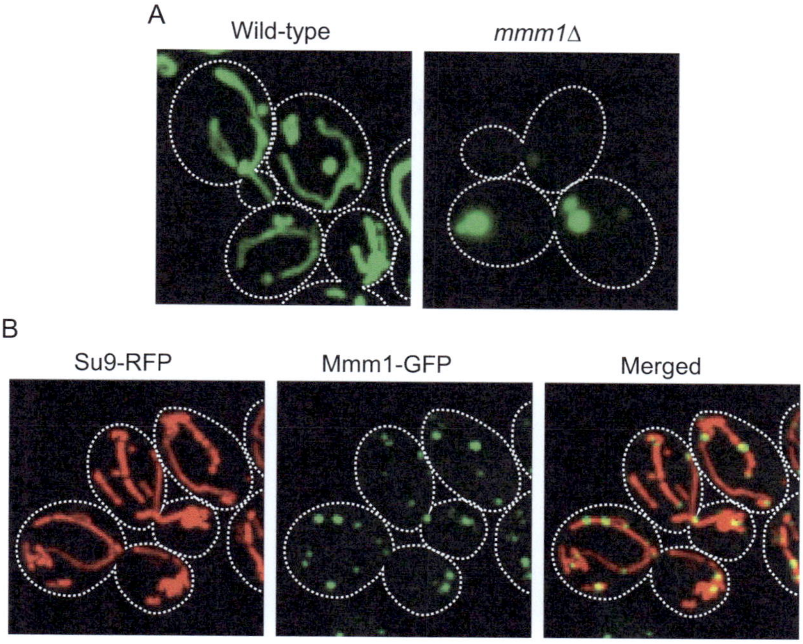

Fig. 9.1 The ERMES complex in yeast. (**a**) Mitochondria are visualized with mitochondria-targeted GFP in wild-type and *mmm1Δ* cells. (**b**) Mitochondria and the ERMES complexes are visualized with mitochondria-targeted RFP (Su9-RFP) and C-terminally GFP-tagged Mmm1, respectively

StAR-related lipid-transfer (StART) domains, respectively (Khafif et al. 2014). Consistent with the predicted functions of the GRAM and VASt domains in lipid binding, Lam6/Ltc1 was shown to facilitate sterol transport (Murley et al. 2015). Both Gem1 and Lam6/Ltc1 also form punctate structures that are co-localized with the ERMES dots. However, although the ERMES foci can no longer form in the absence of any of the ERMES core subunits, they remain intact even when Gem1 or Lam6/Ltc1 is lacking. This indicates that Gem1 and Lam6/Ltc1 are not essential for formation of the ERMES foci, but have a regulatory role in the formation of ERMES foci. Loss of Gem1 results in a slight decrease in the number and increase in the size of ERMES foci in a cell (Kornmann et al. 2011), whereas overexpression of Lam6/Ltc1 enlarges the ERMES region (Elbaz-Alon et al. 2015). Lam6/Ltc1 is localized at several MCSs, not only at the ER–mitochondria but also at the mitochondria–vacuole and ER–vacuole contact sites, likely through interactions with specific receptors. Tom70 and Tom71, the translocase receptors of the TOM (translocase of the mitochondrial OM) complex and a vacuolar protein Vac8, are considered to be responsible for localizations of ER–mitochondria and ER–vacuole membrane contact sites of Lam6/Ltc1, respectively (Fig. 9.2). At vacuoles, Lam6/Ltc1 appears important for formation of sterol-rich domains (Murley et al. 2015).

9.1.2 Mitochondria–ER Tethering 2: The EMC Complex

In addition to the ERMES complex, an additional ER–mitochondria tethering complex, the EMC (conserved ER membrane protein complex), was identified (Lahiri et al. 2014). The EMC, consisting of six ER proteins (Emc1–6), was originally reported as a protein complex important for protein folding in the ER (Jonikas et al. 2009). Subsequently, the EMC was proposed to facilitate phosphatidylserine (PS) transport from the ER to mitochondria by physically tethering the ER and the MOM (Lahiri et al. 2014). Since most phospholipids are synthesized in the ER, they must be distributed to other organelle membranes such as mitochondrial membranes after their syntheses (Fig. 9.3). However, the

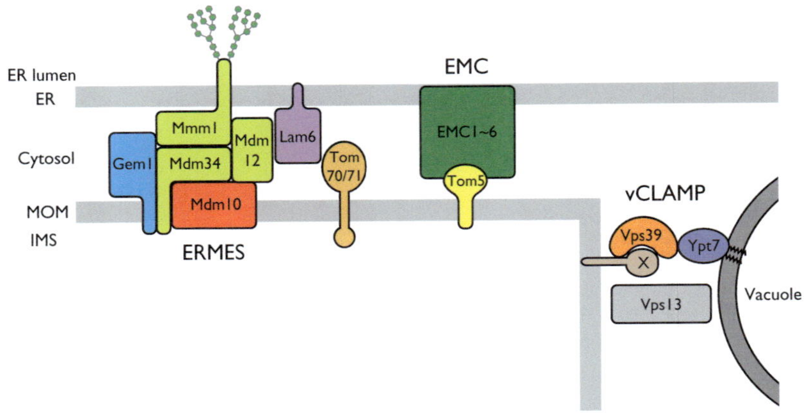

Fig. 9.2 Diagram of inter-mitochondrial tethering complexes. X indicates an unidentified receptor for Vps39

molecular mechanism facilitating transport of phospholipids from the ER to, e.g., mitochondria remained largely unknown.

To identify factors involved in phospholipid transport between the ER and mitochondria, Lahiri et al. (2014) focused on the genetic interaction between the *PSD1* and *CHO2* genes. *PSD1* encodes the mitochondrial PS decarboxylase, which generates phosphatidylethanolamine (PE) from PS, whereas *CHO2* encodes the ER-resident PE methyltransferase, which is responsible for phosphatidylcholine (PC) - production through PE methylations (Clancey et al. 1993; Kodaki and Yamashita, 1989) (Fig. 9.3). Analyses of the synthetic genetic array (SGA) revealed that *PSD1* and *CHO2* genetically interact (Tong et al. 2001); simultaneous deletions of *PSD1* and *CHO2* result in synthetic growth defects. This synthetic growth defect observed in *psd1Δcho2Δ* cells can be rescued by supplement of culture medium with ethanolamine or choline, which results in the production of PE or PC in cells via the Kennedy pathway (Gibellini and Smith. 2010). These results suggest that the gene required for PE production in mitochondria shows a negative genetic interaction with the *cho2*Δ mutation. Since PS synthesized in the ER has to be transported to mitochondria for the production of PE, genes involved in the PS transfer from the ER to mitochondria are also expected to genetically interact with *CHO2*. EMC components, which were identified as genetic interactors with *CHO2*, were thus subjected to analyses of their possible functions in PS transfer and ER–mitochondria tethering. As expected, loss of multiple EMC proteins caused defects in cell growth, the PS to PE conversion rate, as well as formation of the ER–mitochondria contacts (Lahiri et al. 2014). In addition, Lahiri et al. found interactions between EMC proteins such as Emc1, Emc2, and Emc5, and a MOM protein Tom5, which also genetically interacts with *CHO2*. Interestingly, the defects in cell growth and PS to PE conversion due to the loss of EMC can be partially recovered by expressing ChiMERA. These results support the idea that EMC functions as an ER–mitochondria tether like ERMES.

9.1.3 Mitochondria–Vacuole Tethering: vCLAMP

Characterization of Gem1 and Lam6/Ltc1 revealed a dynamic feature of the ER–mitochondria contact; the size and number of ERMES are variable and are probably regulated in response to cellular demand. Such dynamic alteration in the number of ERMES was also uncovered through identification and characterization of vCLAMP (vacuole and mitochondria patch), a mitochondria–vacuole MCS (Elbaz-Alon et al. 2014; Hönscher et al. 2014). Elbaz-Alon et al. utilized the yeast deletion library

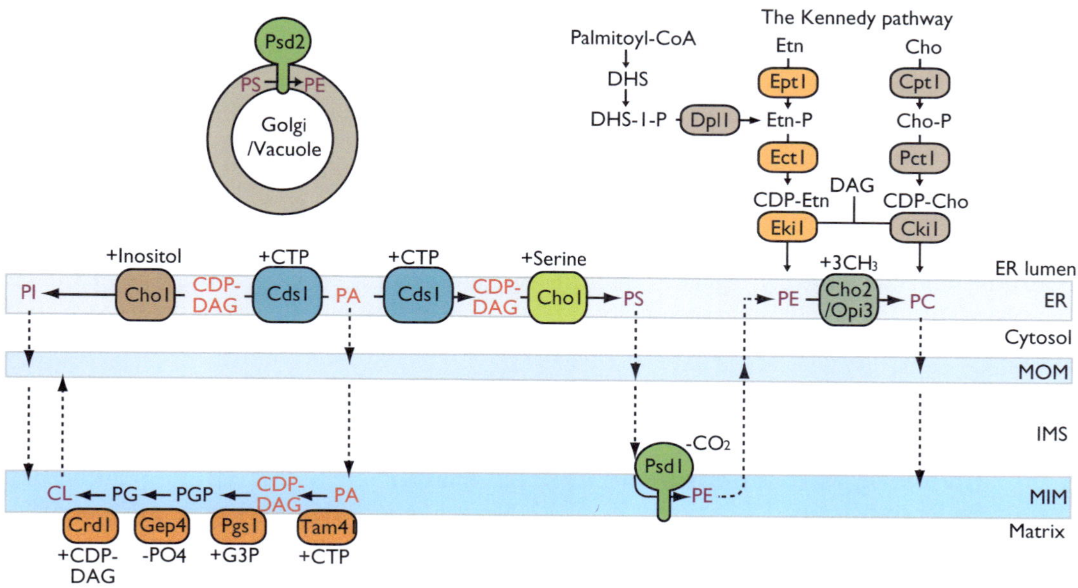

Fig. 9.3 Phospholipid biosynthetic pathways in yeast. Phospholipid syntheses require inter- and intra-organelle phospholipid transport. Broken and solid lines indicate phospholipid movements and conversions, respectively. *DHS* dihydrosphingosine, *DHS-1-P* dihydrosphingosine 1-phosphate, *Etn* ethanolamine, *Etn-P* ethanolamine phosphate, *Cho* choline, *Etn-P* choline phosphate, *PA* phosphatidic acid, *PI* phosphatidylinositol, *PS* phosphatidylserine, *PE* phosphatidylethanolamine, *PC* phosphatidylcholine, *DAG* diacylglycerol, *PGP* phosphatidylglycerol phosphate, *PG* phosphatidylglycerol, *CL* cardiolipin, *G3P* glycerol 3-phosphate

devised to express Mdm34-GFP and screened genes whose loss leads to alteration in the appearance of ERMES dots under a fluorescent microscope. This genome-wide microscopic analysis identified a number of genes that modulate the appearance of ERMES dots. Among them, Vps39, which was previously shown as a component of the homotypic membrane fusion and vacuole protein sorting (HOPS) tethering complex, was found to increase the number of ERMES dots upon its absence (Balderhaar and Ungermann, 2013). Consistently, Vps39 localizes at a specific region of the vacuole that is closely apposed to mitochondria. This vacuole–mitochondria MCS was termed vCLAMP. Hönscher et al. also speculated that the HOPS complex may be involved in the mitochondria–vacuole contact due to its membrane-tethering function upon endosome–vacuole or vacuole–vacuole fusions (Balderhaar and Ungermann, 2013). Interestingly, overexpression of Vps39, but not other HOPS components such as Vps41 or Vps11, enlarges the MCS between mitochondria and the vacuole (Hönscher et al. 2014). Close functional relationships between ERMES and vCLAMP were also suggested by the following observations: (1) simultaneous loss of ERMES and Vps39 is synthetic lethal; (2) overexpression of Vps39 rescues defective growths of ERMES-lacking cells; (3) the absence of ERMES extensively expands the vCLAMP region; (4) phospholipid transport between the ER and mitochondria is compromised in cells lacking both ERMES and vCLAMP (Hönscher et al. 2014; Elbaz-Alon et al. 2014).

Notably, the interrelation between ERMES and vCLAMP may be regulated by the metabolic state of the cell. Since yeast can grow by both fermentation and respiration, it switches between the metabolic states in response to nutrient conditions. When cultivated in glucose-containing media, yeast grow by fermentation, which metabolizes glucose to ethanol and carbon dioxide. On the other hand, when only a non-fermentable carbon source such as ethanol, glycerol, or lactic acid is available in a culture media, yeast cells grow by respiration. The number of ERMES dots increases, while that of vCLAMP decreases, under the respiratory condition. In parallel, Vps39 is phosphorylated at positions S246, S247, S249, and S250 and dissociates from mitochondria under the respiratory condition (Hönscher et al. 2014). Therefore, vCLAMP formation is likely regulated through phosphorylation of Vps39 in response to the growth condition.

9.2 Role of Inter-organelle MCSs Involving Mitochondria in Phospholipid Biogenesis

9.2.1 In Vivo Analysis of ERMES Functions in Phospholipid Transport

Theoretically, close inter-organelle contacts could contribute to facilitating phospholipid transport between the organelles by shortening the physical distances. Kornmann et al. showed that ERMES is important for efficient phospholipid transport between the ER and mitochondria (Kornmann et al. 2009). PS is synthesized in the ER and is converted to PE in mitochondria, where Psd1 PE synthase is present. The synthesized PE then moves back to the ER and is converted to PC through PE methylation, mediated by Cho2/Opi3 PE methyltransferases in the ER. Thus, the fate of pulse-labeled PS can be used as a useful index of the phospholipid transport between the ER and mitochondria in vivo (Fig. 9.3). In vivo pulse-chase experiments revealed that the PS to PC conversion rates are compromised in the absence of Mmm1, Mdm10, Mdm12, or Mdm34, indicating that at least either PS transport from the ER to mitochondria or PE transport from mitochondria to the ER is slowed in the absence of ERMES (Kornmann et al. 2009). Further supporting the lipid-transfer function of ERMES, Mmm1, Mdm34, and Mdm12 possess a SMP (synaptotagmin-like, mitochondrial, and lipid-binding proteins) domain, which could serve as a lipid-binding pocket (Kopec et al. 2011). Mmm1 and Mdm12 indeed bind phospholipids (AhYoung et al. 2015). However, the phospholipid transfer function of ERMES was challenged by other studies. Nguyen et al. and Voss et al. performed similar in vivo pulse-chase experiments using radioisotope (RI)-labeled serine and showed that transport of PS from the ER to mitochondria, which is measured by PS to PE conversion, was not affected by the absence of an ERMES subunit (Nguyen et al. 2012; Voss et al. 2012).

Since the previous in vivo studies focused on different phospholipid transport steps, i.e., only PS transport from the ER to mitochondria (Nguyen et al. 2012; Voss et al. 2012) or transport of both PS and PE between the ER and mitochondria (Kornmann et al. 2009), results from the above studies cannot be directly compared, although the loss of the ERMES subunit may at least not strongly affect phospholipid transport. The moderate effects of the absence of ERMES on phospholipid transport could reflect the presence of redundant phospholipid transport routes distinct from ERMES. As mentioned above, in addition to ERMES, multiple inter-organelle MCSs involving mitochondria, such as EMC and vCLAMP, exist, and these inter-organelle MCSs could compensate for defective phospholipid transport due to functional defects of ERMES. Supporting this idea, EMC was suggested to mediate PS transport from the ER to mitochondria as described above. The vCLAMP region extensively expands in response to the absence of ERMES, suggesting that the vacuole, which is connected to the ER through vesicular transport, may serve as a backup phospholipid source when the ER–mitochondria contact is not fully active. Interestingly, dominant mutants of Vps13, a possible vCLAMP constituting protein, was found to rescue defects in growth, mitochondrial morphology, and maintenance of mitochondrial DNA in cells lacking ERMES (Lang et al. 2015). These results indicate that the functions of ERMES can be bypassed by EMC and/or vCLAMP.

In summary, yeast cells exhibit multiple inter-organelle MCSs involving mitochondria, which may be functionally redundant for inter-organelle phospholipid transport. Such a complementary functional relationship among the multiple MCSs limits the analysis of the direct role of each phospholipid transfer machinery or pathway in inter-organelle phospholipid transport in vivo. Furthermore, strong growth defects in the absence of ERMES could indirectly affect the phospholipid transport in vivo, which would also hamper the straightforward interpretation of the in vivo analyses of MCS functions in lipid transport.

9.2.2 In Vitro Analysis of the Function of ERMES in Phospholipid Transport

To overcome the potential indirect effects of functional defects of MCSs in in vivo lipid transport analysis, an in vitro assay using membrane fractions containing the ER (microsomes) and mitochondria isolated from yeast cells could be developed (Choi et al. 2005). In fact, Nguyen et al. and Voss et al. performed in vitro lipid transport experiments to show that ERMES is dispensable for PS transport from the ER to mitochondria (Nguyen et al. 2012; Voss et al. 2012). However, these previous in vitro experiments were only applied to the PS import from the ER into mitochondria because PC production is inefficient as compared with generation of PS and PE under their in vitro assay conditions (Simbeni et al. 1993; Achleitner et al. 1999; Tamura et al. 2012a). This means that PE export or PC synthesis is inefficient in vitro. Thus, we attempted to optimize the in vitro lipid transport assay suitable for monitoring not only the import of PS into mitochondria from the ER but also the export of PE from mitochondria to the ER, through PS conversions to PE and PC (Kojima et al. 2016). The advantage of this in vitro assay system is that one can directly analyze phospholipid transport between the ER and mitochondria free from possible secondary effects arising from the presence of other organelle membranes, such as endosomes or the vacuole, which could also be the sources of phospholipid supply in vivo.

To examine whether ERMES plays a positive role in phospholipid transport, we prepared membrane fractions from wild-type cells and mutant cells lacking an ERMES subunit, Mmm1, Mdm12, or Mdm34. These cells express Vps13-D716H, a "dominant positive" Vps13 mutant, whose expression rescues the growth defects due to defective ERMES functions, to minimize the secondary effects (Lang et al. 2015). Importantly, we found that the loss of Mmm1, Mdm12, or Mdm34 significantly decelerates the transport of PS from the ER to mitochondria. However in contrast, the transport of PE from mitochondria to the ER, which is estimated from the relative amounts of phosphatidyldimethylethanolamine (PDME) +PC to PE ((PDME+PC)/PE), is less dependent on Mmm1 and Mdm34 (Kojima et al. 2016). While the PE to PDME/PC conversion rate was not significantly altered by the absence of Mdm12, it was even accelerated in the absence of Mmm1 or Mdm34. The *mmm1Δmdm12Δ* membranes essentially showed similar results to the *mmm1Δ* membrane, indicating that *MMM1* is epistatic to *MDM12* (Kojima et al. 2016). In summary, ERMES plays a positive role in PS transport from the ER to mitochondria, whereas it is dispensable for the transport of PE from mitochondria to the ER. In addition, Mmm1 and Mdm34 may perhaps play a negative regulatory role in the unidentified PE transport pathway from mitochondria to the ER.

9.3 Intramitochondrial Membrane Tethering Factors

Factors responsible for formations of inter-organelle MCSs involving mitochondria such as ERMES, EMC, and vCLAMP were found to contribute to inter-organelle phospholipid trafficking. Then intramitochondrial MCSs may also facilitate phospholipid transport between the MOM and the mitochondrial inner membrane (MIM). The surface area of the MIM is much larger than that of the MOM since the MIM forms a number of tubelike or lamellar sheetlike invaginations called cristae. The MIM is thus structurally categorized into two parts: the invaginated cristae membrane (CM) and the flat inner boundary membrane (IBM), which is closely apposed to the MOM. Translocation of mitochondrial proteins takes place at the IBM where the TIM23 complex, the mitochondrial protein translocator complex in the MIM, is mainly present (Vogel et al. 2006; Wurm and Jakobs, 2006). The TIM23 complex directly interacts with the TOM40 complex in the MOM and likely couples protein translocation across the MOM with that across the MIM (Tamura et al. 2009a; Shiota et al. 2011). Therefore, phospholipid transport may well take place between the MOM and the IBM as well. In addition to the TIM23 and TOM40 complexes, the mitochondrial fusion proteins Fzo1 and Ugo1 in the MOM and Mgm1 in the MIM are suggested to physically interact over the membrane, likely participating the formation of the contacts between the MOM and MIM (Sesaki et al. 2003; Wong et al. 2003; Sesaki and Jensen. 2004).

The MICOS (mitochondrial contact site) complex is another MIM protein complex that is proposed to form the MOM–MIM contact site by interacting with MOM proteins such as Sam50, Ugo1, Om45, Por1, and Tom40 (Harner et al. 2011; Hoppins et al. 2011; von der Malsburg et al. 2011). Although the MICOS proteins were first identified as factors responsible for formation of the crista junction, a constricted tubular region connecting the invaginated cristae and flat IBM (Rabl et al. 2009), comprehensive genetic interaction analysis indicated that the MICOS genes genetically interact with the ERMES genes (*MMM1*, *MDM10*, *MDM12*, *MDM34*, and *GEM1*) and with genes involved in cardiolipin (CL) synthesis (*GEP4*, *CRD1*) (Hoppins et al. 2011; Friedman et al. 2015), suggesting that the MICOS may play a role in phospholipid metabolism as well. Indeed, a recent study revealed that the MOM–MIM contact sites formed by the MICOS complex contribute to phospholipid biosynthesis (Aaltonen et al. 2016). The details of the role of MICOS in phospholipid biogenesis are summarized in the following section.

9.4 Phospholipid Transport Between the MOM and MIM and Its Relationship to the Mitochondrial MOM–MIM Contact Sites

Although a number of phospholipids are synthesized in the ER, mitochondria can generate two types of important phospholipids, CL and PE, from the precursor phospholipids PA and PS, respectively (Henry et al. 2012; Horvath and Daum, 2013; Tamura et al. 2014; Tatsuta et al. 2014). Since PA and PS are synthesized in the ER, they are required to be transported to the matrix or intermembrane space (IMS), where CL and PE synthetic enzymes function (Fig. 9.3). Recent studies identified soluble phospholipid transfer proteins in the IMS; evolutionarily conserved mitochondrial protein complexes, Ups1–Mdm35 and Ups2–Mdm35, were found to facilitate PA and PS transfers between the MOM and MIM, respectively (Connerth et al. 2012; Aaltonen et al. 2016; Miyata et al. 2016). Ups1 and Ups2 are homologous proteins and form a complex with the functional partner protein, Mdm35 (Osman et al. 2009; Tamura et al. 2009b). Mdm35 is not a simple binding partner, but rather facilitates vectorial translocations of Ups1 and Ups2 from the cytosol to the IMS, by functioning as a binding trap in the IMS (Potting et al. 2010; Tamura et al. 2010). Lack of Mdm35 thus results in significant decrease in the levels of Ups1 and Ups2 in mitochondria. Structural analyses of Ups1–Mdm35 revealed that the Ups1–Mdm35 complex has a structure similar to lipid-transfer proteins such as CERT, a ceramide transfer protein containing the StART domain (Kudo et al. 2008; Miliara et al. 2015; Watanabe et al. 2015; Yu et al. 2015). Like StART domain-containing proteins, the Ups1–mdm35 complex possesses a deep positively

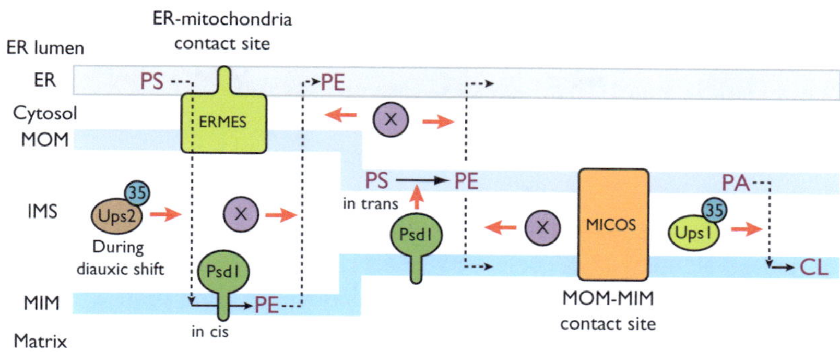

Fig. 9.4 Roles of the ER–mitochondria and MOM–MIM contacts in phospholipid biogenesis. PS transport from the ER to mitochondria depends on ERMES while it is unknown how PE is transported. X indicates unknown factors that mediate PE transport between the ER and mitochondria as well as between the MOM and MIM. The MOM–MIM MCS formed by MICOS contributes to in trans Psd1-dependent PE production, which is not required for the Ups2–Mdm35 complex. The Ups2–Mdm35 complex is indispensable for the Psd1-dependent PE production upon the diauxic shift. The Ups1–Mdm35 is proposed to function in the MOM–MIM MCSs. Broken and solid lines indicate phospholipid movements and conversions, respectively

charged pocket that could accommodate a lipid molecule. Indeed, we could determine the crystal structure of Ups1–Mdm35 with PA, which demonstrated the specific binding of PA to the Ups1 pocket (Miliara et al. 2015; Watanabe et al. 2015; Yu et al. 2015). Ups1–Mdm35 was reported to be enriched in the MOM–MIM contact site in mitochondria, which is consistent with the idea that phospholipid transfer occurs at the MCSs such as the MOM–MIM contact site within mitochondria (Connerth et al. 2012).

Although the PA-transfer function of the Ups1–Mdm35 complex suggested that its homologous Ups2–Mdm35 complex also functions as a lipid-transport machinery, its substrate lipid remained unidentified until recently. While loss of Ups1 results in a decreased level of CL due to the impaired transfer of PA, a precursor lipid for CL, loss of Ups2 causes a decrease in the PE level (Osman et al. 2009; Tamura et al. 2009b). As PS is a precursor phospholipid for PE, the decreased PE levels in the absence of Ups2 may suggest compromised transport of PS, which is synthesized in the ER, to the MIM, where Psd1 PE synthase (PS decarboxylase) is present. Consistent with this speculation, recent studies revealed the PS transfer function of the Ups2–Mdm35 complex (Aaltonen et al. 2016; Miyata et al. 2016). Recombinant Ups2–Mdm35 complex was shown to catalyze PS transport between artificial liposomes by in vitro lipid transport assays. Notably, despite the clear in vitro evidence for the Ups2–Mdm35 complex as a PS transport factor, the Ups2–Mdm35 complex is not essential for PE production in mitochondria in vivo. In vivo pulse-chase experiments using RI-labeled serine showed that the Psd1-dependent PS to PE conversion, which depends on proper PS transfer across the IMS, was only marginally impaired in the absence of Ups2. Besides, the absence of Ups2 did not affect subsequent PE to PC conversion in vivo (Aaltonen et al. 2016; Tamura et al. 2012b), and normal levels of PE are retained even in the absence of Ups2 when yeast cells are cultivated under a fermentable condition (Miyata et al. 2016). However interestingly, PE levels increase approximately two-fold upon the diauxic shift or a metabolic shift from fermentative growth to mitochondria-requiring respiration growth, and this PE increase requires Ups2 (Miyata et al. 2016). This suggests the prominent role of Ups2 in the PS transfer especially when mitochondria proliferate for respiration. These findings in turn raised a question regarding the mechanism responsible for the basal PS transfer under the fermentable condition (Fig. 9.4).

Aaltonen et al. proposed a new role of the MICOS complex in the Ups2-independent PE production. They showed that Psd1, which was reconstituted into liposomes without PS and PE upon cell-free synthesis, can decarboxylate PS in liposomes lacking Psd1 in trans. This indicates that Psd1 in the MIM could catalyze decarboxylation of PS present in the MOM when the MIM is sufficiently close to the MOM, likely at the MOM–MIM contact site (Fig. 9.4). Strikingly, lack of MICOS indeed significantly decelerated the conversions of PS to PE and PC, suggesting that disruption of the MOM–MIM contact site due to the loss of MICOS disturbs Psd1 function in trans. When MICOS is deficient, normal cristae formation and respiration activity become impaired. However, additional loss of Ups2 or Psd1 restores cristae morphogenesis and mitochondrial respiration due to MICOS deficiency likely by limiting the accumulation of PE in the MIM (Aaltonen et al. 2016).

9.5 Future Directions

Although organelle–organelle MCSs have been recognized since the 1950s (Porter and Palade. 1957), their molecular basis remained unknown until recently. As summarized above, recent studies uncovered factors involved in formations of several different inter- and intra-organelle MCSs involving mitochondria and revealed their potential roles in the cellular phospholipid metabolism. Moreover, MCSs appear to be functionally interconnected, and the number and degree of the MCSs are regulated in response to metabolic demand or availability of other MCSs. Therefore, if multiple MCSs provide major routes for phospholipid transport, they would constitute a fail-safe lipid transport system that can minimize the impact of the block of any of the redundant lipid transport routes. This is in contrast to the case of protein targeting to organelles, which usually relies on single specific routes for each protein to the cognate organelle. Thus, a functional analysis of the roles of MCSs would be a key to understanding the molecular mechanisms of intracellular phospholipid transport. Many fundamental questions in regard to the role of MCSs in phospholipid metabolism are open as follows.

First of all, the robust and redundant nature of the phospholipid transport via MCSs raises the question of whether each MCS selects specific phospholipids as substrates to be transported. Our in vitro study suggested that ERMES facilitates PS transport from the ER to mitochondria, whereas PE transport from mitochondria to the ER is not dependent on ERMES (Kojima et al. 2016). This suggests that ERMES specifically selects phospholipids to be transported. Determination of the substrate specificity of phospholipids or fatty acid compositions for each MCS should be important for further understanding of the distinct roles of each MCS.

Although a number of studies have revealed the involvement of ERMES in phospholipid transport between the ER and mitochondria, whether ERMES directly catalyzes phospholipid transport is still not clearly assessed. Similarly, although EMC and vCLAMP were also proposed to facilitate phospholipid transport between the ER and mitochondria and between the vacuole and mitochondria, respectively, how EMC and vCLAMP are involved in the inter-organelle lipid transfer is not clear. Whether EMC and vCLAMP promote lipid transfer by functioning merely as a membrane tether or as an active lipid-transfer machinery is thus an important question to be addressed.

In addition to the ER–mitochondria and vacuole–mitochondria MCSs, MCSs are present between other organelle membranes. Indeed, it was reported that peroxisomes are localized in close proximity of mitochondria, although the relationship of peroxisome with phospholipid metabolism remains unknown (Cohen et al. 2014; Mattiazzi Ušaj et al. 2015). Search for yet-to-be identified MCSs and elucidation of their components and functions would be essential for understanding of the molecular basis of intracellular phospholipid trafficking.

References

Aaltonen MJ, Friedman JR, Osman C, Salin B, di Rago JP, Nunnari J, Langer T, Tatsuta T (2016) MICOS and phospholipid transfer by Ups2–Mdm35 organize membrane lipid synthesis in mitochondria. J Cell Biol 213:525–534

Achleitner G, Gaigg B, Krasser A, Kainersdorfer E, Kohlwein SD, Perktold A, Zellnig G, Daum G (1999) Association between the endoplasmic reticulum and mitochondria of yeast facilitates interorganelle transport of phospholipids through membrane contact. Eur J Biochem 264:545–553

AhYoung AP, Jiang J, Zhang J, Khoi Dang X, Loo JA, Zhou ZH, Egea PF (2015) Conserved SMP domains of the ERMES complex bind phospholipids and mediate tether assembly. Proc Natl Acad Sci USA 112:E3179–E3188

Balderhaar HJK, Ungermann C (2013) CORVET and HOPS tethering complexes- coordinators of endosome and lysosome fusion. J Cell Sci 126:1307–1316

Berger KH, Sogo LF, Yaffe MP (1997) Mdm12p, a component required for mitochondrial inheritance that is conserved between budding and fission yeast. J Cell Biol 136:545–553

Boldogh IR, Nowakowski DW, Yang HC, Chung H, Karmon S, Royes P, Pon LA (2003) A protein complex containing Mdm10p, Mdm12p, and Mmm1p links mitochondrial membranes and DNA to the cytoskeleton-based segregation machinery. Mol Biol Cell 14:4618–4627

Burgess SM, Delannoy M, Jensen RE (1994) *MMM1* encodes a mitochondrial outer membrane protein essential for establishing and maintaining the structure of yeast mitochondria. J Cell Biol 126:1375–1391

Choi JY, Wu WI, Voelker DR (2005) Phosphatidylserine decarboxylases as genetic and biochemical tools for studying phospholipid traffic. Anal Biochem 347:165–175

Clancey CJ, Chang SC, Dowhan W (1993) Cloning of a gene (*PSD1*) encoding phosphatidylserine decarboxylase from *Saccharomyces cerevisiae* by complementation of an *Escherichia coli* mutant. J Biol Chem 268:24580–24590

Cohen Y, Klug YA, Dimitrov L, Erez Z, Chuartzman SG, Elinger D, Yofe I, Soliman K, Gärtner J, Thoms S et al (2014) Peroxisomes are juxtaposed to strategic sites on mitochondria. Mol BioSyst 10:1742–1748

Connerth M, Tatsuta T, Haag M, Klecker T, Westermann B, Langer T (2012) Intramitochondrial transport of phosphatidic acid in yeast by a lipid transfer protein. Science 338:815–818

Dimmer KS, Fritz S, Fuchs F, Messerschmitt M, Weinbach N, Neupert W, Westermann B (2002) Genetic basis of mitochondrial function and morphology in *Saccharomyces cerevisiae*. Mol Biol Cell 13:847–853

Elbaz-Alon Y, Rosenfeld-Gur E, Shinder V, Futerman AH, Geiger T, Schuldiner M (2014) A dynamic interface between vacuoles and mitochondria in yeast. Dev Cell 30:95–102

Elbaz-Alon Y, Eisenberg-Bord M, Shinder V, Stiller SB, Shimoni E, Wiedemann N, Geiger T, Schuldiner M (2015) Lam6 Regulates the Extent of Contacts between Organelles. Cell Rep 12:7–14

Fransson Å, Ruusala A, Aspenström P (2003) Atypical Rho GTPases have roles in mitochondrial homeostasis and apoptosis. J Biol Chem 278:6495–6502

Frederick RL, McCaffery JM, Cunningham KW, Okamoto K, Shaw JM (2004) Yeast Miro GTPase, Gem1p, regulates mitochondrial morphology via a novel pathway. J Cell Biol 167:87–98

Friedman JR, Mourier A, Yamada J, McCaffery JM, Nunnari J (2015) MICOS coordinates with respiratory complexes and lipids to establish mitochondrial inner membrane architecture. eLife 4

Gibellini F, Smith TK (2010) The Kennedy pathway-*De novo* synthesis of phosphatidylethanolamine and phosphatidylcholine. IUBMB Life 62:414–428

Harner M, Körner C, Walther D, Mokranjac D, Kaesmacher J, Welsch U, Griffith J, Mann M, Reggiori F, Neupert W (2011) The mitochondrial contact site complex, a determinant of mitochondrial architecture. EMBO J 30:4356–4370

Henry SA, Kohlwein SD, Carman GM (2012) Metabolism and regulation of glycerolipids in the yeast *Saccharomyces cerevisiae*. Genetics 190:317–349

Hobbs AEA, Srinivasan M, Mccaffery JM, Jensen RE (2001) Mmm1p, a mitochondrial outer membrane protein, is connected to mtDNA nucleoids and required for mtDNA stability. J Cell Biol 152:401–410

Hönscher C, Mari M, Auffarth K, Bohnert M, Griffith J, Geerts W, van der Laan M, Cabrera M, Reggiori F, Ungermann C (2014) Cellular metabolism regulates contact sites between vacuoles and mitochondria. Dev Cell 30:86–94

Hoppins S, Collins SR, Cassidy-Stone A, Hummel E, DeVay RM, Lackner LL, Westermann B, Schuldiner M, Weissman JS, Nunnari J (2011) A mitochondrial-focused genetic interaction map reveals a scaffold-like complex required for inner membrane organization in mitochondria. J Cell Biol 195:323–340

Horvath SE, Daum G (2013) Lipids of mitochondria. Prog Lipid Res 52:590–614

Jonikas MC, Collins SR, Denic V, Oh E, Quan EM, Schmid V, Weibezahn J, Schwappach B, Walter P, Weissman JS et al (2009) Comprehensive characterization of genes required for protein folding in the endoplasmic reticulum. Science 323:1693–1697

Khafif M, Cottret L, Balagué C, Raffaele S (2014) Identification and phylogenetic analyses of VASt, an uncharacterized protein domain associated with lipid-binding domains in Eukaryotes. BMC Bioinformatics 15:222

Kodaki T, Yamashita S (1989) Characterization of the methyltransferases in the yeast phosphatidylethanolamine methylation pathway by selective gene disruption. Eur J Biochem 185:243–251

Kojima R, Endo T, Tamura Y (2016) A phospholipid transfer function of ER-mitochondria encounter structure revealed *in vitro*. Sci Rep 6:30777

Kopec KO, Alva V, Lupas AN (2011) Bioinformatics of the TULIP domain superfamily. Biochem Soc Trans 39:1033–1038
Kornmann B, Currie E, Collins SR, Schuldiner M, Nunnari J, Weissman JS, Walter P (2009) An ER-mitochondria tethering complex revealed by a synthetic biology screen. Science 325:477–481
Kornmann B, Osman C, Walter P (2011) The conserved GTPase Gem1 regulates endoplasmic reticulum-mitochondria connections. Proc Natl Acad Sci USA 108:14151–14156
Kudo N, Kumagai K, Tomishige N, Yamaji T, Wakatsuki S, Nishijima M, Hanada K, Kato R (2008) Structural basis for specific lipid recognition by CERT responsible for nonvesicular trafficking of ceramide. Proc Natl Acad Sci USA 105:488–493
Lahiri S, Chao JT, Tavassoli S, Wong AKO, Choudhary V, Young BP, Loewen CJR, Prinz WA (2014) A conserved endoplasmic reticulum membrane protein complex (EMC) facilitates phospholipid transfer from the ER to mitochondria. PLoS Biol 12: e1001969
Lang AB, Peter ATJ, Walter P, Kornmann B (2015) ER–mitochondrial junctions can be bypassed by dominant mutations in the endosomal protein Vps13. J Cell Biol 210:883–890
Mattiazzi Ušaj M, Brložnik M, Kaferle P, Žitnik M, Wolinski H, Leitner F, Kohlwein SD, Zupan B, Petrovič U (2015) Genome-Wide localization study of yeast Pex11 identifies peroxisome–mitochondria interactions through the ERMES complex. J Mol Biol 427:2072–2087
Meisinger C, Rissler M, Chacinska A, Sanjuán Szklarz LK, Milenkovic D, Kozjak V, Schönfisch B, Lohaus C, Meyer HE, Yaffe MP et al (2004) The mitochondrial morphology protein Mdm10 functions in assembly of the preprotein translocase of the outer membrane. Dev Cell 7:61–71
Meisinger C, Wiedemann N, Rissler M, Strub A, Milenkovic D, Schönfisch B, Müller H, Kozjak V, Pfanner N (2006) Mitochondrial protein sorting: Differentiation of β-barrel assembly by Tom7-mediated segregation of Mdm10. J Biol Chem 281:22819–22826
Meisinger C, Pfannschmidt S, Rissler M, Milenkovic D, Becker T, Stojanovski D, Youngman MJ, Jensen RE, Chacinska A, Guiard B et al (2007) The morphology proteins Mdm12/Mmm1 function in the major β-barrel assembly pathway of mitochondria. EMBO J 26:2229–2239
Miliara X, Garnett JA, Tatsuta T, Abid Ali F, Baldie H, Pérez-Dorado I, Simpson P, Yague E, Langer T, Matthews S (2015) Structural insight into the TRIAP1/PRELI-like domain family of mitochondrial phospholipid transfer complexes. EMBO Rep 16:824–835
Miyata N, Watanabe Y, Tamura Y, Endo T, Kuge O (2016) Phosphatidylserine transport by Ups2–Mdm35 in respiration-active mitochondria. J Cell Biol 214:77–88
Murley A, Sarsam R, Toulmay A, Yamada J, Prinz W, Nunnari J, Murley A, Sarsam R, Toulmay A, Yamada J et al (2015) Ltc1 is an ER-localized sterol transporter and a component of ER–mitochondria and ER–vacuole contacts. J Cell Biol 209:539–548
Nguyen TT, Lewandowska A, Choi JY, Markgraf DF, Junker M, Bilgin M, Ejsing CS, Voelker DR, Rapoport TA, Shaw JM (2012) Gem1 and ERMES do not directly affect phosphatidylserine transport from ER to mitochondria or mitochondrial inheritance. Traffic 13:880–890
Osman C, Haag M, Potting C, Rodenfels J, Dip PV, Wieland FT, Brügger B, Westermann B, Langer T (2009) The genetic interactome of prohibitins: coordinated control of cardiolipin and phosphatidylethanolamine by conserved regulators in mitochondria. J Cell Biol 184:583–596
Porter KR, Palade GE (1957) Studies on the endoplasmic reticulum. J Biophys Biochem Cytol 3:269–300
Potting C, Wilmes C, Engmann T, Osman C, Langer T (2010) Regulation of mitochondrial phospholipids by Ups1/PRELI-like proteins depends on proteolysis and Mdm35. EMBO J 29:2888–2898
Rabl R, Soubannier V, Scholz R, Vogel F, Mendl N, Vasiljev-Neumeyer A, Körner C, Jagasia R, Keil T, Baumeister W et al (2009) Formation of cristae and crista junctions in mitochondria depends on antagonism between Fcj1 and Su e / g. J Cell Biol 185:1047–1063
Sesaki H, Jensen RE (2004) Ugo1p links the Fzo1p and Mgm1p GTPases for mitochondrial fusion. J Biol Chem 279:28298–28303
Sesaki H, Southard SM, Yaffe MP, Jensen RE (2003) Mgm1p, a dynamin-related GTPase, is essential for fusion of the mitochondrial outer membrane. Mol Biol Cell 14:2342–2356
Shiota T, Mabuchi H, Tanaka-Yamano S, Yamano K, Endo T (2011) *In vivo* protein-interaction mapping of a mitochondrial translocator protein Tom22 at work. Proc Natl Acad Sci USA 108:15179–15183
Simbeni R, Tangemann K, Schmidt M, Ceolotto C, Paltauf F, Daum G (1993) Import of phosphatidylserine into isolated yeast mitochondria. Biochim Biophys Acta 1145:1–7
Sogo LF, Yaffe MP (1994) Regulation of mitochondrial morphology and inheritance by Mdm10p, a protein of the mitochondrial outer membrane. J Cell Biol 126:1361–1373
Stroud DA, Oeljeklaus S, Wiese S, Bohnert M, Lewandrowski U, Sickmann A, Guiard B, Van Der Laan M, Warscheid B, Wiedemann N (2011) Composition and topology of the endoplasmic reticulum-mitochondria encounter structure. J Mol Biol 413:743–750
Tamura Y, Harada Y, Shiota T, Yamano K, Watanabe K, Yokota M, Yamamoto H, Sesaki H, Endo T (2009a)

Tim23-Tim50 pair coordinates functions of translocators and motor proteins in mitochondrial protein import. J Cell Biol 184:129–141

Tamura Y, Endo T, Iijima M, Sesaki H (2009b) Ups1p and Ups2p antagonistically regulate cardiolipin metabolism in mitochondria. J Cell Biol 185:1029–1045

Tamura Y, Iijima M, Sesaki H (2010) Mdm35p imports Ups proteins into the mitochondrial intermembrane space by functional complex formation. EMBO J 29:2875–2887

Tamura Y, Onguka O, Itoh K, Endo T, Iijima M, Claypool SM, Sesaki H (2012a) Phosphatidylethanolamine Biosynthesis in Mitochondria: Phosphatidylserine (PS) trafficking in independent of a PS decarboxylase and intermembrane space protiens Ups1p and Ups2p. J Biol Chem 287:43961–43971

Tamura Y, Onguka O, Aiken Hobbs AE, Jensen RE, Iijima M, Claypool SM, Sesaki H (2012b) Role for two conserved intermembrane space proteins, Ups1p and Ups2p, in intra-mitochondrial phospholipid trafficking. J Biol Chem 287:15205–15218

Tamura Y, Sesaki H, Endo T (2014) Phospholipid transport via mitochondria. Traffic 15:933–945

Tatsuta T, Scharwey M, Langer T (2014) Mitochondrial lipid trafficking. Trends Cell Biol 24:44–52

Tong A, Evangelista M, Parsons A, Xu H, Bader G, Pagé N, Robinson M, Raghibizadeh S, Hogue C, Bussey H et al (2001) Systematic genetic analysis with ordered arrays of yeast deletion mutants. Science 294:2364–2368

Vogel F, Bornhövd C, Neupert W, Reichert AS (2006) Dynamic subcompartmentalization of the mitochondrial inner membrane. J Cell Biol 175:237–247

von der Malsburg K, Muller JM, Bohnert M, Oeljeklaus S, Kwiatkowska P, Becker T, Loniewska-Lwowska A, Wiese S, Rao S, Milenkovic D et al (2011) Dual role of mitofilin in mitochondrial membrane organization and protein biogenesis. Dev Cell 21:694–707

Voss C, Lahiri S, Young BP, Loewen CJ, Prinz WA (2012) ER-shaping proteins facilitate lipid exchange between the ER and mitochondria in *S. cerevisiae*. J Cell Sci 125:4791–4799

Watanabe Y, Tamura Y, Kawano S, Endo T (2015) Structural and mechanistic insights into phospholipid transfer by Ups1–Mdm35 in mitochondria. Nat Commun 6:7922

Wong ED, Wagner JA, Scott SV, Okreglak V, Holewinske TJ, Cassidy-Stone A, Nunnari J (2003) The intramitochondrial dynamin-related GTPase, Mgm1p, is a component of a protein complex that mediates mitochondrial fusion. J Cell Biol 160:303–311

Wurm CA, Jakobs S (2006) Differential protein distributions define two sub-compartments of the mitochondrial inner membrane in yeast. FEBS Lett 580:5628–5634

Yamano K, Tanaka-Yamano S, Endo T (2010) Mdm10 as a dynamic constituent of the TOB/SAM complex directs coordinated assembly of Tom40. EMBO Rep 11:187–193

Youngman MJ, Hobbs AEA, Burgess SM, Srinivasan M, Jensen RE (2004) Mmm2p, a mitochondrial outer membrane protein required for yeast mitochondrial shape and maintenance of mtDNA nucleoids. J Cell Biol 164:677–688

Yu F, He F, Yao H, Wang C, Wang J, Li J, Qi X, Xue H, Ding J, Zhang P (2015) Structural basis of intramitochondrial phosphatidic acid transport mediated by Ups1-Mdm35 complex. EMBO Rep 16:813–823

Discovery and Roles of ER-Endolysosomal Contact Sites in Disease

10

William Mike Henne

Abstract

Inter-organelle membrane contact sites (MCSs) serve as unique microenvironments for the sensing and exchange of cellular metabolites and lipids. Though poorly defined, ER-endolysosomal contact sites are quickly becoming recognized as centers for inter-organelle lipid exchange and metabolic decision-making. Here, we review the discovery and current state of knowledge of ER-endolysosomal MCSs with particular focus on the molecular players that establish and/or utilize these contact sites in metabolism. We also discuss associations of ER-endolysosomal MCS-associated proteins in human disease, as well as the therapeutic promise these contact sites hold in modulating cellular physiology.

Keywords

Membrane contact site (MCS) • Inter-organelle • Endoplasmic reticulum • Lysosome • Endosome

10.1 Introduction: ER-Endolysosomal Membrane Contact Sites in Lipid Metabolism and Disease

No organelle is an island, and to maintain cellular homeostasis, organelles must constantly communicate their needs and exchange lipids and other metabolites. The mechanisms that govern this inter-organelle metabolic exchange have, until recently, been poorly defined. Early studies attributed most organelle-organelle lipid exchange through vesicular pathways. Indeed, many lipids produced by the endoplasmic reticulum (ER), the major anabolic organelle of lipid metabolism, can be trafficked along with proteins to other places within the cell such as the plasma membrane (PM) through the formation of coated vesicles (Novick et al. 1980). However, it has been established that even in the absence of ER-derived vesicle trafficking, the transfer of sterol between the ER and PM persists (Baumann et al. 2005).

How, then, does non-vesicular inter-organelle lipid exchange occur? Recent studies suggest that

W.M. Henne (✉)
Department of Cell Biology, UT Southwestern Medical Center, Dallas, TX 75013, USA
e-mail: mike.henne@utsouthwestern.edu

M. Tagaya, T. Simmen (eds.), *Organelle Contact Sites*, Advances in Experimental Medicine and Biology 997, DOI 10.1007/978-981-10-4567-7_10

at least one pathway relies on lipid exchange at sites of close contact between different membrane-bound organelles, sites now established as specific membrane contact sites (MCSs) (reviewed in (Levine and Loewen 2006; Elbaz and Schuldiner 2011; Prinz 2014; Henne et al. 2015a; Phillips and Voeltz 2016)). Often observed but generally ignored, these organelle-organelle contact sites are quickly becoming recognized as hotspots for lipid metabolism and metabolic decision-making. Elegant studies in both budding yeast and mammalian cells have identified key proteins that act as molecular "tethers" connecting different organelles (Manford et al. 2012; Zhang et al. 2012; Giordano et al. 2013; Tavassoli et al. 2013). Perhaps the most characterized of these are contact sites formed between the ER and the PM. Numerous studies demonstrate that ER-PM contact sites serve as major regulators for the exchange of the PM-manufactured phospholipid PtdIns4P with ER-derived phosphatidylserine (PtdSer) (Stefan et al. 2011; Moser von Filseck et al. 2015). Other functions for ER-PM contact sites continue to be revealed, including roles in the regulation of cell signaling and Ca^{2+} signaling (Liou et al. 2007; Omnus et al. 2016).

The PM is constantly internalized via endocytosis, thus initiating the endolysosomal pathway—the network of endosomes and lysosomes that constitutes the principle intracellular trafficking pathway for cell surface proteins and environmentally derived metabolites like cholesterol. In the human body, most cells rely on retrieving extracellular cholesterol through the internalization of LDL particles via their co-internalization with the LDL receptor in so-called receptor-mediated endocytosis (Goldstein 2007) (Fig. 10.1). Following their internalization, LDL particles are trafficked through endosomes to the lysosome, where hydrolases breakdown LDL cholesterol-esters (CE) and triacylglycerides (TAGs) into free cholesterol and fatty acids. Where these lipid metabolites go from here is much less clear. Approximately 80% of LD-derived cholesterol must, eventually, traffic back to the cell surface, and this process requires lysosomal proteins Npc1 and Npc2 (so-called Niemann-Pick proteins) that scavenge cholesterol from the lysosome lumen (Infante et al. 2008).

Following its removal from the lysosome, much of this free cholesterol must also enter the ER, where it is sensed by the sterol regulatory element-binding protein (SREBP) system that serves as the principle nutrient-sensing pathway for cellular sterols (Brown and Goldstein 1997). The SREBP pathway is principally composed of two integral membrane proteins, SREBP-1a and SREBP-2, which function both as sterol sensors and transcriptional regulators of cellular sterol metabolism. When ER-resident cholesterol drops below ~5 mole %, SREBP-2 is efficiently trafficked to the Golgi via COPII vesicles, where it is proteolytically cleaved into a soluble "mature" form by Golgi-resident proteases (Fig. 10.1). This liberates SREBP-2 from membranes, allowing it to translocate into the nucleus where it can initiate a transcriptional response aimed at upregulating sterol biogenesis (Brown and Goldstein 1997).

Despite our understanding of LDL particle internalization and endolysosomal trafficking, how cholesterol exits the lysosome and enters the ER remains largely unclear. However, recent studies implicate ER-lysosome MCSs in the non-vesicular transfer of cholesterol and other lipids between these two organelles, although the proteins and mechanisms that govern this remain poorly defined (Murley et al. 2015; Eden et al. 2016; Hariri et al. 2016). Addressing how cholesterol moves between the ER and endolysosomal system is of major biomedical importance, as numerous genetic and acquired diseases are associated with the pathological accumulation of cholesterol within the lysosome. Most notable is Niemann-Pick type C, a homozygous recessive pediatric neurological disease (one of more than 50 so-called lysosomal storage diseases) associated with loss-of-function alleles in genes *NPC1* and *NPC2*. Both encode lysosomal proteins Npc1 and Npc2 that are essential in the initial steps of cholesterol efflux from the lysosomal lumen (Ko et al. 2003; Infante et al. 2008). Loss of either causes the pathological accumulation of free cholesterol and sphingolipids within lysosomes, eventually blocking endolysosomal trafficking and autophagy and leading to progressive neuronal cell death. Numerous acquired diseases may also be associated with poor

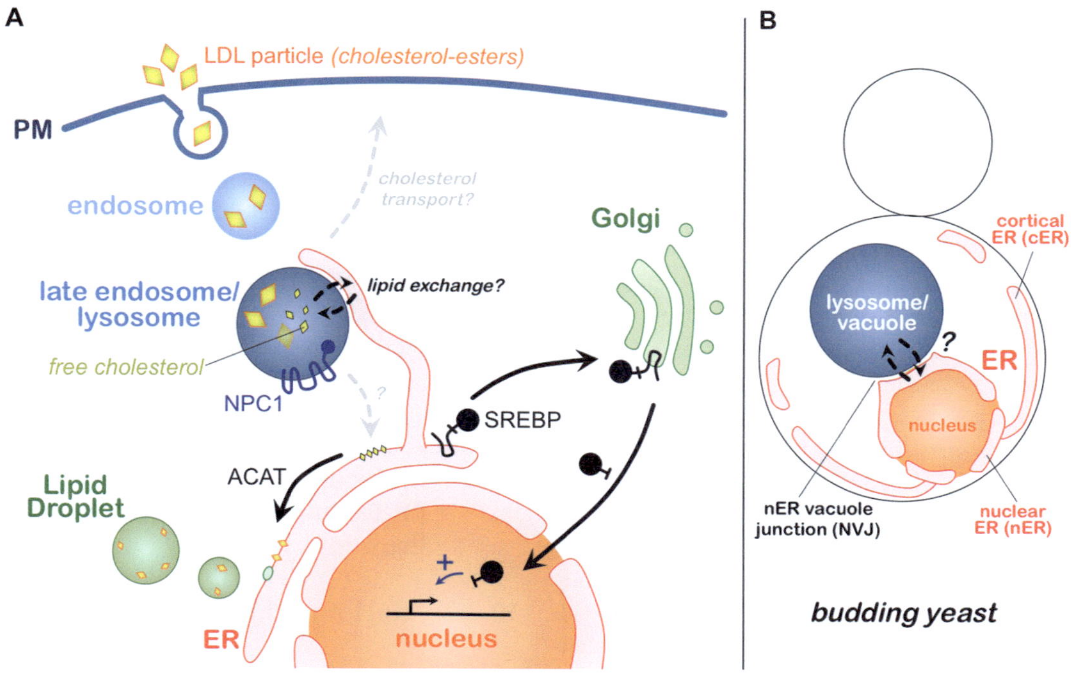

Fig. 10.1 Trafficking pathways in human and yeast ER-endolysosomal systems. (**a**) Receptor-mediated endocytosis of LDL particles (containing sterol-esters) and their trafficking through the endosomal pathway to late endosomes/lysosomes, where cholesterol-esters within LDL particles are degraded into free cholesterol. Free cholesterol must then leave the lysosome lumen in an NPC1-dependent manner and traffic to the plasma membrane (PM) and/or the endoplasmic reticulum (ER). At the ER, the SREBP system senses local cholesterol levels. Low ER cholesterol induces the translocation and proteolytic cleavage of SREBP-2 at the Golgi, producing a soluble transcription factor which enters the nucleus. (**b**) The ER-endolysosomal system of budding yeast *Saccharomyces cerevisiae*. The ER is partitioned into distal cortical ER (cER) and nuclear ER (nER), which makes direct contact with the vacuole/lysosome via the nER vacuole junction (NVJ)

endolysosomal cholesterol homeostasis. Diabetes and general metabolic syndrome are closely tied to cholesterol levels and insulin signaling, which in turn is affected by the cholesterol trafficking burden along the endolysosomal pathway. As such, defects in the retrograde trafficking of cholesterol and other lipids out of lysosomes can contribute to several metabolic syndromes and hypercholesterolemia (high blood cholesterol) that in turn may contribute to pathologies including atherosclerosis and coronary heart disease (for further details, see reference (Reiner et al. 2014)).

The purpose of this chapter is to briefly discuss the current knowledge of ER-endolysosomal MCSs and to outline the importance of understanding this MCS in human health and disease. Many of the discoveries in this field come from elegant studies using both yeast and mammalian model systems, and these will be discussed with respect to their impacts in the basic and biomedical understanding. Finally, we will conclude by providing an assessment of the current trajectory of biomedical research on ER-endolysosomal MCSs, and how current and future studies will pave the way for potentially ground-breaking therapeutic strategies for metabolic diseases.

10.2 Discovery and Imaging of ER-Endolysosomal MCSs

Modern cell biology began in the 1950s with the introduction of transmission electron microscopy (TEM) as a high-resolution visualization tool for biologists. Using this approach, the pioneering work of George Palade and Keith Porter revealed the subcellular architecture of individual cells. Their studies highlighted how crowded cells were, and how organelles often maintained close contact with one another. They also unexpectedly observed the first inter-organelle MCSs including those formed between the ER and PM, as well as the ER and mitochondria (Porter and Palade 1957; Porter and Machado 1960). The ER-PM contact sites of muscle cells were closely noted (so called PM-sarcoplasmic reticulum contact sites that are important for Ca^{2+} regulated muscle contractions).

ER-endosome, ER-late endosome (LE), and ER-lysosome MCSs were not noted during these initial TEM studies, but more recent microscopy work shows them to be highly dynamic and common within mammalian cells. Elegant TEM work from Clare Futter's group showed that the ER makes close contact with LEs and multi-vesicular bodies (MVBs) in mammalian cells (Eden et al. 2010). These studies showed that PTP1B, an ER-localized protein phosphatase, utilized the close inter-organelle contacts provided by ER-LE MCSs to regulate the trafficking and signaling of the EGF receptor, an important oncogene often targeted in anti-cancer therapeutics. Phospho-regulation of EGF receptor trafficking via PTP1B triggered its recognition by the endosomal sorting complex required for transport (ESCRT) pathway and thus its downregulation via delivery to the lysosome for degradation (Eden et al. 2010). Thus the first function for an ER-endolysosomal MCS—receptor regulation—was proposed.

Subsequent work using live-cell imaging further demonstrated that ER-endosome and ER-LE MCSs are highly dynamic and change as an endosome matures. Studies from Gia Voeltz's group showed that the ER makes contact with endosomes shortly after they are formed (Friedman et al. 2013). Intriguingly, the degree of ER-endosome contact increases as early endosomes mature into LEs (monitored by exchanging the early endosome marker Rab5 for Rab7). This suggests that the ER may sense and actively participate in endosomal maturation. This idea was further supported by the observation that ER-LE MCSs appeared to define sites of vesicle scission at the LE from which small tubules sprouted and vesicles budded. This again suggested an active role for the ER in endosomal dynamics and potentially that the ER may contribute mechanical force to the vesicle scission reaction. How this is achieved is currently unclear, but is consistent with the previously proposed role of ER tubules in defining sites of mitochondrial division at ER-mitochondria MCSs (Friedman et al. 2011). An intriguing possibility is that ER-mitochondria and ER-LE MCSs create unique lipid microenvironments that promote spontaneous vesicle fission.

10.3 Molecular Identification of ER-Endolysosomal Tethering Proteins

10.3.1 Nvj1 and the NVJ

Early work on mammalian ER-endolysosomal MCSs is built upon pioneering work using the budding yeast *Saccharomyces cervisiae* to study ER-endolysosomal contact sites. These studies conducted in David Goldfarb's group identified the first proteins that functioned as molecular tethers to establish MCSs between the ER and yeast lysosome, also known as the vacuole (Pan et al. 2000). Notably, the yeast vacuole serves as both a "late endosome" and "lysosome" for the cell, in that it is a Rab7-positive organelle that also contains lumenal hydrolases dedicated to catabolic metabolism. ER-vacuole MCSs may then, by extension, serve as model systems for the study of ER-endosome and ER-lysosome MCSs.

Yeast usually exhibit a single ER-vacuole MCS that forms as a circular "patch" juxtaposed between the vacuole and outer nuclear envelope, a region of ER continuous with the rest of the ER network often referred to as the nuclear ER (nER). Yeast nER-vacuole contacts are thus denoted *n*ER-*v*acuole *j*unctions (NVJs) (Fig. 10.2). Due to their size, and the close physical proximity of the nucleus and vacuole within yeast, NVJs are observed to be stable structures that exist for most of the cell's lifetime. This is in contrast to the dynamic, short-lived ER-endosome MCSs seen in mammalian cells.

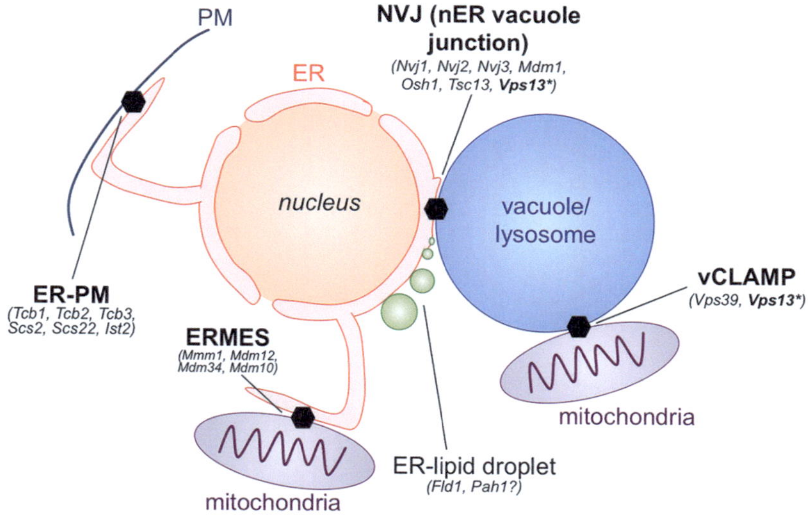

Fig. 10.2 Guide to yeast inter-organelle contact sites and their resident proteins

Pioneering work by David Goldfarb's group revealed the molecular machinery that establishes the NVJ. NVJs form through the direct interaction of ER-resident protein Nvj1 with the vacuole surface protein Vac8, which had previously been identified in vacuolar inheritance screens (Pan et al. 2000). This Nvj1:Vac8 hetero-dimerization thus defines the first molecular "tethers" that establish ER-endolysosomal MCSs.

10.3.2 Other NVJ Proteins

Subsequent studies on the NVJ demonstrated it as a site of lipid metabolism and membrane remodeling. During nitrogen starvation, the NVJ was shown to dramatically invaginate into the vacuole interior where it was eventually engulfed to form intra-lumenal vesicles within the vacuole lumen. Although the reason for this process is still unclear, it appears to represent a specific form of micro-autophagy and was thus termed piecemeal microautophagy of the nucleus (PMN) (Roberts et al. 2003). Other functions for the NVJ, including a site for the production of ceramide via fatty acid processing enzymes like Tsc13, have been proposed for the NVJ (Kvam et al. 2005). Consistent with this, other lipid metabolism proteins such as Nvj2, Osh1, and have also been identified as resident NVJ proteins, suggesting that NVJs function in many different lipid metabolic processes (Fig. 10.3) (Toulmay and Prinz 2012).

Recently, a study by the Will Prinz lab identified a protein domain found to reside specifically at inter-organelle MCSs: the synaptotagmin-like mitochondrial lipid-binding protein (SMP) domain (Toulmay and Prinz 2012). Nvj2 contains a *bone vide* SMP domain. Since a crystal structure of the SMP domain suggests it may physically bind to lipids with a hydrophobic cavity within the domain, it is postulated that SMP domain-containing proteins function in the non-vesicular transfer of lipids between different organelles (AhYoung et al. 2015; Reinisch and De Camilli 2016; Schauder et al. 2014).

10.3.3 Protrudin

Although they are essential for the formation of a proper NVJ, surprisingly, Nvj1 and Vac8 are not conserved in metazoans. How then are the ER and endosomes/lysosomes "tethered" in mammalian cells? Several recent studies have identified potential tethering proteins. Harold Stenmark's group identified Protrudin, an

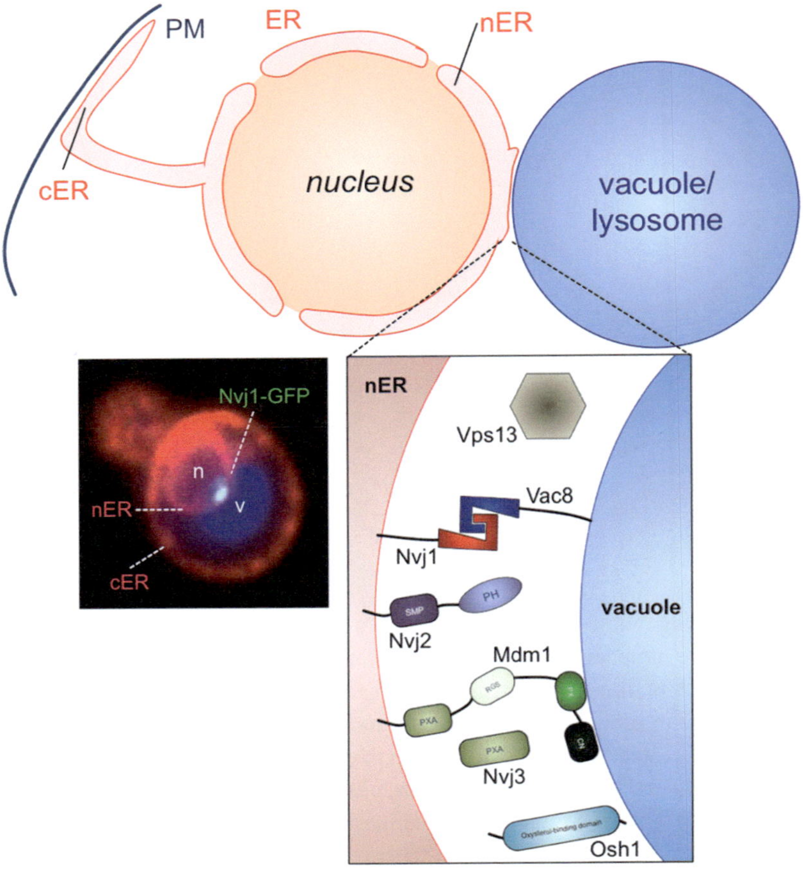

Fig. 10.3 The nER vacuole junction (NVJ) resident proteins and their architectures. The fluorescence micrograph depicts a living yeast with the ER network labeled in red (DsRed-HDEL protein), the vacuole/ lysosome in blue (CMAC dye), and endogenous Nvj1 tagged with GFP (Nvj1-GFP)

integral ER protein with a PtdIns3P-binfing FYVE domain, as a potential tether connecting the ER and Rab7-positive LEs (Raiborg et al. 2015). Protrudin is enriched at ER-endosome MCSs, and its overexpression induced the formation of ER-endosome MCS in mammalian cells. These Protrudin-positive MCSs promoted the migration of LEs to the cell periphery in a microtubule-dependent manner. ER-LE MCSs are thus linked to a novel function: endosomal positioning and migration. This fucntion is supported by previous studies from the Neefjes group, which showed that ORP1L senses endosomal sterol levels and regulates the positioning endosomes within the cytoplasm (Johansson et al. 2007). When endosomal sterol levels are low, ORP1L binding to endosomes is reduced, promoting the formation of ER-LE MCSs in a VAP-dependent manner (Rocha et al. 2009). When endosomal cholesterol levels increase, such as in Niemann-Pick type C, ORP1L associates tightly to endosomes and promotes their trafficking to the perinuclear cell interior. Collectively, Protrudin and ORP1L represent two critical regulators of ER-LE MCSs that govern both endosome positioning and sterol metabolism in mammalian cells.

10.3.4 Mdm1 Family Proteins

Recent work on the yeast NVJ has revealed new, highly conserved proteins that regulate ER-endolysosomal MCSs. Yeast protein Mdm1 and its paralog Nvj3 were recently shown to localize to the NVJ (Henne et al. 2015b). Unlike Nvj1 and Vac8, Mdm1 is highly conserved in metazoans as a member of the Sorting Nexin

(SNX) protein family. The SNX family is defined by the presence of a PtdIns3P-binding Phox homology (PX) domain (Yu and Lemmon 2001). Humans encode four Mdm1 homologs: Snx13, 14, 19, and 25 with near-identical domain architecture to Mdm1. All feature an N-terminal transmembrane region and signal sequence that appears to anchor them in the ER. On their C-terminal half, they contain the PX that directly binds the phospholipid PtdIns3P on the vacuole/lysosome surface, thus establishing ER-vacuole/lysosome tethering (Henne et al. 2015b). Though it requires Mdm1 for its NVJ localization, Nvj3 is soluble and contains a PX-Associated (PXA) that is unique to this protein family. The structure and function of the PXA domain is currently unclear, but implicated in lipid metabolism as its overexpression causes cells to become hypersensitized to drugs that perturb sphingolipid metabolism. In mammalian cells, Mdm1 homolog Snx14 is associated with a form of pediatric cerebellar ataxia and appears phenomenologically similar to lysosomal storage diseases (Thomas et al. 2014; Akizu et al. 2015). Indeed, *SNX14*-deficient cells manifest defects in lysosome homeostasis, suggesting that Snx14 plays some role in lysosome function. A role for Mdm1 family proteins in lipid metabolism is further supported by studies on Snazarus (Snz), the *Drosophila melanogaster* ortholog of Mdm1. These studies link Snz to aging and fat metabolism by showing that fruit flies deficient for Snz displayed extended lifespan (Suh et al. 2008). Intriguingly, Snz was found to be highly expressed in the fly fat body, a metabolic center of insect lipid metabolism akin to both the mammalian liver and adipocyte. Further studies on Snz and its homologs will no doubt continue to reveal roles for this protein family in lipid metabolism and inter-organelle cross talk.

10.3.5 Vps13

In addition to Mdm1 and Nvj3, other yeast proteins with clear metazoan homologs have been implicated in ER-endolysosomal MCSs. Vps13, first identified in the vacuolar protein sorting (*vps*) screen conducted in Scott Emr's group in the 1980s, is 3,144 amino acids in length and one of the largest proteins encoded in the yeast genome (Bankaitis et al. 1986). Several metazoan homologs exist but are poorly characterized. Using elegant yeast genetics, Benoit Kornmann's group recently discovered that gain-of-function Vps13 mutants were capable of rescuing the viability of yeast with a defective ERMES complex, an ER-mitochondria tethering factor required for growth on non-fermentable carbon media (Lang et al. 2015). They then used complementation assays to discover that Vps13 can localize to a previously identified mitochondria-vacuole MCS (also called vacuole and mitochondria patch, vCLAMP) when grown in fermentable (dextrose) conditions (Elbaz-Alon et al. 2014; Honscher et al. 2014). Strikingly, Vps13 re-localized to the NVJ when grown in non-fermentable (glycerol) conditions (Lang et al. 2015). Indeed, Vps13 appears highly dynamic and has been observed to localize at vCLAMPs, ER-endosome, and NVJ MCSs in yeast depending upon growth conditions (Park et al. 2016). This suggests that Vps13 can localize to multiple MCSs, and also that the cell can actively remodel the composition of MCSs in response to changes in nutrient status and demand.

10.3.6 Lam/Ltc Family Proteins

The idea that different MCSs can dynamically change in response to changes in the cell's metabolic status is supported by recent yeast work linking members of the 'lipid transfer protein anchored at membrane contact sites' (Lam) protein family to ER-endolysosomal MCSs. Lam6/Ltc1, the most studied member within this protein family, has been observed to localize to ER-mitochondria (ERMES), NVJ, and vCLAMP MCSs within the same cell, suggesting it may function in multiple metabolic processes (Murley et al. 2015; Elbaz-Alon et al. 2015; Gatta et al. 2015).

What, then, are the functions of Lam proteins at MCSs? At least two functions have been proposed. Overexpression of GFP-Lam6 was found to expand both the NVJ and ERMES MCSs, suggesting Lam proteins may regulate and possibly help extend inter-organelle contact sites in response to changes in cellular nutrient status (Elbaz-Alon et al. 2015). Another potential function is the direct movement of sterols between organelles. This could be achieved via direct binding and shuttling of a sterol molecule within the Lam protein tertiary fold. Lam family members contain a VASt/StART domain that binds sterols and may facilitate this inter-organelle sterol exchange (Murley et al. 2015; Elbaz-Alon et al. 2015). How this occurs, and how these proteins regulate lipid exchange and MCS size remains to be understood.

10.3.7 Annexin A1

Recently, other mammalian ER-endosome "tethers" have been identified that play roles in lipid and sterol metabolism. Using TEM and biochemistry, Emily Eden found that Annexin A1 mediates a specific subtype of ER-LE MCS formed between the ER and MVBs that downregulate the EGF receptor (Eden et al. 2016). Depletion of Annexin A1 reduced ER-MVB MCSs and perturbed EGR receptor signaling. Surprisingly, they find that ER-derived cholesterol is necessary for EGF receptor downregulation at the MVB and suggest that these ER-MVB MCSs function to transfer this cholesterol to the MVB limiting membrane, where cholesterol is required for the formation of intra-lumenal vesicles. How this cholesterol transfer is achieved is still unclear, but appears regulated by ORP1L.

10.4 ER-Endolysosomal MCSs in Human Health and Disease

Numerous ER-endolysosomal MCS proteins are linked to inherited or acquired diseases, indicating important roles in cellular metabolism and development. These connections underlie the important role for ER-endolysosomal cross talk in cellular homoeostasis in metazoans. Refer to Table 10.1 for further details.

Table 10.1 ER-endolysosomal proteins and associated diseases

ER-Endolysosomal contact site proteins in disease				
Protein		Localization(s)	Disease(s) associated	References
Yeast	*Human*			
Vps13	**VPS13A-D**	ER-endosome, ER-vacuole (NVJ), ERMES?	VPS13A: chorea-acanthocytosis (CHAC); VPS13B: Cohen Syndrome (COH)	Ueno et al. (2001) and Kondo et al. (2005)
Scs2/22	**Vap-A,B (Als8)**	ER-PM, ER-vacuole(?)	ALS disease	Teuling et al. (2007)
Tsc13	**TECR**	ER-vacuole (NVJ)	Autosomal recessive non-syndromic intellectual disability	Nolan et al. (2008)
Mdm1	**Snx13, 14, 19, 25**	ER-vacuole (NVJ)	Snx14: autosomal recessive cerebellar ataxia with intellectual disability; Snx19: insulin secretion regulation	Thomas et al. (2014), Akizu et al. (2015) and Harashima et al. (2012)
Osh1	**OSBPs/ORPs**	ER-vacuole (NVJ)	Linked to cholesterol-related metabolic disregulation	Ngo and Ridgway (2009) and Bouchard et al. (2009)
n/a	**Protrudin**	ER-endosome	Heriditary spastic paraplegia (HSP)	Hashimoto et al. (2014)
n/a	**Annexin A1**	ER-late endosome/Multi-vescicular endosome	Links to pancreatic cancer progression	Belvedere et al. (2016)

10.4.1 Neurological Disease

The VAP proteins VAP-A and VAP-B (Scs2 and Scs22 in yeast) have been linked to ER-PM MCSs, but have recently been implicated in controlling ER-endosome contact and endosomal migration via ORP1L (Rocha et al. 2009). A mutant form of VAP-A (containing the P56S mutation) is linked to familial amyotropic lateral sclerosis (ALS) in humans (Teuling et al. 2007). This mutation likely causes the protein to aggregate, resulting in general ER stress and cytotoxicity. Since a number of proteins also interact with VAP proteins via FFAT motifs, this general aggregation likely leads to general protein dysfunction and progressive cell toxicity (Prosser et al. 2008).

Vps13 and Mdm1, both residents of the yeast NVJ, have homologs linked to inherited neurological diseases in humans. Vps13 has at least two human orthologs with clear links to human disease: VPS13A, which is associated with chorea-acanthocytosis (CHAC), and VPS13B, linked to Cohen syndrome (COH) (Ueno et al. 2001; Kondo et al. 2005). Both are progressive neurological conditions associated with muscle dystonia and developmental delay. CHAC is highly variable in phenotype and can include abnormal red blood cell shape, as well as epilepsy. Mdm1 exhibits four human orthologs, two of which have clear mammalian disease links. *SNX14*-deficiency is associated with pediatric cerebellar ataxia and intellectual disability (Thomas et al. 2014; Akizu et al. 2015). A *SNX13*-deficient (also called RGS-PX1) mouse was found to exhibit severe developmental defects, delayed neural tube closure, and embryonic lethality (Zheng et al. 2006). Tsc13, the enoyl-CoA reductase found at the yeast NVJ, is also highly conserved in mammals as TECR, which is linked to an autosomal recessive non-syndromic intellectual disability (Nolan et al. 2008).

The Protrudin-encoding gene *ZFYVE27* is mutated in individuals with hereditary spastic paraplegia (HSP), an inherited disease associated with muscle weakness in the lower half of the body (Hashimoto et al. 2014). This disease is generally associated with ER stress, suggesting that loss-of-function alleles may affect protein stability and trigger general ER stress response.

10.4.2 Cancer

ER-LE MCSs were observed in mammalian cells while studying PTP1B, an ER-localized protein phosphatase that regulates EGF receptor signaling, an oncogene product commonly targeted in anti-cancer therapies (Eden et al. 2010). As ER-LE MCSs may regulate the duration of EGF receptor signaling and other receptor signaling events, this may tie them closely to cancer progression. This is consistent with the proposed roles for Protrudin in cell migration, as cancer cells often display aggressive cell migration during tissue invasion and metastasis (Raiborg et al. 2015). Consistent with this, Annexin A1, which has been proposed as an ER-LE tether, is directly linked to pancreatic cancer progression (Belvedere et al. 2016).

10.4.3 Aging and Metabolism

The *Drosophila* ortholog of yeast Mdm1, Snz, is associated with lifespan extension in insects as well as obesity in mammals, as the expression of SNX13 and SNX14 was shown to increase in *ob/ob* mouse models (Suh et al. 2008). The connection between this aging phenotype and lipid metabolism is currently not clear, but Snz is highly expressed in the fly fat body, the central organ of insect fat metabolism analogous to the liver and adipocytes of mammals. This suggests that Snz may play a role in regulating fat storage and/or mobilization. Numerous connections exist between lipid storage mobilization and lifespan and generally show that caloric restriction may extend lifespan through the inhibition of TORC signaling and subsequent activation of autophagy (McCormick et al. 2011).

Whether Snz plays a role in autophagy and/or lipid mobilization in insects remains to be determined. However, ChIP-seq data from mammalian studies suggested that Snx13, the rodent homolog of Snz, is upregulated in response to cholesterol starvation in an SREBP-2-dependent manner (Seo et al. 2011). Among other genes upregulated were autophagy genes, suggesting Snx13 may be utilized during lipid mobilization in response to cholesterol starvation. In a separate example, human Mdm1 homolog Snx19 has been linked to insulin secretion pancreatic β-cells (Harashima et al. 2012).

10.4.4 Cholesterol Metabolism

The oxysterol-binding protein Osh1 localizes to the NVJ in yeast and is highly conserved in mammals as the oxysterol-binding proteins (OSBPs). OSBP upregulation is correlated with high blood pressure and hypertension, which are risk factors associated with coronary heart disease (Ngo and Ridgway 2009). OSBPL11 is also upregulated in patients exhibiting morbid obesity (Bouchard et al. 2009).

Although the movement of cholesterol between LEs/lysosomes and the ER is not well defined, it must require the Niemann-Pick proteins Npc1 and Npc2. Nps2 resides within the lysosome lumen and binds to the free cholesterol produced from the breakdown of CE within LDL particles. Upon binding, Npc2 transfers this free cholesterol to Npc1, a transmembrane protein that bridging the inner and outlet monolayers of the lysosome membrane (Infante et al. 2008). Through a poorly defined mechanism, Npc1 then transfers free cholesterol from the lysosome interior to the cytoplasmic leaflet of the lysosome. How this free cholesterol leaves the lysosome to travel to other organelles remains a major question in the lipid biology field. However, loss of either Npc1 or Npc2 is associated with Niemann-Pick type C, an autosomal recessive neurological disease characterized by progressive neuronal cell death.

10.5 Closing: The Therapeutic Promise of ER-Endolysosomal MCS Biology

Cholesterol metabolism has arguably been at the center of some of the greatest biopharmaceutical endeavors of the twentieth century. Existing at the crossroads between genetic and acquired disease, modulating human blood cholesterol levels can simultaneously mitigate symptoms associated with obesity, heart disease, diabetes, and an array of inherited genetic diseases afflicting cellular metabolism. Pioneering work by Mike Brown and Joe Goldstein elucidated not only how LDL-derived cholesterol enters cells, but also how the cell senses and responds to cholesterol burden (Brown and Goldstein 1997). These studies revealed the SREBP system: en elaborate sterol-sensing pathway that can modulate cellular cholesterol levels though an elegant feedback mechanism. In times of cholesterol deprivation, when ER-localized cholesterol reaches <5 mole % of the total ER lipid content, the ER protein SREBP-2 is efficiently trafficked to the Golgi, where it is proteolytically cleaved into a soluble transcription factor. It can then traffic into the nucleus and initiate a transcriptional response to this cholesterol need, increasing expression of sterol biosynthesis proteins including HMG-CoA Reductase.

Capitalizing on this metabolic feedback loop, several groups developed statins, which pharmacologically inhibit HMG-CoA reductase, allowing for the acute modulation of cellular cholesterol production, and by extension blood cholesterol levels, thus reducing the risk of heart disease for millions of patients in the Western world. This major biomedical achievement is the product of meticulous basic research coupled with ambitious biomedical application.

One emerging theme of MCS research is that sites of ER-endolysosomal contact may serve as "metabolic platforms" for regulating inter-organelle lipid trafficking. As such, the proteins that function at MCSs may represent a new class of therapeutic targets. MCS-enriched proteins currently fall under at least two general classes: organelle "tethers" (which help establish and maintain MCSs) and lipid "exchangers" (which utilize MCSs for inter-organelle lipid exchange). By pharmacologically targeting either of these two protein classes, it may be possible to modulate rates of lipid exchange and/or production of specific lipid classes within cells. This, by extension, may tune metabolic pathways or the formation of pathological debris that form the basis for pathophysiology in numerous metabolic diseases. Thus, continued focus on the molecular mechanisms that govern MCS formation and regulation are at the center of a new round of impactful biomedical discoveries.

In closing, the field of inter-organelle cross talk is at an exciting time. New proteins and protein complexes that establish organelle-organelle junctions continue to be discovered, providing new and exciting details regarding how MCSs are formed and function in cellular metabolism. Among all MCSs discovered, those formed between the ER and endolysosomal system appear to be highly utilized by cells to control their lipid metabolism. This is not surprising, as ER-endolysosomal MCSs serve as the very interface—both physically and conceptually—that connect catabolic and anabolic metabolism. The ER serves as a major anabolic lipid producer, and endosomes/lysosomes serve as major lipid "sinks" driving the catabolic breakdown of nutrient lipids and proteins received via endocytosis. Mechanisms of cross talk between these two organelle systems will undoubtedly continue to be discovered as we probe the nature of ER-endolysosomal MCSs.

Acknowledgments We would like to thank members of the Henne Lab for their helpful thoughts and suggestions during the preparation of this book chapter. WM Henne is supported by grants from the NIH NIGMS (R35GM119768-01), the Searle Foundation (SSP-2016-1482), The American Federation for Aging Research (AFAR) (A15198), the Welch Foundation (I-1873), and the UT Southwestern Endowed Scholars Program.

References

AhYoung AP, Jiang J, Zhang J, Khoi Dang X, Loo JA, Zhou ZH, Egea PF (2015) Conserved SMP domains of the ERMES complex bind phospholipids and mediate tether assembly. Proc Natl Acad Sci USA 112:E3179–E3188

Akizu N, Cantagrel V, Zaki MS, Al-Gazali L, Wang X, Rosti RO, Dikoglu E, Gelot AB, Rosti B, Vaux KK et al (2015) Biallelic mutations in SNX14 cause a syndromic form of cerebellar atrophy and lysosome-autophagosome dysfunction. Nature genetics 47:528–534

Bankaitis VA, Johnson LM, Emr SD (1986) Isolation of yeast mutants defective in protein targeting to the vacuole. Proc Natl Acad Sci USA 83:9075–9079

Baumann NA, Sullivan DP, Ohvo-Rekilä H, Simonot C, Pottekat A, Klaassen Z, Beh CT, Menon AK (2005) Transport of newly synthesized sterol to the sterol-enriched plasma membrane occurs via nonvesicular equilibration. Biochemistry 44:5816–5826

Belvedere R, Bizzarro V, Forte G, Dal Piaz F, Parente L, Petrella A (2016) Annexin A1 contributes to pancreatic cancer cell phenotype, behaviour and metastatic potential independently of Formyl Peptide Receptor pathway. Sci Rep 6:29660

Bouchard L, Faucher G, Tchernof A, Deshaies Y, Marceau S, Lescelleur O, Biron S, Bouchard C, Pérusse L, Vohl MC (2009) Association of OSBPL11 gene polymorphisms with cardiovascular disease risk factors in obesity. Obesity 17:1466–1472

Brown MS, Goldstein JL (1997) The SREBP pathway: regulation of cholesterol metabolism by proteolysis of a membrane-bound transcription factor. Cell 89:331–340

Eden ER, White IJ, Tsapara A, Futter CE (2010) Membrane contacts between endosomes and ER provide sites for PTP1B-epidermal growth factor receptor interaction. Nat Cell Biol 12:267–272

Eden ER, Sanchez-Heras E, Tsapara A, Sobota A, Levine TP, Futter CE (2016) Annexin A1 Tethers Membrane Contact Sites that Mediate ER to Endosome Cholesterol Transport. Dev Cell 37:473–483

Elbaz Y, Schuldiner M (2011) Staying in touch: the molecular era of organelle contact sites. Trends Biochem Sci 36:616–623

Elbaz-Alon Y, Rosenfeld-Gur E, Shinder V, Futerman AH, Geiger T, Schuldiner M (2014) A dynamic interface between vacuoles and mitochondria in yeast. Dev Cell 30:95–102

Elbaz-Alon Y, Eisenberg-Bord M, Shinder V, Stiller SB, Shimoni E, Wiedemann N, Geiger T, Schuldiner M (2015) Lam6 regulates the extent of contacts between organelles. Cell Rep 12:7–14

Friedman JR, Lackner LL, West M, DiBenedetto JR, Nunnari J, Voeltz GK (2011) ER tubules mark sites of mitochondrial division. Science 334:358–362

Friedman JR, Dibenedetto JR, West M, Rowland AA, Voeltz GK (2013) Endoplasmic reticulum-endosome contact increases as endosomes traffic and mature. Mol Biol Cell 24:1030–1040

Gatta AT, Wong LH, Sere YY, Calderón-Noreña DM, Cockcroft S, Menon AK, Levine TP (2015) A new family of StART domain proteins at membrane contact sites has a role in ER-PM sterol transport. eLife 4:e07253

Giordano F, Saheki Y, Idevall-Hagren O, Colombo SF, Pirruccello M, Milosevic I, Gracheva EO, Bagriantsev SN, Borgese N, De Camilli P (2013) $PI(4,5)P_2$-dependent and Ca^{2+}-regulated ER-PM interactions mediated by the extended synaptotagmins. Cell 153:1494–1509

Goldstein AL (2007) History of the discovery of the thymosins. Ann N Y Acad Sci 1112:1–13

Harashima S, Horiuchi T, Wang Y, Notkins AL, Seino Y, Inagaki N (2012) Sorting nexin 19 regulates the number of dense core vesicles in pancreatic β-cells. J Diabetes Investig 3:52–61

Hariri H, Ugrankar R, Liu Y, Henne WM (2016) Inter-organelle ER-endolysosomal contact sites in metabolism and disease across evolution. Commun Integr Biol 9:e1156278

Hashimoto Y, Shirane M, Matsuzaki F, Saita S, Ohnishi T, Nakayama KI (2014) Protrudin regulates endoplasmic reticulum morphology and function associated with the pathogenesis of hereditary spastic paraplegia. J Biol Chem 289:12946–12961

Henne WM, Liou J, Emr SD (2015a) Molecular mechanisms of inter-organelle ER-PM contact sites. Curr Opin Cell Biol 35:123–130

Henne WM, Zhu L, Balogi Z, Stefan C, Pleiss JA, Emr SD (2015b) Mdm1/Snx13 is a novel ER-endolysosomal interorganelle tethering protein. J Cell Biol 210:541–551

Hönscher C, Mari M, Auffarth K, Bohnert M, Griffith J, Geerts W, van der Laan M, Cabrera M, Reggiori F, Ungermann C (2014) Cellular metabolism regulates contact sites between vacuoles and mitochondria. Dev Cell 30:86–94

Infante RE, Wang ML, Radhakrishnan A, Kwon HJ, Brown MS, Goldstein JL (2008) NPC2 facilitates bidirectional transfer of cholesterol between NPC1 and lipid bilayers, a step in cholesterol egress from lysosomes. Proc Natl Acad Sci USA 105 (40):15287–15292

Johansson M, Rocha N, Zwart W, Jordens I, Janssen L, Kuijl C, Olkkonen VM, Neefjes J (2007) Activation of endosomal dynein motors by stepwise assembly of Rab7-RILP-p150^{Glued}, ORP1L, and the receptor βlll spectrin. J Cell Biol 176:459–471

Ko DC, Binkley J, Sidow A, Scott MP (2003) The integrity of a cholesterol-binding pocket in Niemann-Pick C2 protein is necessary to control lysosome cholesterol levels. Proc Natl Acad Sci USA 100 (5):2518–2525

Kondo I, Shimizu A, Asakawa S, Miyamoto K, Yamagata H, Tabara Y, Shimizu N (2005) COH1 analysis and linkage study in two Japanese families with Cohen syndrome. Clinical Genet 67:270–272

Kvam E, Gable K, Dunn TM, Goldfarb DS (2005) Targeting of Tsc13p to nucleus-vacuole junctions: a role for very-long-chain fatty acids in the biogenesis of microautophagic vesicles. Mol Biol Cell 16:3987–3998

Lang AB, John Peter AT, Walter P, Kornmann B (2015) ER-mitochondrial junctions can be bypassed by dominant mutations in the endosomal protein Vps13. J Cell Biol 210:883–890

Levine T, Loewen C (2006) Inter-organelle membrane contact sites: through a glass, darkly. Curr Opin Cell Biol 18:371–378

Liou J, Fivaz M, Inoue T, Meyer T (2007) Live-cell imaging reveals sequential oligomerization and local plasma membrane targeting of stromal interaction molecule 1 after Ca^{2+} store depletion. Proc Natl Acad Sci USA 104:9301–9306

Manford AG, Stefan CJ, Yuan HL, Macgurn JA, Emr SD (2012) ER-to-plasma membrane tethering proteins regulate cell signaling and ER morphology. Dev Cell 23:1129–1140

McCormick MA, Tsai SY, Kennedy BK (2011) TOR and ageing: a complex pathway for a complex process. Philos Trans R Soc Lond B Biol Sci 366:17–27

Moser von Filseck J, Čopič A, Delfosse V, Vanni S, Jackson CL, Bourguet W, Drin G (2015) Phosphatidylserine transport by ORP/Osh proteins is driven by phosphatidylinositol 4-phosphate. Science 349:432–436

Murley A, Sarsam RD, Toulmay A, Yamada J, Prinz WA, Nunnari J (2015) Ltc1 is an ER-localized sterol transporter and a component of ER-mitochondria and ER-vacuole contacts. J Cell Biol 209:539–548

Ngo M, Ridgway ND (2009) Oxysterol binding protein-related Protein 9 (ORP9) is a cholesterol transfer protein that regulates Golgi structure and function. Mol Biol Cell 20:1388–1399

Nolan DK, Chen P, Das S, Ober C, Waggoner D (2008) Fine mapping of a locus for nonsyndromic mental retardation on chromosome 19p13. Am J Med Genet A Part A 146A:1414–1422

Novick P, Field C, Schekman R (1980) Identification of 23 complementation groups required for post-translational events in the yeast secretory pathway. Cell 21:205–215

Omnus DJ, Manford AG, Bader JM, Emr SD, Stefan CJ (2016) Phosphoinositide kinase signaling controls ER-PM cross-talk. Mol Biol Cell 27:1170–1180

Pan X, Roberts P, Chen Y, Kvam E, Shulga N, Huang K, Lemmon S, Goldfarb DS (2000) Nucleus-vacuole junctions in Saccharomyces cerevisiae are formed through the direct interaction of Vac8p with Nvj1p. M Mol Biol Cell 11:2445–2457

Park JS, Thorsness MK, Policastro R, McGoldrick LL, Hollingsworth NM, Thorsness PE, Neiman AM (2016) Yeast Vps13 promotes mitochondrial function and is localized at membrane contact sites. Mol Biol Cell 27:2435–2449

Phillips MJ, Voeltz GK (2016) Structure and function of ER membrane contact sites with other organelles. Nat Rev Mol Cell Biol 17:69–82

Porter KR, Machado RD (1960) Studies on the endoplasmic reticulum. IV. Its form and distribution during mitosis in cells of onion root tip. J Biophys Biochem Cytol 7:167–180

Porter KR, Palade GE (1957) Studies on the endoplasmic reticulum. III. Its form and distribution in striated muscle cells. J Biophys Biochem Cytol 3:269–300

Prinz WA (2014) Bridging the gap: membrane contact sites in signaling, metabolism, and organelle dynamics. J Cell Biol 205:759–769

Prosser DC, Tran D, Gougeon PY, Verly C, Ngsee JK (2008) FFAT rescues VAPA-mediated inhibition of ER-to-Golgi transport and VAPB-mediated ER aggregation. J Cell Sci121:3052–3061

Raiborg C, Wenzel EM, Pedersen NM, Olsvik H, Schink KO, Schultz SW, Vietri M, Nisi V, Bucci C, Brech A et al (2015) Repeated ER-endosome contacts promote endosome translocation and neurite outgrowth. Nature 520:234–238

Reiner Ž, Guardamagna O, Nair D, Soran H, Hovingh K, Bertolini S, Jones S, Ćorić M, Calandra S, Hamilton J et al (2014) Lysosomal acid lipase deficiency--an under-recognized cause of dyslipidaemia and liver dysfunction. Atherosclerosis 235:21–30

Reinisch KM, De Camilli P (2016) SMP-domain proteins at membrane contact sites: Structure and function. Biochim Biophys Acta 1861:924–927

Roberts P, Moshitch-Moshkovitz S, Kvam E, O'Toole E, Winey M, Goldfarb DS (2003) Piecemeal microautophagy of nucleus in *Saccharomyces cerevisiae*. Mol Biol Cell 14:129–141

Rocha N, Kuijl C, van der Kant R, Janssen L, Houben D, Janssen H, Zwart W, Neefjes J (2009) Cholesterol sensor ORP1L contacts the ER protein VAP to control Rab7-RILP-p150^{Glued} and late endosome positioning. J Cell Biol 185:1209–1225

Schauder CM, Wu X, Saheki Y, Narayanaswamy P, Torta F, Wenk MR, De Camilli P, Reinisch KM (2014) Structure of a lipid-bound extended synaptotagmin indicates a role in lipid transfer. Nature 510:552–555

Seo YK, Jeon TI, Chong HK, Biesinger J, Xie X, Osborne TF (2011) Genome-wide localization of SREBP-2 in hepatic chromatin predicts a role in autophagy. Cell Metab 13:367–375

Stefan CJ, Manford AG, Baird D, Yamada-Hanff J, Mao Y, Emr SD (2011) Osh proteins regulate phosphoinositide metabolism at ER-plasma membrane contact sites. Cell 144:389–401

Suh JM, Stenesen D, Peters JM, Inoue A, Cade A, Graff JM (2008) An RGS-containing sorting nexin controls Drosophila lifespan. PloS one 3:e2152

Tavassoli S, Chao JT, Young BP, Cox RC, Prinz WA, de Kroon AI, Loewen CJ (2013) Plasma membrane--endoplasmic reticulum contact sites regulate phosphatidylcholine synthesis. EMBO Rep 14:434–440

Teuling E, Ahmed S, Haasdijk E, Demmers J, Steinmetz MO, Akhmanova A, Jaarsma D, Hoogenraad CC (2007) Motor neuron disease-associated mutant vesicle-associated membrane protein-associated protein (VAP) B recruits wild-type VAPs into endoplasmic reticulum-derived tubular aggregates. J Neurosci 27:9801–9815

Thomas AC, Williams H, Setó-Salvia N, Bacchelli C, Jenkins D, O'Sullivan M, Mengrelis K, Ishida M, Ocaka L, Chanudet E et al (2014) Mutations in SNX14 cause a distinctive autosomal-recessive cerebellar ataxia and intellectual disability syndrome. Am J Hum Genet 95:611–621

Toulmay A, Prinz WA (2012) A conserved membrane-binding domain targets proteins to organelle contact sites. J Cell Sci 125:49–58

Ueno S, Maruki Y, Nakamura M, Tomemori Y, Kamae K, Tanabe H, Yamashita Y, Matsuda S, Kaneko S, Sano A (2001) The gene encoding a newly discovered protein, chorein, is mutated in chorea-acanthocytosis. Nature Genet 28:121–122

Yu JW, Lemmon MA (2001) All phox homology (PX) domains from Saccharomyces cerevisiae specifically recognize phosphatidylinositol 3-phosphate. J Biol Chem 276:44179–44184

Zhang D, Vjestica A, Oliferenko S (2012) Plasma membrane tethering of the cortical ER necessitates its finely reticulated architecture. Curr Biol 22:2048–2052

Zheng B, Tang T, Tang N, Kudlicka K, Ohtsubo K, Ma P, Marth JD, Farquhar MG, Lehtonen E (2006) Essential role of RGS-PX1/sorting nexin 13 in mouse development and regulation of endocytosis dynamics. Proc Natl Acad Sci USA 103:16776–16781

Alzheimer Disease

11

Estela Area-Gomez and Eric A. Schon

Abstract

The most widely accepted hypothesis to explain the pathogenesis of Alzheimer disease (AD) is the amyloid cascade, in which the accumulation of extraneuritic plaques and intracellular tangles plays a key role in driving the course and progression of the disease. However, there are other biochemical and morphological features of AD, including altered calcium, phospholipid, and cholesterol metabolism and altered mitochondrial dynamics and function that often appear early in the course of the disease, prior to plaque and tangle accumulation. Interestingly, these other functions are associated with a subdomain of the endoplasmic reticulum (ER) called mitochondria-associated ER membranes (MAM). MAM, which is an intracellular lipid raft-like domain, is closely apposed to mitochondria, both physically and biochemically. These MAM-localized functions are, in fact, increased significantly in various cellular and animal models of AD and in cells from AD patients, which could help explain the biochemical and morphological alterations seen in the disease. Based on these and other observations, a strong argument can be made that increased ER-mitochondria connectivity and increased MAM function are fundamental to AD pathogenesis.

Keywords

ApoE • Cholesterol • Cholesteryl esters • Endoplasmic reticulum • Lipid rafts • MAM • Membranes • Mitochondria • Mitochondria-associated ER membranes • Neurodegeneration • Phospholipids

E. Area-Gomez
Department of Neurology, Columbia University Medical Center, 630 West 168th Street, New York, NY 10032, USA

E.A. Schon (✉)
Department of Neurology, Columbia University Medical Center, 630 West 168th Street, New York, NY 10032, USA

Department of Genetics and Development, Columbia University Medical Center, 630 West 168th Street, New York, NY 10032, USA
e-mail: eas3@columbia.edu

M. Tagaya, T. Simmen (eds.), *Organelle Contact Sites*, Advances in Experimental Medicine and Biology 997, DOI 10.1007/978-981-10-4567-7_11

11.1 Introduction

Beginning in primary school and continuing on through secondary school and university, the approach to teaching the structure of eukaryotic cells has been informed by what one might call the "pigeonhole" view. In other words, the cell is described not as a unitary entity, but rather as an object containing various discrete subcellular elements – for example, the nucleus, the endoplasmic reticulum, the Golgi body, mitochondria, peroxisomes, endosomes, and lysosomes – each with its own special place within the cell and each with its own special function. This view is so embedded in our thinking that we have even anthropomorphized many of these functions: the mitochondrion is the "powerhouse of the cell," the nucleus is the cell's "information center," the lysosome is the cell's "garbage disposal and recycling center," and so forth.

Of course, the reality is much more complex. Each subcellular compartment indeed has its own role to play, but to work properly, both spatially and temporally, the function of each organelle has to be coordinated with the function(s) of every other organelle. In addition, organelles can have multiple complementary and/or overlapping functions. For example, the synthesis of cholesterol requires the interplay of at least five organelles – endoplasmic reticulum (ER), the Golgi body, the plasma membrane (PM), mitochondria, and the nucleus – while calcium trafficking requires at least three, ER, mitochondria, and PM.

This interdependence is seen most clearly in the many functions of the ER, which makes physical connections with the nucleus (as the nuclear envelope), the Golgi body (at ER exit sites), the plasma membrane (at plasma membrane-associated membranes, or PAM), peroxisomes (in the "pre-peroxisomal" compartment), and even with lipid droplets (English and Voeltz 2013, Lynes and Simmen 2011). One other important ER connection point, and one that is relevant to the rest of our discussion here, is the association of ER with mitochondria, at mitochondria-associated ER membranes, or MAM. The role of MAM as a highly dynamic entity and its unexpectedly important association with neurodegenerative disease have been revealed only in the last ten years. We will discuss here a hitherto-unsuspected connection between MAM function and the pathogenesis of Alzheimer disease (AD).

11.2 Mitochondria-Associated ER Membranes

As noted elsewhere in this volume, MAM is a dynamic subdomain of the ER that communicates with mitochondria, both biochemically and physically (Csordas et al. 2006, Hayashi et al. 2009, Raturi and Simmen 2013, Rusinol et al. 1994). It is a distinct biochemical/biophysical entity within the overall ER network: as opposed to "free ER," "MAM ER" is a lipid raft-like domain rich in cholesterol and sphingomyelin (Area-Gomez et al. 2012, Hayashi and Fujimoto 2010) and is enriched in approximately 1,000–1,200 proteins, as determined by proteomic analyses of MAM derived from mouse liver (Sala-Vila et al. 2016) and mouse brain (Poston et al. 2013); of these, approximately 165 have been verified in the literature, and of those, mutations in about 65 are associated with human disease. Among the proteins associated with MAM-related functions are those involved in calcium homeostasis (e.g., IP3 receptors (Mendes et al. 2005, Szabadkai et al. 2006)), in phospholipid metabolism (e.g., phosphatidylserine synthase (Stone and Vance 2000, Vance et al. 1997)), in cholesterol metabolism (e.g., acyl-CoA/cholesterol acyltransferase (Rusinol et al. 1994)), in lipid transfer between mitochondria and ER (e.g., fatty acid transfer protein 4 (Jia et al. 2007)), and in the regulation of mitochondrial morphology (e.g., dynamin-related protein 1 and mitochondrial fission factor (Friedman et al. 2011)). MAM is also associated with proteins that regulate and/or stabilize the apposition of mitochondria to ER (at an estimated interorganellar distance of ~10–30 nm (Csordas et al. 2006)), such as mitofusin 2 (de Brito and Scorrano 2008) and phosphofurin acidic cluster sorting protein 2 (Simmen et al. 2005), but the exact "tethering" mechanism is not known.

11.3 Alzheimer Disease

The main histopathological hallmarks of Alzheimer disease (AD), a neurodegenerative disorder characterized by progressive neuronal loss in the cortex and hippocampus, are the accumulation of extracellular neuritic plaques and intracellular neurofibrillary tangles (Querfurth and LaFerla 2010). The plaques are composed of numerous proteins, most prominent among them β-amyloid (Aβ). The tangles consist mainly of hyperphosphorylated forms of a single protein, the microtubule-associated protein tau (Reitz 2012). The majority of AD (>99% of patients) is sporadic (SAD), but genetic variations in *APOE*, encoding apolipoprotein E, a component of circulating lipoproteins, confer an increased risk of developing the disease (Holtzman et al. 2012, Huang 2010). At least three genes have been identified in the far rarer autosomal-dominant familial form (FAD): the amyloid precursor protein (APP), presenilin-1 (PS1), and presenilin-2 (PS2). From a clinical point of view, the two disorders are essentially identical, differing only in the earlier age of onset in FAD (Querfurth and LaFerla 2010).

Disturbances in APP processing play a critical role in both forms of the disease. Full-length APP (which is 695–770 aa in length, depending on the isoform; APP-695 is the predominant isoform in brain) is cleaved near its C-terminus by β-secretase (BACE1) to produce a long soluble N-terminal fragment (sAPPβ) and a shorter membrane-bound C-terminal fragment (APP-C99). APP-C99 is then cleaved by the γ-secretase complex (an aspartyl protease containing PS1 and/or PS2 in its catalytic core; both presenilins are produced as full-length, relatively inactive, precursors that are cleaved autocatalytically to produce the active enzyme) to produce Aβ (~40 aa [$A\beta_{40}$]) and the APP intracellular domain (AICD) peptide (~50 aa). Pathogenic mutations in PS1, PS2, or APP that cause FAD result in the production of aberrantly processed forms of Aβ (and especially an increase in the ratio of $A\beta_{42}/A\beta_{40}$) that accumulate in the neuritic plaques. The accumulated Aβ, and especially $A\beta_{42}$, is toxic to cells, promoting tau hyperphosphorylation. This chain of events has been called the "amyloid cascade" (Hardy and Higgins 1992, Selkoe 2011) and is the most widely accepted hypothesis to explain the pathogenesis of AD.

The amyloid cascade hypothesis helps explain why mutations in both APP and in the presenilins cause FAD. However, the amyloid cascade hypothesis does not address other features of AD that have received less attention in the field (Area-Gomez and Schon 2016, Schon and Area-Gomez 2010, Schon and Area-Gomez 2013). These include altered cholesterol (Stefani and Liguri 2009), glucose (Hoyer et al. 1988, Liu et al. 2009), fatty acid (Fraser et al. 2010), and phospholipid (Pettegrew et al. 2001) metabolism, perturbed calcium homeostasis (Bezprozvanny and Mattson 2008), and mitochondrial dysfunction (Wang et al. 2009). It is notable that these "other" features of AD are the very ones that are implicated in MAM function and that are often associated with proteins enriched in the MAM. This potential connection has given rise to the hypothesis that perturbed MAM function plays a role in the pathogenesis of AD (Area-Gomez and Schon 2016, Schon and Area-Gomez 2010, Schon and Area-Gomez 2013).

11.4 The MAM Connection in AD

In the last few years, a number of groups have found that presenilins and γ-secretase activity itself, while present in the ER (in agreement with the findings of others (Busciglio et al. 1997, Walter et al. 1996)), are *not* present there homogeneously, but rather are enriched heterogeneously in the MAM subcompartment of the ER (Area-Gomez et al. 2009, Newman et al. 2014, Schreiner et al. 2015). The finding that MAM is an intracellular lipid raft (Area-Gomez et al. 2012, Hayashi and Fujimoto 2010) is consistent with the observation that PS1 and γ-secretase activity reside in lipid rafts (Vetrivel et al. 2004) and supports the emerging view that rafts are located not only at the cell surface (Lingwood and Simons 2010, Vieira et al. 2010) but can also be found inside the cell (e.g., at the MAM).

Furthermore, alterations in the processing of APP result in MAM dysfunction, and *vice versa* (Area-Gomez et al. 2012, Hedskog et al. 2013), which links abnormalities in γ-secretase function to the metabolic alterations found early in the course of the disease. On the morphological side, the area of ER-mitochondria appositon is increased significantly in FAD and SAD fibroblasts and in presenilin-mutant cells, compared to controls (Area-Gomez et al. 2012). On the biochemical side, it has long been known that calcium homeostasis, which is in large part a MAM-mediated process (Csordas et al. 2010, Hayashi et al. 2009, Patergnani et al. 2011), is perturbed in AD patients (Gibson et al. 1997, Liang et al. 2015, Mattson 2010, Peterson and Goldman 1986, Sims et al. 1987, Supnet and Bezprozvanny 2010) and in animal models of AD (Sun et al. 2014). Another MAM-mediated process, mitochondrial bioenergetics and dynamics (e.g., organellar localization, fusion, and fission), is also perturbed in AD (Ferrer 2009, Gibson and Huang 2004, Peterson and Goldman 1986, Riemer and Kins 2013, Stokin et al. 2005, Wang et al. 2008).

Another important early feature of AD is disturbed lipid homeostasis (Di Paolo and Kim 2011), which may be behind some of the synaptic alterations seen in the disease (Rohrbough and Broadie 2005). As alluded to above, MAM serves as a regulatory hub for lipid regulation, including that of cholesterol and phospholipids (Vance 2014). Both of these functions are altered in AD (Area-Gomez et al. 2012, Stefani and Liguri 2009, Pettegrew et al. 2001), which can explain the altered lipid profiles seen in the disease (Chan et al. 2012) and the still-controversial connection to cholesterol (Chan et al. 2012).

Early alterations in MAM can also explain the prominent role of ApoE4 as a major genetic risk factor in sporadic AD (Holtzman et al. 2012). As noted above, ApoE is a component of lipoproteins that traffic lipids – mainly cholesterol, cholesteryl esters, and phospholipids – through the circulation, including the brain (where astrocytes, but not neurons, synthesize ApoE). There are a number of naturally occurring variants of ApoE in the population, with the most common being ApoE3 (it has a cysteine at amino acid position 112). ApoE4, with an arginine at that position, confers a significantly increased risk of developing AD compared to that conferred by ApoE3, via a currently unknown mechanism (Holtzman et al. 2012). Notably, ApoE4 has been shown to increase the intracellular concentration of cholesterol compared to the effect of ApoE3 (Heeren et al. 2004).

Consistent with this difference, it was recently shown that lipoproteins containing ApoE4 (but not the free protein) upregulated MAM function to a significantly greater degree than did those containing ApoE3 (Tambini et al. 2016). These results imply that the negative effects of ApoE4 on MAM functionality may well be due to its function in lipoprotein-mediated cholesterol metabolism and trafficking. Thus, the contribution of ApoE4 to the risk of developing AD may be due to the effects of perturbed cholesterol homeostasis on MAM function. This finding is concordant with the discovery of genetic variants in a number of cholesterol metabolism-related genes (Wollmer 2010), such as *ABCA7* (Steinberg et al. 2015), which encodes a cholesterol and phospholipid transport protein (Abe-Dohmae et al. 2004), also predispose to developing AD.

Overall, these data and observations support a view of AD pathogenesis that departs from that afforded by the amyloid cascade hypothesis and is focused less on plaques and tangles and more on altered cellular metabolism as the underlying disturbance in AD. In particular, the "MAM hypothesis" proposes that the development and progression of the AD result from increased communication between ER and mitochondria at the MAM (Area-Gomez and Schon 2016,

Schon and Area-Gomez 2010, Schon and Area-Gomez 2013). This increase affects a panoply of cellular functions, both directly (e.g., alterations in MAM-localized enzymatic functions) and indirectly (e.g., alterations in cellular behavior in response to perturbed MAM behavior). The increased apposition between ER and mitochondria and the concomitant alteration in MAM function are consistent with the perturbed cholesterol homeostasis, the altered phospholipid profiles, the increased calcium trafficking between the two organelles, the changes in mitochondrial bioenergetics and dynamics, and the elevated ratio of $A\beta_{42}/A\beta_{40}$ (Schon and Area-Gomez 2013). Thus, it is possible that the functional cause of AD is an increase in the communication between ER and mitochondria and an associated upregulation in MAM function. What now remains to be elucidated is the biochemical cause of this ER-mitochondria hyperconnectivity and how APP processing plays a role in this process.

In this regard, the finding that ApoE4 impacts on MAM function may provide an important clue. Although the connection between ApoE4 and APP processing is at present unclear, one conceptual connection between the two is perturbed cholesterol homeostasis. Lipoproteins transport cholesterol and cholesteryl esters, and their components are recycled following binding to lipoprotein receptors on the cell surface and internalization into the cell. Interestingly, intracellular lipoprotein-derived cholesterol is recycled poorly in ApoE4-containing cells relative to ApoE3 (Heeren et al. 2004). Thus, one possible connection between APP processing and ApoE in general (and ApoE4 in particular) is the regulation of cholesterol homeostasis (Heeren et al. 2003), for two reasons. First, γ-secretase resides in the MAM (Area-Gomez et al. 2009), a lipid raft rich in cholesterol and sphingomyelin. Second, APP contains a cholesterol-binding domain at its C-terminus (Barrett et al. 2012) that may act as a cholesterol sensor (Beel et al. 2008). Thus, it may well be that in AD cholesterol levels are altered, either at steady state or dynamically (e.g., altered cholesterol turnover). In the case of SAD, this could be the result of aberrant cholesterol trafficking mediated by, for example, ApoE4 or mutated ABCA7. In the case of FAD, it could be the result of aberrant cholesterol sensing or homeostasis due to altered APP structure or amount (as is the case in subjects with Down syndrome, who are at elevated risk for developing AD, likely due to an extra gene dose of *APP*) or in the ability of mutated presenilins to cleave APP properly (Heilig et al. 2010). In either case, altered cholesterol metabolism somehow induces an increase in ER-mitochondria communication that then gives rise to the phenotypes seen in AD (Marquer et al. 2014).

Acknowledgments This work was supported by the US Department of Defense (W911F-15-1-0169), the Ellison Medical Foundation, and the J. Willard and Alice S. Marriott Foundation (to EAS) and by the US National Institutes of Health (K01-AG045335 to E.A.-G.).

References

Abe-Dohmae S, Ikeda Y, Matsuo M, Hayashi M, K-i O, Ueda K, Yokoyama S (2004) Human ABCA7 supports apolipoprotein-mediated release of cellular cholesterol and phospholipid to generate high density lipoprotein. J Biol Chem 279:604–611

Area-Gomez E, Del Carmen Lara Castillo M, Tambini MD, Guardia-Laguarta C, de Groof AJC, Madra M, Ikenouchi J, Umeda M, Bird TD, Sturley SL et al (2012) Upregulated function of mitochondria-associated ER membranes in Alzheimer disease. EMBO J 31:4106–4123

Area-Gomez E, Schon EA (2016) Mitochondria-associated ER membranes and Alzheimer disease. Curr Opin Genet Dev 38:90–96

Area-Gomez E, de Groof AJ, Boldogh I, Bird TD, Gibson GE, Koehler CM, Yu WH, Duff KE, Yaffe MP, Pon LA et al (2009) Presenilins are enriched in endoplasmic reticulum membranes associated with mitochondria. Am J Pathol 175:1810–1816

Barrett PJ, Song Y, Van Horn WD, Hustedt EJ, Schafer JM, Hadziselimovic A, Beel AJ, Sanders CR (2012) The amyloid precursor protein has a flexible transmembrane domain and binds cholesterol. Science 336:1168–1171

Beel AJ, Mobley CK, Kim HJ, Tian F, Hadziselimovic A, Jap B, Prestegard JH, Sanders CR (2008) Structural studies of the transmembrane C-terminal domain of the amyloid precursor protein (APP): does APP function as a cholesterol sensor? Biochemistry 47:9428–9446

Bezprozvanny I, Mattson MP (2008) Neuronal calcium mishandling and the pathogenesis of Alzheimer's disease. Trends Neurosci 31:454–463
Busciglio J, Hartmann H, Lorenzo A, Wong C, Baumann K, Sommer B, Staufenbiel M, Yankner BA (1997) Neuronal localization of presenilin-1 and association with amyloid plaques and neurofibrillary tangles in Alzheimer's disease. J Neurosci 17:5101–5107
Chan RB, Oliveira TG, Cortes EP, Honig LS, Duff KE, Small SA, Wenk MR, Shui G, Di Paolo G (2012) Comparative lipidomic analysis of mouse and human brain with Alzheimer disease. J Biol Chem 287:2678–2688
Csordas G, Renken C, Varnai P, Walter L, Weaver D, Buttle KF, Balla T, Mannella CA, Hajnoczky G (2006) Structural and functional features and significance of the physical linkage between ER and mitochondria. J Cell Biol 174:915–921
Csordas G, Varnai P, Golenar T, Roy S, Purkins G, Schneider TG, Balla T, Hajnoczky G (2010) Imaging interorganelle contacts and local calcium dynamics at the ER-mitochondrial interface. Mol Cell 39:121–132
de Brito OM, Scorrano L (2008) Mitofusin 2 tethers endoplasmic reticulum to mitochondria. Nature 456:605–610
Di Paolo G, Kim TW (2011) Linking lipids to Alzheimer's disease: cholesterol and beyond. Nature reviews Neuroscience 12:284–296
English AR, Voeltz GK (2013) Endoplasmic reticulum structure and interconnections with other organelles. Cold Spring Harb Perspect Biol 5:a013227
Ferrer I (2009) Altered mitochondria, energy metabolism, voltage-dependent anion channel, and lipid rafts converge to exhaust neurons in Alzheimer's disease. J Bioenerg Biomembr 41:425–431
Fraser T, Tayler H, Love S (2010) Fatty acid composition of frontal, temporal and parietal neocortex in the normal human brain and in Alzheimer's disease. Neurochem Res 35:503–513
Friedman JR, Lackner LL, West M, DiBenedetto JR, Nunnari J, Voeltz GK (2011) ER tubules mark sites of mitochondrial division. Science 334:358–362
Gibson GE, Huang H-M (2004) Mitochondrial enzymes and endoplasmic reticulum calcium stores as targets of oxidative stress in neurodegenerative diseases. J Bioenerg Biomembr 36:335–340
Gibson GE, Vestling M, Zhang H, Szolosi S, Alkon D, Lannfelt L, Gandy S, Cowburn RF (1997) Abnormalities in Alzheimer's disease fibroblasts bearing the APP670/671 mutation. Neurobiol Aging 18:573–580
Hardy JA, Higgins GA (1992) Alzheimer's disease: the amyloid cascade hypothesis. Science 256:184–185
Hayashi T, Fujimoto M (2010) Detergent-resistant microdomains determine the localization of sigma-1 receptors to the endoplasmic reticulum-mitochondria junction. Mol Pharmacol 77:517–528
Hayashi T, Rizzuto R, Hajnoczky G, Su TP (2009) MAM: more than just a housekeeper. Trends Cell Biol 19:81–88
Hedskog L, Pinho CM, Filadi R, Ronnback A, Hertwig L, Wiehager B, Larssen P, Gellhaar S, Sandebring A, Westerlund M et al (2013) Modulation of the endoplasmic reticulum-mitochondria interface in Alzheimer's disease and related models. Proc Natl Acad Sci USA 110:7916–7921
Heeren J, Grewal T, Laatsch A, Rottke D, Rinninger F, Enrich C, Beisiegel U (2003) Recycling of apoprotein E is associated with cholesterol efflux and high density lipoprotein internalization. J Biol Chem 278:14370–14378
Heeren J, Grewal T, Laatsch A, Becker N, Rinninger F, Rye K-A, Beisiegel U (2004) Impaired recycling of apolipoprotein E4 is associated with intracellular cholesterol accumulation. J Biol Chem 279:55483–55492
Heilig EA, Xia W, Shen J, Kelleher RJ 3rd (2010) A presenilin-1 mutation identified in familial Alzheimer disease with cotton wool plaques causes a nearly complete loss of γ-secretase activity. J Biol Chem 285:22350–22359
Holtzman DM, Herz J, Bu G (2012) Apolipoprotein E and apolipoprotein E receptors: normal biology and roles in Alzheimer disease. Cold Spring Harb Perspect Med 2:a006312
Hoyer S, Oesterreich K, Wagner O (1988) Glucose metabolism as the site of the primary abnormality in early-onset dementia of Alzheimer type? J Neurol 235:143–148
Huang Y (2010) Aβ-independent roles of apolipoprotein E4 in the pathogenesis of Alzheimer's disease. Trends Mol Med 16:287–294
Jia W, Moulson CL, Pei Z, Miner JH, Watkins PA (2007) Fatty acid transport protein 4 is the principal very long chain fatty acyl-CoA synthetase in skin fibroblasts. J Biol Chem 282:20573–20583
Liang J, Kulasiri D, Samarasinghe S (2015) Ca^{2+} dysregulation in the endoplasmic reticulum related to Alzheimer's disease: a review on experimental progress and computational modeling. Biosystems 134:1–15
Lingwood D, Simons K (2010) Lipid rafts as a membrane-organizing principle. Science 327:46–50
Liu F, Shi J, Tanimukai H, Gu J, Gu J, Grundke-Iqbal I, Iqbal K, Gong C-X (2009) Reduced O-GlcNAcylation links lower brain glucose metabolism and tau pathology in Alzheimer's disease. Brain 132:1820–1832
Lynes EM, Simmen T (2011) Urban planning of the endoplasmic reticulum (ER): how diverse mechanisms segregate the many functions of the ER. Biochim Biophys Acta 1813:1893–1905
Marquer C, Laine J, Dauphinot L, Hanbouch L, Lemercier-Neuillet C, Pierrot N, Bossers K, Le M, Corlier F, Benstaali C et al (2014) Increasing membrane cholesterol of neurons in culture recapitulates Alzheimer's disease early phenotypes. Mol Neurodegen 9:60
Mattson MP (2010) ER calcium and Alzheimer's disease: in a state of flux. Sci Signal 3:pe10
Mendes CC, Gomes DA, Thompson M, Souto NC, Goes TS, Goes AM, Rodrigues MA, Gomez MV, Nathanson MH, Leite MF (2005) The type III inositol

1,4,5-trisphosphate receptor preferentially transmits apoptotic Ca^{2+} signals into mitochondria. J Biol Chem 280:40892–40900
Newman M, Wilson L, Verdile G, Lim A, Khan I, Moussavi Nik SH, Pursglove S, Chapman G, Martins RN, Lardelli M (2014) Differential, dominant activation and inhibition of Notch signalling and APP cleavage by truncations of PSEN1 in human disease. Hum Mol Genet 23:602–617
Patergnani S, Suski JM, Agnoletto C, Bononi A, Bonora M, De Marchi E, Giorgi C, Marchi S, Missiroli S, Poletti F et al (2011) Calcium signaling around mitochondria associated membranes (MAMs). Cell Commun Signal 9:19
Peterson C, Goldman JE (1986) Alterations in calcium content and biochemical processes in cultured skin fibroblasts from aged and Alzheimer donors. Proc Natl Acad Sci USA 83:2758–2762
Pettegrew JW, Panchalingam K, Hamilton RL, McClure RJ (2001) Brain membrane phospholipid alterations in Alzheimer's disease. Neurochem Res 26:771–782
Poston CN, Krishnan SC, Bazemore-Walker CR (2013) In-depth proteomic analysis of mammalian mitochondria-associated membranes (MAM). J Proteomics 79:219–230
Querfurth HW, LaFerla FM (2010) Alzheimer's disease. New Eng J Med 362:329–344
Raturi A, Simmen T (2013) Where the endoplasmic reticulum and the mitochondrion tie the knot: the mitochondria-associated membrane (MAM). Biochim Biophys Acta 1833:213–224
Reitz C (2012) Alzheimer's disease and the amyloid cascade hypothesis: a critical review. Int J Alzheimers Dis 2012:369808
Riemer J, Kins S (2013) Axonal transport and mitochondrial dysfunction in Alzheimer's disease. Neurodegen Dis 12:111–124
Rohrbough J, Broadie K (2005) Lipid regulation of the synaptic vesicle cycle. Nat Rev Neurosci 6:139–150
Rusinol AE, Cui Z, Chen MH, Vance JE (1994) A unique mitochondria-associated membrane fraction from rat liver has a high capacity for lipid synthesis and contains pre-Golgi secretory proteins including nascent lipoproteins. J Biol Chem 269:27494–27502
Sala-Vila A, Navarro-Lerida I, Sanchez-Alvarez M, Bosch M, Calvo C, Lopez JA, Calvo E, Ferguson C, Giacomello M, Serafini A et al (2016) Interplay between hepatic mitochondria-associated membranes, lipid metabolism and caveolin-1 in mice. Sci Rep 6:27351
Schon EA, Area-Gomez E (2010) Is Alzheimer's disease a disorder of mitochondria-associated membranes? J Alzheimers Dis 20:S281–S292
Schon EA, Area-Gomez E (2013) Mitochondria-associated ER membranes in Alzheimer disease. Mol Cell Neurosci 55:26–36
Schreiner B, Hedskog L, Wiehager B, Ankarcrona M (2015) Amyloid-β peptides are generated in mitochondria-associated endoplasmic reticulum membranes. J Alzheimers Dis 43:369–374
Selkoe DJ (2011) Alzheimer's disease. Cold Spring Harb Perspect Biol 3:a004457
Simmen T, Aslan JE, Blagoveshchenskaya AD, Thomas L, Wan L, Xiang Y, Feliciangeli SF, Hung CH, Crump CM, Thomas G (2005) PACS-2 controls endoplasmic reticulum-mitochondria communication and Bid-mediated apoptosis. EMBO J 24:717–729
Sims NR, Finegan JM, Blass JP (1987) Altered metabolic properties of cultured skin fibroblasts in Alzheimer's disease. Ann Neurol 21:451–457
Stefani M, Liguri G (2009) Cholesterol in Alzheimer's disease: unresolved questions. Curr Alz Res 6:15–29
Steinberg S, Stefansson H, Jonsson T, Johannsdottir H, Ingason A, Helgason H, Sulem P, Magnusson OT, Gudjonsson SA, Unnsteinsdottir U et al (2015) Loss-of-function variants in ABCA7 confer risk of Alzheimer's disease. Nat Genet 47:445–447
Stokin GB, Lillo C, Falzone TL, Brusch RG, Rockenstein E, Mount SL, Raman R, Davies P, Masliah E, Williams DS et al (2005) Axonopathy and transport deficits early in the pathogenesis of Alzheimer's disease. Science 307:1282–1288
Stone SJ, Vance JE (2000) Phosphatidylserine synthase-1 and -2 are localized to mitochondria-associated membranes. J Biol Chem 275:34534–34540
Sun S, Zhang H, Liu J, Popugaeva E, Xu N-J, Feske S, White CL 3rd, Bezprozvanny I (2014) Reduced synaptic STIM2 expression and impaired store-operated calcium entry cause destabilization of mature spines in mutant presenilin mice. Neuron 82:79–93
Supnet C, Bezprozvanny I (2010) Neuronal calcium signaling, mitochondrial dysfunction, and Alzheimer's disease. J Alzheimers Dis 20:S487–S498
Szabadkai G, Bianchi K, Varnai P, De Stefani D, Wieckowski MR, Cavagna D, Nagy AI, Balla T, Rizzuto R (2006) Chaperone-mediated coupling of endoplasmic reticulum and mitochondrial Ca^{2+} channels. J Cell Biol 175:901–911
Tambini MD, Pera M, Kanter E, Yang H, Guardia-Laguarta C, Holtzman D, Sulzer D, Area-Gomez E, Schon EA (2016) ApoE4 upregulates the activity of mitochondria-associated ER membranes. EMBO Rep 17:27–36
Vance JE (2014) MAM (mitochondria-associated membranes) in mammalian cells: Lipids and beyond. Biochim Biophys Acta 1841:595–609
Vance DE, Walkey CJ, Cui Z (1997) Phosphatidylethanolamine N-methyltransferase from liver. Biochim Biophys Acta 1348:142–150
Vetrivel KS, Cheng H, Lin W, Sakurai T, Li T, Nukina N, Wong PC, Xu H, Thinakaran G (2004) Association of γ-secretase with lipid rafts in post-Golgi and endosome membranes. J Biol Chem 279:44945–44954
Vieira FS, Correa G, Einicker-Lamas M, Coutinho-Silva R (2010) Host-cell lipid rafts: a safe door for microorganisms? Biol Cell 102:391–407

Walter J, Capell A, Grunberg J, Pesold B, Schindzielorz A, Prior R, Podlisny MB, Fraser P, Hyslop PS, Selkoe DJ et al (1996) The Alzheimer's disease-associated presenilins are differentially phosphorylated proteins located predominantly within the endoplasmic reticulum. Mol Med 2:673–691

Wang X, Su B, Siedlak SL, Moreira PI, Fujioka H, Wang Y, Casadesus G, Zhu X (2008) Amyloid-β overproduction causes abnormal mitochondrial dynamics via differential modulation of mitochondrial fission/fusion proteins. Proc Natl Acad Sci USA 105:19318–19323

Wang X, Su B, Zheng L, Perry G, Smith MA, Zhu X (2009) The role of abnormal mitochondrial dynamics in the pathogenesis of Alzheimer's disease. J Neurochem 109:153–159

Wollmer MA (2010) Cholesterol-related genes in Alzheimer's disease. Biochim Biophys Acta 1801:762–773

Mitochondrial-Associated Membranes in Parkinson's Disease

12

Nobutaka Hattori, Taku Arano, Taku Hatano, Akio Mori, and Yuzuru Imai

Abstract

Parkinson's disease (PD) is a common neurodegenerative disorder, with ageing being a major risk factor. Accordingly, estimates predict an increasing number of PD patients due to our expanding life span. Consequently, developing a true disease-modifying therapy is necessary. In this regard, monogenic PD offers a suitable means for determining pathogenesis. Among monogenic forms of PD, mitochondrial dysfunction may be a major cause and is also likely to be involved in sporadic PD. Thus, mitochondrial impairment may be a common pathway. Recently, mitochondria-associated membranes (MAM) were identified as dynamic sites between mitochondria and endoplasmic reticulum. Indeed, the gene product of α-synuclein is a major component of MAM, with other gene products also involved. This review focuses on the possibility of using MAM as novel therapeutic targets.

Keywords

Parkinson's disease • Familial Parkinson's disease • α-synuclein • Parkin • PINK1 • DJ-1 • Mitochondria-associated membrane

12.1 Introduction

Parkinson's disease (PD) is the second most prevalent neurodegenerative disease after Alzheimer's disease (AD) and is characterized by motor disturbances such as resting tremor, rigidity, bradykinesia, and postural instability. In addition to motor symptoms, nonmotor symptoms are recognized to play an important role in adversely affecting quality of life in patients with PD and may precede formal diagnosis by decades (Berg et al. 2015; Jankovic and Poewe 2012). Defining the epidemiology of PD is confounded by several variables, including diagnostic difficulties and age dependence. PD affects >1% of the population over the age of 60, which equates to more than 127,000

N. Hattori (✉) • T. Arano • T. Hatano • A. Mori • Y. Imai
Department of Neurology, Juntendo University Graduate School of Medicine, 2-1-1 Hongo, Bunkyo, Tokyo 113-8421, Japan
e-mail: nhattori@juntendo.ac.jp

M. Tagaya, T. Simmen (eds.), *Organelle Contact Sites*, Advances in Experimental Medicine and Biology 997, DOI 10.1007/978-981-10-4567-7_12

individuals in the UK or 500,000 individuals in the USA. Further, its prevalence reaches 5% in individuals over the age of 85, highlighting the impact that advancing age has on the risk of developing PD (de Lau and Breteler 2006; Nussbaum and Ellis 2003; Wood-Kaczmar et al. 2006). In a study of the world's ten most populous nations and Western Europe's five most populous nations, it was estimated that the number of people with PD will rise by two times from 4.1–4.6 million in 2005 to 8.7–9.3 million in 2030. Six of the most populous countries are in Asia (China, India, Indonesia, Pakistan, Bangladesh, and Japan), and the number of people with PD in these countries is expected to increase from 2.57 million in 2005 to 6.17 million in 2030. Indeed, the number of people with PD is likely to grow worldwide as a result of our rapidly ageing population and increasing life expectancy (Dorsey et al. 2007). Consequently, determining the pathogenesis of PD and developing new therapies to halt disease progression are crucial. The majority of PD cases are sporadic; however, some (< 10%) are caused by mutations in genes that affect either protein degradation, lysosomal function, or mitochondrial function, including α-synuclein (SNCA) (Park1, 4), Parkin (Park2), PTEN-induced putative kinase 1 (PINK1) (Park6), DJ-1 (Park7), and glucosylceramidase beta (GBA) (Polymeropoulos et al. 1997; Kitada et al. 1998; Kruger et al. 1998; Valente et al. 2004; Bonifati et al. 2003; Multiple-System Atrophy Research 2013). In particular, mitochondrial dysfunction is a key event for pathogenesis of sporadic and familial PD. Emerging evidence shows that mitochondria-associated membrane (MAM) dysfunction plays a prominent role in numerous neurodegenerative diseases, while genes affecting endoplasmic reticulum (ER) and mitochondrial homeostasis are clearly overrepresented in hereditary disorders. Several findings indicate that mitochondria-ER contact sites play crucial roles in neuronal survival and death. Mitochondria in contact with ER are readily observed in brain tissue throughout neurons and at synapses (Hedskog et al. 2013). In addition, choreographed interplay between the ER and mitochondria may be involved in shaping dendritic calcium signals and neuronal activity in hippocampal neurons (Pivovarova et al. 2002). At the synapse, calcium shuttling between the ER and mitochondria is suggested to be essential for exocytosis and synaptic activity (Mironov and Symonchuk 2006). Moreover, MAM signaling is implicated in several neurodegenerative disorders including PD (Cali et al. 2013a; Guardia-Laguarta et al. 2014; Ottolini et al. 2013), AD (Schon and Area-Gomez 2013), amyotrophic lateral sclerosis, frontotemporal dementia (Prause et al. 2013; Stoica et al. 2014), and Charcot–Marie–Tooth disease (CMT) (de Brito and Scorrano 2008). This is underscored by the fact that genes affecting mitochondrial and ER homeostasis are overrepresented in several hereditary neurodegenerative disorders. In contrast, processes affected by mitochondria-ER contact sites are widely implicated in neurodegeneration, including PD. This review will focus on PD and MAM function.

12.2 Mitochondrial Dysfunction in PD

As eukaryotic cells acquired the highly efficient power-generating system of aerobic respiration by incorporating mitochondria into the cytosol, alleviation of oxidative stress by appropriate regulation of mitochondria became an emerging and inevitable issue. Non-dividing cells or tissues with high energy demands in long-lived animals such as humans have a particular need to regulate mitochondrial activity and avoid production of deleterious reactive oxygen species. Currently, mitochondrial dysregulation is implicated in various human diseases including cancer, diabetes, myopathy, and neurodegeneration. Since the discovery of 1-methyl-4-phenyl-1,2,3,6-tetrahydropyridine (MPTP), which appears to specifically target the neurons involved in PD, mitochondrial dysfunction has been proposed as one of the major causes of PD. Indeed, MPTP is used to develop animal models for testing new therapies for PD. Furthermore, investigation of MPTP toxicity has provided insight on the possible pathogenesis of PD. MPTP is amphiphilic and is captured into acidic organelles (mostly lysosomes) in astrocytes. MPTP itself does not appear to be toxic, but its oxidized product, 1-methyl-4-phenylpyridinium (MPP+), is toxic. Although the precise mechanism(s) underlying MPP+ toxicity are unknown, they are suggested to be dependent on concentration in mitochondria via selective uptake. Energy-driven mitochondrial uptake of MPP+ results in sufficiently high toxin concentration, which interferes with mitochondrial respiration. MPP+ appears to act at or near complex I, where several other agents (including rotenone) act to block mitochondrial oxidation. Clinical phenotypes of MPTP-induced parkinsonism are very similar to sporadic forms of PD (Schapira et al. 1988). Moreover, complex I deficiency is observed in autopsied brains, and western blotting and immunohistochemical analyses reveal defects in several complex I molecules (Mizuno et al. 1989; Hattori et al. 1991). Additionally, mitochondrial dysfunction is observed not only in the brain but also skeletal muscle, lymphocytes, and platelets (Yoshino et al. 1992; Benecke et al. 1993). Therefore, mitochondrial dysfunction, especially respiratory chain impairments, is a major cause of PD. Consequently, many neurotoxins have been screened, but none similar to MPTP and rotenone have been discovered so far. Thus, interaction between genetic and environmental factors (e.g., head injuries) may induce PD onset.

12.3 α-Synuclein (SNCA)

More than 20 genes have been identified as causative for familial PD (Table 12.1). Among them, SNCA is mutated or overexpressed in autosomal dominant PD (Polymeropoulos et al. 1997). Several missense mutations and multiplications are reported worldwide (Ibanez et al. 2004; Kiely et al. 2013; Nishioka et al. 2006; Singleton et al. 2003; Chartier-Harlin et al. 2004; Kruger et al. 1998; Pasanen et al. 2014; Proukakis et al.

Table 12.1 Genetics of familial Parkinson's disease PARK1-22 indicates mopnogenic form of PD. GBA is a susceptible gene for developing PD

Gene symbol	Gene locus	Mode	Gene name	Age at onset	Lewy bodies
PARK1 (SNCA), PARK4	4q21	AD	*SNCA*	Approximately 40	+
PARK2	6q252.2-27	AR	*Parkin*	< 40	– (although some patients +)
PARK3	2p13	AD	?	35-89	+
PARK5	4p14	AD	*UCH-L1*	< 50	?
PARK6	1p35-36	AR	*PINK1*	Approximately 50	+
PARK7	1p36	AR	*DJ-1*	27-40	?
PARK8	12q12	AD	*LRRK2*	Approximately 65	+/-
PARK9	1p36	AR	*ATP13A2*	11-16	?
PARK10	1p32	SP	?	Late	?
PARK11	2q36-37	AD	*GIGYF2*	Late	?
PARK12	Xp21-q25	SP	?	Late	?
PARK13	2p12	SP	*HtrA2/Omi*	Late	?
PARK14	22q13.1	AR	*PLA2G6*	20-25	+
PARK15	22q12-q13	AR	*FBXO7*	10-19	?
PARK16	1q32	SP	?	Late	?
PARK17	16q12	AD	*VPS35*	Late	-
PARK18	3q27	AD	*EIF4G1*	Late	+
PARK19	1p31.3	AR	*DNAJC6/HSP40*	10-20	?
PARK20	21q22.11	AR	*SYNJ1*	Early	?
PARK22	7p11.2	AD	*CHCHD2*	Late	?
GBA	1q21	SP	*Glucocerebrosidase*	52 ± 7	+

2013; Zarranz et al. 2004). Yet despite numerous research efforts, the main function of SNCA remains unknown. The majority of SNCA is soluble and resides in the cytoplasm. Interestingly, many studies have shown that upon stimulus by an unidentified factor, SNCA is capable of binding to membranes, and subsequently, changing its N-terminal domain conformation (Eliezer et al. 2001; Jao et al. 2004; Jao et al. 2008). SNCA preferentially binds to anionic phospholipids and high-curvature liposomes in vitro (Davidson et al. 1998; Fortin et al. 2004; Auluck et al. 2010). While in cells, relevant membrane regions include lipid raft domains and detergent-resistant membranes with unique molecular characteristics (Simons and Toomre 2000). Lipid rafts are specialized membrane domains enriched in certain lipids, cholesterol, and proteins. Initially, lipid rafts were believed to form only at the plasma membrane, but there have been many reports showing that these domains also localize intracellularly (Hayashi and Fujimoto 2010). More recently, SNCA was shown to bind to mitochondria (Cole et al. 2008; Li et al. 2007; Devi et al. 2008; Parihar et al. 2008; Zhang et al. 2008), with mitochondrial binding particularly significant in the striatum, substantia nigra, and cortex of PD brains (Devi et al., 2008). Supporting these results, a recent study described an N-terminal sequence in SNCA that may act as a mitochondrial targeting sequence (Devi et al. 2008). Corroboratively, SNCA binding to membranes is favored in the presence of cardiolipin, a lipid specific to the mitochondrial membrane (Zigoneanu et al. 2012).

Mitochondrial presence of SNCA supports involvement of mitochondrial dysfunction and decreased complex I activity in PD patients and cell models with *SNCA* mutations (Hsu et al. 2000; Schon and Przedborski 2011; Devi et al. 2008). Animal models, such as transgenic mice carrying the *SNCA* A53T mutation, reveal not only decreased complex I activity but also damaged mitochondrial DNA and aberrant mitochondrial dynamics (Chinta et al. 2010; Martin et al. 2006; Choubey et al. 2011).

Guardia-Laguarta et al. found that a subpopulation of SNCA resides at MAM (Guardia-Laguarta et al. 2014), which are composed of intracellular lipid rafts. This result is compelling, as previous work has shown that SNCA has an affinity for lipid rafts (Fortin et al. 2004) and negatively-charged membranes (Davidson et al. 1998), and supports SNCA as a mitochondrial protein (Li et al. 2007; Cole et al. 2008; Devi et al. 2008; Parihar et al. 2008; Shavali et al. 2008). Lack of appropriate MAM markers and technical difficulties in fractionating this membrane (because of its ER association) may explain previous results (Area-Gomez et al. 2012). MAM are an ER sub-compartment that is connected to mitochondria (Rusinol et al. 1994; Csordas et al. 2006; Hayashi et al. 2009). Specifically, several proteins related to phospholipid regulation (phosphatidylserine synthase 2), cholesterol metabolism (acyl-CoA cholesterol acyl transferase) (Rusinol et al. 1994), and calcium transport from the ER to mitochondria (type 3 inositol 1,4,5-triphosphate receptor) (Hayashi and Fujimoto 2010) are present in MAM. Notably, mitochondrial distribution and dynamics are influenced by physical connections formed by MAM (Rizzuto et al. 1998; Levine and Rabouille 2005; Csordas et al. 2006; Hayashi et al. 2009; Friedman et al. 2011; Rowland and Voeltz 2012). During mitochondrial fission, ER tubules appear to "embrace" mitochondria and mark mitochondrial division sites (Friedman et al. 2011). In addition, isolated MAM from different tissues are enriched in proteins related to control of mitochondrial dynamics (e.g., FIS1, MFN2, and DRP1). Finally, MAM also contain proteins involved in apoptosis (e.g., voltage-dependent anion channel 1, BAX, and BID) (Garofalo et al. 2005; Ciarlo et al. 2010). Indeed, calcium release at mitochondria-ER contacts, which is important for ATP production, may be responsible for sensitizing mitochondria to apoptosis (Iwasawa et al. 2011; Tabas and Ron 2011). Alteration of mitochondria-ER contacts causes calcium signal deregulation, resulting in inappropriate protein folding, metabolic alterations, and apoptosis (Csordas and Hajnoczky 2009; Bui et al. 2010).

Recent advances in the protein degradation system in PD pathogenesis reveal the importance of both the ubiquitin–proteasome and autophagy–lysosome pathway (Matsuda and Tanaka 2010). Abnormal ubiquitin enrichment in inclusion bodies is found in various neurodegenerative disorders, including PD. This shows that impairment of the protein degradation system may be required for formation of abnormal protein aggregation. Lewy bodies (of which a major component is SNCA) are a pathological diagnostic hallmark. Accordingly, it is crucial to determine SCNA aggregation mechanisms. Compared to wild-type SNCA, mutant SNCA easily forms aggregates in neuronal cells in vitro and in vivo (Winner et al. 2011; Karpinar et al. 2009). The most toxic SNCA species to cells remains unclear, but some studies suggest that mature aggregates are not toxic themselves, but rather an attempt by the cell to clear small toxic oligomers (Cremades et al. 2012). Wild-type SNCA is degraded by both chaperone-mediated autophagy and macroautophagy, while the A30P and A53T mutations involve mainly the latter (Cuervo et al. 2004).

In 1955, Christian de Duve discovered then previously unknown organelles: lysosomes. This discovery significantly contributed to de Duve being awarded a share of the 1974 Nobel Prize in Physiology or Medicine for elucidating "the structural and functional organization of the cell." More recently, Yoshinori Ohsumi won the 2016 Nobel Prize in Physiology or Medicine for discovering and elucidating mechanisms underlying autophagy, a fundamental process for degrading and recycling cellular components. Inclusion bodies are often observed in specific lesions in neurodegenerative disorders such as PD and AD. Hence, autophagy may also be involved in pathogenesis of neurodegenerative disorders.

The term autophagy is used to describe lysosomal-mediated degradation of intracellular contents, which can be divided into three basic mechanisms: (1) chaperone-mediated autophagy, (2) microautophagy, and (3) macroautophagy. First, chaperone-mediated autophagy is initiated by the chaperone, Hsc70, which recognizes one protein at a time. By binding to lysosomal-associated membrane protein, Hsc70 transports proteins to lysosomes (Yabu et al. 2011). Chaperone-mediated autophagy impairment has been reported in PD (Sala et al. 2014). Second, microautophagy involves lysosomal membrane invagination to encircle cellular contents including proteins and lipids. Appropriately, lipids, proteins, or organelles can be degraded through this pathway. Further, it is highly likely that these components are marked by specific modifications to be recognized by lysosomes, although the majority of these modifications have not yet been defined. Third, macroautophagy is the most extensively studied autophagic pathway and involves formation of double-membrane structures that encircle proteins, lipids, and organelles (Klionsky et al. 2012). Degradation of mitochondria through the macroautophagy pathway is termed "mitophagy" and is associated with the Parkin/PINK1 pathway. Similarly, degradation of other cellular structures, such as nuclear fragments, lipid droplets, peroxisomes, ribosomes, and ER, has been named nucleophagy, lipophagy, pexophagy, ribophagy, and reticulophagy, respectively. As already stated, mitochondrial dysregulation is implicated in various human diseases including cancer, diabetes, myopathy, and a variety of neurodegenerative disorders. Therefore, mitochondrial maintenance includes mitophagy as a selective autophagic process to remove abnormal mitochondria. It can easily be speculated that damaged mitochondria fail to provide the cell with life-preserving energy. Subsequently, impaired mitochondria, if not removed, can directly kill cells.

It is well known that mitochondria-ER connections regulate dynamic mitochondrial processes (Csordas et al. 2006; Hayashi et al. 2009; Friedman et al. 2011; Rowland and Voeltz 2012), and mutations in SNCA increase mitochondrial fragmentation (Kamp et al. 2010; Nakamura et al. 2011). Correlating this to MAM dysfunction, Guardia-Laguarta et al. confirmed a fragmented phenotype in mutant cells (Guardia-Laguarta et al. 2014). Nevertheless, it is possible that fragmentation due to SNCA mutations is not due to defects in the fusion/fission machinery of mitochondria, but rather to MAM alterations (Guardia-Laguarta et al. 2014). Several findings support crosstalk between SNCA and MAM, while SNCA aggregation impairs mitochondrial function.

12.4 Parkin and PINK1

The genes for autosomal recessive, young-onset PD, *PINK1* and *Parkin*, encode a kinase and ubiquitin ligase, respectively. Moreover, they have similar clinical phenotypes including age at onset and good levodopa response. Both PINK1 and Parkin gene products are implicated in mitophagy to protect healthy mitochondria. Parkin in cooperation with PINK1 specifically recognizes damaged mitochondria with reduced mitochondrial membrane potential (Δψm) and rapidly isolates them from the mitochondrial network by eliminating them through the ubiquitin–proteasome and autophagy–lysosome pathways. Parkin expressed in cultured cells mainly localizes to the cytosol; therefore, how Parkin regulates mitochondrial function remains unknown. Narendra et al. reported that Δψm disruption in cultured mammalian cells using mitochondrial-damaging reagents such as carbonyl cyanide *m*-chlorophenylhydrazone (CCCP) causes Parkin translocation to mitochondria with low Δψm (Narendra et al. 2008). Further, this translocation induces microtubule-associated protein light chain 3 (LC3)-mediated autophagic elimination of damaged mitochondria (i.e., mitophagy) (Fig. 12.1). After Parkin translocation, mitochondrial accumulation of proteins mainly polyubiquitinated with Lys63-linked polyubiquitin and a small proportion of Lys48 linkage (Chan et al. 2011; Lee et al. 2010; Olzmann et al. 2007) recruits the ubiquitin- and LC3-binding adaptor protein, p62/sequestosome-1 (Geisler et al. 2010; Narendra et al. 2010; Okatsu et al. 2010), and the ubiquitin-binding deacetylase, histone deacetylase 6 (HDAC6) (Lee et al. 2010) (Fig. 12.2).

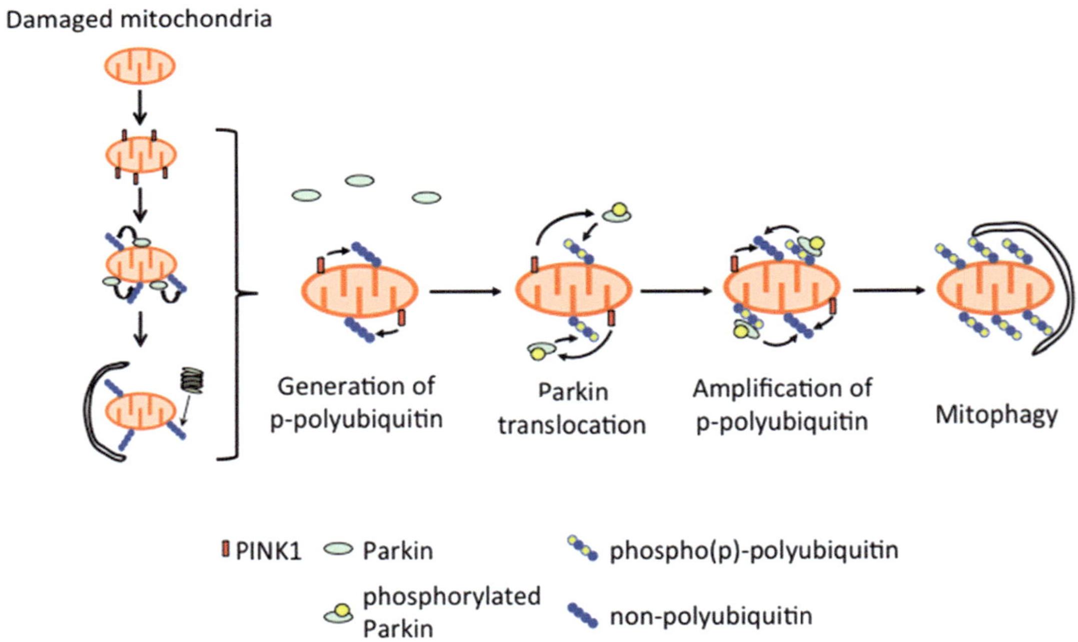

Fig. 12.1 PINK1-Parkin-mediated mitophagy. (1) Mitochondrial DNA (mtDNA) mutations, reactive oxygen species overproduction, and misfolded protein accumulation cause a reduction in mitochondrial membrane potential. (2) PINK1 is constitutively degraded under steady-state conditions and accumulates on damaged mitochondria. (3) Accumulated PINK1 activates itself, eliciting mitochondrial translocation and Parkin activation via phosphorylation. Activated Parkin ubiquitinates mitochondrial substrates. (4) Polyubiquitinated proteins on the outer mitochondrial membrane are degraded by the proteasome, with damaged mitochondria concurrently eliminated by mitophagy

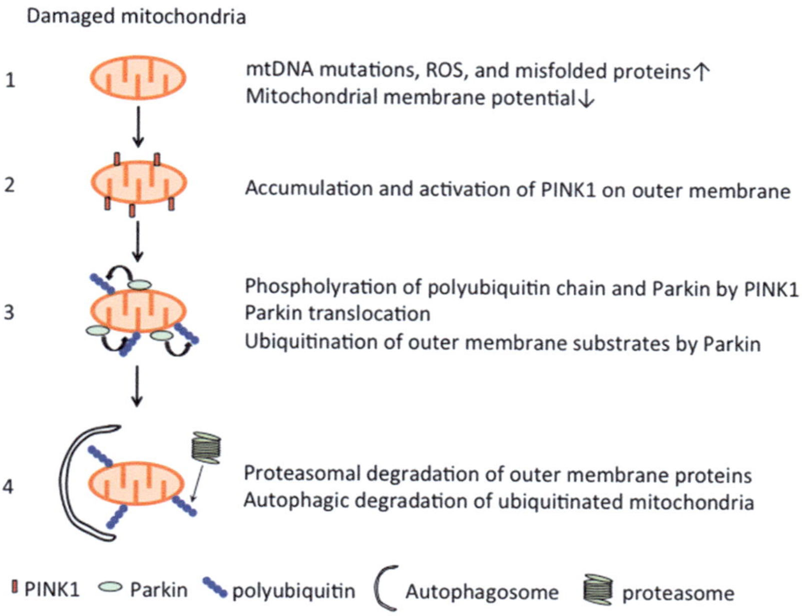

Fig. 12.2 Amplification of phospho-polyubiquitin chain production achieves rapid Parkin translocation and activation. Phospho-polyubiquitin chains on mitochondria are produced in collaboration with PINK1 and Parkin. The mechanism responsible for initial formation of ubiquitin chains on mitochondria remains unresolved. Ubiquitin chains may be attached to outer membrane proteins via mitochondrial ubiquitin ligases (other than Parkin). Alternatively, Parkin, which is activated by phospho-monoubiquitin in the cytosol, may attach ubiquitin chains to mitochondrial proteins. Phosphorylation of polyubiquitin chains by PINK1 promotes further Parkin activation and relocation to the mitochondrial outer membrane. In turn this amplifies generation of phospho-polyubiquitin chains and subsequently recruits autophagy machinery to ubiquitinated mitochondria (Arano and Imai 2015)

Depolarized mitochondria treated with CCCP or paraquat accumulate in the perinuclear compartment in a p62/SQSTM1-dependent manner (Geisler et al. 2010; Narendra et al. 2010; Okatsu et al. 2010), followed by activation of autophagic isolation and subsequent lysosomal degradation of mitochondria (Narendra et al. 2008) (Fig. 12.1). Clustering of ubiquitinated mitochondria by p62 and HDAC6 is reminiscent of sequestration of ubiquitinated proteins known as aggresomes (Komatsu et al. 2007; Pandey et al. 2007) and may facilitate isolation of damaged mitochondria from the healthy mitochondrial network and their efficient removal.

In mammalian tissue, Parkin translocation from the cytosol to mitochondria requires intact PINK1 with kinase activity and is an initial step in the mitophagy process (Geisler et al. 2010; Kawajiri et al. 2010; Matsuda and Tanaka 2010; Narendra et al. 2010; Vives-Bauza et al. 2010). Most pathogenic mutations in PINK1 and Parkin involve Parkin's mitochondrial translocation activity, which have been partly confirmed using neuronal cultures derived from induced pluripotent stem cells (iPSC) obtained from *PINK1*-linked PD cases (Seibler et al. 2011). Importantly, iPSC-derived neurons with Parkin mutations, but not fibroblasts or iPSCs with the same Parkin mutations, exhibit abnormal mitochondrial morphology and impaired mitochondrial homeostasis (Imaizumi et al. 2012). Thus, Parkin-mediated mitophagy may be closely associated with young-onset PD etiology caused by both *PINK1* and *Parkin* mutations.

A series of *Drosophila* genetic and cell biology studies clearly show that PINK1 is required for Parkin-mediated mitochondrial maintenance. Mitophagy of damaged mitochondria is a well-characterized event in which PINK1 and Parkin are involved, yet how PINK1 regulates Parkin is largely unclear. Recently, Ser65 in the ubiquitin-like domain of endogenous Parkin was shown to be phosphorylated in an activated PINK1-dependent manner (Shiba-Fukushima et al. 2012). Therefore, PINK1 is required for Parkin phosphorylation to activate mitophagy, while lack of phosphorylation in fibroblasts from a PARK6 patient shows its relevance to PD pathogenesis.

In *Drosophila*, mitochondria with accumulated misfolded proteins are also removed by Parkin (Pimenta de Castro et al. 2012). This phenomenon is reminiscent of ER-associated protein degradation (ERAD), which eliminates misfolded or unassembled proteins from the ER (Meyer et al. 2012). An AAA+ ATPase, p97/valosin-containing protein, is required for ERAD along with proteasome activity and involved in elimination of proteins on the outer mitochondrial membrane during mitophagy (Tanaka et al. 2010). Hence, the pathways for protein quality control and mitochondrial maintenance are likely to be linked. Although there is increasing evidence that with respect to PD, autophagy and mitophagy are critically linked to pathogenic mechanisms, at least under experimental conditions, the molecular mechanisms by which PINK1 and Parkin signal a commitment to elimination remain largely unclear. In particular, an important question currently being raised is whether mitophagy performs a protective role or whether dysregulation is in fact a common underlying cause of disease initiation. This is an important issue and its resolution may lead to identification of new genes associated with neurodegenerative disorders, as well as etiological understanding of idiopathic PD and *PINK1*- and *Parkin*-linked PD.

Upon loss of ΔΨm, cytosolic Parkin is recruited to mitochondria by PINK1 through an uncharacterized mechanism, an initial step that triggers sequential events for autophagic activation in mitophagy. In our previous study, we found that Ser65 in the ubiquitin-like domain of Parkin is phosphorylated by activated PINK1 upon ΔΨm depolarization and required for Parkin translocation to mitochondria. We also provided evidence that Ser65 phosphorylation is required, but not sufficient, for Parkin translocation by showing that pathogenic mutant forms of Parkin lacking mitochondrial translocation activity are still phosphorylated by PINK1 (Shiba-Fukushima et al. 2014). Our study partly uncovered the molecular mechanisms underlying mitochondrial translocation of Parkin in a PINK1-dependent manner as an initial step in mitophagy. Moreover, ubiquitin is the genuine substrate of PINK1. PINK1 phosphorylates ubiquitin at Ser65 both in vitro and in cells, while a Ser65 phosphopeptide derived from endogenous ubiquitin is only detected in cells in the presence of PINK1 following decreased ΔΨm. Furthermore, phosphomimetic ubiquitin accelerates discharge of a thioester conjugate formed by ubiquitin-conjugating enzyme E2 L3 (UBCH7) (also known as UBE2L3) and ubiquitin (UBCH7∼ubiquitin) in the presence of Parkin in vitro, indicating that it acts allosterically. Phosphorylation-dependent interaction between ubiquitin and Parkin suggests that phosphorylated ubiquitin unlocks autoinhibition of the catalytic cysteine. Koyano et al. showed that PINK1-dependent phosphorylation of both Parkin and ubiquitin is sufficient for full activation of Parkin E3 activity (Koyano et al. 2014). Although various models for the recruitment process have been proposed, none adequately explain the accumulating data, and thus the molecular basis for PINK1 recruitment of Parkin remains to be fully elucidated. Koyano et al. demonstrated that a linear ubiquitin chain of phosphomimetic tetra-ubiquitin (S65D) recruits Parkin to energized mitochondria in the absence of PINK1, whereas a wild-type tetra-ubiquitin chain does not. Under more physiologically relevant conditions, a lysosomal phosphorylated polyubiquitin chain recruits phosphomimetic Parkin to lysosomes. A cellular ubiquitin replacement system confirmed that ubiquitin phosphorylation is indeed essential for Parkin translocation to mitochondria (Okatsu et al. 2015). Furthermore, physical interactions between phosphomimetic Parkin and phosphorylated polyubiquitin chain were detected by immunoprecipitation from cells and in vitro reconstitution using recombinant proteins. Okatsu et al. proposed that the phosphorylated ubiquitin chain functions as the genuine Parkin receptor for recruitment to depolarized mitochondria (Okatsu et al. 2015).

In this regards, both molecules (Parkin and PINK1) collaborate for mitophagy, and it is possible that both molecules might function at MAM.

Cali et al. reported that Parkin overexpression in model cells physically and functionally enhances ER–mitochondria coupling, favoring calcium transfer from the ER to mitochondria following cell stimulation with an agonist that generates 1,4,5 inositol trisphosphate (InsP3), and increases agonist-induced ATP production (Cali et al. 2013b). Overexpression of a Parkin mutant lacking the ubiquitin-like domain failed to enhance mitochondrial calcium transients, thus highlighting the importance of the N-terminal ubiquitin-like domain in the observed phenotype. siRNA-mediated Parkin silencing causes mitochondrial fragmentation, impaired mitochondrial calcium handling, and reduced mitochondrial-ER tethering. These data support a novel role for Parkin in regulation of mitochondrial homeostasis, calcium signaling, and energy metabolism under physiological conditions. Moreover, no direct involvement for PINK1 in mitochondrial-ER crosstalk has been reported. However, it does play a role in mitochondrial motility, regulation of which also takes place at this interface. In addition to local calcium levels, mitochondrial motility is regulated through proteasomal breakdown of the transport complex, as PINK1 promotes proteasomal degradation of mitochondrial Rho GTPase (Miro)-1/2 and mitofusin (MFN)-1/2 in a Parkin-dependent manner (Gegg et al. 2010; Saotome et al. 2008). Intriguingly, loss of Miro was shown to rescue the phenotype caused by expression of PD-causing PINK1 mutations in *Drosophila* and facilitate clearance of damaged mitochondria in HeLa cells, suggesting this process plays a significant role in PD (Tsai et al. 2014). Further research is needed to investigate how contact between mitochondria and the ER is involved in this pathway and whether ubiquitination of MFN2 by Parkin can modulate these contacts, as reported for MITOL, a mitochondrial ubiquitin ligase (Gegg et al. 2010; Kim et al. 2015). DJ-1 may also be linked to MAM, as DJ-1 overexpression promotes MAM function and mitochondrial interaction (Ottolini et al. 2013).

12.5 Conclusions

The monogenic form of PD provides a hint toward elucidating PD pathogenesis. Among them, SNCA, Parkin, PINK1, and DJ-1 are directly linked to mitochondrial function and MAM. In addition, CHCHD2 (Funayama et al. 2015), a novel gene product associated with PD, is also involved in mitochondrial homeostasis. Accumulating evidence suggests that these PD-associated genes have multifaceted roles in mitochondria, including regulation of mitochondrial motility and quality as well as redox and respiration regulation.

References

Arano T, Imai Y (2015) Mitophagy regulated by the PINK1-Parkin pathway, vol Chapter 6. Cell death-Autophagy, Apotosis and Necrosis. InTech, Chapters published

Area-Gomez E, Del Carmen Lara Castillo M, Tambini MD, Guardia-Laguarta C, de Groof AJ, Madra M, Ikenouchi J, Umeda M, Bird TD, Sturley SL et al (2012) Upregulated function of mitochondria-associated ER membranes in Alzheimer disease. EMBO J 31:4106–4123

Auluck PK, Caraveo G, Lindquist S (2010)α-Synuclein: membrane interactions and toxicity in Parkinson's disease. Annu Rev Cell Dev Biol 26:211–233

Benecke R, Strumper P, Weiss H (1993) Electron transfer complexes I and IV of platelets are abnormal in Parkinson's disease but normal in Parkinson-plus syndromes. Brain 116:1451–1463

Berg D, Postuma RB, Adler CH, Bloem BR, Chan P, Dubois B, Gasser T, Goetz CG, Halliday G, Joseph L et al (2015) MDS research criteria for prodromal Parkinson's disease. Mov Disord 30:1600–1611

Bonifati V, Rizzu P, van Baren MJ, Schaap O, Breedveld GJ, Krieger E, Dekker MC, Squitieri F, Ibanez P, Joosse M et al (2003) Mutations in the DJ-1 gene associated with autosomal recessive early-onset parkinsonism. Science 299:256–259

Bui M, Gilady SY, Fitzsimmons RE, Benson MD, Lynes EM, Gesson K, Alto NM, Strack S, Scott JD, Simmen

T (2010) Rab32 modulates apoptosis onset and mitochondria-associated membrane (MAM) properties. J Biol Chem 285:31590–31602
Cali T, Ottolini D, Brini M (2013a) Calcium and endoplasmic reticulum-mitochondria tethering in neurodegeneration. DNA Cell Biol 32:140–146
Cali T, Ottolini D, Negro A, Brini M (2013b) Enhanced parkin levels favor ER-mitochondria crosstalk and guarantee Ca^{2+} transfer to sustain cell bioenergetics. Biochim Biophys Acta 1832:495–508
Chan NC, Salazar AM, Pham AH, Sweredoski MJ, Kolawa NJ, Graham RL, Hess S, Chan DC (2011) Broad activation of the ubiquitin-proteasome system by Parkin is critical for mitophagy. Hum Mol Genet 20:1726–1737
Chartier-Harlin MC, Kachergus J, Roumier C, Mouroux V, Douay X, Lincoln S, Levecque C, Larvor L, Andrieux J, Hulihan M et al (2004) α-synuclein locus duplication as a cause of familial Parkinson's disease. Lancet 364:1167–1169
Chinta SJ, Mallajosyula JK, Rane A, Andersen JK (2010) Mitochondrial α-synuclein accumulation impairs complex I function in dopaminergic neurons and results in increased mitophagy *in vivo*. Neurosci Lett 486:235–239
Choubey V, Safiulina D, Vaarmann A, Cagalinec M, Wareski P, Kuum M, Zharkovsky A, Kaasik A (2011) Mutant A53T α-synuclein induces neuronal death by increasing mitochondrial autophagy. J Biol Chem 286:10814–10824
Ciarlo L, Manganelli V, Garofalo T, Matarrese P, Tinari A, Misasi R, Malorni W, Sorice M (2010) Association of fission proteins with mitochondrial raft-like domains. Cell Death Differ 17:1047–1058
Cole NB, Dieuliis D, Leo P, Mitchell DC, Nussbaum RL (2008) Mitochondrial translocation of α-synuclein is promoted by intracellular acidification. Exp Cell Res 314:2076–2089
Cremades N, Cohen SI, Deas E, Abramov AY, Chen AY, Orte A, Sandal M, Clarke RW, Dunne P, Aprile FA et al (2012) Direct observation of the interconversion of normal and toxic forms of α-synuclein. Cell 149:1048–1059
Csordas G, Hajnoczky G (2009) SR/ER-mitochondrial local communication: calcium and ROS. Biochim Biophys Acta 1787:1352–1362
Csordas G, Renken C, Varnai P, Walter L, Weaver D, Buttle KF, Balla T, Mannella CA, Hajnoczky G (2006) Structural and functional features and significance of the physical linkage between ER and mitochondria. J Cell Biol 174:915–921
Cuervo AM, Stefanis L, Fredenburg R, Lansbury PT, Sulzer D (2004) Impaired degradation of mutant α-synuclein by chaperone-mediated autophagy. Science 305:1292–1295
Davidson WS, Jonas A, Clayton DF, George JM (1998) Stabilization of α-synuclein secondary structure upon binding to synthetic membranes. J Biol Chem 273:9443–9449
de Brito OM, Scorrano L (2008) Mitofusin 2 tethers endoplasmic reticulum to mitochondria. Nature 456:605–610
de Lau LM, Breteler MM (2006) Epidemiology of Parkinson's disease. Lancet Neurol 5:525–535
Devi L, Raghavendran V, Prabhu BM, Avadhani NG, Anandatheerthavarada HK (2008) Mitochondrial import and accumulation of α-synuclein impair complex I in human dopaminergic neuronal cultures and Parkinson disease brain. J Biol Chem 283:9089–9100
Dorsey ER, Constantinescu R, Thompson JP, Biglan KM, Holloway RG, Kieburtz K, Marshall FJ, Ravina BM, Schifitto G, Siderowf A et al (2007) Projected number of people with Parkinson disease in the most populous nations, 2005 through 2030. Neurology 68:384–386
Eliezer D, Kutluay E, Bussell R Jr, Browne G (2001) Conformational properties of α-synuclein in its free and lipid-associated states. J Mol Biol 307:1061–1073
Fortin DL, Troyer MD, Nakamura K, Kubo S, Anthony MD, Edwards RH (2004) Lipid rafts mediate the synaptic localization of α-synuclein. J Neurosci 24:6715–6723
Friedman JR, Lackner LL, West M, DiBenedetto JR, Nunnari J, Voeltz GK (2011) ER tubules mark sites of mitochondrial division. Science 334:358–362
Funayama M, Ohe K, Amo T, Furuya N, Yamaguchi J, Saiki S, Li Y, Ogaki K, Ando M, Yoshino H et al (2015) CHCHD2 mutations in autosomal dominant late-onset Parkinson's disease: a genome-wide linkage and sequencing study. Lancet Neurol 14:274–282
Garofalo T, Giammarioli AM, Misasi R, Tinari A, Manganelli V, Gambardella L, Pavan A, Malorni W, Sorice M (2005) Lipid microdomains contribute to apoptosis-associated modifications of mitochondria in T cells. Cell Death Differ 12:1378–1389
Gegg ME, Cooper JM, Chau KY, Rojo M, Schapira AH, Taanman JW (2010) Mitofusin 1 and mitofusin 2 are ubiquitinated in a PINK1/parkin-dependent manner upon induction of mitophagy. Hum Mol Genet 19:4861–4870
Geisler S, Holmstrom KM, Skujat D, Fiesel FC, Rothfuss OC, Kahle PJ, Springer W (2010) PINK1/Parkin-mediated mitophagy is dependent on VDAC1 and p62/SQSTM1. Nat Cell Biol 12:119–131
Guardia-Laguarta C, Area-Gomez E, Rub C, Liu Y, Magrane J, Becker D, Voos W, Schon EA, Przedborski S (2014) α-Synuclein is localized to mitochondria-associated ER membranes. J Neurosci 34:249–259
Hattori N, Tanaka M, Ozawa T, Mizuno Y (1991) Immunohistochemical studies on complexes I, II, III, and IV of mitochondria in Parkinson's disease. Ann Neurol 30:563–571
Hayashi T, Fujimoto M (2010) Detergent-resistant microdomains determine the localization of sigma-1 receptors to the endoplasmic reticulum-mitochondria junction. Mol Pharmacol 77:517–528
Hayashi T, Rizzuto R, Hajnoczky G, Su TP (2009) MAM: more than just a housekeeper. Trends Cell Biol 19:81–88

Hedskog L, Pinho CM, Filadi R, Ronnback A, Hertwig L, Wiehager B, Larssen P, Gellhaar S, Sandebring A, Westerlund M et al (2013) Modulation of the endoplasmic reticulum-mitochondria interface in Alzheimer's disease and related models. Proc Natl Acad Sci USA 110:7916–7921

Hsu LJ, Sagara Y, Arroyo A, Rockenstein E, Sisk A, Mallory M, Wong J, Takenouchi T, Hashimoto M, Masliah E (2000) α-synuclein promotes mitochondrial deficit and oxidative stress. Am J Pathol 157:401–410

Ibanez P, Bonnet AM, Debarges B, Lohmann E, Tison F, Pollak P, Agid Y, Durr A, Brice A (2004) Causal relation between α-synuclein gene duplication and familial Parkinson's disease. Lancet 364:1169–1171

Imaizumi Y, Okada Y, Akamatsu W, Koike M, Kuzumaki N, Hayakawa H, Nihira T, Kobayashi T, Ohyama M, Sato S et al (2012) Mitochondrial dysfunction associated with increased oxidative stress and α-synuclein accumulation in PARK2 iPSC-derived neurons and postmortem brain tissue. Mol Brain 5:35

Iwasawa R, Mahul-Mellier AL, Datler C, Pazarentzos E, Grimm S (2011) Fis1 and Bap31 bridge the mitochondria-ER interface to establish a platform for apoptosis induction. EMBO J 30:556–568

Jankovic J, Poewe W (2012) Therapies in Parkinson's disease. Curr Opin Neurol 25:433–447

Jao CC, Der-Sarkissian A, Chen J, Langen R (2004) Structure of membrane-bound α-synuclein studied by site-directed spin labeling. Proc Natl Acad Sci USA 101:8331–8336

Jao CC, Hegde BG, Chen J, Haworth IS, Langen R (2008) Structure of membrane-bound α-synuclein from site-directed spin labeling and computational refinement. Proc Natl Acad Sci USA 105:19666–19671

Kamp F, Exner N, Lutz AK, Wender N, Hegermann J, Brunner B, Nuscher B, Bartels T, Giese A, Beyer K et al (2010) Inhibition of mitochondrial fusion by α-synuclein is rescued by PINK1, Parkin and DJ-1. EMBO J 29:3571–3589

Karpinar DP, Balija MB, Kugler S, Opazo F, Rezaei-Ghaleh N, Wender N, Kim HY, Taschenberger G, Falkenburger BH, Heise H et al (2009) Pre-fibrillar α-synuclein variants with impaired β-structure increase neurotoxicity in Parkinson's disease models. EMBO J 28:3256–3268

Kawajiri S, Saiki S, Sato S, Sato F, Hatano T, Eguchi H, Hattori N (2010) PINK1 is recruited to mitochondria with parkin and associates with LC3 in mitophagy. FEBS Letters 584:1073–1079

Kiely AP, Asi YT, Kara E, Limousin P, Ling H, Lewis P, Proukakis C, Quinn N, Lees AJ, Hardy J et al (2013) α-Synucleinopathy associated with G51D SNCA mutation: a link between Parkinson's disease and multiple system atrophy? Acta Neuropathol 125:753–769

Kim HJ, Nagano Y, Choi SJ, Park SY, Kim H, Yao TP, Lee JY (2015) HDAC6 maintains mitochondrial connectivity under hypoxic stress by suppressing MARCH5/MITOL dependent MFN2 degradation. Biochem Biophys Res Commun 464:1235–1240

Kitada T, Asakawa S, Hattori N, Matsumine H, Yamamura Y, Minoshima S, Yokochi M, Mizuno Y, Shimizu N (1998) Mutations in the parkin gene cause autosomal recessive juvenile parkinsonism. Nature 392:605–608

Klionsky DJ, Abdalla FC, Abeliovich H, Abraham RT, Acevedo-Arozena A, Adeli K, Agholme L, Agnello M, Agostinis P, Aguirre-Ghiso JA et al (2012) Guidelines for the use and interpretation of assays for monitoring autophagy. Autophagy 8:445–544

Komatsu M, Waguri S, Koike M, Sou YS, Ueno T, Hara T, Mizushima N, Iwata J, Ezaki J, Murata S et al (2007) Homeostatic levels of p62 control cytoplasmic inclusion body formation in autophagy-deficient mice. Cell 131:1149–1163

Koyano F, Okatsu K, Kosako H, Tamura Y, Go E, Kimura M, Kimura Y, Tsuchiya H, Yoshihara H, Hirokawa T et al (2014) Ubiquitin is phosphorylated by PINK1 to activate parkin. Nature 510:162–166

Kruger R, Kuhn W, Muller T, Woitalla D, Graeber M, Kosel S, Przuntek H, Epplen JT, Schols L, Riess O (1998) Ala30Pro mutation in the gene encoding α-synuclein in Parkinson's disease. Nat Genet 18:106–108

Lee JY, Nagano Y, Taylor JP, Lim KL, Yao TP (2010) Disease-causing mutations in parkin impair mitochondrial ubiquitination, aggregation, and HDAC6-dependent mitophagy. J Cell Biol 189:671–679

Levine T, Rabouille C (2005) Endoplasmic reticulum: one continuous network compartmentalized by extrinsic cues. Curr Opin Cell Biol 17:362–368

Li WW, Yang R, Guo JC, Ren HM, Zha XL, Cheng JS, Cai DF (2007) Localization of α-synuclein to mitochondria within midbrain of mice. Neuroreport 18:1543–1546

Martin LJ, Pan Y, Price AC, Sterling W, Copeland NG, Jenkins NA, Price DL, Lee MK (2006) Parkinson's disease α-synuclein transgenic mice develop neuronal mitochondrial degeneration and cell death. J Neurosci 26:41–50

Matsuda N, Tanaka K (2010) Does impairment of the ubiquitin-proteasome system or the autophagy-lysosome pathway predispose individuals to neurodegenerative disorders such as Parkinson's disease? J Alzheimers Dis 19:1–9

Meyer H, Bug M, Bremer S (2012) Emerging functions of the VCP/p97 AAA-ATPase in the ubiquitin system. Nat Cell Biol 14:117–123

Mironov SL, Symonchuk N (2006) ER vesicles and mitochondria move and communicate at synapses. J Cell Sci 119:4926–4934

Mizuno Y, Ohta S, Tanaka M, Takamiya S, Suzuki K, Sato T, Oya H, Ozawa T, Kagawa Y (1989) Deficiencies in complex I subunits of the respiratory chain in Parkinson's disease. Biochem Biophys Res Commun 163:1450–1455

Multiple-System Atrophy Research C (2013) Mutations in COQ2 in familial and sporadic multiple-system atrophy. N Engl J Med 369:233–244

Nakamura K, Nemani VM, Azarbal F, Skibinski G, Levy JM, Egami K, Munishkina L, Zhang J, Gardner B, Wakabayashi J et al (2011) Direct membrane association drives mitochondrial fission by the Parkinson disease-associated protein α-synuclein. J Biol Chem 286:20710–20726

Narendra D, Tanaka A, Suen DF, Youle RJ (2008) Parkin is recruited selectively to impaired mitochondria and promotes their autophagy. J Cell Biol 183:795–803

Narendra D, Kane LA, Hauser DN, Fearnley IM, Youle RJ (2010) p62/SQSTM1 is required for Parkin-induced mitochondrial clustering but not mitophagy; VDAC1 is dispensable for both. Autophagy 6:1090–1106

Nishioka K, Hayashi S, Farrer MJ, Singleton AB, Yoshino H, Imai H, Kitami T, Sato K, Kuroda R, Tomiyama H et al (2006) Clinical heterogeneity of α-synuclein gene duplication in Parkinson's disease. Ann Neurol 59:298–309

Nussbaum RL, Ellis CE (2003) Alzheimer's disease and Parkinson's disease. N Engl J Med 348:1356–1364

Okatsu K, Saisho K, Shimanuki M, Nakada K, Shitara H, Sou YS, Kimura M, Sato S, Hattori N, Komatsu M et al (2010) p62/SQSTM1 cooperates with Parkin for perinuclear clustering of depolarized mitochondria. Genes Cells 15:887–900

Okatsu K, Koyano F, Kimura M, Kosako H, Saeki Y, Tanaka K, Matsuda N (2015) Phosphorylated ubiquitin chain is the genuine Parkin receptor. J Cell Biol 209:111–128

Olzmann JA, Li L, Chudaev MV, Chen J, Perez FA, Palmiter RD, Chin LS (2007) Parkin-mediated K63-linked polyubiquitination targets misfolded DJ-1 to aggresomes via binding to HDAC6. J Cell Biol 178:1025–1038

Ottolini D, Cali T, Negro A, Brini M (2013) The Parkinson disease-related protein DJ-1 counteracts mitochondrial impairment induced by the tumour suppressor protein p53 by enhancing endoplasmic reticulum-mitochondria tethering. Hum Mol Genet 22:2152–2168

Pandey UB, Nie Z, Batlevi Y, McCray BA, Ritson GP, Nedelsky NB, Schwartz SL, DiProspero NA, Knight MA, Schuldiner O et al (2007) HDAC6 rescues neurodegeneration and provides an essential link between autophagy and the UPS. Nature 447:859–863

Parihar MS, Parihar A, Fujita M, Hashimoto M, Ghafourifar P (2008) Mitochondrial association of α-synuclein causes oxidative stress. Cell Mol Life Sci 65:1272–1284

Pasanen P, Myllykangas L, Siitonen M, Raunio A, Kaakkola S, Lyytinen J, Tienari PJ, Poyhonen M, Paetau A (2014) Novel α-synuclein mutation A53E associated with atypical multiple system atrophy and Parkinson's disease-type pathology. Neurobiol Aging 35(2180):e1–e5

Pimenta de Castro I, Costa AC, Lam D, Tufi R, Fedele V, Moisoi N, Dinsdale D, Deas E, Loh SH, Martins LM (2012) Genetic analysis of mitochondrial protein misfolding in *Drosophila melanogaster*. Cell Death Differ 19:1308–1316

Pivovarova NB, Pozzo-Miller LD, Hongpaisan J, Andrews SB (2002) Correlated calcium uptake and release by mitochondria and endoplasmic reticulum of CA3 hippocampal dendrites after afferent synaptic stimulation. J Neurosci 22:10653–10661

Polymeropoulos MH, Lavedan C, Leroy E, Ide SE, Dehejia A, Dutra A, Pike B, Root H, Rubenstein J, Boyer R et al (1997) Mutation in the α-synuclein gene identified in families with Parkinson's disease. Science 276:2045–2047

Prause J, Goswami A, Katona I, Roos A, Schnizler M, Bushuven E, Dreier A, Buchkremer S, Johann S, Beyer C et al (2013) Altered localization, abnormal modification and loss of function of Sigma receptor-1 in amyotrophic lateral sclerosis. Hum Mol Genet 22:1581–1600

Proukakis C, Dudzik CG, Brier T, MacKay DS, Cooper JM, Millhauser GL, Houlden H, Schapira AH (2013) A novel α-synuclein missense mutation in Parkinson disease. Neurology 80:1062–1064

Rizzuto R, Pinton P, Carrington W, Fay FS, Fogarty KE, Lifshitz LM, Tuft RA, Pozzan T (1998) Close contacts with the endoplasmic reticulum as determinants of mitochondrial Ca^{2+} responses. Science 280:1763–1766

Rowland AA, Voeltz GK (2012) Endoplasmic reticulum-mitochondria contacts: function of the junction. Nat Rev Mol Cell Biol 13:607–625

Rusinol AE, Cui Z, Chen MH, Vance JE (1994) A unique mitochondria-associated membrane fraction from rat liver has a high capacity for lipid synthesis and contains pre-Golgi secretory proteins including nascent lipoproteins. J Biol Chem 269:27494–27502

Sala G, Stefanoni G, Arosio A, Riva C, Melchionda L, Saracchi E, Fermi S, Brighina L, Ferrarese C (2014) Reduced expression of the chaperone-mediated autophagy carrier hsc70 protein in lymphomonocytes of patients with Parkinson's disease. Brain Res 1546:46–52

Saotome M, Safiulina D, Szabadkai G, Das S, Fransson A, Aspenstrom P, Rizzuto R, Hajnoczky G (2008) Bidirectional Ca^{2+}-dependent control of mitochondrial dynamics by the Miro GTPase. Proc Natl Acad Sci USA 105:20728–20733

Schapira AH, Cooper JM, Morgan-Hughes JA, Patel SD, Cleeter MJ, Ragan CI, Clark JB (1988) Molecular basis of mitochondrial myopathies: polypeptide analysis in complex-I deficiency. Lancet 1:500–503

Schon EA, Area-Gomez E (2013) Mitochondria-associated ER membranes in Alzheimer disease. Mol Cell Neurosci 55:26–36

Schon EA, Przedborski S (2011) Mitochondria: the next (neurode)generation. Neuron 70:1033–1053

Seibler P, Graziotto J, Jeong H, Simunovic F, Klein C, Krainc D (2011) Mitochondrial Parkin recruitment is impaired in neurons derived from mutant PINK1 induced pluripotent stem cells. J Neurosci 31:5970–5976

Shavali S, Brown-Borg HM, Ebadi M, Porter J (2008) Mitochondrial localization of α-synuclein protein in α-synuclein overexpressing cells. Neurosci Lett 439:125–128

Shiba-Fukushima K, Imai Y, Yoshida S, Ishihama Y, Kanao T, Sato S, Hattori N (2012) PINK1-mediated phosphorylation of the Parkin ubiquitin-like domain primes mitochondrial translocation of Parkin and regulates mitophagy. Sci Rep 2:1002

Shiba-Fukushima K, Arano T, Matsumoto G, Inoshita T, Yoshida S, Ishihama Y, Ryu KY, Nukina N, Hattori N, Imai Y (2014) Phosphorylation of mitochondrial polyubiquitin by PINK1 promotes Parkin mitochondrial tethering. PLoS Genet 10:e1004861

Simons K, Toomre D (2000) Lipid rafts and signal transduction. Nat Rev Mol Cell Biol 1:31–39

Singleton AB, Farrer M, Johnson J, Singleton A, Hague S, Kachergus J, Hulihan M, Peuralinna T, Dutra A, Nussbaum R et al (2003) α-Synuclein locus triplication causes Parkinson's disease. Science 302:841

Stoica R, De Vos KJ, Paillusson S, Mueller S, Sancho RM, Lau KF, Vizcay-Barrena G, Lin WL, Xu YF, Lewis J et al (2014) ER-mitochondria associations are regulated by the VAPB-PTPIP51 interaction and are disrupted by ALS/FTD-associated TDP-43. Nat Commun 5:3996

Tabas I, Ron D (2011) Integrating the mechanisms of apoptosis induced by endoplasmic reticulum stress. Nat Cell Biol 13:184–190

Tanaka A, Cleland MM, Xu S, Narendra DP, Suen DF, Karbowski M, Youle RJ (2010) Proteasome and p97 mediate mitophagy and degradation of mitofusins induced by Parkin. J Cell Biol 191:1367–1380

Tsai PI, Course MM, Lovas JR, Hsieh CH, Babic M, Zinsmaier KE, Wang X (2014) PINK1-mediated phosphorylation of Miro inhibits synaptic growth and protects dopaminergic neurons in Drosophila. Sci Rep 4:6962

Valente EM, Abou-Sleiman PM, Caputo V, Muqit MM, Harvey K, Gispert S, Ali Z, Del Turco D, Bentivoglio AR, Healy DG et al (2004) Hereditary early-onset Parkinson's disease caused by mutations in PINK1. Science 304:1158–1160

Vives-Bauza C, de Vries RL, Tocilescu M, Przedborski S (2010) PINK1/Parkin direct mitochondria to autophagy. Autophagy 6:315–316

Winner B, Jappelli R, Maji SK, Desplats PA, Boyer L, Aigner S, Hetzer C, Loher T, Vilar M, Campioni S et al (2011) In vivo demonstration that α-synuclein oligomers are toxic. Proc Natl Acad Sci USA 108:4194–4199

Wood-Kaczmar A, Gandhi S, Wood NW (2006) Understanding the molecular causes of Parkinson's disease. Trends Mol Med 12:521–528

Yabu T, Imamura S, Mohammed MS, Touhata K, Minami T, Terayama M, Yamashita M (2011) Differential gene expression of HSC70/HSP70 in yellowtail cells in response to chaperone-mediated autophagy. FEBS J 278:673–685

Yoshino H, Nakagawa-Hattori Y, Kondo T, Mizuno Y (1992) Mitochondrial complex I and II activities of lymphocytes and platelets in Parkinson's disease. J Neural Transm Park Dis Dement Sect 4:27–34

Zarranz JJ, Alegre J, Gomez-Esteban JC, Lezcano E, Ros R, Ampuero I, Vidal L, Hoenicka J, Rodriguez O, Atares B et al (2004) The new mutation, E46K, of α-synuclein causes Parkinson and Lewy body dementia. Ann Neurol 55:164–173

Zhang L, Zhang C, Zhu Y, Cai Q, Chan P, Ueda K, Yu S, Yang H (2008) Semi-quantitative analysis of α-synuclein in subcellular pools of rat brain neurons: an immunogold electron microscopic study using a C-terminal specific monoclonal antibody. Brain Res 1244:40–52

Zigoneanu IG, Yang YJ, Krois AS, Haque E, Pielak GJ (2012) Interaction of α-synuclein with vesicles that mimic mitochondrial membranes. Biochim Biophys Acta 1818:512–519

13 Role of Endoplasmic Reticulum-Mitochondria Communication in Type 2 Diabetes

Jennifer Rieusset

Abstract

Although mitochondrial dysfunction, endoplasmic reticulum (ER) stress, and disrupted lipid and calcium (Ca^{2+}) homeostasis are classically associated with both insulin resistance and β-cell dysfunction in type 2 diabetes mellitus (T2DM), the interplay between these metabolic stresses is less known. Both organelles interact through contact points known as mitochondria-associated membranes (MAM), in order to exchange both lipids and Ca^{2+} and regulate cellular homeostasis. Recent evidences suggest that MAM could be an important hub for hormonal and nutrient signaling in the liver and that ER-mitochondria miscommunication could participate to hepatic insulin resistance, highlighting the importance of MAM in the control of glucose homeostasis. Here, we specifically discuss the role of MAM in hormonal and nutrient-regulated signaling pathways supporting a role in the control of glucose homeostasis and analyze the evidences pointing a role of ER-mitochondria miscommunication in T2DM. Collectively, these data suggest that targeting MAM structure and function might be a novel strategy for the treatment of T2DM.

Keywords

Mitochondria • Endoplasmic reticulum (ER) • Organelle communication • Mitochondria-associated membranes (MAM) • Calcium homeostasis • Glucose homeostasis • Insulin resistance • β-Cell dysfunction • Type 2 diabetes mellitus (T2DM)

13.1 Introduction

Type 2 diabetes mellitus (T2DM) is a complex endocrine and metabolic disorder associating peripheral insulin resistance and altered insulin

J. Rieusset (✉)
Faculté de médecine Lyon-Sud, UMR INSERM U1060, CarMeN Laboratory, Lyon 1 University, INRA U1397, 165 chemin du grand Revoyet, BP12, F-69921 Oullins Cedex, France
e-mail: jennifer.rieusset@univ-lyon1.fr

M. Tagaya, T. Simmen (eds.), *Organelle Contact Sites*, Advances in Experimental Medicine and Biology 997, DOI 10.1007/978-981-10-4567-7_13

secretion by pancreatic β-cells, thus leading to hyperglycemia. Its incidence has increased dramatically during the last decades, and now T2DM is considered as a global epidemic with an estimated worldwide prevalence of 415 million people in 2015 and an estimated rise to 642 million people by 2040 (Kharroubi and Darwish 2015). The epidemic of obesity is a major risk factor for the development of insulin resistance and T2DM (Kahn et al. 2006). Therefore, T2DM is responsible for enormous health and social and economic burdens, thus representing a major challenge for health-care systems all around the world. Peripheral insulin resistance precedes by several years the development of T2DM (De Fronzo and Tripathy 2009). Then, T2DM emerges when β-cells can no longer secrete enough insulin to overcome peripheral insulin resistance. Therefore, a better understanding of molecular mechanisms underlying defects in both insulin action and secretion is crucial for proposing new preventive and/or therapeutic approaches. The challenge for the coming years is to identify molecular targets shared by both peripheral tissues and pancreatic islets, in order to uncover more efficient treatments targeting the two components of T2DM.

Alterations of intracellular organelles such as the endoplasmic reticulum (ER) (Salvadó et al. 2015) and mitochondria (Montgomery and Turner 2015), as well as disruption of both lipid (Stinkens et al. 2015) and calcium (Ca^{2+}) (Arruda and Hotamisligil 2015) homeostasis, have been commonly associated with dysfunctions of both peripheral tissues and pancreas, highlighting the importance of organelle stress in the physiopathology of T2DM. Since ER and mitochondria are known to interact at sites named mitochondria-associated membranes (MAM) in order to exchange lipids and Ca^{2+} and to regulate cellular homeostasis (Giorgi et al. 2015), miscommunication between both organelles may be a new mechanism underlying T2DM.

In this review, we briefly remind the metabolic and cellular pathways controlling glucose homeostasis and their alterations in the context of T2DM. Next, we highlight the emerging role of ER-mitochondria contacts in hormonal and cellular signaling required for glucose homeostasis and discuss recent data pointing ER-mitochondria miscommunication as a major contributor of T2DM.

13.2 Control of Glucose Homeostasis and Disruption in T2DM

13.2.1 Hormonal Regulation of Glucose Homeostasis

Glucose is an essential metabolic substrate of all mammalian cells, which represents an important source of energy for most tissues and organs. Some of them, such as the brain, depend solely on glucose for fuel and require a steady supply of glucose at all times. The process of maintaining blood glucose at a steady-state level (4–6 mM) is called glucose homeostasis and is finely orchestrated by hormonal regulations, exemplified by insulin and glucagon (Hers 1990). Following a meal, the increase of circulating glucose levels induces insulin secretion in pancreatic β-cells. Then, insulin controls glucose homeostasis by inhibiting hepatic glucose production and stimulating glucose utilization by skeletal muscle and white adipose tissue (WAT). In addition, insulin inhibits triglyceride hydrolysis in WAT, preventing the liberation of circulating free fatty acids (FFA). At the opposite, during fasting periods, glucagon secretion by pancreatic α-cells stimulates gluconeogenesis and glycogenolysis leading to hepatic glucose release.

13.2.2 Altered Glucose Homeostasis in T2DM

Hyperglycemia in T2DM results from both peripheral insulin resistance and altered insulin secretion by pancreatic β-cells. Insulin resistance is defined as the failure of cells to respond normally to insulin's glucose-lowering effects. In skeletal muscle, insulin resistance induces a reduction of glucose uptake (Carnagarin et al. 2015), whereas, in the liver, altered insulin signaling leads to greater glycogenolysis and gluconeogenesis (Roden and Bernroider 2003), thereby raising blood glucose levels. In WAT, insulin resistance leads to increased lipolysis, resulting in elevated circulating FFA levels (De Fronzo 2004). Increased FFA levels favor ectopic lipid accumulation in the liver and skeletal muscle, further exacerbating insulin resistance, and induce hepatic triglyceride synthesis favoring hyperlipidemia (Samuel and Shulman 2016). In order to compensate peripheral insulin resistance, pancreatic β-cells secrete more and more insulin leading progressively to an exhaustion of the pancreas. Therefore, disruption of glucose-stimulated insulin secretion by β-cells ultimately leads to disrupted glucose homeostasis.

13.2.2.1 Alterations of Insulin Action in Peripheral Tissues

At the cellular level, insulin mediates its action by binding to its plasma membrane receptor and activating a cascade of protein phosphorylation. Briefly, insulin binding to its receptor promotes receptor autophosphorylation and activation. The activated receptor is then able to phosphorylate the insulin receptor substrates 1 and 2 (IRS1 and 2) that induces the recruitment and activation of the phosphatidylinositol 3-kinase (PI3K). Subsequent activation of protein kinase B (PKB), atypical protein kinases C (PKCs), and mammalian target of rapamycin (mTOR) pathways mediates then the anabolic actions of insulin.

Molecular mechanisms of insulin resistance are complex and rather unclear, and even some common threads have been highlighted (Samuel and Shulman 2016). Particularly, an increased serine phosphorylation of IRS1/IRS2 is implicated in the inhibition of insulin signaling (Copps and White 2012). This serine phosphorylation may reduce the ability of IRS proteins to interact either with the cytosolic domain of the insulin receptor (Paz et al. 1997) or with the downstream PI3K (Mothe and Van Obberghen 1996), thus disrupting insulin signals. In agreement, mutations of key serines of IRS1 (Ser302, Ser307, and Ser612) prevented the development of diet-induced muscle insulin resistance in mice (Morino et al. 2008). Phosphorylation on serine residues of IRS1/IRS2 is mediated by several serine/threonine kinases, including mTOR complex 1, c-JUN N-terminal kinase, IkB kinase, protein kinase C, and the double-stranded RNA-activated protein kinase. Several metabolic stresses in peripheral tissues are able to activate these serine/threonine kinases, including lipid accumulation, mitochondrial dysfunction, oxidative stress, ER stress, and inflammation (Samuel and Shulman 2016).

13.2.2.2 Alterations of the Insulin Secretion by β-Cells of the Pancreas

Insulin release from pancreatic β-cells is crucial to maintain normal glucose homeostasis and depends on both β-cell mass and function. Pancreatic β-cell mass is regulated by the balance of neogenesis, proliferation, hypertrophy, and apoptosis (Ackermann and Gannon 2006). In healthy β-cells, glucose stimulates insulin secretion through mechanisms involving the regulation of intracellular energy level and the regulation of ion channels (ATP-sensitive K^+ channels, voltage-dependent Ca^{2+} channels). In T2DM, both pancreatic β-cell loss and abnormal insulin secretion are involved in altered glucose homeostasis (Halban et al. 2014). Multiple pathways underlie decreased β-cell function and mass, including inflammatory stress, ER stress, oxidative stress, amyloid stress, and loss of islet cell integrity (Halban et al. 2014).

13.2.2.3 Organelle Stress in T2DM

Mitochondrial dysfunction and ER stress have been largely and independently associated with the loss of both insulin action and secretion in T2DM. These independent roles of each organelle stress in T2DM have been extensively reviewed and will not be covered in detail (for review, see Wang et al. 2015; Chang et al. 2015; Rieusset 2015; Salvadó et al. 2015; Hasnain et al. 2016). Briefly, mitochondria dysfunction alters insulin action and secretion, mainly via subsequent oxidative stress and/or lipid accumulation, whereas ER stress mediates these effects through the activation of the unfolded response (UPR). In addition, both organelles are able to activate inflammatory pathways (Hummasti and Hotamisligil 2010, Chaudhari et al. 2014), further inhibiting insulin action and secretion through inflammatory responses.

Importantly, ER and mitochondria are no longer considered as individual organelles in the cell as they are structurally and functionally linked through contact points defined as MAM (for review, see Giorgi et al. 2015; Marchi et al. 2014; van Vliet et al. 2014; Naon and Scorrano 2014). Therefore, our vision of organelle stress involvement in T2DM needs to evolve in order to integrate this subcellular communication. In this context, ER-mitochondria miscommunication might be a unifying mechanism promoting mitochondria dysfunction, ER stress, altered lipid, and Ca^{2+} homeostasis and leading to alterations of both insulin action and secretion in T2DM (Fig. 13.1). Recently, a new function of MAM in metabolic health has emerged (Lopez-Crisosto et al. 2015), further confirming the importance of MAM in metabolic diseases. Consequently, targeting MAM structure and function might be a new and interesting strategy to improve glucose homeostasis in T2DM.

13.3 Role of ER-Mitochondria Communication in Hormonal and Cellular Signaling

Through their connection, ER and mitochondria reciprocally transmit signals and control their function. Beyond organelle homeostasis, MAM also control several cellular signaling pathways (van Vliet et al. 2014), further highlighting the importance of ER-mitochondria communication in cellular homeostasis. Particularly, new functions of MAM in hormonal and nutrient-regulated signaling pathways have recently emerged in the liver. In this section, after briefly introducing MAM structure, we focus on these emerged roles of MAM in metabolic pathways and their relevance in the control of glucose homeostasis (Fig. 13.2).

13.3.1 Organization and Function of ER-Mitochondria Contact Sites

The physical interactions between ER and mitochondria are rather frequent in the cell and can be dynamically regulated (Rizzuto et al. 1998). The space between the ER and the outer mitochondrial membranes is small, up to 10 nm at the smooth ER and 25 nm at the rough ER (Csordás et al. 2006). Organelle contact site does not involve membrane fusion, but is mediated through protein bridges (Csordás et al. 2006). The proteinaceous nature of MAM is illustrated by the interaction between the voltage-dependent anion channel (VDAC) at the outer mitochondrial membrane and the inositol 1,4,5-trisphosphate receptor (IP3R) at the ER through the molecular chaperone glucose-regulated protein 75 (Grp75), allowing Ca^{2+} transfer from the ER to mitochondria (Szabadkai et al. 2006). In addition, MAM are enriched in phospholipid enzymes supporting their role as a phospholipid exchange platform (Vance 2014). Several ER chaperones, such as sigma-1 receptor (SigR1) (Hayashi and Su 2007) and calnexin (Lynes et al. 2012), are enriched in MAM, and the recent discovery that cyclophilin D (CypD) is also present at MAM interface where it regulates inter-organelle Ca^{2+} exchange (Paillard et al. 2013; Rieusset et al. 2016) supports a role of mitochondrial chaperones in Ca^{2+} signaling at MAM interface. Nevertheless, the list of MAM proteins and their widely varied functions has increased dramatically in recent years, highlighting the new role of MAM in cellular signaling (van Vliet et al. 2014). In agreement, recent proteomic analysis of MAM fractions has highlighted the huge number of proteins in this fraction and their involvement in diverse cellular processes (Poston et al. 2011; Poston et al. 2013; Horner et al. 2015). They have also pointed the tissue specificity of MAM composition as only 30% of proteins are in common between MAM from the brain and liver (Poston et al. 2013). Lastly, caveolin-1 was recently revealed as an integral component of hepatic MAM, which controls

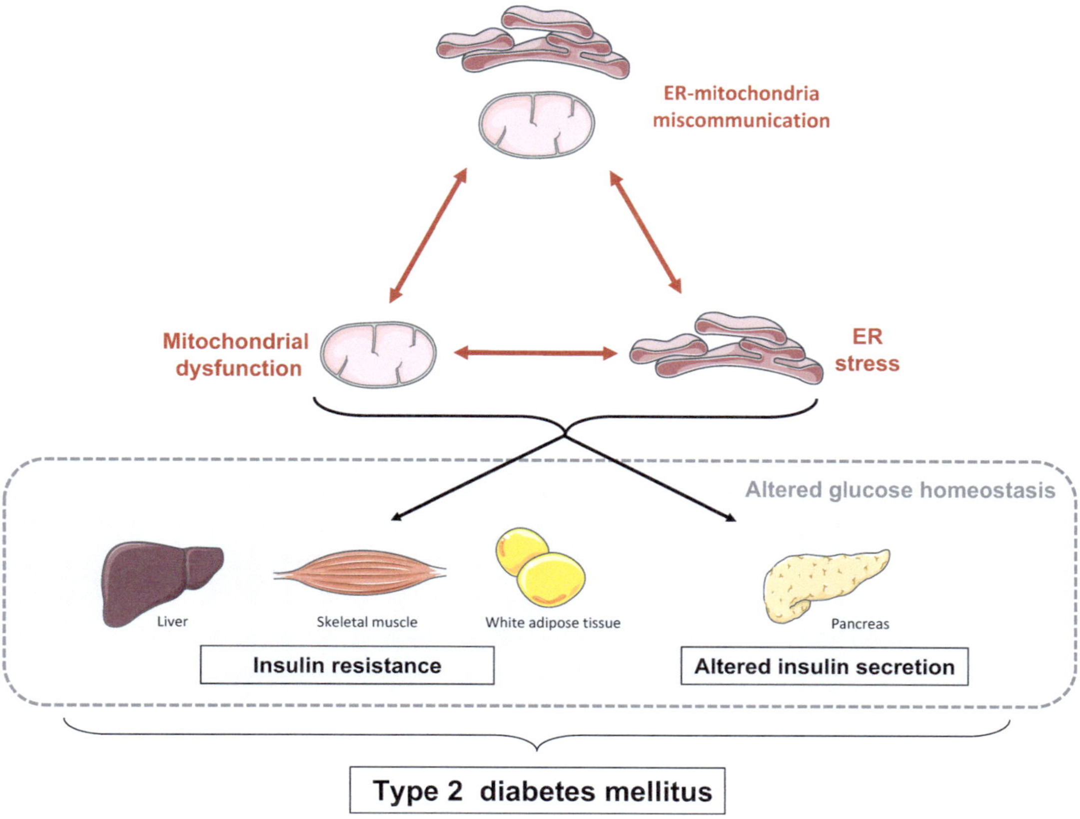

Fig. 13.1 Proposed model for the role of ER-mitochondria miscommunication in T2DM. A strong interplay between ER-mitochondria interactions and their respective dysfunction are involved in the alterations of both insulin action and secretion, leading to disrupted glucose homeostasis and T2DM

their relative cholesterol content (Sala-Vila et al. 2016), pointing at a role of MAM in lipid metabolism beyond phospholipids. Until now, the most described functions of ER-mitochondria interactions are lipid biosynthesis, Ca^{2+} transfer, and response to cellular stress. These functions have been largely reviewed (Giorgi et al. 2015; Marchi et al. 2014, van Vliet et al. 2014; Naon and Scorrano 2014)) and will not be described here. More interestingly in the context of metabolic diseases is their new function in hormonal and nutrient signaling, described below.

13.3.2 Role of MAM in Insulin Signaling

It has been known for a long time that Ca^{2+} homeostasis controls hepatic metabolism (Bartlett et al. 2014). Recently, it was further shown that IP3R was both a modulator and modulated by insulin and hormonal pathways (Wang et al. 2012), supporting a potential role of MAM in hormonal signaling. In agreement, several proteins of insulin signaling pathway are located at the MAM interface, including the protein kinase Akt, the protein phosphatase 2A (PP2A), the mTOR complex 2 (mTORC2), and the phosphatase and tensin homolog (PTEN). Akt is present at the ER-mitochondria interface (Betz et al. 2013; Giorgi et al. 2010; Tubbs et al. 2014), where it phosphorylates and inhibits IP3R, thus reducing Ca^{2+} release and preventing apoptosis (Giorgi et al. 2010). Importantly, insulin stimulation increases Akt phosphorylation at MAM interface (Tubbs et al. 2014), whereas it is actually unclear whether Akt is phosphorylated directly at MAM or localized at MAM after its

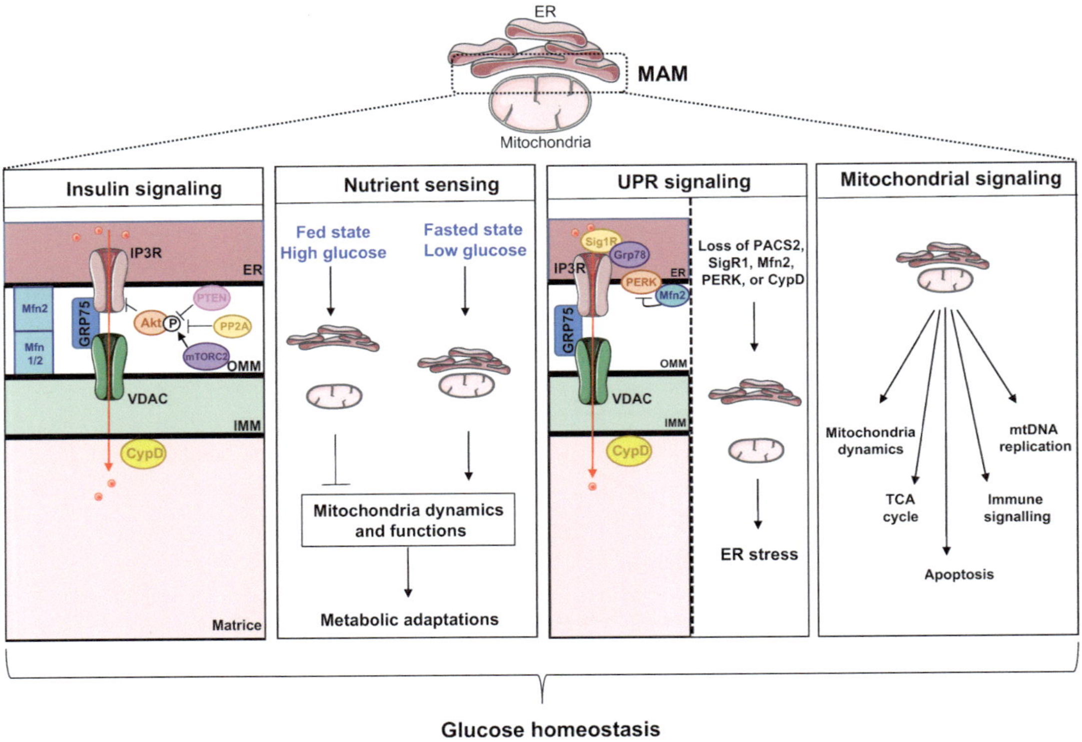

Fig. 13.2 Connections between MAM functions and metabolism. The roles of ER-mitochondria interactions in insulin signaling, nutrient sensing, UPR signaling, and in mitochondria physiology highlight the importance of MAM in metabolism and glucose homeostasis

phosphorylation in the cytoplasm. Its activity at MAM interface is controlled by PP2A, which is also present in MAM fractions (Giorgi et al. 2010). The dephosphorylation of Akt by PP2A counteracts its inhibitory action on IP3R, thus increasing Ca^{2+} transfer from ER to mitochondria (Giorgi et al. 2010). Furthermore, mTORC2, a kinase phosphorylating and activating Akt (Hresko and Mueckler 2005), was also found at MAM interface, and its amount was increased in response to growth factors or insulin stimulation (Betz et al. 2013). Interestingly, mTORC2 at MAM controls Akt and its targets to ultimately control MAM integrity, Ca^{2+} release, and mitochondrial physiology (Betz et al. 2013). Lastly, the tumor suppressor PTEN is also enriched in MAM, where it sensitizes cells to apoptosis by counteracting Akt-mediated phosphorylation and inhibition of IP3R, thus restoring ER Ca^{2+} release (Bononi et al. 2013). Nevertheless, whereas MAM appear as an important hub for Akt signaling, the involvement of ER-mitochondria coupling in the control of insulin action was unknown. Recently, our laboratory demonstrated that MAM integrity is required for insulin signaling in the liver (Tubbs et al. 2014). Experimental disruption of MAM alters insulin signaling and action in hepatocytes, whereas reinforcement of MAM by overexpressing MAM proteins enhances it. In agreement, mitochondrial Ca^{2+} uptake is critical for effective insulin signaling in both skeletal myocytes (del campo et al. 2014) and cardiomyocytes (Gutiérrez et al. 2014). Alternatively, cellular knockout (KO) of key proteins of insulin signaling, such as Akt or mTORC2 (Betz et al. 2013), or disruption of insulin signaling by palmitate treatment (Tubbs et al. 2014) disrupted MAM integrity. This suggests that the relationship between MAM integrity and insulin signaling is reciprocal, even if the molecular mechanism is actually unknown. Whether

MAM control the canonical insulin signaling pathways in the cytosol or whether insulin signaling needs to transit by MAM interface to regulate metabolism should be determined in the future.

13.3.3 Role of MAM in Sensing Nutrient Shift

Whereas both mitochondria (Gao et al. 2014) and ER (Mandl et al. 2009) are independently considered as nutrient sensors regulating cellular metabolism, recent data highlight the fact that MAM are dynamic structures that respond to the energy state of the cell. Starvation-induced autophagy is a cellular catabolic process degrading cellular constituents in order to generate energy in period of scarcity (Kaur and Debnath 2015). Interestingly, disruption of MAM inhibits starvation-induced autophagy by inhibiting the phosphatidylserine transfer from ER to mitochondria-derived autophagosomes (Hailey et al. 2010). Recently, it was further demonstrating that autophagosomes form at MAM and that different proteins of autophagy were enriched at MAM after starvation (Hamasaki et al. 2013), supporting the importance of ER-mitochondria communication in starvation-induced processes. In addition, it was recently shown that ER-mitochondria doubled in length in the liver when nutrients become limiting (Sood et al. 2014), suggesting that the liver could adapt to metabolic transitions through a mechanism dependent of MAM. In agreement, our laboratory recently demonstrated that MAM integrity is regulated in the liver during nutritional transition, with a reduction of ER-mitochondria interactions after feeding (Theurey et al. 2016). Furthermore, we demonstrated both in vitro and in vivo that increasing glucose levels can reproduce the effects of feeding on MAM, pointing toward glucose as a major regulator of MAM integrity during nutritional transition. At the molecular level, we revealed that high glucose levels disrupt MAM integrity and function through the activation of the pentose phosphate (PP)-PP2A pathway (Theurey et al. 2016). The presence of PP2A at MAM interface (Giorgi et al. 2010) supports the importance of PP2A in the control of MAM integrity/function. Therefore, glucose-mediated regulation of ER-mitochondria interactions could depend on phosphorylation state of yet to be identified substrates of PP2A on the MAM. Importantly, we found that the glucose sensing by MAM is crucial for the regulation of mitochondrial dynamics and function in the liver. Indeed, glucose-mediated reduction of MAM induced mitochondria fission and impaired respiration. Altogether, these data point to MAM as a glucose sensor adapting to cellular bioenergetics, likely contributing to the adaptive fuel partitioning during nutritional transition. This capacity to switch from lipid to glucose oxidation during the transition from a fasted to a fed state is called metabolic flexibility. As metabolic flexibility is important to avoid ectopic lipid accumulation and the development of insulin resistance (Galgani et al. 2008), the capacity of MAM to connect energy sensing to mitochondria physiology could be important for the control of glucose homeostasis.

13.3.4 Role of MAM in UPR Signaling

The close relationship between MAM and UPR signaling was first suggested by the observation that loss of the MAM protein PACS2 disrupted ER-mitochondria interactions and induced ER stress (Simmen et al. 2005). In agreement, disturbance in MAM integrity following loss of other MAM proteins, such as SigR1 (Hayashi and Su 2007), Mfn2 (Sebastián et al. 2012), or CypD (Rieusset et al. 2016), also induced ER stress. Conversely, it was demonstrated that early phases of ER stress increased ER-mitochondria coupling to promote mitochondrial respiration and bioenergetics, whereas massive and/or prolonged mitochondrial Ca^{2+} accumulation induced swelling and dysfunction of the organelles (Bravo et al. 2011). PERK, an important ER stress sensor in the UPR pathway, was shown to be localized at MAM and to physically increase the contact between both organelles

(Verfaillie et al. 2012). Furthermore, PERK activity is regulated by mitofusin 2 (Mfn2) (Muñoz et al. 2013), an important tether at MAM interface (de Brito and Scorrano 2008) which has been associated with ER stress responses (Ngoh et al. 2012; Sebastián et al. 2012). Lastly, the ER protein Sig-1R also forms a complex with another protein of UPR, called GRP78; upon ER Ca^{2+} depletion, dissociation of this complex was shown to prolong Ca^{2+} signaling from the ER to mitochondria via IP3R at MAM (Hayashi and Su 2007). Increasing Sig-1R expression in cells counteracts ER stress, thus inhibiting apoptosis (Hayashi and Su 2007). As UPR signaling controls insulin action (Salvadó et al. 2015), MAM could therefore regulate glucose homeostasis through its effects on UPR signaling.

13.3.5 Role of MAM in Mitochondria Signaling

Several aspects of mitochondria function are controlled by ER-mitochondria interactions, including oxidative metabolism, mitochondria dynamics, apoptosis, and antiviral signaling, which could consequently impact glucose homeostasis. Most of these functions are intimately connected to the Ca^{2+} status of mitochondria, since IP3R-mediated Ca^{2+} transfer from ER to mitochondria regulates TCA cycle enzyme activity (Denton 2009), the activity of the mitochondrial fission protein DRP1 (Hoppins and Nunnari 2012), apoptosis (Rizzuto et al. 1998), and NLRP3 inflammasome activation (Lee et al. 2012). Beyond Ca^{2+}, ER-mitochondria contact sites also support ROS-mediated signals that influence mitochondria function. Indeed, loss of PERK induced MAM disruption and reduced ROS transfer from ER to mitochondria, protecting cells from ROS-mediated mitochondrial apoptosis (Verfaillie et al. 2012). Accordingly, MAM are particularly enriched in ER folding chaperones and oxidoreductases, which regulate Ca^{2+} flux from the ER through reversible, calcium- and redox-dependent interactions with IP3R and therefore signaling with mitochondria (Gilady et al. 2010; Simmen et al. 2010). Very recently, ER-mitochondria contacts were shown to coordinate the licensing of mtDNA replication with division to distribute newly replicated nucleoids to daughter mitochondria, highlighting a new role of MAM in mitochondria physiology (Lewis et al. 2016). However, its involvement in the control of glucose homeostasis is currently unknown.

13.4 Miscommunication Between ER and Mitochondria in Type 2 Diabetes

Recent evidences suggest a role of ER-mitochondria miscommunication in hepatic insulin resistance, but it is actually unknown whether it could also participate to alterations of insulin action in other tissues or to dysfunction of β-cells. In this section, we mainly focus on the role of MAM disruption in hepatic metabolic alterations, trying to generalize to other tissues (Fig. 13.3).

13.4.1 Disruption of Calcium Homeostasis in Hepatic Insulin Resistance

MAM is an important hub of Ca^{2+} signaling, and altered Ca^{2+} homeostasis in the context of hepatic insulin resistance supports a potential role of ER-mitochondria miscommunication in T2DM. It has been shown that obesity-induced dysfunctional ER Ca^{2+} transport is an important cause of ER stress in the liver (Fu et al. 2011). Furthermore, an increase in cytosolic Ca^{2+} is also associated with obesity and insulin resistance and is responsible for Ca^{2+}/calmodulin-dependent protein kinase II (CaMKII) hyperphosphorylation, JNK activation, inflammatory signaling, and increased hepatic glucose production (Ozcan et al. 2012). CaMKII also impacts insulin signaling by modulating PERK and subsequently inhibiting insulin-mediated Akt phosphorylation (Ozcan et al. 2013). Interestingly, genetic and

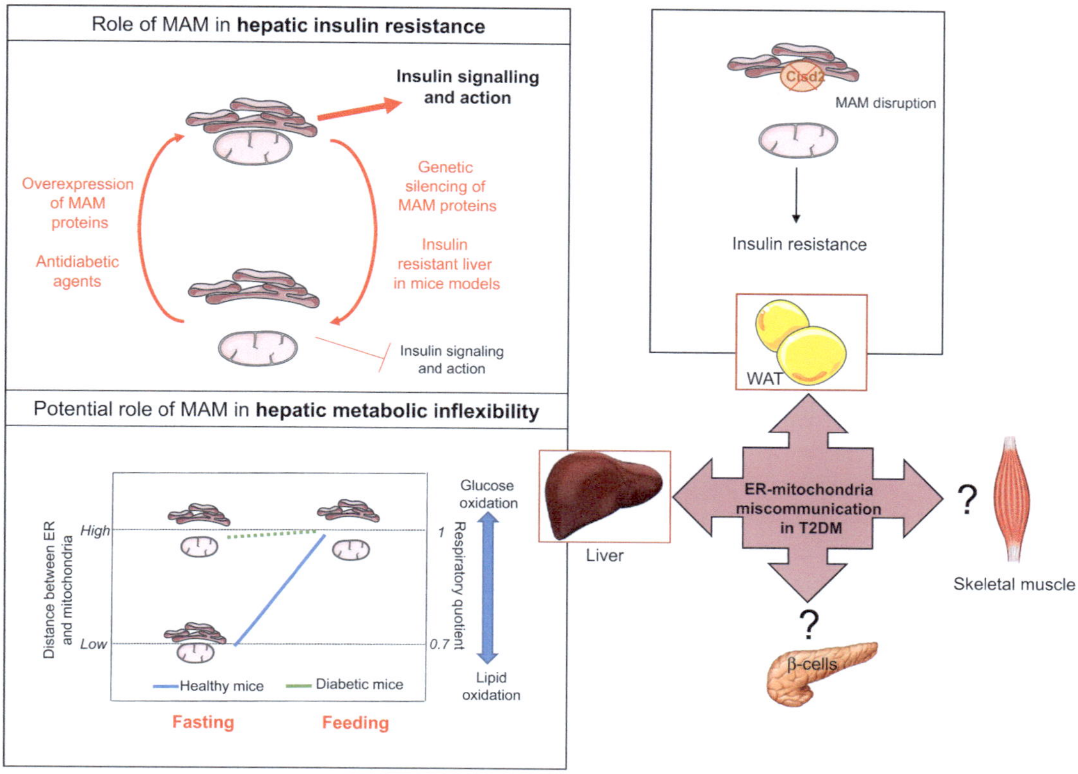

Fig. 13.3 ER-mitochondria miscommunication in T2DM. In liver, MAM control insulin signaling and potentially metabolic flexibility. Chronic disruption of ER-mitochondria interactions participates to hepatic insulin resistance and increased gluconeogenesis, whereas reinforcement of MAM integrity improves hepatic insulin sensitivity. As MAM allow the coupling of nutrient sensing to metabolic homeostasis, the loss of MAM regulation by nutritional status in insulin-resistant liver could participate to metabolic inflexibility associated with T2DM. In white adipose tissue (WAT), loss of cisd2 induces disruption of ER-mitochondria interactions and Ca^{2+} exchange and reduces insulin sensitivity. The state of ER-mitochondria interactions in both skeletal muscle and pancreatic islets is unknown in T2DM

pharmacological inhibition of CaMKII in obese mice results in a marked improvement of glucose homeostasis (Ozcan et al. 2013). In line with these evidences, loss of Ca^{2+}/calmodulin-dependent protein kinase kinase 2, an enzyme activating CaMKII, protects mice from diet-induced obesity and insulin resistance, whereas liver-specific loss of this enzyme improves glucose tolerance (Anderson et al. 2012). Taken together, these data directly link altered Ca^{2+} homeostasis with hepatic insulin resistance, supporting a role of disrupted MAM in metabolic diseases. In agreement, mutation of IP3R1 in mice was associated with glucose intolerance (Ye et al. 2011), further pointing at the importance of IP3R-mediated Ca^{2+} signaling in the control of glucose homeostasis.

13.4.2 Disruption of MAM in Hepatic Insulin Resistance

An interplay between ER stress and mitochondria dysfunction initially suggested a potential role of MAM in hepatic insulin resistance. Experimental mitochondrial dysfunction induced ER stress through an elevation of cytosolic-free Ca^{2+} and led subsequently to aberrant insulin signaling and increased hepatic gluconeogenesis (Lim et al. 2009). Similarly, liver-specific KO of the mitochondrial Mfn2 (Sebastián et al. 2012) or loss of the mitochondrial CypD (Rieusset et al. 2016) induced hepatic ER stress and insulin resistance.

Recently, our laboratory investigated the involvement of MAM disruption in hepatic insulin resistance (Tubbs et al. 2014). We found that MAM integrity is altered in palmitate-induced insulin-resistant HuH7 cells, as well as in the liver of different models of obese and diabetic mice (*ob/ob* and high-fat diet (HFD)-fed mice). Importantly, disruption of MAM integrity by genetic or pharmacological inhibition of CypD induced insulin resistance in mice and disrupted insulin signaling in human primary hepatocytes. Interestingly, treatment of diabetic mice with antidiabetic drugs (HFD mice with rosiglitazone or CypD-KO mice with metformin) improved insulin sensitivity and restored organelle communication. Conversely, the rescue of MAM integrity in primary hepatocytes of *ob/ob* or HFD mice by adenoviral overexpression of CypD improved insulin action. Therefore, our data demonstrate for the first time a key role of disrupted MAM in hepatic insulin resistance (Tubbs et al. 2014). In agreement, cellular loss of mTORC2 disrupted ER-mitochondria interactions and mitochondria alterations (Betz et al. 2013), whereas mice with liver-specific KO of rictor, a mTORC2 subunit, showed impaired glucose homeostasis (Hagiwara et al. 2012). Mechanistically, we suggest that a loss of Ca^{2+} transfer from ER to mitochondria links MAM disruption to hepatic insulin resistance (Rieusset et al. 2016). In the case of CypD-KO mice, the disruption of Ca^{2+} exchange between the two organelles leads to organelle stress, lipid accumulation, activation of PKCε, and alteration of insulin-stimulated PKB phosphorylation (Rieusset et al. 2016). Whether this mechanism can be generalized to genetic and nutritional models of obesity and T2DM has to be determined. Interestingly, ER-mitochondria contacts are also reduced in POMC neurons of HFD mice (Schneeberger et al. 2013), suggesting a miscommunication between ER and mitochondria in multiple tissues in the context of T2DM. A link between disrupted MAM and insulin resistance was also found in WAT of mice deficient in cisd2, an iron-sulfur protein localized at MAM interface (Chen et al. 2009, Wang et al. 2014). However, another group recently reported that MAM content is increased in the liver of obese mice, leading to mitochondrial Ca^{2+} overload and mitochondrial dysfunction (Arruda et al. 2014). They further showed that reinforcing hepatic MAM by IP3R1 or PACS2 overexpression induced insulin resistance, whereas reducing the expression of these proteins in the liver of obese mice improved insulin sensitivity (Arruda et al. 2014). The discrepancy between studies is actually unclear but could be related to differences in mice metabolic status, environmental housing conditions, or experimental analysis. Furthermore, as none of MAM proteins are specific of this subcellular compartment, we cannot exclude that modulating their expression could have nonspecific effects, further participating to the discrepancy between studies. Along these lines, reduced or excessive ER-mitochondria contacts, likely depending on the timing of the adaptive response upon a metabolic challenge, could represent a new and important mechanism contributing to hepatic mitochondrial dysfunction and insulin resistance. Future studies in which MAM will be dynamically studied are required to clarify this element.

13.4.3 Disruption of Dynamic Modulation of MAM by Nutritional Transition in Insulin-Resistant Liver

T2DM is classically associated with metabolic inflexibility (Galgani et al. 2008). Importantly, we found that chronic disruption of MAM in the liver of insulin-resistant mice is associated with a loss of MAM regulation by change in nutritional state (Theurey et al. 2016). Furthermore, we found that the regulation of MAM by glucose levels is lost in the liver of *ob/ob* and CypD-KO mice, characterized by chronic disruption of MAM integrity, mitochondrial fission, and altered mitochondrial respiration. Therefore, chronic disruption of MAM may participate to both hepatic-metabolic inflexibility and mitochondrial dysfunction associated with hepatic insulin resistance. In line with these evidences, ER-mitochondria interactions are controlled by PP2A (Theurey et al. 2016), and hyperactivation of PP2A was associated with insulin resistance (Kowluru and Matti 2012). Therefore, increased PP2A activity could participate to disruption of MAM in the liver of insulin-resistant mice. Future studies are required to understand the molecular mechanisms of MAM disruption in the context of hepatic metabolic diseases.

13.4.4 Is There a Role of MAM Disruption in Other Tissues?

Ca^{2+} also play an essential role in other metabolic tissues, such as in skeletal muscle where it affects contraction and metabolism (Contreras-Ferrat et al. 2014), in WAT where it controls adipogenesis and triglyceride accumulation (Shi et al. 2000), and in pancreas where it is indispensable for insulin secretion (Gilon et al. 2014). Disruption of Ca^{2+} homeostasis has been linked to T2DM (Guerrero-Hernandez and Verkhratsky 2014). Furthermore, mitochondria and ER functions are vital for muscle insulin action and secretion, and their dysfunction has been associated with T2DM (Rieusset 2015). Therefore, it is conceivable to imagine that MAM integrity could affect the function of these tissues and that disruption of MAM could participate to both muscle and adipose tissue insulin resistance or even to β-cell dysfunction. Nevertheless, it has to be precisely investigated in the future. From now, evidence suggests it could be possible in adipose tissue, since WAT-specific loss of cisd2 disrupts ER-mitochondria interactions and reduces insulin-stimulated glucose uptake in adipose tissue (Wang et al. 2014). For skeletal muscle and β-cells, the role of ER-mitochondria miscommunication in their functional alterations in the context of T2DM remains to be investigated. Nevertheless, the fact that blocking Ca^{2+} through ryanodine receptors (RyR) lowers the ATP/ADP ratio in pancreatic β-cells (Dror et al. 2008) suggests that RyR-mediated Ca^{2+} flux from the ER to mitochondria regulates basal mitochondrial metabolism to maintain beta cell survival (Johnson et al. 2012). Similarly, the fact that a IP3R-mediated Ca^{2+} release is required for GLUT4 translocation and glucose uptake in skeletal muscle (Contreras-Ferrat et al. 2014) supports a potential role of MAM in the control of muscle glucose homeostasis.

13.5 Targeting Organelle Stress and Communication to Treat Type 2 Diabetes

As organelle function and communication play an important role in the control of glucose homeostasis, their targeting could represent an efficient way to improve T2DM. In agreement, several antioxidants alleviating mitochondrial stress were shown to improve insulin action and/or secretion (Abdali et al. 2015; Cheng et al. 2014). Similarly, drugs targeting ER stress enhanced hepatic insulin sensitivity and glucose homeostasis in HFD mice and in obese patients (Engin and Hotamisligil 2010). So far, it is unknown whether targeting MAM could help to fight against metabolic diseases. For that, we need to learn more about the physiological regulation of MAM before finding pharmacological and/or nutritional strategies to modulate ER-mitochondria contacts and improve T2DM. Currently, we only know that some hormones (Tubbs et al. 2014) and growth factors (Betz et al. 2013), as well as some nutrients (Theurey et al. 2016), can regulate organelle proximity. Furthermore, the expression level (Simmen et al. 2005; de Brito and Scorrano 2008; Tubbs et al. 2014) or the posttranscriptional modifications (Lynes et al. 2012) of some MAM proteins are able to regulate organelle crosstalk. However, no proteins are exclusively expressed in the MAM, and targeting their expression could have side effects. The specific targeting of some proteins at MAM interface could be an alternative strategy. However, it is actually unknown whether specific targeting sequences to address proteins in the MAM exist. It was suggested that a mitochondrial targeting sequence found in diacylglycerol O-acyltransferase 2 was required for its targeting to MAM (Stone et al. 2009), whereas a functional lipid transfer domain was required for the localization of oxysterol-binding protein-related proteins ORP5 and ORP8 at MAM interface (Galmes et al. 2016). However, it is not known whether these findings can be generalized to other MAM proteins. Lastly, we recently identified PP2A as an important negative regulator of ER-mitochondria interactions (Theurey et al. 2016), suggesting that inhibiting its activity into the cell should increase organelle communication. Nevertheless, PP2A is not a good intracellular target to improve organelle cross-talk, as it has pleiotropic effects and its inhibition is pro-cancerous (Sangodkar et al. 2016). This issue notwithstanding, these data indicate that phosphorylation status of some MAM proteins should be important for the crosstalk between the two organelles. Therefore, the next challenge is to identify either interacting partners or targets of PP2A. Whereas the interacting partners of PP2A in total cell start to emerge (Zhang et al. 2016), those at MAM interface are less known. Nevertheless, cumulative evidence indicate that some MAM proteins are good candidates including Akt (Giorgi et al. 2010), IP3R (DeSouza et al. 2002), and Drp1 (Dickey and Stracks 2011; Cho et al. 2012). Therefore, a better understanding of the role of phosphorylated state of MAM proteins, as well as the identity of PP2A-regulated proteins at MAM, will be required.

13.6 Conclusion

MAM are at the crossroad of several important hormonal and nutrient-regulated signaling pathways in the liver, suggesting that ER-mitochondria miscommunication could be involved in hepatic metabolic diseases. Therefore, targeting MAM might be a novel strategy for the treatment of T2DM, especially if ER-miscommunication is also involved in impaired insulin action in other tissues or in β-cell dysfunction. However, the dynamic properties of MAM should be taken into account, and future studies are required to uncover the molecular mechanisms underlying MAM regulation and to better understand the role of MAM in the control of glucose homeostasis.

References

Abdali D, Samson SE, Grover AK (2015) How effective are antioxidant supplements in obesity and diabetes? Med Princ Pract 24:201–215

Ackermann AM, Gannon M (2006) Molecular regulation of pancreatic β-cell mass development, maintenance, and expansion. J Mol Endocrinol 38:193–206

Anderson KA, Lin F, Ribar TJ, Stevens RD, Muehlbauer MJ, Newgard CB, Means AR (2012) Deletion of CaMKK2 from the liver lowers blood glucose and improves whole-body glucose tolerance in the mouse. Mol Endocrinol 26:281–291

Arruda AP, Hotamisligil GS (2015) Calcium homeostasis and organelle function in the pathogenesis of obesity and diabetes. Cell Metab 22:381–397

Arruda AP, Pers BM, Parlakgül G, Güney E, Inouye K, Hotamisligil GS (2014) Chronic enrichment of hepatic endoplasmic reticulum-mitochondria contact leads to mitochondrial dysfunction in obesity. Nat Med 20:1427–1435

Bartlett PJ, Gaspers LD, Pierobon N, Thomas AP (2014) Calcium-dependent regulation of glucose homeostasis in the liver. Cell Calcium 55:306–316

Betz C, Stracka D, Prescianotto-Baschong C, Frieden M, Demaurex N, Hall MN (2013) Feature article: mTOR complex 2-Akt signaling at mitochondria-associated endoplasmic reticulum membranes (MAM) regulates mitochondrial physiology. Proc Natl Acad Sci USA 110:12526–12534

Bononi A, Bonora M, Marchi S, Missiroli S, Poletti F, Giorgi C, Pandolfi PP, Pinton P (2013) Identification of PTEN at the ER and MAMs and its regulation of Ca^{2+} signaling and apoptosis in a protein phosphatase-dependent manner. Cell Death Differ 20:1631–1643

Bravo R, Vicencio JM, Parra V, Troncoso R, Munoz JP, Bui M, Quiroga C, Rodriguez AE, Verdejo HE, Ferreira J et al (2011) Increased ER-mitochondrial coupling promotes mitochondrial respiration and bioenergetics during early phases of ER stress. J Cell Sci 124:2143–2152

Carnagarin R, Dharmarajan AM, Dass CR (2015) Molecular aspects of glucose homeostasis in skeletal muscle – a focus on the molecular mechanisms of insulin resistance. Mol Cell Endocrinol 417:52–62

Chang YC, Hee SW, Hsieh ML, Jeng YM, Chuang LM (2015) The role of organelle stresses in diabetes mellitus and obesity: implication for treatment. Anal Cell Pathol (Amst) 2015:972891

Chaudhari N, Talwar P, Parimisetty A, Lefebvre d'Hellencourt C, Ravanan P (2014) A molecular web: endoplasmic reticulum stress, inflammation, and oxidative stress. Front Cell Neurosci 8:213

Chen YF, Kao CH, Chen YT, Wang CH, Wu CY, Tsai CY, Liu FC, Yang CW, Wei YH, Hsu MT et al (2009) Cisd2 deficiency drives premature aging and causes mitochondria-mediated defects in mice. Genes Dev 23:1183–1194

Cheng Z, Schmelz EM, Liu D, Hulver MW (2014) Targeting mitochondrial alterations to prevent type 2 diabetes – evidence from studies of dietary redox-active compounds. Mol Nutr Food Res 58:1739–1749

Cho MH, Kim DH, Choi JE, Chang EJ, Seung-Yongyoon (2012) Increased phosphorylation of dynamin-related protein 1 and mitochondrial fission in okadaic acid-treated neurons. Brain Res 1454:100–110

Contreras-Ferrat A, Lavandero S, Jaimovich E, Klip A (2014) Calcium signaling in insulin action on striated muscle. Cell Calcium 56:390–396

Copps KD, White MF (2012) Regulation of insulin sensitivity by serine/threonine phosphorylation of insulin receptor substrate proteins IRS1 and IRS2. Diabetologia 55:2565–2582

Csordás G, Renken C, Várnai P, Walter L, Weaver D, Buttle KF, Balla T, Mannella CA, Hajnóczky G (2006) Structural and functional features and significance of the physical linkage between ER and mitochondria. J Cell Biol 174:915–921

de Brito OM, Scorrano L (2008) Mitofusin 2 tethers endoplasmic reticulum to mitochondria. Nature 456:605–610

De Fronzo RA (2004) Dysfunctional fat cells, lipotoxicity and type 2 diabetes. Int J Clin Pract 143:9–21

DeFronzo RA, Tripathy D (2009) Skeletal muscle insulin resistance is the primary defect in type 2 diabetes. Diabetes Care 32:S157–S163

Del Campo A, Parra V, Vásquez-Trincado C, Gutiérrez T, Morales PE, López-Crisosto C, Bravo-Sagua R, Navarro-Marquez MF, Verdejo HE, Contreras-Ferrat A et al (2014) Mitochondrial fragmentation impairs insulin-dependent glucose uptake by modulating Akt activity through mitochondrial Ca^{2+} uptake. Am J Physiol Endocrinol Metab 306:E1–E13

Denton RM (2009) Regulation of mitochondrial dehydrogenases by calcium ions. Biochim Biophys Acta 1787:1309–1316

DeSouza N, Reiken S, Ondrias K, Yang YM, Matkovich S, Marks AR (2002) Protein kinase A and two phosphatases are components of the inositol 1,4,5-trisphosphate receptor macromolecular signaling complex. J Biol Chem 277:39397–39400

Dickey AS, Strack S (2011) PKA/AKAP1 and PP2A/Bβ2 regulate neuronal morphogenesis via Drp1 phosphorylation and mitochondrial bioenergetics. J Neurosci 31:15716–15726

Dror V, Kalynyak TB, Bychkivska Y, Frey MH, Tee M, Jeffrey KD, Nguyen V, Luciani DS, Johnson JD (2008) Glucose and endoplasmic reticulum calcium channels regulate HIF-1β via presenilin in pancreatic β-cells. J Biol Chem 283:9909–9916

Engin F, Hotamisligil GS (2010) Restoring endoplasmic reticulum function by chemical chaperones: an emerging therapeutic approach for metabolic diseases. Diabetes Obes Metab 12:108–115

Fu S, Yang L, Li P, Hofmann O, Dicker L, Hide W, Lin X, Watkins SM, Ivanov AR, Hotamisligil GS (2011) Aberrant lipid metabolism disrupts calcium homeostasis causing liver endoplasmic reticulum stress in obesity. Nature 473:528–531

Galgani JE, Moro C, Ravussin E (2008) Metabolic flexibility and insulin resistance. Am J Physiol Endocrinol Metab 295:E1009–E1017

Galmes R, Houcine A, van Vliet AR, Agostinis P, Jackson CL, Giordano F (2016) ORP5/ORP8 localize to endoplasmic reticulum-mitochondria contacts and are involved in mitochondrial function. EMBO Rep 17:800–810

Gao AW, Cantó C, Houtkooper RH (2014) Mitochondrial response to nutrient availability and its role in metabolic disease. EMBO Mol Med 6:580–589

Gilady SY, Bui M, Lynes EM, Benson MD, Watts R, Vance JE, Simmen T (2010) Ero1α requires oxidizing and normoxic conditions to localize to the mitochondria-associated membrane (MAM). Cell Stress Chaperones 15:619–629

Gilon P, Chae HY, Rutter GA, Ravier MA (2014) Calcium signaling in pancreatic β-cells in health and in Type 2 diabetes. Cell Calcium 56:340–361

Giorgi C, Ito K, Lin HK, Santangelo C, Wieckowski MR, Lebiedzinska M, Bononi A, Bonora M, Duszynski J, Bernardi R et al (2010) PML regulates apoptosis at endoplasmic reticulum by modulating calcium release. Science 330:1247–1251

Giorgi C, Missiroli S, Patergnani S, Duszynski J, Wieckowski MR, Pinton P (2015) Mitochondria-associated membranes: composition, molecular mechanisms, and physiopathological implications. Antioxid Redox Signal 22:995–1019

Guerrero-Hernandez A, Verkhratsky A (2014) Calcium signalling in diabetes. Cell Calcium 56:297–301

Gutiérrez T, Parra V, Troncoso R, Pennanen C, Contreras-Ferrat A, Vasquez-Trincado C, Morales PE, Lopez-Crisosto C, Sotomayor-Flores C, Chiong M et al (2014) Alteration in mitochondrial Ca^{2+} uptake disrupts insulin signaling in hypertrophic cardiomyocytes. Cell Commun Signal 12:68

Hagiwara A, Cornu M, Cybulski N, Polak P, Betz C, Trapani F, Terracciano L, Heim MH, Rüegg MA, Hall MN (2012) Hepatic mTORC2 activates glycolysis and lipogenesis through Akt, glucokinase, and SREBP1c. Cell Metab 15:725–738

Hailey DW, Rambold AS, Satpute-Krishnan P, Mitra K, Sougrat R, Kim PK, Lippincott-Schwartz J (2010) Mitochondria supply membranes for autophagosome biogenesis during starvation. Cell 141:656–667

Halban PA, Polonsky KS, Bowden DW, Hawkins MA, Ling C, Mather KJ, Powers AC, Rhodes CJ, Sussel L, Weir GC (2014) β-cell failure in type 2 diabetes: postulated mechanisms and prospects for prevention and treatment. Diabetes Care 37:1751–1758

Hamasaki M, Furuta N, Matsuda A, Nezu A, Yamamoto A, Fujita N, Oomori H, Noda T, Haraguchi T, Hiraoka Y et al (2013) Autophagosomes form at ER-mitochondria contact sites. Nature 495:389–393

Hasnain SZ, Prins JB, McGuckin MA (2016) Oxidative and endoplasmic reticulum stress in β-cell dysfunction in diabetes. J Mol Endocrinol 56:R33–R54
Hayashi T, Su TP (2007) Sigma-1 receptor chaperones at the ER-mitochondrion interface regulate Ca^{2+} signaling and cell survival. Cell 13:596–610
Hers HG (1990) Mechanisms of blood glucose homeostasis. J Inherit Metab Dis 13(4):395–410
Hoppins S, Nunnari J (2012) Cell biology. Mitochondrial dynamics and apoptosis – the ER connection. Science 337:1052–1054
Horner SM, Wilkins C, Badil S, Iskarpatyoti J, Gale M Jr (2015) Proteomic analysis of mitochondrial-associated ER membranes (MAM) during RNA virus infection reveals dynamic changes in protein and organelle trafficking. PLoS One 10:e0117963
Hresko RC, Mueckler M (2005) mTOR.RICTOR is the Ser473 kinase for Akt/protein kinase B in 3T3-L1 adipocytes. J Biol Chem 280:40406–40416
Hummasti S, Hotamisligil GS (2010) Endoplasmic reticulum stress and inflammation in obesity and diabetes. Circ Res 107:579–591
Johnson JD, Bround MJ, White SA, Luciani DS (2012) Nanospaces between endoplasmic reticulum and mitochondria as control centres of pancreatic β-cell metabolism and survival. Protoplasma 249:S49–S58
Kahn SE, Hull RL, Utzschneider KM (2006) Mechanisms linking obesity to insulin resistance and type 2 diabetes. Nature 444:840–846
Kaur J, Debnath J (2015) Autophagy at the crossroads of catabolism and anabolism. Nat Rev Mol Cell Biol 16:461–472
Kharroubi AT, Darwish HM (2015) Diabetes mellitus: the epidemic of the century. World J Diabetes 6:850–867
Kowluru A, Matti A (2012) Hyperactivation of protein phosphatase 2A in models of glucolipotoxicity and diabetes: potential mechanisms and functional consequences. Biochem Pharmacol 84:591–597
Lee GS, Subramanian N, Kim AI, Aksentijevich I, Goldbach-Mansky R, Sacks DB, Germain RN, Kastner DL, Chae JJ (2012) The calcium-sensing receptor regulates the NLRP3 inflammasome through Ca^{2+} and cAMP. Nature 492:123–127
Lewis SC, Uchiyama LF, Nunnari J (2016) ER-mitochondria contacts couple mtDNA synthesis with mitochondrial division in human cells. Science 353:aaf5549
Lim JH, Lee HJ, Ho Jung M, Song J (2009) Coupling mitochondrial dysfunction to endoplasmic reticulum stress response: a molecular mechanism leading to hepatic insulin resistance. Cell Signal 21:169–177
López-Crisosto C, Bravo-Sagua R, Rodriguez-Peña M, Mera C, Castro PF, Quest AF, Rothermel BA, Cifuentes M, Lavandero S (2015) ER-to-mitochondria miscommunication and metabolic diseases. Biochim Biophys Acta 1852:2096–2105
Lynes EM, Bui M, Yap MC, Benson MD, Schneider B, Ellgaard L, Berthiaume LG, Simmen T (2012) Palmitoylated TMX and calnexin target to the mitochondria-associated membrane. EMBO J 31:457–470
Mandl J, Mészáros T, Bánhegyi G, Hunyady L, Csala M (2009) Endoplasmic reticulum: nutrient sensor in physiology and pathology. Trends Endocrinol Metab 20:194–201
Marchi S, Patergnani S, Pinton P (2014) The endoplasmic reticulum-mitochondria connection: one touch, multiple functions. Biochim Biophys Acta 1837:461–469
Montgomery MK, Turner N (2015) Mitochondrial dysfunction and insulin resistance: an update. Endocr Connect 4:R1–R15
Morino K, Neschen S, Bilz S, Sono S, Tsirigotis D, Reznick RM, Moore I, Nagai Y, Samuel V, Sebastian D et al (2008) Muscle-specific IRS-1 Ser->Ala transgenic mice are protected from fat-induced insulin resistance in skeletal muscle. Diabetes 57:2644–2651
Mothe I, Van Obberghen E (1996) Phosphorylation of insulin receptor substrate-1 on multiple serine residues, 612, 632, 662, and 731, modulates insulin action. J Biol Chem 271:11222–11227
Muñoz JP, Ivanova S, Sánchez-Wandelmer J, Martínez-Cristóbal P, Noguera E, Sancho A, Díaz-Ramos A, Hernández-Alvarez MI, Sebastián D, Mauvezin C et al (2013) Mfn2 modulates the UPR and mitochondrial function via repression of PERK. EMBO J 32:2348–2361
Naon D, Scorrano L (2014) At the right distance: ER-mitochondria juxtaposition in cell life and death. Biochim Biophys Acta 1843:2184–2194
Ngoh GA, Papanicolaou KN, Walsh K (2012) Loss of mitofusin 2 promotes endoplasmic reticulum stress. J Biol Chem 287:20321–20332
Ozcan L, Wong CC, Li G, Xu T, Pajvani U, Park SK, Wronska A, Chen BX, Marks AR, Fukamizu A et al (2012) Calcium signaling through CaMKII regulates hepatic glucose production in fasting and obesity. Cell Metab 15:739–751
Ozcan L, Cristina de Souza J, Harari AA, Backs J, Olson EN, Tabas I (2013) Activation of calcium/calmodulin-dependent protein kinase II in obesity mediates suppression of hepatic insulin signaling. Cell Metab 18:803–815
Paillard M, Tubbs E, Thiebaut PA, Gomez L, Fauconnier J, Da Silva CC, Teixeira G, Mewton N, Belaidi E, Durand A et al (2013) Depressing mitochondria-reticulum interactions protects cardiomyocytes from lethal hypoxia-reoxygenation injury. Circulation 128:1555–1565
Paz K, Hemi R, LeRoith D, Karasik A, Elhanany E, Kanety H, Zick Y (1997) A molecular basis for insulin resistance. Elevated serine/threonine phosphorylation of IRS-1 and IRS-2 inhibits their binding to the juxtamembrane region of the insulin receptor and impairs their ability to undergo insulin-induced tyrosine phosphorylation. J Biol Chem 272:29911–29918
Poston CN, Duong E, Cao Y, Bazemore-Walker CR (2011) Proteomic analysis of lipid raft-enriched

membranes isolated from internal organelles. Biochem Biophys Res Commun 415:355–360

Poston CN, Krishnan SC, Bazemore-Walker CR (2013) In-depth proteomic analysis of mammalian mitochondria-associated membranes (MAM). J Proteomics 79:219–230

Rieusset J (2015) Contribution of mitochondria and endoplasmic reticulum dysfunction in insulin resistance: distinct or interrelated roles? Diabetes Metab 41:358–368

Rieusset J, Fauconnier J, Paillard M, Belaidi E, Tubbs E, Chauvin MA, Durand A, Bravard A, Teixeira G, Bartosch B et al (2016) Disruption of calcium transfer from ER to mitochondria links alterations of mitochondria-associated ER membrane integrity to hepatic insulin resistance. Diabetologia 59:614–623

Rizzuto R, Pinton P, Carrington W, Fay FS, Fogarty KE, Lifshitz LM, Tuft RA, Pozzan T (1998) Close contacts with the endoplasmic reticulum as determinants of mitochondrial Ca^{2+} responses. Science 280:1763–1766

Roden M, Bernroider E (2003) Hepatic glucose metabolism in humans – its role in health and disease. Best Pract Res Clin Endocrinol Metab 17:365–383

Sala-Vila A, Navarro-Lérida I, Sánchez-Alvarez M, Bosch M, Calvo C, López JA, Calvo E, Ferguson C, Giacomello M, Serafini A, Scorrano L, Enriquez JA, Balsinde J, Parton RG, Vázquez J, Pol A, Del Pozo MA (2016) Interplay between hepatic mitochondria-associated membranes, lipid metabolism and caveolin-1 in mice. Sci Rep 6:27351

Salvadó L, Palomer X, Barroso E, Vázquez-Carrera M (2015) Targeting endoplasmic reticulum stress in insulin resistance. Trends Endocrinol Metab 26:438–448

Samuel VT, Shulman GI (2016) The pathogenesis of insulin resistance: integrating signaling pathways and substrate flux. J Clin Invest 126:12–22

Sangodkar J, Farrington CC, McClinch K, Galsky MD, Kastrinsky DB, Narla G (2016) All roads lead to PP2A: exploiting the therapeutic potential of this phosphatase. FEBS J 283:1004–1024

Schneeberger M, Dietrich MO, Sebastián D, Imbernón M, Castaño C, Garcia A, Esteban Y, Gonzalez-Franquesa A, Rodríguez IC, Bortolozzi A et al (2013) Mitofusin 2 in POMC neurons connects ER stress with leptin resistance and energy imbalance. Cell 155:172–187

Sebastián D, Hernández-Alvarez MI, Segalés J, Sorianello E, Muñoz JP, Sala D, Waget A, Liesa M, Paz JC, Gopalacharyulu P et al (2012) Mitofusin 2 (Mfn2) links mitochondrial and endoplasmic reticulum function with insulin signaling and is essential for normal glucose homeostasis. Proc Natl Acad Sci USA 109:5523–5528

Shi H, Halvorsen YD, Ellis PN, Wilkison WO, Zemel MB (2000) Role of intracellular calcium in human adipocyte differentiation. Physiol Genomics 3:75–82

Simmen T, Aslan JE, Blagoveshchenskaya AD, Thomas L, Wan L, Xiang Y, Feliciangeli SF, Hung CH, Crump CM, Thomas G (2005) PACS-2 controls endoplasmic reticulum-mitochondria communication and Bid-mediated apoptosis. EMBO J 24:717–729

Simmen T, Lynes EM, Gesson K, Thomas G (2010) Oxidative protein folding in the endoplasmic reticulum: tight links to the mitochondria-associated membrane (MAM). Biochim Biophys Acta 1798:1465–1473

Sood A, Jeyaraju DV, Prudent J, Caron A, Lemieux P, McBride HM, Laplante M, Tóth K, Pellegrini L (2014) A Mitofusin-2-dependent inactivating cleavage of Opa1 links changes in mitochondria cristae and ER contacts in the postprandial liver. Proc Natl Acad Sci USA 111:16017–16022

Stinkens R, Goossens GH, Jocken JW, Blaak EE (2015) Targeting fatty acid metabolism to improve glucose metabolism. Obes Rev 16:715–757

Stone SJ, Levin MC, Zhou P, Han J, Walther TC, Farese RV Jr (2009) The endoplasmic reticulum enzyme DGAT2 is found in mitochondria-associated membranes and has a mitochondrial targeting signal that promotes its association with mitochondria. J Biol Chem 284:5352–5361

Szabadkai G, Bianchi K, Várnai P, De Stefani D, Wieckowski MR, Cavagna D, Nagy AI, Balla T, Rizzuto R (2006) Chaperone-mediated coupling of endoplasmic reticulum and mitochondrial Ca^{2+} channels. J Cell Biol 175:901–911

Theurey P, Tubbs E, Vial G, Jacquemetton J, Bendridi N, Chauvin MA, Alam MR, Le Romancer M, Vidal H, Rieusset J (2016) Mitochondria-associated endoplasmic reticulum membranes allow adaptation of mitochondrial metabolism to glucose availability in the liver. J Mol Cell Biol 8:129–143

Tubbs E, Theurey P, Vial G, Bendridi N, Bravard A, Chauvin MA, Ji-Cao J, Zoulim F, Bartosch B, Ovize M et al (2014) Mitochondria-associated endoplasmic reticulum membrane (MAM) integrity is required for insulin signaling and is implicated in hepatic insulin resistance. Diabetes 63:3279–3294

van Vliet AR, Verfaillie T, Agostinis P (2014) New functions of mitochondria associated membranes in cellular signaling. Biochim Biophys Acta 1843:2253–2262

Vance JE (2014) MAM (mitochondria-associated membranes) in mammalian cells: lipids and beyond. Biochim Biophys Acta 1841:595–609

Verfaillie T, Rubio N, Garg AD, Bultynck G, Rizzuto R, Decuypere JP, Piette J, Linehan C, Gupta S, Samali A et al (2012) PERK is required at the ER-mitochondrial contact sites to convey apoptosis after ROS-based ER stress. Cell Death Differ 19:1880–1891

Wang Y, Li G, Goode J, Paz JC, Ouyang K, Screaton R, Fischer WH, Chen J, Tabas I, Montminy M (2012) Inositol-1,4,5-trisphosphate receptor regulates hepatic gluconeogenesis in fasting and diabetes. Nature 485:128–132

Wang CH, Chen YF, Wu CY, Wu PC, Huang YL, Kao CH, Lin CH, Kao LS, Tsai TF, Wei YH (2014) Cisd2 modulates the differentiation and functioning of adipocytes by regulating intracellular Ca^{2+} homeostasis. Hum Mol Genet 23:4770–4785

Wang CH, Tsai TF, Wei YH (2015) Role of mitochondrial dysfunction and dysregulation of Ca^{2+} homeostasis in insulin insensitivity of mammalian cells. Ann N Y Acad Sci 1350:66–76

Ye R, Ni M, Wang M, Luo S, Zhu G, Chow RH, Lee AS (2011) Inositol 1,4,5-trisphosphate receptor 1 mutation perturbs glucose homeostasis and enhances susceptibility to diet-induced diabetes. J Endocrinol 210:209–217

Zhang X, Damacharla D, Ma D, Qi Y, Tagett R, Draghici S, Kowluru A, Yi Z (2016) Quantitative proteomics reveals novel protein interaction partners of PP2A catalytic subunit in pancreatic β-cells. Mol Cell Endocrinol 424:1–11

14 Mitochondria–Endoplasmic Reticulum Contact Sites Mediate Innate Immune Responses

Takuma Misawa, Michihiro Takahama, and Tatsuya Saitoh

Abstract

Mitochondria and the endoplasmic reticulum (ER) are fundamental organelles that coordinate high-order cell functions. Mitochondria are centers of energy production, whereas the ER is responsible for folding, transport, and degradation of proteins. In addition to their specific functions, mitochondria and ER actively communicate with each other to promote a variety of cellular events, such as material transfer and signal transduction. Recent studies have shown the critical involvement of these organelles in regulation of the innate immune system, which functions in host defense. The innate immune system utilizes a wide range of germline-encoded pattern recognition receptors (PRRs) to recognize pathogen-associated molecular patterns (PAMPs) and damage-associated molecular patterns (DAMPs) and induces inflammatory and antiviral responses. Contact sites between mitochondria and the ER function in assembly of the NLR family pyrin domain containing 3 (NLRP3)-inflammasome to promote the inflammatory response. The NLRP3-inflammasome is a protein complex composed of the receptor NLRP3 on the ER side and the adaptor apoptosis-associated speck-like protein containing a CARD on the mitochondrial side; it induces caspase-1-dependent maturation of proinflammatory cytokines such as interleukin (IL)-1β and IL-18. Furthermore, ER–mitochondria contact sites function in initiation and mediation of signal transduction pathways downstream of intracellular PRRs, such as retinoic acid-inducible gene I-like receptor and cyclic GMP-AMP synthase, to promote the antiviral response. Therefore, ER–mitochondria contact sites, also known as mitochondria-associated membranes, play key roles in regulation of innate immune responses.

T. Misawa
Center for the Genetics of Host Defense, University of Texas Southwestern Medical Center, Dallas, TX 75390-8505, USA

M. Takahama • T. Saitoh (✉)
Division of Inflammation Biology, Institute for Enzyme Research, Tokushima University, Tokushima 770-8503, Japan
e-mail: saitohtatsuya@tokushima-u.ac.jp

M. Tagaya, T. Simmen (eds.), *Organelle Contact Sites*, Advances in Experimental Medicine and Biology 997, DOI 10.1007/978-981-10-4567-7_14

Keywords
Cytokine • Host defense • Inflammasome • Inflammatory disease • Innate immune response • Interferon • Macrophage • Microbe • Signal transduction • Sterilized inflammation

14.1 Introduction

14.1.1 The Mitochondria-Associated Membrane: A Functional Organellar Contact Site

Organelles are fundamental components of living organisms. Each organelle performs specific functions to maintain cellular homeostasis. Accumulating evidence has revealed that organelles not only function independently but also actively interact with each other to mediate a variety of cellular events (Fig. 14.1). ER–mitochondrial contact is mediated by a specific membrane compartment, known as the mitochondria-associated membrane (MAM). Although the concept of the MAM was first described more than four decades ago (Franke and Kartenbeck 1971; Morré et al. 1971), its unique functions have only recently been recognized. Many proteins in the MAM are shared with the ER and are evolutionarily conserved among a wide range of species, from yeast to mammals (Vance 1990). Marker proteins of mitochondria, the Golgi apparatus, lysosomes, the nucleus, or the plasma membrane are seldom identified in the MAM. The MAM plays a pivotal role in calcium and lipid exchange between mitochondria and the ER (Vance 2014). Moreover, the MAM serves as an optimal platform for the processes of apoptosis, mitochondrial fission and fusion, autophagy, and certain types of immune responses (Vance 2014). Here, we describe the involvement of the MAM in regulation of innate immune responses.

14.1.2 Innate Immune System

The host defense system comprises the innate and adaptive arms of the immune system. The innate immune system has long been thought of as the "primitive" immune system, which detects invading pathogens in a non-specific manner and assists in subsequent activation of the adaptive immune system. However, discovery of the germ-line-encoded receptors known as pattern recognition receptors (PRRs) has completely changed the concept of the innate immune system. PRRs include Toll-like receptors (TLRs),

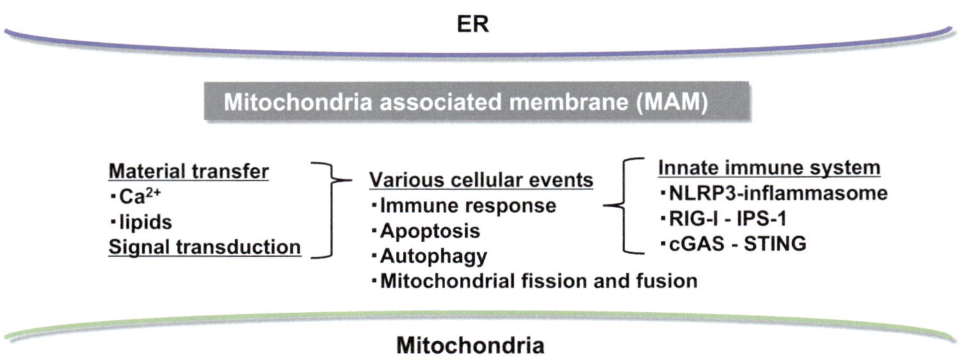

Fig. 14.1 Functions of the mitochondria-associated membrane. The mitochondria-associated membrane (MAM) is an organelle contact site that mediates the interplay between mitochondria and the ER. The MAM promotes transfer of materials and interaction of molecules between mitochondria and the ER. The MAM regulates the progression of various cellular events, including the innate immune response

retinoic acid-inducible gene-I (RIG-I)-like receptors (RLRs), C-type lectin receptors, NOD-like receptors (NLRs), and cytosolic DNA sensors (Pandey et al. 2014). These PRRs specifically recognize conserved microbial structures known as pathogen-associated molecular patterns (PAMPs). Some PRRs are also responsible for detecting endogenous molecules released from damaged or dying cells, termed as damage-associated molecular patterns (DAMPs).

14.1.3 Spatial Regulation of TLRs: Sensors for Extracellular PAMPs and DAMPs

TLRs are well-studied PRRs that sense extracellular PAMPs and DAMPs on cell surfaces and in endolysosomes. TLRs were initially identified in *Drosophila*, as a factor that induces host defense responses against fungal infections (Lemaitre et al. 1996). Later, the mammalian homolog, TLR4, was identified as a specific sensor of lipopolysaccharide (LPS), a component of the outer membranes of gram-negative bacteria; TLR4 accelerates further inflammatory responses (Medzhitov et al. 1997; Poltorak et al. 1998; Hoshino et al. 1999). In mammals, 13 TLRs have been identified to date. Upon recognizing their specific ligands, TLRs transmit signals to downstream components using the adaptors myeloid differentiation primary response gene 88 (Myd88) and TIR-domain-containing adapter-inducing interferon-β (TRIF). Signaling through either the Myd88 or TRIF pathway will eventually activate transcription factors such as nuclear factor kappa B (NF-kB), activator protein-1, or interferon regulatory factors (e.g., IRF3 and IRF7), leading to production of proinflammatory cytokines or type I interferons (Takeuchi and Akira 2010).

Proper positioning of TLRs on specific organelles is important for their activation. TLR1, TLR2, TLR4, TLR5, TLR6, and TLR10 are present on the plasma membrane and are responsible for detecting bacterial or viral components. When TLR4 detects LPS, the membrane-resident molecule CD14 induces translocation of TLR4 from the plasma membrane to an endosome. This translocation of TLR4 further activates TRIF-mediated inflammatory responses (Kagan et al. 2008; Zanoni et al. 2011). TLR3, TLR7, and TLR9 recognize nucleic acids derived from pathogens or host endolysosomes. UNC93B1 delivers TLR3, TLR7, and TLR9 from the endoplasmic reticulum (ER) to endolysosome translocation (Tabeta et al. 2006). PRAT4A delivers TLR1, TLR4, and TLR5 from the ER to the cell surface and also performs relocation of TLR9 (Takahashi et al. 2007; Shibata et al. 2012). When TLR9 reaches an endolysosome, TLR9 is cleaved by cathepsins, thereby acquiring the ability to recognize its ligand, unmethylated CpG DNA (Ewald et al. 2008; Park et al. 2008). Molecules derived from organelles also assist in TLR activation. Mitochondrial reactive oxygen species (ROS) mediate the signaling pathway downstream of TLRs (West et al. 2011). Mitochondrial DNA released from damaged mitochondria is capable of stimulating TLR9, because mitochondrial DNA contains CpG motifs (Oka et al. 2012). Moreover, succinate, an intermediate of the tricarboxylic acid cycle, stabilizes the transcription factor hypoxia-inducible factor 1-α and activates LPS-induced inflammatory responses (Tannahill et al. 2013). These findings indicate the essential roles of organelles in activation of TLRs, sensors for extracellular PAMPs and DAMPs. Furthermore, recent studies have revealed the involvement of organellar crosstalk in regulation of intracellular PRR-mediated innate immune responses.

14.2 MAM and Intracellular PRR-Mediated Inflammatory Responses

14.2.1 The NLRP3-Inflammasome

Inflammasomes are components of the innate immune system that activate the protease caspase-1 and induce subsequent maturation of proinflammatory cytokines such as IL-1β and IL-18 (de Zoete et al. 2014). PRRs that belong to the NLR family, or absent in melanoma 2, form inflammasomes in response to different types of PAMPs or DAMPs. NLR family pyrin domain containing 3 (NLRP3; also known as cryopyrin)-containing inflammasomes are a well-characterized type of inflammasome. NLRP3 recruits caspase-1 via the adaptor apoptosis-associated speck-like protein containing a CARD (ASC; also known as PYCARD) to form the NLRP3-inflammasome.

The NLRP3-inflammasome can detect infection by diverse pathogens and induce production of cytokines such as IL-1β and IL-18 (Hise et al. 2009; Ichinohe et al. 2010; Broz et al. 2010; Kim et al. 2010; Dorhoi et al. 2012). Thus, the NLRP3-inflammasome plays a pivotal role in host defense responses against infectious diseases. However, excessive activation of the NLRP3-inflammasome, caused by host-derived metabolites or environmental irritants, results in development of severe inflammatory disorders, such as gout, type 2 diabetes, atherosclerosis, Alzheimer's disease, and silicosis. In humans, mutations in NLRP3 cause autoinflammatory disorders such as familial cold autoinflammatory syndrome, Muckle–Wells syndrome, neonatal-onset multisystem inflammatory disease, and cryopyrin-associated periodic syndromes (Hoffman et al. 2001; Feldmann et al. 2002; Aksentijevich et al. 2002; Neven et al. 2004; Agostini et al. 2004). In patients developing these diseases, the IL-1β level is significantly elevated, probably due to inappropriate regulation of the NLRP3-inflammasome. Therefore, the NLRP3-inflammasome has two faces, both protecting and damaging us.

14.2.2 Triggers of NLRP3-Inflammasome Activation

NLRP3 is a unique PRR that responds to a variety of activators. Distinct from other PRR family members, the NLRP3-inflammasome is more likely to respond to cellular stress, so-called danger signals, rather than sensing specific ligands (Fig. 14.2). Most NLRP3-inflammasome activators induce severe mitochondrial damage. Thus, mitochondrial damage is the major source of the danger signals. ROS generated by damaged mitochondria could potently activate the NLRP3-inflammasome (Naik and Dixit 2011). In addition, mitochondrial DNA or the mitochondrial inner membrane component cardiolipin, which is released from damaged mitochondria, can directly activate the NLRP3-inflammasome (Shimada et al. 2012; Iyer et al. 2013). Autophagy is an intracellular clearance system that promotes turnover of aged organelles and elimination of damaged organelles. Thus, autophagy-deficient macrophages show enhanced production of IL-1β and IL-18 in response to NLRP3-inflammasome activators, due to the defective clearance of damaged mitochondria (Saitoh et al. 2008; Nakahira et al. 2011; Saitoh and Akira 2016). Consistently, the mitochondria-specific autophagy system, mitophagy, may also regulate NLRP3-inflammasome activation (Zhou et al. 2011).

Dramatic changes in intracellular cation levels is another danger signal that can activate the NLRP3-inflammasome. Release of adenosine triphosphate (ATP) from dying cells or the bacterial toxin nigericin, which induces efflux of intracellular potassium, can potently activate the NLRP3-inflammasome (Mariathasan et al. 2006; Petrilli et al. 2007; Franchi et al. 2007; Munoz-Planillo et al. 2013). The intracellular calcium level is also critical for NLRP3-inflammasome activation. Transient receptor potential (TRP) cation channels such as TRPM2, TRPM7, and TRPV2 induce a calcium influx in macrophages in response to NLRP3-inflammasome activators (Zhong et al. 2013; Compan et al. 2012). G-protein coupled receptors (GPR) such as calcium-sensing receptor (CASR) and GPR class C group 6 member A (GPRC6A) sense high levels of extracellular calcium and activate phospholipase C-mediated generation of inositol 1,4,5-triphosphate (IP3). This triggers release of calcium from the ER to the cytoplasmic space via IP3 receptors (Murakami et al. 2012; Rossol et al. 2012; Lee et al. 2012). Mitochondria uptake calcium to buffer the cytosolic calcium level. However, an overload of calcium leads to mitochondrial dysfunction and accelerates NLRP3-inflammasome activation (Triantafilou et al. 2013). This

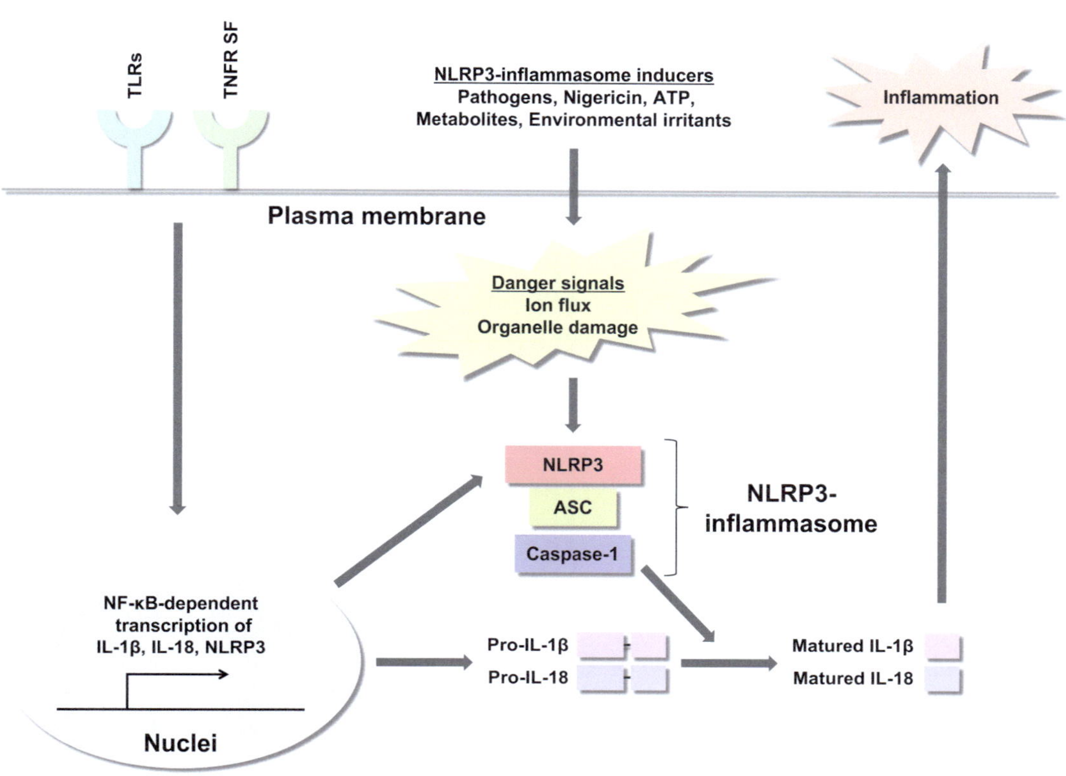

Fig. 14.2 NLRP3-inflammasome activation pathways. Two processes are required for maximal activation of the NLRP3-inflammasome. The first is induced by TLR signaling and induces activation of transcription factors, such as NF-κB. This upregulates expression of IL-1β and IL-18, which encode precursor forms of these cytokines. The second is the assembly and activation of the NLRP3-inflammasome, triggered by PAMPs or DAMPs. The NLRP3-inflammasome induces processing of pro-IL-1β and pro-IL-18 by caspase-1. Mature IL-β and IL-18 are released to the extracellular space, where they mediate a variety of inflammatory responses

pathway could be regulated by voltage-dependent anion channels (VDAC), which are indispensable for transferring calcium into mitochondria. Consistently, knockdown of VDAC in the human monocyte cell line THP-1 suppresses mitochondrial ROS production in response to NLRP3-inflammasome activators and subsequent maturation of IL-1β (Zhou et al. 2011).

Other organelles, such as lysosomes, the ER, and the Golgi, are also involved in activation of the NLRP3-inflammasome. Host-derived metabolites (monosodium urate crystals, cholesterol crystals), amyloid toxic oligomers (amyloid β, human islet amyloid polypeptides), and environmental irritants (silica crystals, asbestos, diesel particulates) are well-known activators of the NLRP3-inflammasome (Martinon et al. 2006; Eisenbarth et al. 2008; Dostert et al. 2008; Hornung et al. 2008; Duewell et al. 2010; Masters et al. 2010; Vandanmagsar et al. 2011; Wen et al. 2011; Heneka et al. 2012). When these crystals or particles are engulfed by macrophages, they induce lysosomal rupture, leading to leakage of acidic lysosomal contents and subsequent activation of the NLRP3-inflammasome. Moreover, since lysosomes contain a high level of calcium, lysosomal rupture elevates the cytosolic calcium level, followed by activation of a series of mitogen-activated protein kinases and the NLRP3-inflammasome (Okada et al. 2014). The ER stress sensor inositol-requiring enzyme 1α (IRE1α) has been shown to promote transcription of thioredoxin-interacting protein, an amplifier of the NLRP3-inflammasome (Zhou et al. 2010; Bronner et al. 2015). Influenza A virus-encoded M2 ion channels localize to the host trans-Golgi network (TGN) and induce Golgi stress. M2 ion channels will then accelerate H^+ release from the TGN to the cytosol. This ionic imbalance triggers NLRP3-inflammasome activation (Ichinohe et al. 2010). Lysosomal rupture, ER stress, and influenza A viral infection cause mitochondrial damage, leading to production of ROS (Misawa et al. 2013; Lupfer et al. 2013; Bronner et al. 2015). Because BAX and BAK-1, mitochondrial executioners of apoptosis, are dispensable for NLRP3-inflammasome activation, NLRP3-inflammasome activators cause mitochondrial damage by undefined mechanisms (Allam et al. 2014).

14.2.3 Assembly of the NLRP3-Inflammasome at the MAM

Spatiotemporal regulation of organelles plays a pivotal role in activating the NLRP3-inflammasome. Mitochondrial damage induced by NLRP3-inflammasome activators leads to accumulation of NLRP3 and its adaptor ASC, at the MAM (Zhou et al. 2011). ROS are fragile and thus can only be mobilized over short distances. Therefore, it is reasonable that the NLRP3-inflammasome would be formed at the MAM, for efficient activation by danger signals derived from damaged mitochondria.

Interferon-beta promoter stimulator 1 (IPS-1; also known as MAVS, VISA, or Cardif) is critical for recruiting NLRP3 to mitochondria and subsequent assembly of the NLRP3-inflammasome (Subramanian et al. 2013). IPS-1 is a mitochondria-associated molecule that is best characterized for its role in antiviral responses mediated by RLRs. IPS-1 deficient macrophages show impaired IL-1β production in response to ATP or nigericin. On the other hand, IPS-1 is dispensable for NLRP3-inflammasome activation induced by particles such as MSU or alum. Thus, particle-induced NLRP3-inflammasome activation could occur in an IPS-1 independent manner, via an undefined mechanism.

Mitochondrial dynamics mediated by the protein mitofusin (MFN) also affect NLRP3-inflammasome activation. MFN1 and MFN2 are GTPases localized on the outer membrane of mitochondria that regulate mitochondrial fusion. MFN2 is also expressed on the MAM to promote mitochondria–ER tethering. Ichinohe et al. have shown that MFN2, but not MFN1, is indispensable for activating the NLRP3-inflammasome during RNA virus infection (Ichinohe et al. 2013). When cells are infected with an RNA virus, MFN2 directly interact with NLRP3 and IPS-1, in a mitochondrial membrane potential-dependent manner. Thus, MFN2 may serve as a scaffold for formation and activation of the NLRP3-inflammasome at the MAM. Further studies are required to clarify whether a similar pathway could also be induced by other NLRP3-inflammasome activators.

Microtubule-driven mitochondria–ER contact promotes assembly of the NLRP3-inflammasome (Misawa et al. 2013; Misawa et al. 2015). Under resting conditions, endogenous ASC localizes on mitochondria, in the nucleus, and the cytoplasm, while NLRP3 mainly localizes in the ER in macrophages. Mitochondrial damage induced by NLRP3-inflammasome inducers causes reduction of intracellular NAD^+ levels, followed by inactivation of the NAD^+-dependent deacetylase SIRT2. This enhances accumulation of acetylated α-tubulin. Dynein, a molecular motor, transports mitochondria and the ER along acetylated microtubules, resulting in contact between ASC on mitochondria and NLRP3 on the ER. Danger signals, such as ROS or potassium efflux, then amplify NLRP3-inflammasome signaling (Fig. 14.3).

14.3 The MAM and the Intracellular PRR-Mediated Antiviral Response

14.3.1 Involvement of the MAM in the RIG-I-IPS-1 Signaling Axis

The MAM not only serves as a platform for NLRP3-inflammasome assembly but also for function of the antiviral innate immune system. RIG-I is a PRR that senses cytosolic double-stranded (ds) RNA and induces an antiviral

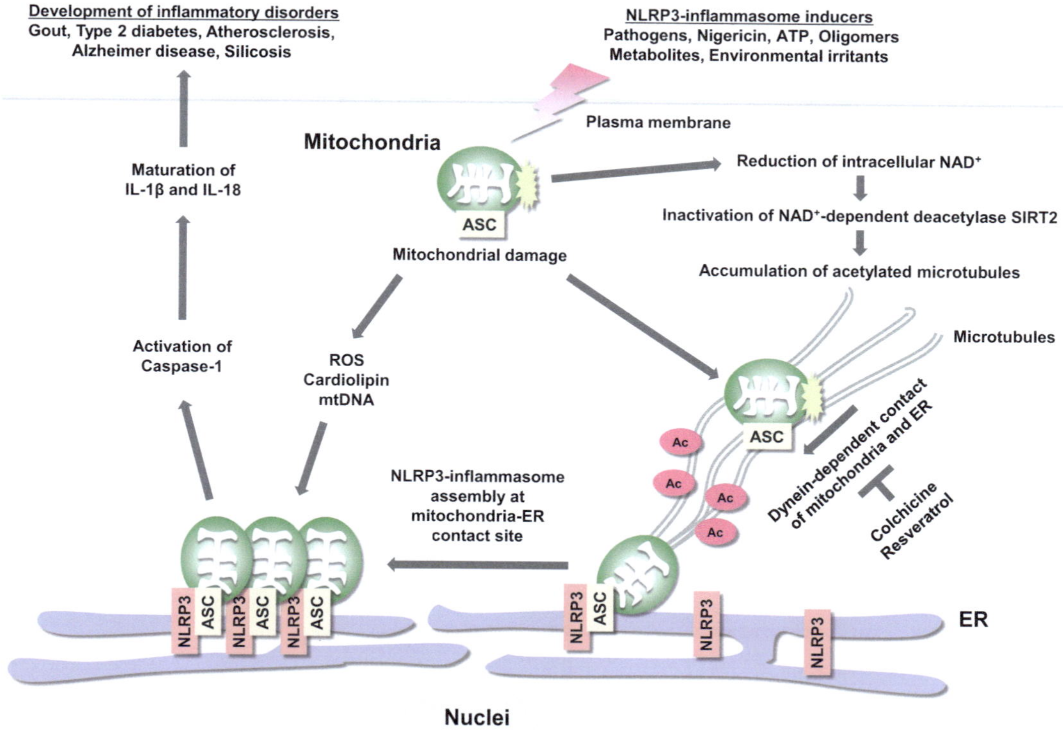

Fig. 14.3 Acetylated microtubules drive contact between mitochondria and the ER to promote NLRP3-inflammasome assembly. NLRP3-inflammasome activators severely damage mitochondria, followed by reduction of the intracellular NAD^+ level. Subsequent inactivation of the NAD^+-dependent deacetylase SIRT2 leads to accumulation of acetylated α-tubulin. Acetylated microtubules accelerate dynein-dependent contact between mitochondria and the ER, resulting in assembly of ASC on mitochondria and NLRP3 on the ER. Danger signals, such as potassium efflux or mitochondrial ROS, induce subsequent NLRP3-inflammasome activation, leading to production of IL-1β and IL-18

response against RNA viruses (Yoneyama et al. 2015). RIG-I is an RNA helicase that recognizes viral RNA and interacts with the adaptor IPS-1 to induce expression of type I interferons and antiviral genes. The MAM functions as a site for assembly of RIG-I and IPS-1 (Horner et al. 2011). Under resting conditions, RIG-I localizes in the cytosol, whereas IPS-1 localizes on mitochondria, the MAM, and peroxisomes. During hepatitis C virus infection, RIG-I preferentially accumulates at the MAM, to facilitate interaction with IPS-1. Interestingly, hepatitis C virus cleaves IPS-1 from the MAM by its NS3/4A protease, to escape RIG-I-mediated antiviral responses. MFN1 promotes activation of the RIG-I-dependent antiviral response, whereas MFN2 limits the response (Yasukawa et al. 2009; Onoguchi et al. 2010; Castanier et al. 2010; Horner et al. 2011) (Fig. 14.4).

14.3.2 Involvement of the MAM in the cGAS-STING Signaling Axis

The nucleotidyltransferase family member cyclic GMP-AMP synthase (cGAS) is a recently identified PRR that senses cytosolic dsDNA and induces an antiviral response against DNA viruses (Cai et al. 2014). Upon binding to dsDNA, cGAS synthesizes cyclic-GMP-AMP (cGAMP) from ATP and GTP. Stimulator of interferon genes (STING), an ER-resident sensor, detects cGAMP to induce expression of type

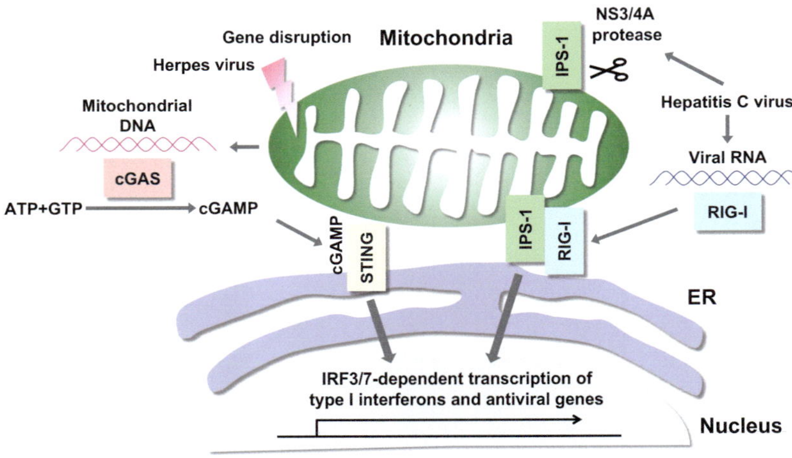

Fig. 14.4 The MAM functions as a platform for the antiviral innate immune system. Upon hepatitis C virus infection, RIG-I detects viral RNA and is recruited to the MAM to interact with its adaptor IPS-1, leading to activation of downstream signaling pathways. Herpes virus infection and mitochondrial dysfunction accelerate release of mitochondrial DNA into the cytosol. cGAS detects cytosolic mitochondrial DNA and is recruited to the MAM to produce cGAMP, which in turn activates STING, an ER-resident downstream signal transducer, leading to activation of downstream signaling pathways

I interferons and antiviral genes. Herpes virus infection and mitochondrial dysfunction induce release of mitochondrial DNA, which in turn activates cGAS (West et al. 2015). At the MAM, binding of cGAS to mitochondrial DNA and STING on the ER membrane are temporally clustered. Therefore, it is possible that the MAM facilitates detection of cGAMP via STING to promote an antiviral response.

14.4 Conclusion

The MAM plays pivotal roles in regulation of intracellular PRR-triggered innate immune responses. The MAM serves as an optimal platform for the innate immune system and promotes host defense against invading pathogens. However, inappropriate activation of a MAM-mediated innate immune response often results in development of diverse inflammatory disorders. Therefore, strict regulation of the MAM-mediated innate immune response is essential. The unique features of the MAM have only been revealed in the past few decades, and its functions are not completely understood. Future investigation is needed to determine the molecular basis of the MAM and to further illuminate its pathophysiological role in a variety of cellular events, including innate immune responses.

References

Agostini L, Martinon F, Burns K, McDermott MF, Hawkins PN, Tschopp J (2004) NALP3 forms an IL-1-β-processing inflammasome with increased activity in Muckle-Wells autoinflammatory disorder. Immunity 20:319–325

Aksentijevich I, Nowak M, Mallah M, Chae JJ, Watford WT, Hofmann SR, Stein L, Russo R, Goldsmith D, Dent P et al (2002) De novo CIAS1 mutations, cytokine activation, and evidence for genetic heterogeneity in patients with neonatal-onset multisystem inflammatory disease (NOMID): a new member of the expanding family of pyrin-associated autoinflammatory diseases. Arthritis Rheum 46:3340–3348

Allam R, Lawlor KE, Yu EC, Mildenhall AL, Moujalled DM, Lewis RS, Ke F, Mason KD, White MJ, Stacey KJ et al (2014) Mitochondrial apoptosis is dispensable for NLRP3 inflammasome activation but non-apoptotic caspase-8 is required for inflammasome priming. EMBO Rep 15:982–990

Bronner DN, Abuaita BH, Chen X, Fitzgerald KA, Nuñez G, He Y, Yin XM, O'Riordan MX (2015) Endoplasmic reticulum stress activates the inflammasome via NLRP3- and caspase-2-driven mitochondrial damage. Immunity 43:451–462

Broz P, Newton K, Lamkanfi M, Mariathasan S, Dixit VM, Monack DM (2010) Redundant roles for inflammasome receptors NLRP3 and NLRC4 in host defense against Salmonella. J Exp Med 207:1745–1755

Cai X, Chiu YH, Chen ZJ (2014) The cGAS-cGAMP-STING pathway of cytosolic DNA sensing and signaling. Mol Cell 54:289–296

Castanier C, Garcin D, Vazquez A, Arnoult D (2010) Mitochondrial dynamics regulate the RIG-I-like receptor antiviral pathway. EMBO Rep 11:133–138

Compan V, Baroja-Mazo A, López-Castejón G, Gomez AI, Martínez CM, Angosto D, Montero MT, Herranz AS, Bazán E, Reimers D et al (2012) Cell volume regulation modulates NLRP3 inflammasome activation. Immunity 37:487–500

de Zoete MR, Palm NW, Zhu S, Flavell RA (2014) Inflammasomes. Cold Spring Harb Perspect Biol 6: a016287

Dorhoi A, Nouailles G, Jörg S, Hagens K, Heinemann E, Pradl L, Oberbeck-Müller D, Duque-Correa MA, Reece ST, Ruland J et al (2012) Activation of the NLRP3 inflammasome by Mycobacterium tuberculosis is uncoupled from susceptibility to active tuberculosis. Eur J Immunol 42:374–384

Dostert C, Pétrilli V, Van Bruggen R, Steele C, Mossman BT, Tschopp J (2008) Innate immune activation through Nalp3 inflammasome sensing of asbestos and silica. Science 320:674–677

Duewell P, Kono H, Rayner KJ, Sirois CM, Vladimer G, Bauernfeind FG, Abela GS, Franchi L, Nuñez G, Schnurr M et al (2010) NLRP3 inflammasomes are required for atherogenesis and activated by cholesterol crystals. Nature 464:1357–1361

Eisenbarth SC, Colegio OR, O'Connor W, Sutterwala FS, Flavell RA (2008) Crucial role for the Nalp3 inflammasome in the immunostimulatory properties of aluminium adjuvants. Nature 453:1122–1126

Ewald SE, Lee BL, Lau L, Wickliffe KE, Shi GP, Chapman HA, Barton GM (2008) The ectodomain of Toll-like receptor 9 is cleaved to generate a functional receptor. Nature 456:658–662

Feldmann J, Prieur AM, Quartier P, Berquin P, Certain S, Cortis E, Teillac-Hamel D, Fischer A, de Saint BG (2002) Chronic infantile neurological cutaneous and articular syndrome is caused by mutations in CIAS1, a gene highly expressed in polymorphonuclear cells and chondrocytes. Am J Hum Genet 71:198–203

Franchi L, Kanneganti TD, Dubyak GR, Núñez G (2007) Differential requirement of P2X7 receptor and intracellular K^+ for caspase-1 activation induced by intracellular and extracellular bacteria. J Biol Chem 282:18810–18818

Franke WW, Kartenbeck J (1971) Outer mitochondrial membrane continuous with endoplasmic reticulum. Protoplasma 73:35–41

Heneka MT, Kummer MP, Stutz A, Delekate A, Schwartz S, Vieira-Saecker A, Griep A, Axt D, Remus A, Tzeng TC et al (2012) NLRP3 is activated in Alzheimer's disease and contributes to pathology in APP/PS1 mice. Nature 493:674–678

Hise AG, Tomalka J, Ganesan S, Patel K, Hall BA, Brown GD, Fitzgerald KA (2009) An essential role for the NLRP3 inflammasome in host defense against the human fungal pathogen Candida albicans. Cell Host Microbe 5:487–497

Hoffman HM, Mueller JL, Broide DH, Wanderer AA, Kolodner RD (2001) Mutation of a new gene encoding a putative pyrin-like protein causes familial cold autoinflammatory syndrome and Muckle-Wells syndrome. Nat Genet 29:301–305

Horner SM, Liu HM, Park HS, Briley J, Gale M Jr (2011) Mitochondrial-associated endoplasmic reticulum membranes (MAM) form innate immune synapses and are targeted by hepatitis C virus. Proc Natl Acad Sci USA 108:14590–14595

Hornung V, Bauernfeind F, Halle A, Samstad EO, Kono H, Rock KL, Fitzgerald KA, Latz E (2008) Silica crystals and aluminum salts activate the NALP3 inflammasome through phagosomal destabilization. Nat Immunol 9:847–856

Hoshino K, Takeuchi O, Kawai T, Sanjo H, Ogawa T, Takeda Y, Takeda K, Akira S (1999) Toll-Like Receptor 4 (TLR4)-deficient mice are hyporesponsive to lipopolysaccharide: Evidence for TLR4 as the Lps gene product. J Immunol 162:3749–3752

Ichinohe T, Pang IK, Iwasaki A (2010) Influenza virus activates inflammasomes via its intracellular M2 ion channel. Nat Immunol 11:404–410

Ichinohe T, Yamazaki T, Koshiba T, Yanagi Y (2013) Mitochondrial protein mitofusin 2 is required for NLRP3 inflammasome activation after RNA virus infection. Proc Natl Acad Sci USA 110:17963–17968

Iyer SS, He Q, Janczy JR, Elliott EI, Zhong Z, Olivier AK, Sadler JJ, Knepper-Adrian V, Han R, Qiao L et al (2013) Mitochondrial cardiolipin is required for Nlrp3 inflammasome activation. Immunity 39:311–323

Kagan JC, Su T, Horng T, Chow A, Akira S, Medzhitov R (2008) TRAM couples endocytosis of Toll-like receptor 4 to the induction of interferon-β. Nat Immunol 9:361–368

Kim S, Bauernfeind F, Ablasser A, Hartmann G, Fitzgerald KA, Latz E, Hornung V (2010) Listeria monocytogenes is sensed by the NLRP3 and AIM2 inflammasome. Eur J Immunol 40:1545–1551

Lee GS, Subramanian N, Kim AI, Aksentijevich I, Goldbach-Mansky R, Sacks DB, Germain RN, Kastner DL, Chae JJ (2012) The calcium-sensing receptor regulates the NLRP3 inflammasome through Ca^{2+} and cAMP. Nature 492:123–127

Lemaitre B, Nicolas E, Michaut L, Reichhart JM, Hoffmann JA (1996) The dorsoventral regulatory gene cassette spätzle/Toll/cactus controls the potent sntifungal response in *drosophila* adults. Cell 86:973–983

Lupfer C, Thomas PG, Anand PK, Vogel P, Milasta S, Martinez J, Huang G, Green M, Kundu M, Chi H et al (2013) Receptor interacting protein kinase 2-mediated mitophagy regulates inflammasome activation during virus infection. Nat Immunol 14:480–488

Mariathasan S, Weiss DS, Newton K, McBride J, O'Rourke K, Roose-Girma M, Lee WP, Weinrauch Y, Monack DM, Dixit VM (2006) Cryopyrin activates the inflammasome in response to toxins and ATP. Nature 440:228–232

Martinon F, Pétrilli V, Mayor A, Tardivel A, Tschopp J (2006) Gout-associated uric acid crystals activate the NALP3 inflammasome. Nature 440:237–241

Masters SL, Dunne A, Subramanian SL, Hull RL, Tannahill GM, Sharp FA, Becker C, Franchi L, Yoshihara E, Chen Z et al (2010) Activation of the NLRP3 inflammasome by islet amyloid polypeptide provides a mechanism for enhanced IL-1β in type 2 diabetes. Nat Immunol 11:897–904

Medzhitov R, Preston-Hurlburt P, Janeway CA Jr (1997) A human homologue of the Drosophila Toll protein signals activation of adaptive immunity. Nature 388:394–397

Misawa T, Takahama M, Kozaki T, Lee H, Zou J, Saitoh T, Akira S (2013) Microtubule-driven spatial arrangement of mitochondria promotes activation of the NLRP3 inflammasome. Nat Immunol 14:454–460

Misawa T, Saitoh T, Kozaki T, Park S, Takahama M, Akira S (2015) Resveratrol inhibits the acetylated α-tubulin-mediated assembly of NLRP3-inflammasome. Int Immunol 27:425–434

Morré DJ, Merritt WD, Lembi CA (1971) Connections between mitochondria and endoplasmic reticulum in rat liver and onion stem. Protoplasma 73:43–49

Muñoz-Planillo R, Kuffa P, Martínez-Colón G, Smith BL, Rajendiran TM, Núñez G (2013) K^+ efflux is the common trigger of NLRP3 inflammasome activation by bacterial toxins and particulate matter. Immunity 38:1142–1153

Murakami T, Ockinger J, Yu J, Byles V, McColl A, Hofer AM, Horng T (2012) Critical role for calcium mobilization in activation of the NLRP3 inflammasome. Proc Natl Acad Sci USA 109:11282–11287

Naik E, Dixit VM (2011) Mitochondrial reactive oxygen species drive proinflammatory cytokine production. J Exp Med 208:417–420

Nakahira K, Haspel JA, Rathinam VA, Lee SJ, Dolinay T, Lam HC, Englert JA, Rabinovitch M, Cernadas M, Kim HP et al (2011) Autophagy proteins regulate innate immune responses by inhibiting the release of mitochondrial DNA mediated by the NALP3 inflammasome. Nat Immunol 12:222–230

Neven B, Callebaut I, Prieur AM, Feldmann J, Bodemer C, Lepore L, Derfalvi B, Benjaponpitak S, Vesely R, Sauvain MJ et al (2004) Molecular basis of the spectral expression of CIAS1 mutations associated with phagocytic cell-mediated autoinflammatory disorders CINCA/NOMID, MWS, and FCU. Blood 103:2809–2815

Oka T, Hikoso S, Yamaguchi O, Taneike M, Takeda T, Tamai T, Oyabu J, Murakawa T, Nakayama H, Nishida K et al (2012) Mitochondrial DNA that escapes from autophagy causes inflammation and heart failure. Nature 485:251–255

Okada M, Matsuzawa A, Yoshimura A, Ichijo H (2014) The lysosome rupture-activated TAK1-JNK pathway regulates NLRP3 inflammasome activation. J Biol Chem 289:32926–32936

Onoguchi K, Onomoto K, Takamatsu S, Jogi M, Takemura A, Morimoto S, Julkunen I, Namiki H, Yoneyama M, Fujita T (2010) Virus-infection or 5'ppp-RNA activates antiviral signal through redistiribution of IPS-1 mediated by MFN1. PLoS Pathog 6:e1001012

Pandey S, Kawai T, Akira S (2014) Microbial sensing by Toll-like receptors and intracellular nucleic acid sensors. Cold Spring Harb Perspect Biol 7:a016246

Park B, Brinkmann MM, Spooner E, Lee CC, Kim YM, Ploegh HL (2008) Proteolytic cleavage in an endolysosomal compartment is required for activation of Toll-like receptor 9. Nat Immunol 9:1407–1414

Pétrilli V, Papin S, Dostert C, Mayor A, Martinon F, Tschopp J (2007) Activation of the NALP3 inflammasome is triggered by low intracellular potassium concentration. Cell Death Differ 14:1583–1589

Poltorak A, He X, Smirnova I, Liu MY, Van Huffel C, Du X, Birdwell D, Alejos E, Silva M, Galanos C et al (1998) Defective LPS signaling in C3H/HeJ and C57BL/10ScCr mice: mutations in Tlr4 gene. Science 282:2085–2088

Rossol M, Pierer M, Raulien N, Quandt D, Meusch U, Rothe K, Schubert K, Schöneberg T, Schaefer M, Krügel U et al (2012) Extracellular Ca^{2+} is a danger signal activating the NLRP3 inflammasome through G protein-coupled calcium sensing receptors. Nat Commun 3:1329

Saitoh T, Akira S (2016) Regulation of inflammasomes by autophagy. J Allergy Clin Immunol 138:28–36

Saitoh T, Fujita N, Jang MH, Uematsu S, Yang BG, Satoh T, Omori H, Noda T, Yamamoto N, Komatsu M et al (2008) Loss of the autophagy protein Atg16L1 enhances endotoxin-induced IL-1β production. Nature 456:264–268

Shibata T, Takemura N, Motoi Y, Goto Y, Karuppuchamy T, Izawa K, Li X, Akashi-Takamura S, Tanimura N, Kunisawa J et al (2012) PRAT4A-dependent expression of cell surface TLR5 on neutrophils, classical monocytes and dendritic cells. Int Immunol 24:613–623

Shimada K, Crother TR, Karlin J, Dagvadorj J, Chiba N, Chen S, Ramanujan VK, Wolf AJ, Vergnes L, Ojcius DM et al (2012) Oxidized mitochondrial DNA activates the NLRP3 inflammasome during apoptosis. Immunity 36:401–414

Subramanian N, Natarajan K, Clatworthy MR, Wang Z, Germain RN (2013) The adaptor MAVS promotes NLRP3 mitochondrial localization and inflammasome activation. Cell 153:348–361

Tabeta K, Hoebe K, Janssen EM, Du X, Georgel P, Crozat K, Mudd S, Mann N, Sovath S, Goode J et al (2006) The Unc93b1 mutation 3d disrupts exogenous antigen presentation and signaling via Toll-like receptors 3, 7 and 9. Nat Immunol 7:156–164

Takahashi K, Shibata T, Akashi-Takamura S, Kiyokawa T, Wakabayashi Y, Tanimura N, Kobayashi T, Matsumoto F, Fukui R, Kouro T et al (2007) A protein associated with Toll-like receptor (TLR) 4 (PRAT4A) is required for TLR-dependent immune responses. J Exp Med 204:2963–2976

Takeuchi O, Akira S (2010) Pattern recognition receptors and inflammation. Cell 140:805–820

Tannahill GM, Curtis AM, Adamik J, Palsson-McDermott EM, McGettrick AF, Goel G, Frezza C, Bernard NJ, Kelly B, Foley NH et al (2013) Succinate is an inflammatory signal that induces IL-1β through HIF-1α. Nature 496:238–242

Triantafilou K, Hughes TR, Triantafilou M, Morgan BP (2013) The complement membrane attack complex triggers intracellular Ca^{2+} fluxes leading to NLRP3 inflammasome activation. J Cell Sci 126:2903–2913

Vance JE (1990) Phospholipid synthesis in a membrane fraction associated with mitochondria. J Biol Chem 265:7248–7256

Vance JE (2014) MAM (mitochondria-associated membranes) in mammalian cells: lipids and beyond. Biochim Biophys Acta 1841:595–609

Vandanmagsar B, Youm YH, Ravussin A, Galgani JE, Stadler K, Mynatt RL, Ravussin E, Stephens JM, Dixit VD (2011) The NLRP3 inflammasome instigates obesity-induced inflammation and insulin resistance. Nat Med 7:179–188

Wen H, Gris D, Lei Y, Jha S, Zhang L, Huang MT, Brickey WJ, Ting JP (2011) Fatty acid-induced NLRP3-ASC inflammasome activation interferes with insulin signaling. Nat Immunol 12:408–415

West AP, Brodsky IE, Rahner C, Woo DK, Erdjument-Bromage H, Tempst P, Walsh MC, Choi Y, Shadel GS, Ghosh S (2011) TLR signalling augments macrophage bactericidal activity through mitochondrial ROS. Nature 472:476–480

West AP, Khoury-Hanold W, Staron M, Tal MC, Pineda CM, Lang SM, Bestwick M, Duguay BA, Raimundo N, MacDuff DA et al (2015) Mitochondrial DNA stress primes the antiviral innate immune response. Nature 520:553–557

Yasukawa K, Ohiumi H, Takeda M, Ishihara N, Yanagi Y, Seya T, Kawabata S, Koshiba T (2009) Mitofusin 2 inhibits mitochondrial antiviral signaling. Sci Signal 2:ra47

Yoneyama M, Onomoto K, Jogi M, Akaboshi T, Fujita T (2015) Viral RNA detection by RIG-I-like receptors. Curr Opin Immunol 32:48–53

Zanoni I, Ostuni R, Marek LR, Barresi S, Barbalat R, Barton GM, Granucci F, Kagan JC (2011) CD14 controls the LPS-induced endocytosis of Toll-like receptor 4. Cell 147:868–880

Zhong Z, Zhai Y, Liang S, Mori Y, Han R, Sutterwala FS, Qiao L (2013) TRPM2 links oxidative stress to NLRP3 inflammasome activation. Nat Commun 4:1611

Zhou R, Tardivel A, Thorens B, Choi I, Tschopp J (2010) Thioredoxin-interacting protein links oxidative stress to inflammasome activation. Nat Immunol 11:136–140

Zhou R, Yazdi AS, Menu P, Tschopp J (2011) A role for mitochondria in NLRP3 inflammasome activation. Nature 469:221–225

Hepatitis C Virus Replication

15

Tetsuro Suzuki

Abstract

Viruses use synthetic mechanism and organelles of the host cells to facilitate their replication and make new viruses. Host's ATP provides necessary energy. Hepatitis C virus (HCV) is a major cause of liver disease. Like other positive-strand RNA viruses, the HCV genome is thought to be synthesized by the replication complex, which consists of viral- and host cell-derived factors, in tight association with structurally rearranged vesicle-like cytoplasmic membranes. The virus-induced remodeling of subcellular membranes, which protect the viral RNA from nucleases in the cytoplasm, promotes efficient replication of HCV genome. The assembly of HCV particle involves interactions between viral structural and nonstructural proteins and pathways related to lipid metabolisms in a concerted fashion. Association of viral core protein, which forms the capsid, with lipid droplets appears to be a prerequisite for early steps of the assembly, which are closely linked with the viral genome replication. This review presents the recent progress in understanding the mechanisms for replication and assembly of HCV through its interactions with organelles or distinct organelle-like structures.

Keywords

Hepatitis C virus • Replication factory • Endoplasmic reticulum • Membrane vesicle • Mitochondria • ATP • Virus-host interaction

15.1 Introduction

Hepatitis C virus (HCV) is a major etiologic agent of human liver diseases, infecting approximately 2% of the population worldwide, and typically establishes a chronic infection in the liver, leading to chronic hepatitis, liver cirrhosis,

T. Suzuki (✉)
Department of Virology and Parasitology, Hamamatsu University School of Medicine, Hamamatsu, Shizuoka 431-3192, Japan
e-mail: tesuzuki@hama-med.ac.jp

M. Tagaya, T. Simmen (eds.), *Organelle Contact Sites*, Advances in Experimental Medicine and Biology 997, DOI 10.1007/978-981-10-4567-7_15

and hepatocellular carcinoma (Wasley and Alter 2000; Pawlotsky 2006). HCV belongs to the genus *Hepacivirus*, a member of the family *Flaviviridae* (as reviewed in Lindenbach et al. 2007). The viral particle consists of a nucleocapsid, surrounded by a lipid envelope containing two viral glycoproteins, E1 and E2. HCV has a positive-sense plus-stranded RNA genome, which is approximately 9.6 kb in length and consists of an open reading frame encoding a polyprotein of ~3000 amino acids and untranslated regions (UTRs) located at the 5′ and 3′ termini. The UTRs are highly structured sequences encompassing critical *cis*-active RNA elements essential for genome replication and translation. The large precursor polyprotein is cleaved into 10 different proteins, the structural proteins core, E1, E2, and p7 and the nonstructural (NS) proteins NS2, NS3, NS4A, NS4B, NS5A, and NS5B (Fig. 15.1). The structural components are released from the precursor by cellular proteases, whereas the mature NS proteins are produced by virus-encoded proteases. NS3 to NS5B proteins are presumably not incorporated into viral particles but instead assemble into a membrane-associated HCV RNA replicase complex catalyzing RNA replication (Gosert et al. 2003). The N-terminal side of NS3 harbors a serine protease domain, which contributes to cleavage of the four downstream NS protein junctions (Bartenschlager et al. 1993; Bartenschlager et al. 1995). The C-terminal portion of NS3 possesses activities of RNA helicase and NTPase that are essential for the viral replication (Suzich et al. 1993; Kim et al. 1995). NS4A is a transmembrane protein and serves as a cofactor for NS3 and is involved in targeting NS3 to the ER membrane (Wolk et al. 2000). NS4B contains multiple transmembrane segments and plays a role in the remodeling of host cell membranes (Egger et al. 2002; Lundin et al. 2003; Ferraris et al. 2010), potentially to generate a site for assembly of the viral replication complex (Egger et al. 2002; Lundin et al. 2003; Gouttenoire et al. 2010; Romero-Brey et al. 2012; Paul et al. 2013). NS5A is a membrane-anchored phosphoprotein with no enzyme activity but is important for the viral genome replication (Ross-Thrieplend and Harris 2015; Masaki and Suzuki 2015). NS5B functions as an RNA-dependent RNA polymerase with a C-terminal membrane-anchoring segment (Schmidt-Mende et al. 2001; Moradpour et al. 2004). In addition, the roles of HCV NS proteins such as NS2, NS3, and NS5A in assembly of the infectious particles have been shown (Suzuki 2012; Lindenbach 2013).

15.2 Cytoplasmic Membrane Reorganization for Supporting Replication of the HCV Genome

To facilitate RNA replication, HCV interacts with numerous host molecules through protein–protein, RNA–protein, and protein–lipid interactions. These interactions lead to the formation of viral replication complexes, which produce new viral RNA progeny in host cells. While the HCV genomic RNA is translated on the endoplasmic reticulum (ER), the viral genome is thought to replicate within a complex network of structurally rearranged vesicle-like ER-derived membranes, which are often called replication factory (Fig. 15.1). Like other positive-sense RNA viruses, HCV replication or expression of certain viral proteins induces the altered membrane structures as observed ultrastructurally in cultured cells (Egger et al. 2002; Mottola et al. 2002; Gosert et al. 2003; El-Hage and Luo 2003; Shi et al. 2003; Ferraris et al. 2010). It is generally considered for positive-sense RNA viruses that the virus-induced remodeling of subcellular membranes potentially provides a framework for RNA synthesis by facilitating the concentration and cooperation of viral macromolecules on a dedicated membrane surface (Paul and Bartenschlager 2013). The membrane structures may protect the viral genomes from nucleases in the cytoplasm of the host cell, aid in shielding the double-stranded RNA intermediates of virus replication from the innate immune system in host cells, and/or contribute to the coordination of the viral life cycle in time and space. It may also be likely that such

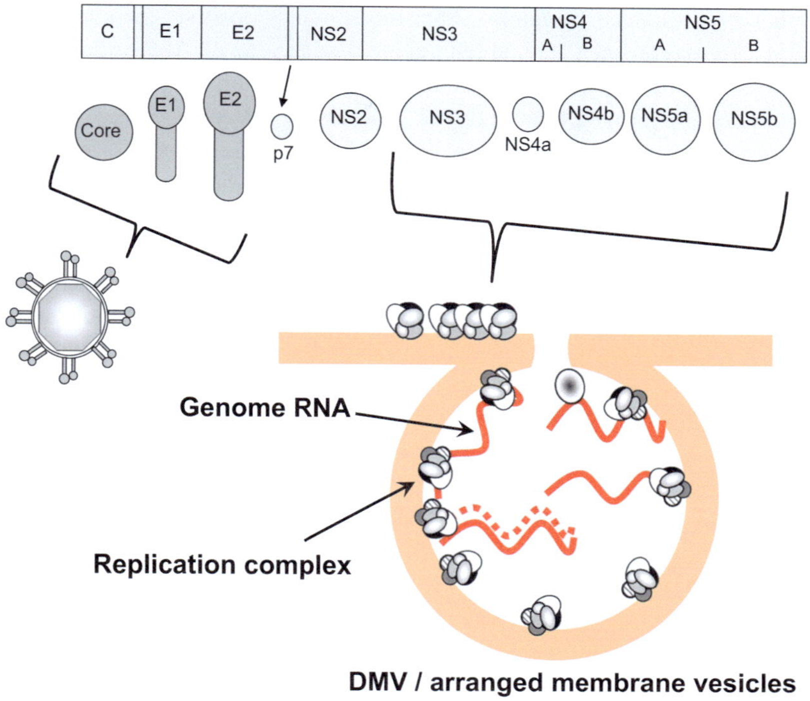

Fig. 15.1 Gene organization of HCV, processing of precursor polyprotein, and maturation of ten viral proteins (*upper*). Hypothetical model of membrane-associated replication of HCV genome (*lower*)

a membranous environment yet allows influx of nucleotide triphosphates and exit of newly synthesized genomic RNA molecules.

With regard to HCV replication, it has been demonstrated that upon HCV infection, accumulation of double-membrane vesicles (DMVs) with an average diameter of 150 nm was initially observed. The kinetics of DMV accumulation correlated with replication of HCV RNA; at later time points of infection, after the peak of HCV replication, multi-membrane vesicles with larger diameter of ~330 nm became more predominant (Romero-Brey et al. 2012). Three-dimensional reconstructions of electron microscopic tomography imaging indicated that most of the DMVs were tightly opposed to ER membranes and some of them were identified as protrusions from the ER membrane into the cytosol, suggesting that these membrane vesicles originate from ER membranes (Romero-Brey et al. 2012). Although the general mechanism for accumulation of virus-induced membrane vesicles and generation of membrane-associated replication complexes has been shown, hypothetical model of formation of HCV replication complex is proposed as follows. HCV NS proteins are translated and processed on the ER membrane, followed by colocalization of a fraction of NS3, NS4A, NS4B, NS5A, and NS5B proteins at certain sites on the membrane through the membrane anchoring of individual NS proteins as well as interactions between NS proteins. NS4B, a 27-kDa integral membrane protein, is thought to play an essential role in remodeling of the intracellular membrane and biogenesis of membranous web by self-oligomerization, providing a platform for the viral replication complex (Egger et al. 2002; Lundin et al. 2003; Ferraris et al. 2010; Gouttenoire et al. 2010; Romero-Brey et al. 2012; Paul et al. 2013). In case of the protrusion and detachment model, part of the ER cisterna nearby viral integrated membrane protein(s) starts to bend and pinches off, leading to sealing formation of DMV contained with the viral replication complex. Interactions between the lumenal domains of viral membrane proteins could mediate the tight apposition of the two bilayers and induce curvature. Other models for DMV formation such as double-budding model, in which a single-membrane vesicle buds into the lumen of the ER and then buds out again, may operate simultaneously in the viral-infected cells.

The potential role of cellular host factors in formation of HCV replication complex has also been investigated. HCV NS5A and NS5B have been shown to interact with vesicle-associated membrane protein-associated protein (VAP). VAP-A and VAP-B localize primarily at the cytoplasmic face of the ER and Golgi apparatus and are involved in maintaining ER homeostasis and vesicular trafficking, including the regulation of COPI-mediated transport. Recent studies have shown that VAP-A and VAP-B play roles at the contact of the ER with Golgi, endosomes, and mitochondria (also with plasma membrane in the case of yeast) (Wakana et al. 2015; Dong et al. 2016; Paillusson et al. 2016). In HCV-replicating cells, VAP-A can be detected in the detergent-resistant membrane fraction that contains the viral RC. Interaction between VAP-A and NS5A is required for the efficient replication of HCV genomic RNA (Gao et al. 2004). VAP-B may also be involved in viral replication through the formation of homodimers and/or heterodimers with VAP-A (Hamamoto et al. 2005). VAP-A and VAP-B form either hetero- or homodimers through their transmembrane regions and interact with NS5A and NS5B. Further studies showed that VAP-A and VAP-B are enriched in the cytoplasmic membrane compartment, the formation of which is induced by HCV RNA replication (Paul et al. 2013) and potentially bind NS4A (Ramage et al. 2015). In addition, VAP was shown to interact with the cellular protein NIR2, known to be implicated in membrane trafficking, and thus remodel the structure of the ER. The cellular lipid kinase PI4KA (also known as phosphatidylinositol 4-kinase III alpha, PI4KIIIα, and PIK4CA) was identified as a key factor for HCV replication.

Inhibition of PI4KA, either by RNA interference or by pharmacologic inhibitors (Tai et al. 2009; Reiss et al. 2011; Berger et al. 2011; Tai and Salloum 2011; Bianco et al. 2012), leads to accumulation of large "clusters" of NS5A-positive membranes at the light microscopic level. At the ultrastructural level, these "clusters" correspond to clusters of DMVs with reduced diameter, suggesting that PI4KA is essential for the proper formation and/or integrity of membrane rearrangement for the viral replication.

15.3 ATP Requirement During Virus Life Cycles

Adenosine 5′-triphosphate (ATP) is the major energy currency molecule of the cell and is essential for the maintenance of energy homeostasis of cells. Viruses are intracellular parasites that hijack cellular mechanisms including energy homeostasis to sponsor their replication. Several key steps during the viral life cycle, such as genome synthesis, duplex unwinding of viral genes, and assembly of virus particles via interaction with ATP-dependent chaperones, require high-energy phosphoryl groups.

It has been shown that ATP is required for the formation of a preinitiation complex for de novo RNA synthesis by RNA-dependent RNA polymerase (RdRp) of RNA viruses (Klumpp et al. 1998; Nomaguchi et al. 2003; Vreede et al. 2008). Dengue virus is a member of the *Flavivirus* genus of positive-strand RNA viruses. The dengue virus is causative of dengue fever and the most prevalent tropical infectious agent today. NS5, the largest among 10 viral proteins, contains conserved motif of RdRp consistent with the corresponding proteins encoded by several positive-strand RNA viruses. By using an in vitro viral RNA replicase assay, the components of the preinitiation complex for de novo synthesis of negative-strand RNA are defined as ATP, a high concentration of GTP, NS5 RdRp, and the template viral RNA. CTP and UTP are required only for the reaction of RNA elongation.

RNA helicases function in myriad aspects of RNA metabolism in cells and are also encoded by virus genomes (Frick and Lam 2006). HCV NS3 possesses helicase activity, which is essential for the viral genome replication (Suzich et al. 1993; Kim et al. 1995). NS3 has a DExH-box domain and potentially binds to a single-/double-strand RNA or DNA junction, leading to ATP-independent, local melting of the duplex region and formation of a complex capable of ATP-dependent unwinding. Members of the processive DExH group often translocate along single-strand RNA and displace paired strands in their path, presumably through consuming energy from ATP hydrolysis.

There is abundant evidence that heat shock proteins 70 (Hsp70) participate in the life cycles of various viruses. Hsp70 family members facilitate assembly and disassembly of oligomeric protein complexes as well as their folding and intracellular transport in an energy-dependent fashion (Mayer and Bukau 2005). Hsp70 proteins are upregulated by a variety of stresses including heat shock and virus infection. The proteins have three functional domains. The N-terminal ATPase domain binds ATP and hydrolyzes it to ADP. This exchange of ATP drives conformational changes in the other two domains, substrate-binding domain and the C-terminal domain rich in alpha-helical structure. Hsp70 is known to exist inside purified virions such as rabies virus and HIV and to be involved in the assembly of capsid protein complexes of adenovirus, enterovirus, and polyomavirus (Lahaye et al. 2012; Li et al. 2009). In case of HIV, the viral Gag polyprotein, which is required for assembly and release of enveloped virion from infected cells, is sufficient for Hsp70 incorporation. Further findings obtained with ATPgammaS, which is resistant to hydrolysis by Hsp70, support a model in which the catalytic activity of virion-associated Hsp70 is required to maintain structural integrity of virion core of retroviruses (Gurer et al. 2005).

15.4 ATP Imaging in HCV Replication

Although virus replication is known to be one of the physiological mechanisms that require energy for operation such as nucleic acid synthesis and its unwinding, it has been difficult to comprehend how the ATP level is regulated inside single living cells where virus replicates, since a method by which to monitor ATP production in real time and by which to quantify ATP in individual cells and subcellular compartments is lacking and average ATP values in cell extracts are only available from existing methods for ATP measurement. Thus, studies aimed at elucidating precise mechanisms for controlling the viral replication energized by ATP have not been carried out for several years. Imamura et al. have developed a method for measuring ATP levels in individual living cells using a genetically encoded Förster resonance energy transfer (FRET)-based indicator for ATP, called ATeam, composed of a small bacterial protein sandwiched between two fluorescent proteins (Imamura et al. 2009). This ATP indicator employs the epsilon subunit of a bacterial F_0F_1-ATPase that has advantages of highly specific binding to ATP and conformational change upon ATP binding. The ATeam enables us to examine the subcellular compartmentation of ATP as well as time-dependent changes in cellular ATP levels under various physiological conditions.

To visualize and monitor ATP levels in living cells during replication of the viral genome, we first introduced the original ATeam-expressing plasmids into cells and found that cytoplasmic ATP levels in cells undergoing HCV RNA replication were less than those in parental cells. The finding agreed with the results of CE-TOF MS analysis, demonstrating that cells involved in viral RNA replication actively consume ATP, thereby reducing cytoplasmic ATP levels (Ando et al. 2012). Next, NS5A-ATeam fusion constructs, in which the ATeam gene was introduced into the C-terminal end of the NS5A coding region, and SGR-ATeam constructs containing HCV subgenomic replicon within the NS5A-ATeam fused sequence were engineered (Fig. 15.2). It is of interest that FRET imaging using SGR-ATeam constructs provided evidence for the formation of ATP-enriched foci within the cytoplasm of HCV-replicating cells but not in non-replicating cells (Ando et al. 2012). FRET signal detection followed by indirect immunofluorescence

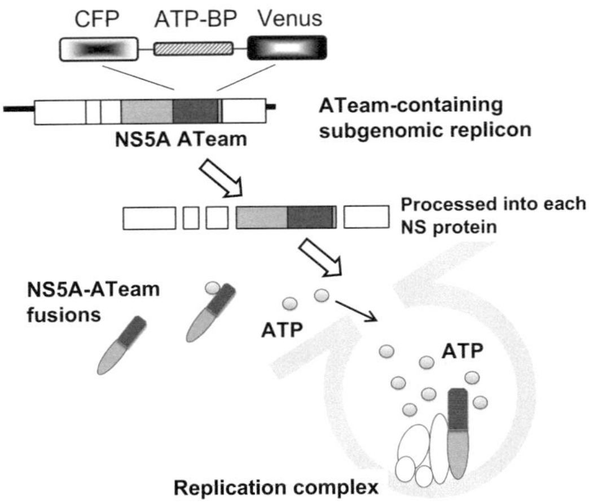

Fig. 15.2 Development of ATeam-containing HCV replicon to enable real-time monitoring of ATP. CFP, cyan fluorescent protein; ATP-BP, ATP-binding protein; Venus, yellow fluorescent protein (See the reference (Ando et al. 2012) for further description)

allowed us to visualize colocalization of viral NS proteins at sites of ATP accumulation in cells, suggesting that these membrane-associated ATP-enriched foci likely represent sites of HCV RNA replication. The obtained FRET signals allow us to estimate ATP concentrations within the viral replicating cells, which approximate 5–10 mM at possible replicating sites and ~1 mM at their peripheral sites that do not appear to be involved in HCV replication. In contrast, cytoplasmic ATP levels in non-replicating HuH-7 cells are ~2 mM (Ando et al. 2012).

The mechanism by which ATP accumulates at potential sites of HCV RNA replication is unknown. We demonstrated that creatine kinase B (CKB), which is critical for the maintenance and regulation of cellular energy stores, accumulates in the HCV replication complex-rich fraction of the viral replicating cells. It is likely that CKB is directed to the HCV replication machinery through interaction with NS4A and that the enzyme functions as a positive regulator of the viral genome replication by providing ATP (Hara et al. 2009). Creatine kinases catalyze the reversible transfer of high-energy phosphate groups between ATP and phosphocreatine, thereby playing a storage and distribution role in cellular energetics. CKB is a brain-type creatine kinase isoenzyme and is also detected in a variety of other tissues, including human liver. It can be hypothesized that recruitment of the ATP generating machinery into the membrane-associated site or DMVs, through its interaction with viral proteins comprising the replication complex, is at least in part linked with elevated concentrations of ATP at a particular site. Another possibility may be the implication of communication between mitochondria and DMVs associated with HCV replication complex in ATP transport through membrane-to-membrane contact. Our preliminary analysis demonstrated that putative sites of the viral RNA replication with high FRET signals were mainly localized proximal to mitochondria (Ando et al. 2012).

It is well known that the ER and mitochondria join together at multiple contact sites to form specific domains, such as mitochondria-associated (ER) membranes, with distinct biochemical properties in cells. However, to our knowledge, connection between membrane vesicles associated with the viral replication complex and mitochondria has not been characterized to date.

15.5 Role of Lipid Droplets in Assembly of HCV Particles

HCV core, E1, and E2 proteins are structural components of the viral particles. HCV particles likely consist of a nucleocapsid and outer envelope composed of a host cell-derived lipid membrane and envelope E1 and E2. After translation and processing of precursor polyprotein, these structural proteins are targeted to distinct places within the cell, which is considered that assembly of HCV particles is potentially regulated by the coordinated subcellular localization of these proteins.

Core protein interacts with cellular membranes mainly through two amphipathic helices located within its central domain. This mode of peripheral membrane interaction allows matured core, which is generated through processing by signal peptidase and signal peptide peptidase, to migrate to the surface of lipid droplets (LDs) (McLauchlan et al. 2002; Schwer et al. 2004). LDs are storage organelles used as a source of neutral lipid for metabolism and membrane synthesis. LDs are composed of a core of triacylglycerol and cholesterol ester, surrounded by a phospholipid monolayer derived from the outer leaflet of the ER, which in turn is bounded by a proteinaceous coat. The ~120 N-terminal residues of core contain multiple positively charged residues that are implicated in RNA binding and homo-oligomerization. It is therefore likely that this domain is a prerequisite for assembly of the HCV nucleocapsid (Kunkel et al. 2001; Klein et al. 2004; Majeau et al. 2004; Klein et al. 2005). With regard to the molecular mechanism for nucleocapsid formation, we have recently identified the 3′UTR as a *cis*-acting element that was sufficient for packaging of HCV genome (Shi et al. 2016). Mutagenesis analyses, together with a binding assay with core, demonstrated that, whereas the best encapsidation occurs with the entire 3′ UTR, the two loop sequences at the 3′ end appear to be essential for encapsidation.

In addition to core, three viral proteins have been shown to interact with the viral RNA: NS5A, the NS3 helicase domain, and the RNA polymerase NS5B. Structural analyses have shown that the N-terminal region of NS5A forms a “clawlike” dimer, which might accommodate RNA and interact with viral and cellular proteins and membranes (Huang et al. 2005; Tellinghuisen et al. 2005). It may be likely that newly synthesized HCV genomic RNA bound to NS5A is released from the replication complex containing DMV or replication factory, leading to recruitment of the NS5A-RNA complex to LDs, thereby enabling interaction with core and encapsidation (Fig. 15.3). Although both basal phosphorylation and hyperphosphorylation of NS5A appear to be required for an efficient interaction between NS5A and core essential for virion assembly, its hyperphosphorylation mediated by CKI-α contributes to the recruitment of NS5A from the ER membranes to LDs or LD-associated membranes (Masaki et al. 2014). DGAT1, one of the two enzymes mediating triglyceride synthesis, binds to both NS5A and core, and this interaction also mediates recruitment to LDs (Herker et al. 2010; Camus et al. 2013).

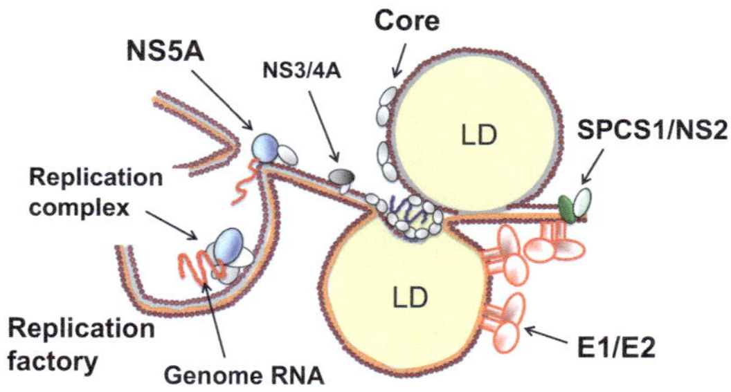

Fig. 15.3 Hypothetical model for assembly of HCV particles. Newly synthesized HCV genome at the replication factory is recruited to the surface of lipid droplet (LD), where ER membrane is associated, possibly via interaction between NS5A and core. The viral genome thus enables to associate with core, thereby nucleocapsid formation. Formation of the complex, E1/E2 and signal peptidase complex subunit 1 (SPCS1) and NS2, in tight association with the ER, may be critical for the viral envelopment

Likewise, it has been shown that NS2 brings together the E1-E2 complex, p7, and the NS3-4A complex (Phan et al. 2009; Jirasko et al. 2010; Boson et al. 2011; Ma et al. 2011; Popescu et al. 2011; Stapleford and Lindenbach 2011), and the interaction between p7-NS2 and NS3-4A is important for recruiting core to sites of virus assembly (Boson et al. 2011; Counihan et al. 2011). Signal peptidase complex subunit 1 (SPCS1), a component of the microsomal signal peptidase complex that is responsible for cleavage of the signal peptides of many secreted and membrane-associated proteins, has been identified as an NS2-binding host factor (Suzuki et al. 2013). SPCS1 interacts with both NS2 and E2, forming an E2-SPCS1-NS2 complex at the cytoplasmic membranes (Fig. 15.3). SPCS1 thus plays a critical role in the virion assembly of HCV (Suzuki et al. 2013), as interaction between E2 and NS2 is important for infectious viral particle assembly, as indicated above.

For the final maturation of the particles, HCV hijacks the pathway responsible for secretion of the VLDL lipoprotein particles, and virions are considered to be released into the extracellular space as low-density lipoviroparticles, whose lipids are presumably derived from LDs.

Thus, VLDL-associated proteins, including apolipoprotein B (ApoB), ApoE, and microsomal triglyceride transfer protein, have been shown to play crucial roles in the virion formation (Gastaminza et al. 2008; Jiang and Luo 2009; Andre et al. 2002). A recent study has demonstrated the redundant roles of ApoB and the exchangeable apolipoproteins ApoA1, ApoA2, ApoC1, ApoC2, ApoC3, and ApoE in the assembly of infectious HCV particles (Fukuhara et al. 2014; Fukuhara et al. 2015).

15.6 Future Perspectives

There is accumulating evidence regarding the involvement of host cell organelles or organelle-like structures, such as ER, membrane vesicle-associated replication factories, and LDs, in the genome replication and the virion assembly of HCV particles. Mitochondria may also be implicated in the regulation of the viral replication. To advance our understanding of the regulatory mechanisms of the virus lifecycle, further researches to address key questions such as the dynamics of biogenesis of the viral replication factory, its biochemical and structural features, and contact mechanisms of the replication factory with mitochondria and LDs should be required.

Acknowledgments The author thanks Ms. Tomoko Mochizuki for her technical assistance. Financial support came from the Japan Agency for Medical Research and Development, AMED, and from the Ministry of Education, Culture, Sports, Science and Technology, Japan.

References

Ando T, Imamura H, Suzuki R, Aizaki H, Watanabe T, Wakita T, Suzuki T (2012) Visualization and measurement of ATP levels in living cells replicating hepatitis C virus genome RNA. PLoS Pathog 8: e1002561

Andre P, Komurian-Pradel F, Deforges S, Perret M, Berland JL, Sodoyer M, Pol S, Brechot C, Paranhos-Baccala G, Lotteau V (2002) Characterization of low- and very-low-density hepatitis C virus RNA-containing particles. J Virol 76:6919–6928

Bartenschlager R, Ahlborn-Laake L, Mous J, Jacobsen H (1993) Nonstructural protein 3 of the hepatitis C virus encodes a serine-type proteinase required for cleavage at the NS3/4 and NS4/5 junctions. J Virol 67:3835–3844

Bartenschlager R, Ahlborn-Laake L, Yasargil K, Mous J, Jacobsen H (1995) Substrate determinants for cleavage in cis and in trans by the hepatitis C virus NS3 proteinase. J Virol 69:198–205

Berger KL, Kelly SM, Jordan TX, Tartell MA, Randall G (2011) Hepatitis C virus stimulates the phosphatidylinositol 4-kinase III alpha-dependent phosphatidylinositol 4-phosphate production that is essential for its replication. J Virol 85:8870–8883

Bianco A, Reghellin V, Donnici L, Fenu S, Alvarez R, Baruffa C, Peri F, Pagani M, Abrignani S, Neddermann P et al (2012) Metabolism of phosphatidylinositol 4-kinase IIIα-dependent PI4P Is subverted by HCV and is targeted by a 4-anilino quinazoline with antiviral activity. PLoS Pathog 8:e1002576

Boson B, Granio O, Bartenschlager R, Cosset FL (2011) A concerted action of hepatitis C virus p7 and nonstructural protein 2 regulates core localization at the endoplasmic reticulum and virus assembly. PLoS Pathog 7:e1002144

Camus G, Herker E, Modi AA, Haas JT, Ramage HR, Farese RV Jr, Ott M (2013) Diacylglycerol acyltransferase-1 localizes hepatitis C virus NS5A protein to lipid droplets and enhances NS5A interaction with the viral capsid core. J Biol Chem 288:9915–9923

Counihan NA, Rawlinson SM, Lindenbach BD (2011) Trafficking of hepatitis C virus core protein during virus particle assembly. PLoS Pathog 7:e1002302

Dong R, Saheki Y, Swarup S, Lucast L, Harper JW, De Camilli P (2016) Endosome-ER contacts control actin nucleation and retromer function through VAP-dependent regulation of PI4P. Cell 166:408–423

Egger D, Wolk B, Gosert R, Bianchi L, Blum HE, Moradpour D, Bienz K (2002) Expression of hepatitis C virus proteins induces distinct membrane alterations including a candidate viral replication complex. J Virol 76:5974–5984

El-Hage N, Luo G (2003) Replication of hepatitis C virus RNA occurs in a membrane-bound replication complex containing nonstructural viral proteins and RNA. J Gen Virol 84:2761–2769

Ferraris P, Blanchard E, Roingeard P (2010) Ultrastructural and biochemical analyses of hepatitis C virus-associated host cell membranes. J Gen Virol 91:2230–2237

Frick DN, Lam AM (2006) Understanding helicases as a means of virus control. Curr Pharm Des 12:1315–1338

Fukuhara T, Wada M, Nakamura S, Ono C, Shiokawa M, Yamamoto S, Motomura T, Okamoto T, Okuzaki D, Yamamoto M et al (2014) Amphipathic α-helices in apolipoproteins are crucial to the formation of infectious hepatitis C virus particles. PLoS Pathog 10: e1004534

Fukuhara T, Ono C, Puig-Basagoiti F, Matsuura Y (2015) Roles of Lipoproteins and Apolipoproteins in Particle Formation of Hepatitis C Virus. Trends Microbiol 23:618–629

Gao L, Aizaki H, He JW, Lai MM (2004) Interactions between viral nonstructural proteins and host protein hVAP-33 mediate the formation of hepatitis C virus RNA replication complex on lipid raft. J Virol 78:3480–3488

Gastaminza P, Cheng G, Wieland S, Zhong J, Liao W, Chisari FV (2008) Cellular determinants of hepatitis C virus assembly, maturation, degradation, and secretion. J Virol 82:2120–2129

Gosert R, Egger D, Lohmann V, Bartenschlager R, Blum HE, Bienz K, Moradpour D (2003) Identification of the hepatitis C virus RNA replication complex in Huh-7 cells harboring subgenomic replicons. J Virol 77:5487–5492

Gouttenoire J, Penin F, Moradpour D (2010) Hepatitis C virus nonstructural protein 4B: a journey into unexplored territory. Rev Med Virol 20:117–129

Gurer C, Hoglund A, Hoglund S, Luban J (2005) ATPgammaS disrupts human immunodeficiency virus type 1 virion core integrity. J Virol 79:5557–5567

Hamamoto I, Nishimura Y, Okamoto T, Aizaki H, Liu M, Mori Y, Abe T, Suzuki T, Lai MM, Miyamura T et al (2005) Human VAP-B is involved in hepatitis C virus replication through interaction with NS5A and NS5B. J Virol 79:13473–13482

Hara H, Aizaki H, Matsuda M, Shinkai-Ouchi F, Inoue Y, Murakami K, Shoji I, Kawakami H, Matsuura Y, Lai MM et al (2009) Involvement of creatine kinase B in hepatitis C virus genome replication through interaction with the viral NS4A protein. J Virol 83:5137–5147

Herker E, Harris C, Hernandez C, Carpentier A, Kaehlcke K, Rosenberg AR, Farese RV Jr, Ott M (2010) Efficient hepatitis C virus particle formation requires diacylglycerol acyltransferase-1. Nat Med 16:1295–1298

Huang L, Hwang J, Sharma SD, Hargittai MR, Chen Y, Arnold JJ, Raney KD, Cameron CE (2005) Hepatitis C virus nonstructural protein 5A (NS5A) is an RNA-binding protein. J Biol Chem 280:36417–36428

Imamura H, Nhat KP, Togawa H, Saito K, Iino R, Kato-Yamada Y, Nagai T, Noji H (2009) Visualization of ATP levels inside single living cells with fluorescence resonance energy transfer-based genetically encoded indicators. Proc Natl Acad Sci USA 106:15651–15656

Jiang J, Luo G (2009) Apolipoprotein E but not B is required for the formation of infectious hepatitis C virus particles. J Virol 83:12680–12691

Jirasko V, Montserret R, Lee JY, Gouttenoire J, Moradpour D, Penin F, Bartenschlager R (2010) Structural and functional studies of nonstructural protein 2 of the hepatitis C virus reveal its key role as organizer of virion assembly. PLoS Pathog 6:e1001233

Kim DW, Gwack Y, Han JH, Choe J (1995) C-terminal domain of the hepatitis C virus NS3 protein contains an RNA helicase activity. Biochem Biophys Res Commun 215:160–166

Klein KC, Polyak SJ, Lingappa JR (2004) Unique features of hepatitis C virus capsid formation revealed by de novo cell-free assembly. J Virol 78:9257–9269

Klein KC, Dellos SR, Lingappa JR (2005) Identification of residues in the hepatitis C virus core protein that are critical for capsid assembly in a cell-free system. J Virol 79:6814–6826

Klumpp K, Ford MJ, Ruigrok RW (1998) Variation in ATP requirement during influenza virus transcription. J Gen Virol 79:1033–1045

Kunkel M, Lorinczi M, Rijnbrand R, Lemon SM, Watowich SJ (2001) Self-assembly of nucleocapsid-like particles from recombinant hepatitis C virus core protein. J Virol 75:2119–2129

Lahaye X, Vidy A, Fouquet B, Blondel D (2012) Hsp70 protein positively regulates rabies virus infection. J Virol 86:4743–4751

Li PP, Itoh N, Watanabe M, Shi Y, Liu P, Yang HJ, Kasamatsu H (2009) Association of simian virus 40 vp1 with 70-kilodalton heat shock proteins and viral tumor antigens. J Virol 83:37–46

Lindenbach BD (2013) Virion assembly and release. Curr Top Microbiol Immunol 369:199–218

Lindenbach BD, Thiel HJ, Rice CM (2007) The viruses and their replication. In: Fields virology. Lippincott-Raven, Philadelphia, pp 1101–1152

Lundin M, Monne M, Widell A, Von Heijne G, Persson MA (2003) Topology of the membrane-associated hepatitis C virus protein NS4B. J Virol 77:5428–5438

Ma Y, Anantpadma M, Timpe JM, Shanmugam S, Singh SM, Lemon SM, Yi M (2011) Hepatitis C virus NS2 protein serves as a scaffold for virus assembly by interacting with both structural and nonstructural proteins. J Virol 85:86–97

Majeau N, Gagne V, Boivin A, Bolduc M, Majeau JA, Ouellet D, Leclerc D (2004) The N-terminal half of the core protein of hepatitis C virus is sufficient for nucleocapsid formation. J Gen Virol 85:971–981

Masaki T, Suzuki T (2015) NS5A phosphorylation: its functional role in the life cycle of hepatitis C virus. Future Virology 10:751–762

Masaki T, Matsunaga S, Takahashi H, Nakashima K, Kimura Y, Ito M, Matsuda M, Murayama A, Kato T, Hirano H et al (2014) Involvement of hepatitis C virus

NS5A hyperphosphorylation mediated by casein kinase I-α in infectious virus production. J Virol 88:7541–7555
Mayer MP, Bukau B (2005) Hsp70 chaperones: cellular functions and molecular mechanism. Cell Mol Life Sci 62:670–684
McLauchlan J, Lemberg MK, Hope G, Martoglio B (2002) Intramembrane proteolysis promotes trafficking of hepatitis C virus core protein to lipid droplets. EMBO J 21:3980–3988
Moradpour D, Brass V, Bieck E, Friebe P, Gosert R, Blum HE, Bartenschlager R, Penin F, Lohmann V (2004) Membrane association of the RNA-dependent RNA polymerase is essential for hepatitis C virus RNA replication. J Virol 78:13278–13284
Mottola G, Cardinali G, Ceccacci A, Trozzi C, Bartholomew L, Torrisi MR, Pedrazzini E, Bonatti S, Migliaccio G (2002) Hepatitis C virus nonstructural proteins are localized in a modified endoplasmic reticulum of cells expressing viral subgenomic replicons. Virology 293:31–43
Nomaguchi M, Ackermann M, Yon C, You S, Padmanabhan R (2003) De novo synthesis of negative-strand RNA by Dengue virus RNA-dependent RNA polymerase in vitro: nucleotide, primer, and template parameters. J Virol 77:8831–8842
Paillusson S, Stoica R, Gomez-Suaga P, Lau DH, Mueller S, Miller T, Miller CC (2016) There's something wrong with my MAM; the ER-mitochondria axis and neurodegenerative diseases. Trends Neurosci 39:146–157
Paul D, Bartenschlager R (2013) Architecture and biogenesis of plus-strand RNA virus replication factories. World J Virol 2:32–48
Paul D, Hoppe S, Saher G, Krijnse-Locker J, Bartenschlager R (2013) Morphological and biochemical characterization of the membranous hepatitis C virus replication compartment. J Virol 87:10612–10627
Pawlotsky JM (2006) Hepatitis C virus population dynamics during infection. Curr Top Microbiol Immunol 299:261–284
Phan T, Beran RK, Peters C, Lorenz IC, Lindenbach BD (2009) Hepatitis C virus NS2 protein contributes to virus particle assembly via opposing epistatic interactions with the E1-E2 glycoprotein and NS3-NS4A enzyme complexes. J Virol 83:8379–8395
Popescu CI, Callens N, Trinel D, Roingeard P, Moradpour D, Descamps V, Duverlie G, Penin F, Heliot L, Rouille Y et al (2011) NS2 protein of hepatitis C virus interacts with structural and non-structural proteins towards virus assembly. PLoS Pathog 7: e1001278
Ramage HR, Kumar GR, Verschueren E, Johnson JR, Von Dollen J, Johnson T, Newton B, Shah P, Horner J, Krogan NJ et al (2015) A combined proteomics/genomics approach links hepatitis C virus infection with nonsense-mediated mRNA decay. Mol Cell 57:329–340
Reiss S, Rebhan I, Backes P, Romero-Brey I, Erfle H, Matula P, Kaderali L, Poenisch M, Blankenburg H, Hiet MS et al (2011) Recruitment and activation of a lipid kinase by hepatitis C virus NS5A is essential for integrity of the membranous replication compartment. Cell Host Microbe 9:32–45
Romero-Brey I, Merz A, Chiramel A, Lee JY, Chlanda P, Haselman U, Santarella-Mellwig R, Habermann A, Hoppe S, Kallis S et al (2012) Three-dimensional architecture and biogenesis of membrane structures associated with hepatitis C virus replication. PLoS Pathog 8:e1003056
Ross-Thriepland D, Harris M (2015) Hepatitis C virus NS5A: enigmatic but still promiscuous 10 years on! J Gen Virol 96:727–738
Schmidt-Mende J, Bieck E, Hugle T, Penin F, Rice CM, Blum HE, Moradpour D (2001) Determinants for membrane association of the hepatitis C virus RNA-dependent RNA polymerase. J Biol Chem 276:44052–44063
Schwer B, Ren S, Pietschmann T, Kartenbeck J, Kaehlcke K, Bartenschlager R, Yen TS, Ott M (2004) Targeting of hepatitis C virus core protein to mitochondria through a novel C-terminal localization motif. J Virol 78:7958–7968
Shi ST, Lee KJ, Aizaki H, Hwang SB, Lai MM (2003) Hepatitis C virus RNA replication occurs on a detergent-resistant membrane that cofractionates with caveolin-2. J Virol 77:4160–4168
Shi G, Ando T, Suzuki R, Matsuda M, Nakashima K, Ito M, Omatsu T, Oba M, Ochiai H, Kato T et al (2016) Involvement of the 3′ Untranslated Region in Encapsidation of the Hepatitis C Virus. PLoS Pathog 12:e1005441
Stapleford KA, Lindenbach BD (2011) Hepatitis C virus NS2 coordinates virus particle assembly through physical interactions with the E1-E2 glycoprotein and NS3-NS4A enzyme complexes. J Virol 85:1706–1717
Suzich JA, Tamura JK, Palmer-Hill F, Warrener P, Grakoui A, Rice CM, Feinstone SM, Collett MS (1993) Hepatitis C virus NS3 protein polynucleotide-stimulated nucleoside triphosphatase and comparison with the related pestivirus and flavivirus enzymes. J Virol 67:6152–6158
Suzuki T (2012) Morphogenesis of infectious hepatitis C virus particles. Front Microbiol 3:38
Suzuki R, Matsuda M, Watashi K, Aizaki H, Matsuura Y, Wakita T, Suzuki T (2013) Signal peptidase complex subunit 1 participates in the assembly of hepatitis C virus through an interaction with E2 and NS2. PLoS Pathog 9:e1003589
Tai AW, Salloum S (2011) The role of the phosphatidylinositol 4-kinase PI4KA in hepatitis C virus-induced host membrane rearrangement. PLoS ONE 6:e26300
Tai AW, Benita Y, Peng LF, Kim SS, Sakamoto N, Xavier RJ, Chung RT (2009) A functional genomic

screen identifies cellular cofactors of hepatitis C virus replication. Cell Host Microbe 5:298–307

Tellinghuisen TL, Marcotrigiano J, Rice CM (2005) Structure of the zinc-binding domain of an essential component of the hepatitis C virus replicase. Nature 435:374–379

Vreede FT, Gifford H, Brownlee GG (2008) Role of initiating nucleoside triphosphate concentrations in the regulation of influenza virus replication and transcription. J Virol 82:6902–6910

Wakana Y, Kotake R, Oyama N, Murate M, Kobayashi T, Arasaki K, Inoue H, Tagaya M (2015) CARTS biogenesis requires VAP-lipid transfer protein complexes functioning at the endoplasmic reticulum-Golgi interface. Mol Biol Cell 26:4686–4699

Wasley A, Alter MJ (2000) Epidemiology of hepatitis C: geographic differences and temporal trends. Semin Liver Dis 20:1–16

Wolk B, Sansonno D, Krausslich HG, Dammacco F, Rice CM, Blum HE, Moradpour D (2000) Subcellular localization, stability, and trans-cleavage competence of the hepatitis C virus NS3-NS4A complex expressed in tetracycline-regulated cell lines. J Virol 74:2293–2304

Hijacking of Membrane Contact Sites by Intracellular Bacterial Pathogens

16

Isabelle Derré

Abstract

Intracellular bacterial pathogens have evolved sophisticated mechanisms to hijack host cellular processes to promote their survival and replication inside host cells. Over the past two decades, much attention has been given to the strategies employed by these pathogens to manipulate various vesicular trafficking pathways. But in the past 5 years, studies have brought to light that intracellular bacteria also target non-vesicular trafficking pathways. Here we review how three vacuolar pathogens, namely, *Legionella*, *Chlamydia*, and *Coxiella* hijack components of cellular MCS with or without the formation of stable MCS. A common theme in the manipulation of MCS by intracellular bacteria is the dependence on the secretion of bacterial effector proteins. During the early stages of the *Legionella* life cycle, the bacteria connects otherwise unrelated cellular pathways (i.e., components of ER-PM MCS, PI4KIIIα, and Sac1 and the early secretory pathway) to remodel its nascent vacuole into an ER-like compartment. *Chlamydia* and *Coxiella* vacuoles establish direct MCS with the ER and target lipid transfer proteins that contain a FFAT motif, CERT, and ORP1L, respectively, suggesting a common mechanism of VAP-dependent lipid acquisition. *Chlamydia* also recruits STIM1, an ER calcium sensor involved in store-operated calcium entry (SOCE) at ER-PM MCS, and elucidating the role of STIM1 at ER-*Chlamydia* inclusion MCS may uncover additional role for these contacts. Altogether, the manipulation of MCS by intracellular bacterial pathogens has open a new and exciting area of research to investigate the molecular mechanisms supporting pathogenesis.

I. Derré (✉)
Department of Microbiology, Immunology and Cancer Biology, University of Virginia, Charlottesville, VA, USA
e-mail: id8m@virginia.edu

M. Tagaya, T. Simmen (eds.), *Organelle Contact Sites*, Advances in Experimental Medicine and Biology 997, DOI 10.1007/978-981-10-4567-7_16

Keywords
Legionella • PI4KIIIα • Sac1 • *Chlamydia* • IncD • CERT • VAP • STIM1 • *Coxiella* • ORP1L

16.1 Introduction

Intracellular bacterial pathogens have evolved sophisticated mechanisms to hijack host cellular processes to promote their survival and replication inside host cells. *Legionella pneumophila*, *Chlamydia trachomatis*, and *Coxiella burnetii* are intracellular vacuolar pathogens that replicate inside a large membrane-bound compartment. Although their respective lifestyles are quite different, *Legionella* resides in an endoplasmic reticulum (ER)-like compartment; the nature of the membrane surrounding *Chlamydia* is unknown and *Coxiella* replicates in phagolysosomes; all three bacteria have evolved sophisticated secretion systems to inject effector molecules into the cytosol of the host cell and promote the selective interaction of their vacuole with cellular organelles. Over the past two decades, research focused on the strategies employed by these pathogens to manipulate various vesicular trafficking pathways in order to create their replication niche. However, in the past 5 years, studies have brought to light the notion that intracellular bacteria not only target vesicular trafficking pathways but also manipulate non-vesicular pathways via interaction with components of cellular membrane contact sites (MCS) or through the formation of MCS between their vacuole and cellular organelles. In this chapter, we review how *Legionella* hijacks components of ER-plasma membrane (PM) MCS to promote the remodeling of its nascent vacuole, how *Chlamydia* initiates the formation of hybrid MCS-harboring components of both ER-Golgi and ER-PM MCS, and how *Coxiella* establishes MCS that bear resemblance to ER-endosome MCS.

16.2 *Legionella* Hijacks Components of ER-PM MCS to Promote the Remodeling of Its Nascent Vacuole into an ER-Like Compartment

Legionella pneumophila is a water-borne bacterium that can lead to a potentially fatal pneumoniae, known as Legionnaire's disease, in the elderly and in immunocompromised persons (Marston et al. 1994). Infection occurs by inhalation of contaminated aerosols (Nguyen et al. 2006) and subsequent phagocytosis of the bacteria by monocytes and alveolar macrophages (Horwitz and Silverstein 1980). The membrane of the nascent *Legionella*-containing vacuole (LCV) originally derives from the plasma membrane, but it is rapidly remodeled to avoid fusion with lysosomes. In addition, trafficking from the ER to the Golgi is redirected toward the LCV, and fusion of secretory vesicles with the LCV generates an ER-like compartment favorable to bacterial replication.

Legionella participates actively to the remodeling of the membrane of the LCV. The bacteria encode a type IVB secretion system (T4SS), known as the Dot/Icm system (Segal et al. 1998; Vogel et al. 1998), which allows the bacteria to secrete up to 300 bacterial effector proteins into the cytosol of the host cell (Zhu et al. 2011). The specific function of these effectors is for the most part unknown, but some of them have been shown to play a role in the subversion of host cell signaling and secretory pathways. In particular, Dot/Icm effectors implicated in the manipulation of the eukaryotic small GTPase Rab1, which supports vesicular trafficking in the early secretory pathway, have been extensively characterized over the past decade (Goody and Itzen 2013).

Two Dot/Icm effectors regulate the nucleotide state of Rab1. DrrA (also known as SidM) acts as a Rab1 guanine nucleotide exchange factor (GEF) and plays a critical role in the activation and recruitment of Rab1 to the LCV (Machner and Isberg 2006; Murata et al. 2006). LepB on the other end acts as a Rab1-GTPase-activating protein (GAP) and contributes to Rab1 inactivation and release from the LCV (Ingmundson et al. 2007). In addition, *Legionella* Dot/Icm effectors also regulate Rab1 activity through reversible posttranslational modifications. Through adenosine monophosphorylation (AMPylation) activity (Muller et al. 2010) and phosphocholine transferase activity (Mukherjee et al. 2011; Tan et al. 2011), DrrA and AnkX, respectively, keep Rab1 in its GTP-bound active state by preventing access to LepB. The de-AMPylase SidD (Neunuebel et al. 2011; Tan and Luo 2011) and the dephosphocholinase Lem3 (Tan et al. 2011) reverse this effect.

The presence of phosphatidylinositol 4-phosphate (PI4P) in the LCV membrane is important for DrrA association with the LCV. Although PI4P is mostly present at the Golgi, a pool of PI4P is generated at the PM by the host PI 4-kinase type IIIα (PI4KIIIα, SttA in yeast) (Balla et al. 2005; Baird et al. 2008; Hammond et al. 2012; Nakatsu et al. 2012) (Fig. 16.1). In yeast, points of intimate contact between the ER and the PM, known as ER-PM MCS, constitute a hot spot for regulating PI4P metabolism at the PM. PI4P turnover requires the oxysterol-binding homology (Osh) protein, Osh3, which

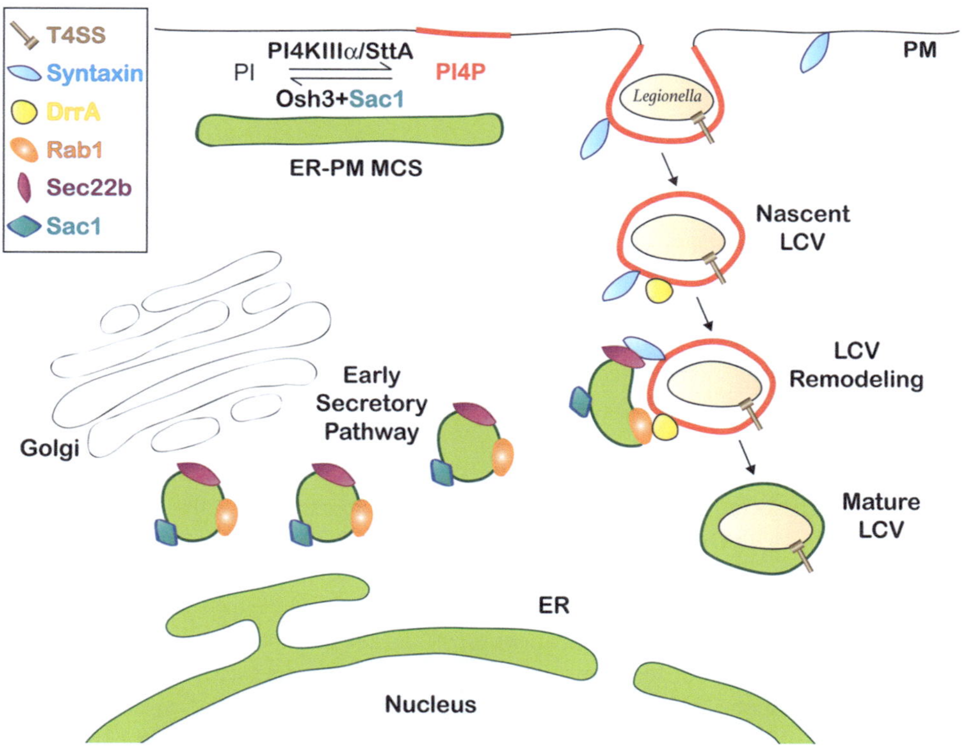

Fig. 16.1 ***Legionella*** **hijacks components of ER-PM MCS to temporally position effector proteins involved in vacuole maturation.** Upon phagocytosis by macrophages, *Legionella* resides in a PM-derived vacuole. The PI4P produced at the PM by the host PI4KIIIα accumulates in the membrane of the nascent LCV allowing for DrrA, a Dot/Icm effector containing a PI4P-binding domain, to associate with the LCV. DrrA recruits and activates the small GTPase Rab1, thereby promoting the tethering and fusion of ER-derived vesicles with the LCV by non-canonical pairing of the PM t-SNARE syntaxin and the ER v-SNARE Sec22b. Upon delivery to the LCV, the ER-associated phosphatase, Sac1, hydrolyzes the PI4P present on the LCV-promoting DrrA and subsequently Rab1 dissociation from the LCV

acts as a PI4P sensor. When the PM levels of PI4P are high, Osh3 localizes to ER-PM MCS and stimulates the phosphatase activity of the ER-localized phosphoinositide phosphatase, Sac1, therefore reducing the PI4P level of the PM (Stefan et al. 2011).

The nascent LCV is enriched in PI4P (Hubber et al. 2014), and a PI4P-binding domain is located in the carboxy-terminal of the DrrA protein (Murata et al. 2006; Brombacher et al. 2009; Schoebel et al. 2009; Zhu et al. 2010; Hammond et al. 2014). In agreement with the notion that DrrA association with membranes relies on PI4P, point mutations in the PI4P-binding domain of DrrA affect its localization to the LCV and prevent Rab1 recruitment (Hubber et al. 2014). PI4KIIIα plays an important role in generating the PI4P that mediates DrrA association with membrane. In a permeabilized cell system or when DrrA is expressed in eukaryotic cells, the PI4P-binding domain of DrrA drives its localization to the PM (Murata et al. 2006; Hubber et al. 2014). However in mouse embryonic fibroblasts (MEFs) derived from PI4KIIIα knockout mice, the localization of DrrA to the PM is reduced (Hubber et al. 2014), suggesting that PI4KIIIα is important for generating the PI4P that mediates DrrA recruitment to the PM. The importance of PI4KIIIα during *Legionella* infection is further exemplified by the defect in Rab1 recruitment to the LCV and the reduced bacterial replication in PI4KIIIα-deficient macrophages (Hubber et al. 2014).

The DrrA/Rab1 association with the LCV is temporally regulated, with DrrA and Rab1 being present on the surface of the LCV only during the first 4 hours of infection prior to remodeling of the LCV into an ER-like compartment (Ingmundson et al. 2007). PI4P turnover on the LCV is most likely responsible for the dissociation of DrrA from the LCV. Interestingly the overexpression of Sac1 in *Legionella*-infected cells led to a decrease in Rab1 association with the LCV (Hubber et al. 2014), suggesting a role for Sac1 in PI4P turnover at the LCV.

Altogether the data gathered so far is compatible with a model in which *Legionella* manipulates components of ER-PM MCSs, PI4KIIIα, and Sac1, to modulate PI4P dynamics on the LCV and control effector localization (Fig. 16.1). The PI4P- and DrrA-dependent activation of Rab1 on the nascent LCV is sufficient to promote the tethering and fusion of ER-derived vesicles with the LCV. This process involves the non-canonical pairing of the PM t-SNARE Syntaxins and the ER v-SNARE Sec22b (Arasaki and Roy 2010; Arasaki et al. 2012) and ultimately results in converting the PM-derived LCV into an ER-like compartment (Tilney et al. 2001; Robinson and Roy 2006).

16.3 *Chlamydia* Initiates the Formation of Hybrid MCS-Harboring Components of Both ER-Golgi and ER-PM MCS

The *Chlamydiae* phylum regroups obligate intracellular bacterial pathogens infecting a wide range of hosts (rodents, cattle, swine, human, etc.) and responsible for various diseases (pneumoniae, trachoma, and sexually transmitted diseases) (Schachter 1999). Of the human pathogen *C. trachomatis*, ocular serovars are associated with non-congenital blindness in developing countries, and genital serovars are the leading cause of bacteria-associated sexually transmitted diseases in the developed world. In women, genital *C. trachomatis* infections are often asymptomatic but nevertheless manifest with inflammation of the upper reproductive tract, which if left untreated, leads to long-term sequelae such as chronic pelvic pain, ectopic pregnancy, and infertility (Malhotra et al. 2013). Despite the availability of diagnostic tests and antibiotic treatments, cases are increasing each year, and in the absence of a vaccine, *Chlamydia* infections remain a burden to human health.

In vivo and in vitro, *Chlamydia* has a tropism for epithelial cells in which the bacteria undergo a biphasic developmental cycle that occurs exclusively in a membrane-bound compartment termed the inclusion (Moulder 1991). The bacteria alternate between an infectious but non-replicating form (elementary body, EB) and a replicative form (reticulate body, RB). Under favorable conditions, a single EB differentiates into the RB form and replicates to high numbers. Toward mid-cycle, the RBs transition back to EBs. Once the cycle is completed, the EBs are released from the host cell.

During coevolution with the mammalian host, *Chlamydia* genome has undergone drastic reduction and is now limited to ~1000 genes (Stephens et al. 1998) (for comparison *E. coli* K12 genome encodes ~4000 ORFs (Riley et al. 2006)). As a result, multiple metabolic pathways are incomplete, and *Chlamydia* relies on the host cells for essential nutrients. While the inclusion offers a perfect niche for the bacteria to replicate, the inclusion membrane separates the bacteria from essential building blocks. To circumvent this problem, *Chlamydia* is equipped with a bacterial type III secretion system (T3SS) and uses this syringe-like apparatus to deliver effector proteins across the inclusion membrane (reviewed by (Mueller et al. 2014)). One family of T3SS effectors, the Inc proteins, is inserted into the inclusion membrane (Dehoux et al. 2011; Lutter et al. 2012) and is therefore strategically positioned to promote interaction between the inclusion and cellular organelles (reviewed by (Elwell et al. 2016)).

Chlamydia incorporates host cellular lipids such as phosphatidylcholine, phosphatidylinositol, cholesterol, and sphingomyelin into their cell wall (Hackstadt et al. 1995, 1996; Wylie et al. 1997; Hatch and McClarty 1998; Carabeo et al. 2003). Moreover, sphingomyelin acquisition is necessary for bacterial replication (van Ooij et al. 2000). The bulk of sphingomyelin is synthesized within the Golgi by the sphingomyelin synthase SMS1 (Huitema et al. 2004), but its precursor, ceramide, is synthetized in the ER. At ER-Golgi MCSs, the lipid-transfer protein CERT allows for the transfer of ceramide from the ER to the Golgi (Hanada et al. 2003) (Fig. 16.2). CERT contains three domains that are critical for this process, a pleckstrin homology (PH) domain (Lemmon 2008) that binds PI4P and potentially Arf1 at the Golgi membrane (Balla and Balla 2006; Hanada et al. 2009), a FFAT (two phenylalanines in an acidic tract) motif (Loewen et al. 2003) interacting with the VAPA and VAPB (vesicle-associated membrane protein-associated protein A and B, respectively) proteins in the ER (Lev et al. 2008), and a START (steroidogenic acute regulatory protein-related lipid transfer) domain that binds ceramide (Ponting and Aravind 1999). Once synthesized in the Golgi, Sphingomyelin-containing vesicles bud from the Golgi and are delivered to the plasma membrane. In *Chlamydia*-infected cells, the bacteria hijack this vesicular trafficking pathway, and the sphingomyelin-laden vesicles are redirected toward the inclusion membrane (Hackstadt et al. 1995, 1996). Sphingomyelin is incorporated into the inclusion membrane and the cell wall of the bacteria (Fig. 16.2).

An alternative pathway of sphingomyelin acquisition has been proposed through the formation MCS (Fig. 16.2). During *Chlamydia* infection, the inclusion membrane establishes intimate contact with the endoplasmic reticulum, leading to the formation of ER-inclusion MCS (Derré et al. 2011). CERT and VAP localize to these contacts (Derré et al. 2011; Elwell et al. 2011). The *Chlamydia* inclusion membrane protein IncD promotes CERT association with the inclusion membrane via the direct interaction of IncD with the PH domain of CERT (Derré et al. 2011; Agaisse and Derré 2014). Similar to ER-Golgi MCS, the FFAT motif of CERT mediates the interaction between CERT and VAP at ER-inclusion MCS (Agaisse and Derré 2014). The importance of CERT and VAP for the completion of *Chlamydial* developmental cycle is illustrated by the formation of smaller inclusions containing less infectious particles upon depletion of CERT or VAP (Derré et al. 2011; Elwell et al. 2011). The small molecule HPA-12 (Yasuda et al. 2001; Kudo et al. 2010), a synthetic analog of ceramide that inhibits CERT-mediated transfer of ceramide, leads to a similar phenotype but also to a reduction in sphingomyelin acquisition by the bacteria (Elwell et al. 2011). The association of the sphingomyelin synthase SMS2, an enzyme involved in sphingomyelin synthesis at the PM (Tafesse et al. 2007), with the inclusion

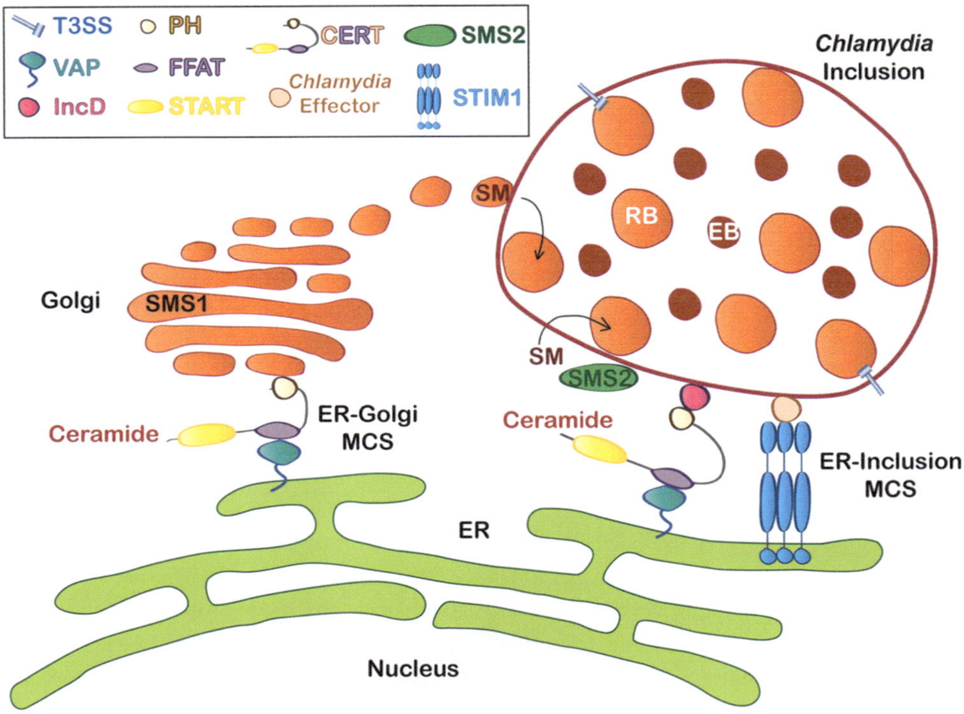

Fig. 16.2 Vesicular and non-vesicular trafficking of lipids to *Chlamydia* inclusion. At ER-Golgi MCS, the lipid-transfer protein CERT binds to the ER-resident proteins VAPA and VAPB via its FFAT motif and to the Golgi membrane via its PH domain. The START domain of CERT promotes the transfer of ceramide from the ER to the Golgi-localized sphingomyelin synthase SMS1 leading to sphingomyelin (SM) synthesis in the Golgi. In *Chlamydia*-infected cells, the trafficking of Golgi-derived vesicles containing SM en route to the PM is redirected to the *Chlamydia* inclusion. Upon fusion of these vesicles with the inclusion membrane, SM is incorporated into the inclusion membrane and into the cell wall of the bacteria. MCS between the ER and the inclusion membrane may constitute an alternative pathway of SM acquisition. At ER-inclusion MCS, CERT binds to VAPA and VAPB on the ER via its FFAT motif and to the inclusion membrane via direct interaction of its PH domain with the *Chlamydia* inclusion membrane protein IncD. The transfer of ceramide to SMS2 localized on the inclusion membrane may lead to direct SM synthesis at the inclusion membrane. STIM1, an ER calcium sensor involved in store-operated calcium entry (SOCE) at ER-PM MCS, also localized to the ER-inclusion MCS. Its function is unknown, and it has been proposed that a *Chlamydia* factor is involved in STIM1 association with the inclusion membrane

membrane and the reduced bacterial replication in cells depleted of SMS2 (Elwell et al. 2011), has led to propose a model in which, at ER-inclusion MCS, the CERT-dependent transfer of ceramide to SMS2 allows for synthesis of sphingomyelin directly at the inclusion membrane (Fig. 16.2).

MCS between the ER and various organelles differs in their molecular composition, providing each type of MCS with a specific molecular signature and function (Prinz 2014). ER-inclusion MCS, however, appears as hybrid MCS-harboring components found at ER-Golgi MCS, such as CERT and VAP, but also STIM1 (Agaisse and Derré 2015), a key component of ER-PM MCS. STIM1 is an ER-localized calcium (Ca^{2+}) sensor that plays an essential role during store-operated calcium entry (SOCE) (reviewed by (Prakriya and Lewis 2015)). When the ER Ca^{2+} store are depleted, STIM1 oligomerizes and relocates to ER-PM MCS where it binds and activates the calcium channel Orai, allowing for Ca^{2+} influx and replenishment of the ER Ca^{2+} store. Throughout *Chlamydial* developmental cycle, STIM1, but not Orai, is highly enriched at ER-inclusion MCS (Agaisse and Derré 2015) (Fig. 16.2). STIM1 association with the inclusion depends on the CAD domain of STIM1, a domain that is known to interact with and activate Orai. It was proposed that, as shown for IncD and CERT, a *Chlamydia* T3SS effector mediates the association of STIM1 with the inclusion membrane, possibly through molecular mimicry of Orai. Unlike CERT and VAP, however, STIM1 depletion does not affect *Chlamydia* replication, leaving the role of STIM1 at ER-inclusion MCS unresolved.

Beside a potential role in lipid acquisition, the function of ER-inclusion MCS remains unclear. Structures resembling ER-inclusion MCS, but referred to as pathogen synapse, where ordered arrays of *Chlamydia* T3SS connect *Chlamydia* with the inclusion membrane and the apposed rough ER have been proposed to represent special sites for effector translocation (Dumoux et al. 2012). In addition, a role in stimulator of interferon genes (STING)-dependent pathogen sensing has been proposed (Barker et al. 2013) and studies in *Chlamydia pneumoniae* and in the environmental *Chlamydia*, *Simkania negevensis*, also suggest a role in the modulation of the ER stress response (Mehlitz et al. 2014; Shima et al. 2015).

16.4 *Coxiella* Establishes MCS That Bear Resemblance to ER-Endosome MCS

Coxiella burnetii is a gram-negative bacterium that causes the zoonotic disease Q fever. Ruminants are the primary reservoir. Infections are mostly asymptomatic in these animals, but *Coxiella* is present in the milk, urine, and feces, which contributes to contamination of the local environment. In addition, *Coxiella* infection can trigger abortion, and the very high bacterial load in birth products, especially the placenta, is the major cause of environmental contamination (Delsing et al. 2011). The low infectious dose (Moos and Hackstadt 1987) and the environmental stability of the bacteria contribute to human infections via inhalation of contaminated aerosols. Although *Coxiella* infections are often asymptomatic in human, Q fever can manifest as acute or chronic diseases (debilitating flue like illness and endocarditis, respectively).

Coxiella has tropism for alveolar macrophages in vivo (Khavkin and Tabibzadeh 1988) but has the ability to infect phagocytic and non-phagocytic cells in vitro (reviewed by (Voth and Heinzen 2007)). The bacteria replicate into a membrane-bound compartment termed the *Coxiella*-containing vacuole (CCV). The CCV traffics through the endocytic pathway and eventually fuses with lysosomes (Heinzen et al. 1996). Surprisingly, the acidic pH environment and the plethora of acidic hydrolases, which constitute a hostile environment that most intracellular pathogens have evolved to avoid, do not affect *Coxiella*. On the contrary, *Coxiella* thrives and replicates to high numbers in this environment.

The CCV is highly fusogenic. The expansion of the CCV precedes bacterial replication and is driven by the bacteria and relies on the recruitment of cellular vesicles that fuse with the CCV (Howe et al. 2003). The *Coxiella* genome encodes a type IV secretion system (T4SS) related to *Legionella pneumophila* Dot/Icm system (Seshadri et al. 2003). *Coxiella* T4SS is essential for bacterial replication but not for survival within the lysosomal environment (Beare et al. 2011; Carey et al. 2011). In fact, *Coxiella* passively enters the cell through the endocytic pathway, and it is not until the CCV fuses with lysosomes that translocation of T4SS effector begins (Newton et al. 2013). The development of an axenic culture condition (Omsland et al. 2009) and of genetic tools to study *Coxiella* pathogenesis (Beare 2012; Beare and Heinzen 2014) has been instrumental in elucidating and linking the function of *Coxiella* T4SS effectors to cellular processes important in the remodeling of the CCV (reviewed by (Voth and Heinzen 2007; van Schaik et al. 2013; Moffatt et al. 2015)).

One hallmark of the CCV membrane is its high sterol content (Howe and Heinzen 2006). The importance of sterol during *Coxiella* infection is further illustrated by the increased expression of genes involved in the uptake and synthesis of cholesterol in *Coxiella* infected cells and the reduced bacterial replication in cells treated with pharmacological inhibitors of cholesterol uptake and biosynthesis (Howe and Heinzen 2006).

Beside its role in maintaining cellular membrane integrity and fluidity, cholesterol is also involved in endosome positioning (Mukherjee and Maxfield 2004). Cholesterol-rich endosomes migrate toward the minus end of microtubules and fuse with lysosomes, whereas cholesterol-poor endosomes establish MCS with the ER (Fig. 16.3). This process is regulated through conformational changes of the oxysterol-binding protein (OSBP)-related protein L1, ORP1L (Rocha et al. 2009). ORP1L is a multi-domain protein. It contains N-terminal ankyrin repeats that binds Rab7 (Johansson et al. 2005), a PH domain that binds phosphoinositides (Johansson et al. 2005), a FFAT motif that interacts with VAPA and VAPB (Loewen et al. 2003; Loewen and Levine 2005; Weber-Boyvat et al. 2015), and a C-terminal OSBP-related domain (ORD) that binds sterols (Im et al. 2005; Suchanek et al. 2007). ORP1L binds to cholesterol-rich endosomes via the ORD domain and the ankyrin repeats and adopts a conformation that allows the

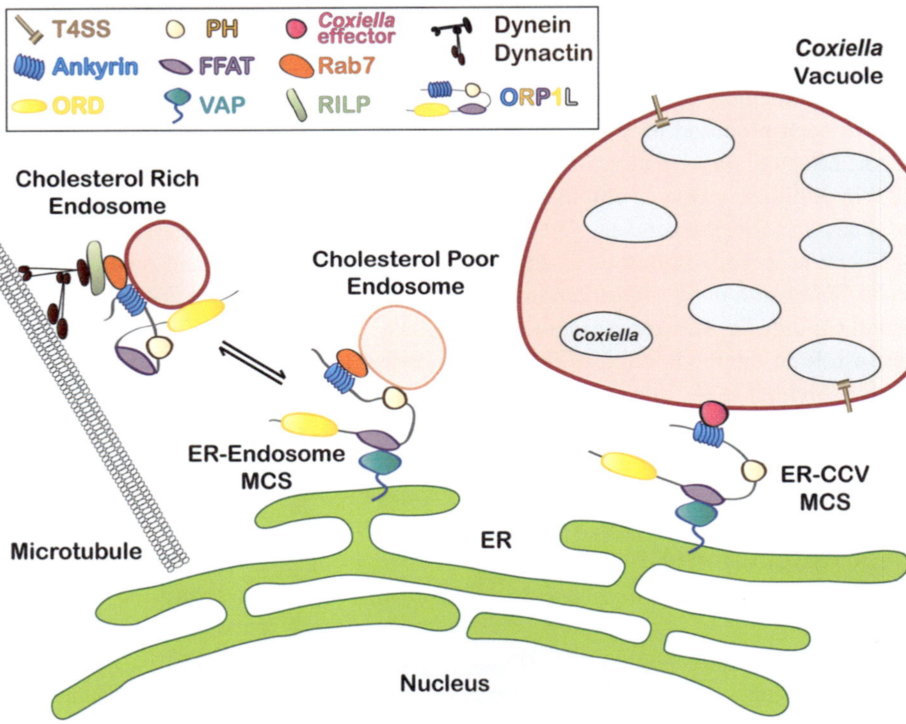

Fig. 16.3 ***Coxiella*** **vacuole establishes MCS resembling ER-endosome MCS.** The cholesterol content of endosomes dictates whether they travel along microtubules or form ER-endosome MCS. This phenomenon relies on conformational change of the sterol-binding protein ORP1L. ORP1L binds to cholesterol-rich endosomes via its ankyrin repeats and its ORD domain allowing for Rab7 to interact with RILP, which in turn interact with the motor dynein driving the minus-end movement of endosomes on microtubules. On cholesterol-poor endosome, ORP1L adopts an open conformation that prevents the recruitment of RIPL and dynein but instead allows for ORP1L interaction with the ER-resident VAP protein and the formation of ER-endosome MCS. In *Coxiella*-infected cells, ORP1L is found at ER-CCV MCS. The ankyrin repeats of ORP1L mediate its association with the sterol-rich membrane of the CCV, potentially by binding a *Coxiella* effector present in the CCV membrane. In addition, it was proposed that ORP1L associates with the ER through FFAT motif-mediated interaction with VAP

Rab7 effector RILP (Rab-interacting lysosomal protein) to interact with the motor dynein and the tethering factor HOPS (homotypic fusion and protein sorting) leading to endosome migration toward the minus end of microtubules (dynein dependent) and fusion with lysosomes (HOPS dependent). When the endosomes contain low level of cholesterol, the ankyrin repeats and the PH domain facilitate the association of ORP1L with the endosomal membrane. The ORD domain no longer interacts with the endosomal membrane modifying ORP1L conformation. This conformation prevents the interaction of RILP with dynein or the tethering factor HOPS and therefore endosome migration and fusion with lysosomes and instead favors ORP1L interaction with VAP at ER-endosome MCSs.

In *Coxiella*-infected cells, ORP1L localize to the CCV, but despite the high sterol content of the CCV, ORP1L association with the CCV does not require the ORD domain (Justis et al. 2016). Instead, the ankyrin repeats of ORP1L are necessary and sufficient for ORP1L binding to the CCV, suggesting that protein-protein interaction is involved. In addition, a functional T4SS is required for ORP1L recruitment to the CCV. Several T4SS effectors localize to the CCV membrane (Larson et al. 2013), and future work will determine if these effectors are directly involved in the recruitment of ORP1L to the CCV. Alternately, *Coxiella* T4SS effectors could play a role in ORP1L association with the CCV by modulating Rab7 activity. ORP1L associates with both the CCV membrane and the ER in a reticulate pattern that is lost when ORP1L FFAT motif is mutated (Justis et al. 2016). Together with the presence of ER tubules in close apposition of the CCV membrane, it is proposed that by binding to an unknown bacterial factor on the CCV membrane and to the VAP proteins in the ER, ORP1L is involved in the formation of ER-CCV MCS during *Coxiella* infection (Fig. 16.3). The role of ORP1L or ER-CCV MCS during *Coxiella* infection is still unclear. ORP1L depletion leads to smaller CCV suggesting a potential role in CCV expansion, and future studies will focus on a potential role in sterol transfer to the CCV (Justis et al. 2016).

16.5 Conclusion and Perspectives

The hijacking of components of cellular MCS with or without the formation of stable MCS is emerging as a new mechanism allowing intracellular bacterial pathogens to establish their replication niche.

The early stages of the *Legionella* life cycle illustrate how the bacteria is capable of connecting otherwise unrelated cellular pathways and machineries, namely, PI4KIIIα and Sac1, two components of ER-PM MCS involved in PI4P homeostasis at the PM, and the early secretory pathway, to remodel the lipid composition of the nascent LCV and convert it into an ER-like compartment.

The vacuoles harboring *Chlamydia* or *Coxiella* establish direct MCS with the ER. *Chlamydia* creates MCS hybrid between ER-Golgi and ER-PM MCS, while *Coxiella* MCS resembles ER-endosome MCS. These two pathogens have in common the targeting of lipid-transfer proteins that contain a FFAT motif, CERT, and ORP1L, respectively, suggesting similar mechanisms of lipid acquisition that rely on the VAP proteins. It will be interesting to determine if any of the other lipid-transfer proteins that localize to various cellular MCS, especially the one harboring a FFAT motif (Prinz 2014), are also targeted by *Chlamydia* and/or *Coxiella*.

Another common theme among the manipulation of MCS by intracellular bacteria is the dependence on the secretion of bacterial effector proteins. As exemplified by the IncD/CERT interaction, one could envision that through molecular mimicry, these effectors allow the bacteria to redirect cellular components to their vacuole.

How MCS between the pathogen-containing vacuole and the ER are formed is still unclear. As shown for ER-mitochondria MCS (Kornmann 2013) or ER-PM (Giordano et al. 2013), there could be specific tethering factors of bacterial and/or host origin that bring and maintain the membranes in close apposition. Alternately, it is possible that the interaction between bacterial effectors and their interacting partners, such as the IncD/CERT/VAP trio, is sufficient to maintain the contact between the two compartments. As for their function, although a common theme of lipid acquisition is emerging, elucidating the role of the ER calcium sensor STIM1 at ER-*Chlamydia* inclusion MCS may uncover additional role for these contacts.

Altogether, the link between MCS and intracellular bacterial pathogens has opened a new and exciting area of research to investigate the molecular mechanisms supporting pathogenesis. In their quest to better understand the formation and the function of MCS in the context of bacterial infection, bacteriologists will greatly benefits from years of research and discoveries made by cell biologists in the field of MCS. One can only hope that bacterial pathogens will give back by shedding light on unknown cellular processes and contribute to the expansion of our knowledge.

References

Agaisse H, Derré I (2014) Expression of the effector protein IncD in *Chlamydia trachomatis* mediates recruitment of the lipid transfer protein CERT and the endoplasmic reticulum-resident protein VAPB to the inclusion membrane. Infect Immun 82:2037–2047

Agaisse H, Derré I (2015) STIM1 is a novel component of ER-*Chlamydia trachomatis* inclusion membrane contact sites. PLoS One 10:e0125671

Arasaki K, Roy CR (2010) *Legionella pneumophila* promotes functional interactions between plasma membrane syntaxins and Sec22b. Traffic 11:587–600

Arasaki K, Toomre DK, Roy CR (2012) The *Legionella pneumophila* effector DrrA is sufficient to stimulate SNARE-dependent membrane fusion. Cell Host Microbe 11:46–57

Baird D, Stefan C, Audhya A, Weys S, Emr SD (2008) Assembly of the PtdIns 4-kinase Stt4 complex at the plasma membrane requires Ypp1 and Efr3. J Cell Biol 183:1061–1074

Balla A, Balla T (2006) Phosphatidylinositol 4-kinases: old enzymes with emerging functions. Trends Cell Biol 16:351–361

Balla A, Tuymetova G, Tsiomenko A, Varnai P, Balla T (2005) A plasma membrane pool of phosphatidylinositol 4-phosphate is generated by phosphatidylinositol 4-kinase type-III alpha: studies with the PH domains of the oxysterol binding protein and FAPP1. Mol Biol Cell 16:1282–1295

Barker JR, Koestler BJ, Carpenter VK, Burdette DL, Waters CM, Vance RE, Valdivia RH (2013) STING-dependent recognition of cyclic di-AMP mediates type I interferon responses during *Chlamydia trachomatis* infection. MBio 4:e00018–e00013

Beare PA (2012) Genetic manipulation of *Coxiella burnetii*. Adv Exp Med Biol 984:249–271

Beare PA, Heinzen RA (2014) Gene inactivation in *Coxiella burnetii*. Methods Mol Biol 1197:329–345

Beare PA, Gilk SD, Larson CL, Hill J, Stead CM, Omsland A, Cockrell DC, Howe D, Voth DE, Heinzen RA (2011) Dot/Icm type IVB secretion system requirements for *Coxiella burnetii* growth in human macrophages. MBio 2:e00175–e00111

Brombacher E, Urwyler S, Ragaz C, Weber SS, Kami K, Overduin M, Hilbi H (2009) Rab1 guanine nucleotide exchange factor SidM is a major phosphatidylinositol 4-phosphate-binding effector protein of *Legionella pneumophila*. J Biol Chem 284:4846–4856

Carabeo RA, Mead DJ, Hackstadt T (2003) Golgi-dependent transport of cholesterol to the *Chlamydia trachomatis* inclusion. Proc Natl Acad Sci USA 100:6771–6776

Carey KL, Newton HJ, Luhrmann A, Roy CR (2011) The *Coxiella burnetii* Dot/Icm system delivers a unique repertoire of type IV effectors into host cells and is required for intracellular replication. PLoS Pathog 7: e1002056

Dehoux P, Flores R, Dauga C, Zhong G, Subtil A (2011) Multi-genome identification and characterization of *chlamydiae*-specific type III secretion substrates: the Inc proteins. BMC Genomics 12:109

Delsing CE, Warris A, Bleeker-Rovers CP (2011) Q fever: still more queries than answers. Adv Exp Med Biol 719:133–143

Derré I, Swiss R, Agaisse H (2011) The lipid transfer protein CERT interacts with the *Chlamydia* inclusion protein IncD and participates to ER-*Chlamydia* inclusion membrane contact sites. PLoS Pathog 7: e1002092

Dumoux M, Clare DK, Saibil HR, Hayward RD (2012) *Chlamydiae* assemble a pathogen synapse to hijack the host endoplasmic reticulum. Traffic 13:1612–1627

Elwell CA, Jiang S, Kim JH, Lee A, Wittmann T, Hanada K, Melancon P, Engel JN (2011) *Chlamydia trachomatis* co-opts GBF1 and CERT to acquire host sphingomyelin for distinct roles during intracellular development. PLoS Pathog 7:e1002198

Elwell C, Mirrashidi K, Engel J (2016) *Chlamydia* cell biology and pathogenesis. Nat Rev Microbiol 14:385–400

Giordano F, Saheki Y, Idevall-Hagren O, Colombo SF, Pirruccello M, Milosevic I, Gracheva EO, Bagriantsev SN, Borgese N, De Camilli P (2013) PI(4,5)P(2)-dependent and Ca^{2+}-regulated ER-PM interactions mediated by the extended synaptotagmins. Cell 153:1494–1509

Goody RS, Itzen A (2013) Modulation of small GTPases by *Legionella*. Curr Top Microbiol Immunol 376:117–133

Hackstadt T, Scidmore MA, Rockey DD (1995) Lipid metabolism in *Chlamydia trachomatis*-infected cells: directed trafficking of Golgi-derived sphingolipids to the chlamydial inclusion. Proc Natl Acad Sci USA 92:4877–4881

Hackstadt T, Rockey DD, Heinzen RA, Scidmore MA (1996) *Chlamydia trachomatis* interrupts an exocytic pathway to acquire endogenously synthesized sphingomyelin in transit from the Golgi apparatus to the plasma membrane. EMBO J 15:964–977

Hammond GR, Fischer MJ, Anderson KE, Holdich J, Koteci A, Balla T, Irvine RF (2012) PI4P and PI(4,5) P2 are essential but independent lipid determinants of membrane identity. Science 337:727–730

Hammond GR, Machner MP, Balla T (2014) A novel probe for phosphatidylinositol 4-phosphate reveals multiple pools beyond the Golgi. J Cell Biol 205:113–126

Hanada K, Kumagai K, Yasuda S, Miura Y, Kawano M, Fukasawa M, Nishijima M (2003) Molecular machinery for non-vesicular trafficking of ceramide. Nature 426:803–809

Hanada K, Kumagai K, Tomishige N, Yamaji T (2009) CERT-mediated trafficking of ceramide. Biochim Biophys Acta 1791:684–691

Hatch GM, McClarty G (1998) Phospholipid composition of purified *Chlamydia trachomatis* mimics that of the eucaryotic host cell. Infect Immun 66:3727–3735

Heinzen RA, Scidmore MA, Rockey DD, Hackstadt T (1996) Differential interaction with endocytic and exocytic pathways distinguish parasitophorous vacuoles of *Coxiella burnetii* and *Chlamydia trachomatis*. Infect Immun 64:796–809

Horwitz MA, Silverstein SC (1980) Legionnaires' disease bacterium (*Legionella pneumophila*) multiples intracellularly in human monocytes. J Clin Invest 66:441–450

Howe D, Heinzen RA (2006) *Coxiella burnetii* inhabits a cholesterol-rich vacuole and influences cellular cholesterol metabolism. Cell Microbiol 8:496–507

Howe D, Melnicakova J, Barak I, Heinzen RA (2003) Maturation of the *Coxiella burnetii* parasitophorous vacuole requires bacterial protein synthesis but not replication. Cell Microbiol 5:469–480

Hubber A, Arasaki K, Nakatsu F, Hardiman C, Lambright D, De Camilli P, Nagai H, Roy CR (2014) The machinery at endoplasmic reticulum-plasma membrane contact sites contributes to spatial regulation of multiple *Legionella* effector proteins. PLoS Pathog 10:e1004222

Huitema K, van den Dikkenberg J, Brouwers JF, Holthuis JC (2004) Identification of a family of animal sphingomyelin synthases. EMBO J 23:33–44

Im YJ, Raychaudhuri S, Prinz WA, Hurley JH (2005) Structural mechanism for sterol sensing and transport by OSBP-related proteins. Nature 437:154–158

Ingmundson A, Delprato A, Lambright DG, Roy CR (2007) *Legionella pneumophila* proteins that regulate Rab1 membrane cycling. Nature 450:365–369

Johansson M, Lehto M, Tanhuanpaa K, Cover TL, Olkkonen VM (2005) The oxysterol-binding protein homologue ORP1L interacts with Rab7 and alters functional properties of late endocytic compartments. Mol Biol Cell 16:5480–5492

Justis AV, Hansen B, Beare PA, King KB, Heinzen RA, Gilk SD (2016) Interactions between the *Coxiella burnetii* parasitophorous vacuole and the endoplasmic reticulum involve the host protein ORP1L. Cell Microbiol

Khavkin T, Tabibzadeh SS (1988) Histologic, immunofluorescence, and electron microscopic study of infectious process in mouse lung after intranasal challenge with *Coxiella burnetii*. Infect Immun 56:1792–1799

Kornmann B (2013) The molecular hug between the ER and the mitochondria. Curr Opin Cell Biol 25:443–448

Kudo N, Kumagai K, Matsubara R, Kobayashi S, Hanada K, Wakatsuki S, Kato R (2010) Crystal structures of the CERT START domain with inhibitors provide insights into the mechanism of ceramide transfer. J Mol Biol 396:245–251

Larson CL, Beare PA, Howe D, Heinzen RA (2013) *Coxiella burnetii* effector protein subverts clathrin-mediated vesicular trafficking for pathogen vacuole biogenesis. Proc Natl Acad Sci USA 110:E4770–E4779

Lemmon MA (2008) Membrane recognition by phospholipid-binding domains. Nat Rev Mol Cell Biol 9:99–111

Lev S, Ben Halevy D, Peretti D, Dahan N (2008) The VAP protein family: from cellular functions to motor neuron disease. Trends Cell Biol 18:282–290

Loewen CJ, Levine TP (2005) A highly conserved binding site in vesicle-associated membrane protein-associated protein (VAP) for the FFAT motif of lipid-binding proteins. J Biol Chem 280:14097–14104

Loewen CJ, Roy A, Levine TP (2003) A conserved ER targeting motif in three families of lipid binding proteins and in Opi1p binds VAP. EMBO J 22:2025–2035

Lutter EI, Martens C, Hackstadt T (2012) Evolution and conservation of predicted inclusion membrane proteins in *chlamydiae*. Comp Funct Genomics 2012:362104

Machner MP, Isberg RR (2006) Targeting of host Rab GTPase function by the intravacuolar pathogen *Legionella pneumophila*. Dev Cell 11:47–56

Malhotra M, Sood S, Mukherjee A, Muralidhar S, Bala M (2013) Genital *Chlamydia trachomatis*: an update. Ind J Med Res 138:303–316

Marston BJ, Lipman HB, Breiman RF (1994) Surveillance for Legionnaires' disease. Risk factors for morbidity and mortality. Arch Intern Med 154:2417–2422

Mehlitz A, Karunakaran K, Herweg JA, Krohne G, van de Linde S, Rieck E, Sauer M, Rudel T (2014) The chlamydial organism *Simkania negevensis* forms ER vacuole contact sites and inhibits ER-stress. Cell Microbiol 16:1224–1243

Moffatt JH, Newton P, Newton HJ (2015) *Coxiella burnetii*: turning hostility into a home. Cell Microbiol 17:621–631

Moos A, Hackstadt T (1987) Comparative virulence of intra- and interstrain lipopolysaccharide variants of *Coxiella burnetii* in the guinea pig model. Infect Immun 55:1144–1150

Moulder JW (1991) Interaction of *chlamydiae* and host cells in vitro. Microbiol Rev 55:143–190

Mueller KE, Plano GV, Fields KA (2014) New frontiers in type III secretion biology: the *Chlamydia* perspective. Infect Immun 82:2–9

Mukherjee S, Maxfield FR (2004) Lipid and cholesterol trafficking in NPC. Biochim Biophys Acta 1685:28–37

Mukherjee S, Liu X, Arasaki K, McDonough J, Galan JE, Roy CR (2011) Modulation of Rab GTPase function by a protein phosphocholine transferase. Nature 477:103–106

Muller MP, Peters H, Blumer J, Blankenfeldt W, Goody RS, Itzen A (2010) The *Legionella* effector protein DrrA AMPylates the membrane traffic regulator Rab1b. Science 329:946–949

Murata T, Delprato A, Ingmundson A, Toomre DK, Lambright DG, Roy CR (2006) The *Legionella pneumophila* effector protein DrrA is a Rab1 guanine nucleotide-exchange factor. Nat Cell Biol 8:971–977

Nakatsu F, Baskin JM, Chung J, Tanner LB, Shui G, Lee SY, Pirruccello M, Hao M, Ingolia NT, Wenk MR et al (2012) PtdIns4P synthesis by PI4KIIIα at the plasma membrane and its impact on plasma membrane identity. J Cell Biol 199:1003–1016

Neunuebel MR, Chen Y, Gaspar AH, Backlund PS Jr, Yergey A, Machner MP (2011) De-AMPylation of the small GTPase Rab1 by the pathogen *Legionella pneumophila*. Science 333:453–456

Newton HJ, McDonough JA, Roy CR (2013) Effector protein translocation by the *Coxiella burnetii* Dot/Icm type IV secretion system requires endocytic maturation of the pathogen-occupied vacuole. PLoS One 8: e54566

Nguyen TM, Ilef D, Jarraud S, Rouil L, Campese C, Che D, Haeghebaert S, Ganiayre F, Marcel F, Etienne J et al (2006) A community-wide outbreak of legionnaires disease linked to industrial cooling towers – how far can contaminated aerosols spread? J Infect Dis 193:102–111

Omsland A, Cockrell DC, Howe D, Fischer ER, Virtaneva K, Sturdevant DE, Porcella SF, Heinzen RA (2009) Host cell-free growth of the Q fever bacterium *Coxiella burnetii*. Proc Natl Acad Sci USA 106:4430–4434

Ponting CP, Aravind L (1999) START: a lipid-binding domain in StAR, HD-ZIP and signalling proteins. Trends Biochem Sci 24:130–132

Prakriya M, Lewis RS (2015) Store-operated calcium channels. Physiol Rev 95:1383–1436

Prinz WA (2014) Bridging the gap: membrane contact sites in signaling, metabolism, and organelle dynamics. J Cell Biol 205:759–769

Riley M, Abe T, Arnaud MB, Berlyn MK, Blattner FR, Chaudhuri RR, Glasner JD, Horiuchi T, Keseler IM, Kosuge T et al (2006) *Escherichia coli* K-12: a cooperatively developed annotation snapshot – 2005. Nucleic Acids Res 34:1–9

Robinson CG, Roy CR (2006) Attachment and fusion of endoplasmic reticulum with vacuoles containing *Legionella pneumophila*. Cell Microbiol 8:793–805

Rocha N, Kuijl C, van der Kant R, Janssen L, Houben D, Janssen H, Zwart W, Neefjes J (2009) Cholesterol sensor ORP1L contacts the ER protein VAP to control Rab7-RILP-p150 Glued and late endosome positioning. J Cell Biol 185:1209–1225

Schachter J (1999) Infection and disease epidemiology. In: Stephens RS (ed) Chlamydia: intracellular biology, pathogenesis, and immunity. American Society for Microbiology, Wahsington, DC, pp 139–170

Schoebel S, Oesterlin LK, Blankenfeldt W, Goody RS, Itzen A (2009) RabGDI displacement by DrrA from *Legionella* is a consequence of its guanine nucleotide exchange activity. Mol Cell 36:1060–1072

Segal G, Purcell M, Shuman HA (1998) Host cell killing and bacterial conjugation require overlapping sets of genes within a 22-kb region of the *Legionella pneumophila* genome. Proc Natl Acad Sci USA 95:1669–1674

Seshadri R, Paulsen IT, Eisen JA, Read TD, Nelson KE, Nelson WC, Ward NL, Tettelin H, Davidsen TM, Beanan MJ et al (2003) Complete genome sequence of the Q-fever pathogen *Coxiella burnetii*. Proc Natl Acad Sci USA 100:5455–5460

Shima K, Klinger M, Schutze S, Kaufhold I, Solbach W, Reiling N, Rupp J (2015) The role of ER-related BiP/GRP78 in IFN-gamma induced persistent *Chlamydia pneumoniae* infection. Cell Microbiol 17 (7):923–934

Stefan CJ, Manford AG, Baird D, Yamada-Hanff J, Mao Y, Emr SD (2011) Osh proteins regulate phosphoinositide metabolism at ER-plasma membrane contact sites. Cell 144:389–401

Stephens RS, Kalman S, Lammel C, Fan J, Marathe R, Aravind L, Mitchell W, Olinger L, Tatusov RL, Zhao Q et al (1998) Genome sequence of an obligate intracellular pathogen of humans: *Chlamydia trachomatis*. Science 282:754–759

Suchanek M, Hynynen R, Wohlfahrt G, Lehto M, Johansson M, Saarinen H, Radzikowska A, Thiele C, Olkkonen VM (2007) The mammalian oxysterol-binding protein-related proteins (ORPs) bind 25-hydroxycholesterol in an evolutionarily conserved pocket. Biochem J 405:473–480

Tafesse FG, Huitema K, Hermansson M, van der Poel S, van den Dikkenberg J, Uphoff A, Somerharju P, Holthuis JC (2007) Both sphingomyelin synthases SMS1 and SMS2 are required for sphingomyelin homeostasis and growth in human HeLa cells. J Biol Chem 282:17537–17547

Tan Y, Luo ZQ (2011) *Legionella pneumophila* SidD is a deAMPylase that modifies Rab1. Nature 475:506–509

Tan Y, Arnold RJ, Luo ZQ (2011) *Legionella pneumophila* regulates the small GTPase Rab1 activity by reversible phosphorylcholination. Proc Natl Acad Sci USA 108:21212–21217

Tilney LG, Harb OS, Connelly PS, Robinson CG, Roy CR (2001) How the parasitic bacterium *Legionella pneumophila* modifies its phagosome and transforms it into rough ER: implications for conversion of plasma membrane to the ER membrane. J Cell Sci 114:4637–4650

van Ooij C, Kalman L, van Ijzendoorn, Nishijima M, Hanada K, Mostov K, Engel JN (2000) Host cell-derived sphingolipids are required for the intracellular growth of *Chlamydia trachomatis*. Cell Microbiol 2:627–637

van Schaik EJ, Chen C, Mertens K, Weber MM, Samuel JE (2013) Molecular pathogenesis of the obligate intracellular bacterium *Coxiella burnetii*. Nat Rev Microbiol 11:561–573

Vogel JP, Andrews HL, Wong SK, Isberg RR (1998) Conjugative transfer by the virulence system of *Legionella pneumophila*. Science 279:873–876

Voth DE, Heinzen RA (2007) Lounging in a lysosome: the intracellular lifestyle of *Coxiella burnetii*. Cell Microbiol 9:829–840

Weber-Boyvat M, Kentala H, Peranen J, Olkkonen VM (2015) Ligand-dependent localization and function of ORP-VAP complexes at membrane contact sites. Cell Mol Life Sci 72:1967–1987

Wylie JL, Hatch GM, McClarty G (1997) Host cell phospholipids are trafficked to and then modified by *Chlamydia trachomatis*. J Bacteriol 179:7233–7242

Yasuda S, Kitagawa H, Ueno M, Ishitani H, Fukasawa M, Nishijima M, Kobayashi S, Hanada K (2001) A novel inhibitor of ceramide trafficking from the endoplasmic reticulum to the site of sphingomyelin synthesis. J Biol Chem 276:43994–44002

Zhu Y, Hu L, Zhou Y, Yao Q, Liu L, Shao F (2010) Structural mechanism of host Rab1 activation by the bifunctional *Legionella* type IV effector SidM/DrrA. Proc Natl Acad Sci USA 107:4699–4704

Zhu W, Banga S, Tan Y, Zheng C, Stephenson R, Gately J, Luo ZQ (2011) Comprehensive identification of protein substrates of the Dot/Icm type IV transporter of *Legionella pneumophila*. PLoS One 6: e17638

Alterations in Ca^{2+} Signalling via ER-Mitochondria Contact Site Remodelling in Cancer

17

Martijn Kerkhofs, Carlotta Giorgi, Saverio Marchi, Bruno Seitaj, Jan B. Parys, Paolo Pinton, Geert Bultynck, and Mart Bittremieux

Abstract

Inter-organellar contact sites establish microdomains for localised Ca^{2+}-signalling events. One of these microdomains is established between the ER and the mitochondria. Importantly, the so-called mitochondria-associated ER membranes (MAMs) contain, besides structural proteins and proteins involved in lipid exchange, several Ca^{2+}-transport systems, mediating efficient Ca^{2+} transfer from the ER to the mitochondria. These Ca^{2+} signals critically control several mitochondrial functions, thereby impacting cell metabolism, cell death and survival, proliferation and migration. Hence, the MAMs have emerged as critical signalling hubs in physiology, while their dysregulation is an important factor that drives or at least contributes to oncogenesis and tumour progression. In this book chapter, we will provide an overview of the role of the MAMs in cell function and how alterations in the MAM composition contribute to oncogenic features and behaviours.

Keywords

Calcium signaling • Ca^{2+}-transport systems • IP_3 receptors • Voltage-dependent anion channels • Chaperones • Cell death and survival • Mitochondrial bioenergetics • Autophagy • Mitochondria-associated ER membranes (MAMs) • Cancer

M. Kerkhofs • B. Seitaj • J.B. Parys • G. Bultynck (✉) • M. Bittremieux
Laboratory Molecular and Cellular Signaling, Department of Cellular and Molecular Medicine and Leuven Kanker Instituut (LKI), KU Leuven, Campus Gasthuisberg O&N 1 Box 802, Herestraat 49, 3000 Leuven, Belgium
e-mail: geert.bultynck@kuleuven.be

C. Giorgi • S. Marchi • P. Pinton
Laboratory for Technologies of Advanced Therapies (LTTA), Department of Morphology, Surgery and Experimental Medicine, Section of Pathology, Oncology and Experimental Biology, University of Ferrara, Ferrara, Italy

M. Tagaya, T. Simmen (eds.), *Organelle Contact Sites*, Advances in Experimental Medicine and Biology 997, DOI 10.1007/978-981-10-4567-7_17

17.1 Introduction

Eukaryotic cells are faced with many challenges to sustain life (Chen and Silver 2012). To tackle these challenges, nature has come up with cellular compartmentalisation resulting in different organelles including the nucleus, endoplasmic reticulum (ER), mitochondria, peroxisomes and Golgi apparatus (Diekmann and Pereira-Leal 2013). The ability of eukaryotic cells to restrict processes to a subcellular localisation brings about numerous advantages (Chen and Silver 2012). However, to ensure the smooth orchestration of cellular processes, communication between the organelles is critical.

One way of inter-organellar communication is Ca^{2+} signalling. The fact that Ca^{2+} is not homogeneously distributed throughout the cell makes it possible to use Ca^{2+} as a messenger (Clapham 2007). Ca^{2+} signals typically arise from the ER and the lysosomes (Berridge et al. 2000; Berridge et al. 2003). These Ca^{2+} signals impact the other organelles via microdomains established by membrane contact sites (La Rovere et al. 2016; Raffaello et al. 2016). Important target organelles of ER-derived Ca^{2+} signals are the mitochondria, which help to maintain Ca^{2+} homeostasis in the cell (Rizzuto et al. 2012). Ca^{2+} uptake in the mitochondria is driven by the negative mitochondrial membrane potential (Rizzuto et al. 2000). The outer mitochondrial membrane (OMM) is freely permeable to Ca^{2+} due to the presence of porins, like the type 1 voltage-gated anion channel (VDAC1) (Gincel et al. 2001; Rapizzi et al. 2002). While Ca^{2+} easily reaches the intermembrane space, a more intriguing problem was the entry of Ca^{2+} into the mitochondrial matrix. Initial research pointed to the existence of a mitochondrial uniporter, albeit one with low affinity for Ca^{2+} (Carafoli 2012; Marchi and Pinton 2014). However, the low affinity of this uniporter could not easily be matched to observations that Ca^{2+} was dynamically and efficiently exchanged between the cytosol and the mitochondria, considering that cytosolic $[Ca^{2+}]$ is typically in the low-to-middle nanomolar range (Rizzuto et al. 1998; Csordás et al. 1999; Carafoli 2012).

The mechanism underlying this seemingly paradoxical observation is the presence of close contacts between ER and mitochondria, which favour quasi-synaptic Ca^{2+} transfer from the ER to the mitochondria (Rizzuto et al. 1998; Csordás et al. 1999). These sites can be isolated biochemically as mitochondria-associated ER membranes (MAMs). MAMs are parts of the ER membrane that are in close proximity to the mitochondrial membrane and are tethered to it (Szabadkai et al. 2006; van Vliet et al. 2014). At the MAMs, the distance between the ER and mitochondria is believed to be approximately 10 to 25 nm (Rizzuto et al. 1998; Csordás et al. 2006; Marchi et al. 2014b), allowing proteins situated on the ER membrane and OMM to interact and enabling efficient Ca^{2+}-based communication between the ER and mitochondria (Decuypere et al. 2011; Rowland and Voeltz 2012; Marchi et al. 2014b). In this way, MAMs provide a microdomain in which the $[Ca^{2+}]$ is several folds higher than in the bulk cytosol (estimated to be >10 μM at the ER-mitochondrial interface) (Csordás et al. 2010), alleviating the problem of the paradoxical low affinity of the mitochondrial Ca^{2+} uniporter (Rizzuto et al. 1993, 1998). Apart from Ca^{2+} homeostasis, MAMs are implicated in several processes critical to cell function, e.g. lipid transport, ER stress, apoptosis, autophagy, inflammation and anti-viral response (Vance 2014; van Vliet et al. 2014).

Due to their involvement in these specific functions, MAMs contain a select protein population. The inositol 1,4,5-trisphosphate (IP_3) receptor (IP_3R) and VDAC1 channels are present in this sub-organellar domain (Várnai et al. 2005; Szabadkai et al. 2006), underlying the prominent role of the MAMs in ER-mitochondrial Ca^{2+} signalling. Furthermore, proteins indirectly involved in ER-mitochondrial Ca^{2+} flux can be found in and around the MAMs as well. These include glucose-regulated protein 75 (GRP75) (Szabadkai et al. 2006), mitofusin-2 (Mfn2) (de Brito and Scorrano 2008), phosphofurin acidic cluster sorting protein-2 (PACS-2) (Myhill et al. 2008), promyelocytic leukaemia protein (PML) (Pinton et al. 2011), sigma-1 receptor (Sig-1R) (Hayashi and Su 2007) and protein kinase RNA-like endoplasmic reticulum kinase (PERK) (Verfaillie et al. 2012), amongst others. Since Ca^{2+} signalling fulfils an important role in several cell physiological processes and considering its dysregulation in pathophysiological conditions, the expression of these proteins is often altered, and their functional activity is converted to promote tumour growth, proliferation, migration, apoptosis resistance and changes in cellular metabolism (Urra et al. 2016; Marchi and Pinton 2016). Furthermore, the MAMs harbour an increasing number of oncogenes and tumour suppressors that functionally impact ER-mitochondrial Ca^{2+} transfer and oncogenic features (Marchi et al. 2014a; Giorgi et al. 2015b; Bittremieux et al. 2016).

In this chapter, we will discuss (1) the MAM components, including the Ca^{2+}-transport systems, chaperones and structural proteins that are present, (2) how MAM components impact ER-mitochondrial Ca^{2+} transfer and their structural organisation and (3) how alterations in the function of these MAM components drive oncogenesis and tumour progression. An overview of the MAM components to be discussed in this chapter can be found in Fig. 17.1.

17.2 MAM Components

17.2.1 The Ca^{2+}-Signalling Machinery at the MAMs

17.2.1.1 IP_3R

The IP_3R is an intracellular Ca^{2+}-release channel, present in the ER membrane (Ashby and Tepikin 2001; Choe and Ehrlich 2006). IP_3Rs are opened by IP_3 (Foskett et al. 2007; Parys and De Smedt 2012; Fedorenko et al. 2014), a second messenger released into the cytosol after phosphatidylinositol 4,5-bisphosphate cleavage by phospholipase C. IP_3R activity is tightly controlled by cytosolic [Ca^{2+}] in a biphasic manner (Iino 1990; Finch et al. 1991; Bezprozvanny et al. 1991; Parys et al. 1992). The Ca^{2+}-flux properties of the IP_3R are also regulated by other cellular factors, including ATP, regulatory proteins (Choe and Ehrlich 2006; Foskett et al. 2007; Parys and De Smedt 2012) and protein kinases and phosphatases (Vanderheyden et al. 2009). Structurally, the IP_3R consists of three domains: an N-terminal ligand-binding domain, which is subdivided in a suppressor region and the IP_3-binding core, a central modulatory domain and a pore-forming region in the C-terminal tail (Yoshikawa et al. 1999). The IP_3R comes in three different isoforms (IP_3R1, IP_3R2 and IP_3R3) which are encoded by different genes (*ITPR1*, *ITPR2* and *ITPR3*) and display 60–80% homology at the level of the amino acid sequence (Mikoshiba 2007; Foskett et al. 2007). Sensitivity towards their ligand IP_3 as well as regulation by Ca^{2+}, ATP and phosphorylation appears to be isoform specific (Newton et al. 1994; Miyakawa et al. 1999; Tu et al. 2005; Khan et al. 2006; Betzenhauser et al. 2008; Wagner et al. 2008; Vervloessem et al. 2015).

A subset of IP_3Rs is present at the MAMs, where it is responsible for ER-mitochondrial Ca^{2+} flux (Várnai et al. 2005; Mendes et al. 2005). By varying the spatio-temporal pattern of Ca^{2+} release from the ER, Ca^{2+} signalling can convey messages, which are differentially decoded at the subcellular level (Berridge et al. 2000, 2003). For example, constitutive low-level IP_3R-mediated ER-mitochondrial Ca^{2+} transfer maintains mitochondrial bioenergetics through stimulation of mitochondrial respiration and ATP production (Cárdenas et al. 2010; Kaufman and Malhotra 2014), whereas excessive Ca^{2+} release from the ER triggers apoptotic cell death due to mitochondrial Ca^{2+} overload (Giorgi et al. 2012; Morciano et al. 2015). Hence, IP_3Rs located at the MAMs play an important role in determining cell fate. Especially IP_3R3 seems to be abundant at the MAMs, since it has been proposed that pro-apoptotic ER-mitochondrial Ca^{2+} transfers preferentially occur via IP_3R3 (Blackshaw et al. 2000; Mendes et al. 2005). However, also the other IP_3R isoforms have been implicated in Ca^{2+}-mediated cell death (Gutstein and Marks 1997;

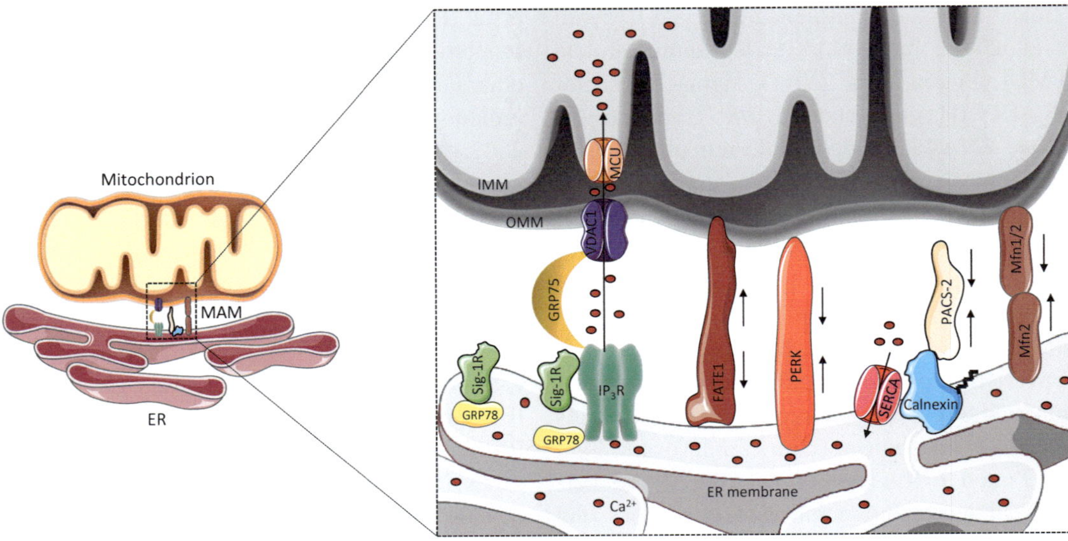

Fig. 17.1 MAM components playing a role in ER-mitochondrial Ca^{2+} signalling. The MAMs harbour a specific protein population consisting of Ca^{2+}-transport proteins and chaperones, as well as of proteins that enable their structural organisation. The principal components of the ER-mitochondrial Ca^{2+} exchange at the MAMs are the IP_3R and VDAC1, which are physically coupled by the chaperone protein GRP75. When Ca^{2+} is released from the ER by the IP_3R, it freely permeates the OMM via VDAC1, to be transported to the mitochondrial matrix by the MCU, located in the IMM. The chaperone Sig-1R is able to modify IP_3R-mediated Ca^{2+} signalling. Sig-1R is held inactive by binding to GRP78, but under ER stress binding to GRP78 is disrupted and Sig-1R interacts with the IP_3R, stabilising the IP_3R and enabling proficient Ca^{2+} signalling even under conditions of ER stress. The efficiency of Ca^{2+} exchange between ER and mitochondria is influenced by the presence and action of tethering proteins like PERK and Mfn2 and anti-tethering proteins like FATE1. The functional effect of tethering and anti-tethering proteins at the MAMs is indicated by *arrows* facing each other or *arrows pointing* in opposite directions, respectively. Besides its function as a tethering protein, PACS-2 also contributes to MAM organisation, while simultaneously having a role in the enrichment of the chaperone calnexin at the MAMs. Furthermore, calnexin is enriched at the MAMs by palmitoylation, a process that switches calnexin function from quality control/protein folding to ER Ca^{2+}-signalling control by enhancing SERCA activity. The interaction between calnexin and SERCA2b appears to be counteracted by the thiol reductase TMX1 (not shown for clarity reasons), which inhibits SERCA2b activity (For more details, please see text)

Jayaraman and Marks 1997; Assefa et al. 2004; Li et al. 2009; Akl et al. 2013) and thus may reside in the MAMs in certain cell types or systems.

17.2.1.2 VDAC1

As described above, Ca^{2+} released by the IP_3R is able to cross the OMM through VDAC1, a weakly anion-selective channel that is Ca^{2+} permeable and which is enriched at the MAMs (Hajnóczky et al. 2002; Shoshan-Barmatz and Gincel 2003; Colombini 2012). Apart from its role in mitochondrial Ca^{2+} transport, VDAC allows for substrates of the electron transport chain, like malate, succinate and nicotinamide adenine dinucleotide, to access the mitochondria (Shoshan-Barmatz et al. 2010). Moreover, VDAC's channel properties permit ATP, produced by oxidative phosphorylation, as well as other mitochondrial products like reactive oxygen species (ROS) to diffuse into the cytosol (Shoshan-Barmatz et al. 2010). Additionally, VDAC1 oligomers have been implicated in the release of cytochrome c into the cytosol (Weisthal et al. 2014). As it were, VDAC functions as the channel that allows the mitochondria to communicate with their subcellular environment.

VDAC is able to switch between an open and a closed state in a voltage-dependent manner. While the channel is stable in the open state at low voltages, high voltages cause VDAC to switch to the closed state (Hodge and Colombini

1997; Gincel et al. 2000). Interestingly, the open state shows a weak selectivity towards anions, as opposed to the closed state, which blocks the passage of large anionic molecules, while it has been proposed to be selective for cations (Gincel et al., 2000; Schein et al., 1976; Shoshan-Barmatz et al., 2010). At the structural level, the N-terminus of the protein, which forms an α-helix, is important for its voltage-dependent gating (Abu-Hamad et al. 2009). Different mechanistic models have been proposed, albeit the exact mechanism has not been established yet (Shoshan-Barmatz et al. 2010), and this voltage-dependency has only been observed in vitro (Shoshan-Barmatz et al. 2010). Apart from voltage-dependent gating, the closed or open state of the channel is affected by modulators interacting with VDAC1. Examples include Bcl-XL (Vander Heiden et al. 2000, 2001), hexokinase (Azoulay-Zohar et al. 2004), tubulin (Rostovtseva et al. 2008), mitochondrial membrane lipids (Rostovtseva et al. 2006) and Ca^{2+} (Báthori et al. 2006).

In mammals, there are three known isoforms of VDAC: VDAC1, VDAC2 and VDAC3, with VDAC1 being the isoform that is expressed at the highest level and consequently has been studied most extensively (Messina et al. 2012). Recombinant expression of VDAC1 enhanced the Ca^{2+} transfer to the mitochondria (Rapizzi et al. 2002), yet this property seems not to be unique for VDAC1, as also the other isoforms display it (De Stefani et al. 2012). However, there is a unique role for VDAC1, but not for VDAC2 nor VDAC3, in conveying pro-apoptotic Ca^{2+} signals to the mitochondria (De Stefani et al. 2012).

17.2.2 The Chaperones

17.2.2.1 GRP75

GRP75 is a chaperone protein belonging to the heat shock 70 kDa (HSP70) protein family (Wadhwa et al. 2002a). GRP75 has been found at different subcellular localisations, e.g. the cytosol, the mitochondria, the ER and the Golgi apparatus (Wadhwa et al. 1995; Ran et al. 2000). Moreover, GRP75 is a pleiotropic protein. For example, GRP75 located in the mitochondrial matrix helps to import unfolded proteins into the matrix in an ATP-dependent manner in cooperation with Tim44 (Scherer et al. 1992; Kronidou et al. 1994; Schneider et al. 1996; Voos and Röttgers 2002). Furthermore, GRP75 is thought to play a role in endocytosis as well as exocytosis (Flachbartová and Kovacech 2013). At the MAMs, GRP75 plays an important role in Ca^{2+} signalling, as it forms a physical link between the IP_3R and VDAC1 (Szabadkai et al. 2006; Betz et al. 2013; Rieusset et al. 2016), thereby increasing the efficiency of ER-mitochondrial Ca^{2+} signalling.

17.2.2.2 Sig-1R

Sig-1R is a chaperone, which was first mistakenly identified as an opioid receptor subtype (Su 1982; Hayashi and Su 2007; Tagashira et al. 2014). Sig-1R mainly resides in the ER, specifically at the MAMs, under resting conditions (Hayashi and Su 2007). There, Sig-1R is bound to an ER lumen chaperone, glucose-regulated protein 78 (GRP78, also known as BiP) (Hayashi and Su 2007). Upon ER stress, however, this association is broken, and Sig-1R gains its chaperone function, targeting its client proteins like the ER stress sensor inositol-requiring enzyme 1 (IRE1) and the IP_3R, thereby regulating ER-mitochondrial Ca^{2+} signalling (cfr. infra) (Hayashi and Su 2007; Mori et al. 2013). Interestingly, Shioda et al. (2012) reported the existence of a truncated splice form of Sig-1R, which did not bind to the IP_3R. Overexpression of this splice variant decreased mitochondrial Ca^{2+} uptake, while promoting IP_3R degradation, as opposed to the non-truncated Sig-1R (Shioda et al. 2012). Furthermore, Sig-1R can undergo translocation to various subcellular localisations upon stimulation by agonists. These locations include the plasma membrane and the nuclear envelope (Su et al. 2010; Mavlyutov et al. 2015; Tsai et al. 2015; Chu and Ruoho 2016).

17.2.2.3 Calnexin

Calnexin is another ER chaperone that is enriched at the MAMs and this in a palmitoylation-dependent manner (Lynes et al. 2012). This chaperone interacts with glycoproteins that are monoglucosylated (Hebert et al. 1995). Functionally, calnexin improves efficiency of ER protein folding and helps retaining misfolded proteins in the ER (Lamriben et al. 2016). Furthermore, calnexin modifies Ca^{2+} signalling in the cell through its functional interaction with sarco/endoplasmic reticulum Ca^{2+}-ATPase 2b (SERCA2b) (Roderick et al. 2000). When the cytosolic domain of calnexin is phosphorylated, the interaction inhibits SERCA2b activity, diminishing intracellular Ca^{2+} oscillations (Roderick et al. 2000). In further work, it was shown that calnexin interaction with SERCA was critically dependent on calnexin palmitoylation and was impaired upon ER stress induction (Lynes et al. 2013). The binding of calnexin to SERCA2b appeared to promote its activity, given the higher ER Ca^{2+}-store content of cells overexpressing wild-type calnexin, but not a palmitoylation-deficient calnexin mutant (Lynes et al. 2013). Interestingly, the binding of calnexin to SERCA2b is influenced by the thioredoxin-related transmembrane protein (TMX1), which is also targeted to the MAMs through palmitoylation (Lynes et al. 2012), since knockout of TMX1 improved binding of calnexin to SERCA2b (Krols et al. 2016; Raturi et al. 2016). TMX1 and calnexin thus may target overlapping binding sites in SERCA2b. As such, the stimulatory effect of calnexin on SERCA activity may be partially related to its ability to reduce the binding of TMX1, which inhibits SERCA activity, to SERCA2b (Krols et al. 2016; Raturi et al. 2016).

17.2.3 Proteins Defining MAM Structure

17.2.3.1 Mfn2

Mfn2, a GTPase protein situated in the OMM, was first studied for its function as a mitochondrial fusion protein, together with its homologue Mfn1 (Ranieri et al. 2013). Interestingly, in 2008 it was found that Mfn2 is also located on the ER membrane and enriched at the MAMs, where it acts as a tether that links mitochondria to the ER and supports efficient Ca^{2+} signalling between the two organelles (de Brito and Scorrano 2008). This tethering function of Mfn2 is realised by the interaction of Mfn2 in the ER membrane with Mfn1 or Mfn2 localised in the OMM. Recently, however, a discussion has risen about the role of Mfn2 as a mitochondrial tether in the MAMs. Ultrastructural analyses as well as functional, biochemical and genetic approaches showed that Mfn2 antagonised ER-mitochondrial tethering (Cosson et al. 2012; Filadi et al. 2015), since ablation of Mfn2 resulted in an increased inter-organellar proximity. It was proposed that Mfn2 functions as an anti-tether that maintains a correct, non-toxic distance between both the ER and the mitochondria (Filadi et al. 2015). However, very recently, the role of Mfn2 as a *bona fide* ER-mitochondrial tether has been confirmed in a series of experiments aiming to critically reappraise its function (Naon et al. 2016).

17.2.3.2 PACS-2

PACS-2 is an ER-associated protein involved in retrograde ER-Golgi trafficking of multiple proteins (Youker et al. 2009). This sorting protein was initially studied for its role in mitochondrial network and MAM organisation (Simmen et al. 2005). PACS-2 regulates communication between the ER and the mitochondria by controlling contact sites between the two organelles (Simmen et al. 2005). In this way, PACS-2 mediates apoptosis (cfr. infra) and ER homeostasis, while promoting transfer of lipids between the ER and mitochondria (Simmen et al. 2005). Interestingly, PACS-2 can also assist in calnexin enrichment at the MAMs in concert with the coat protein complex COPI (Myhill et al. 2008). Also mechanistic target of rapamycin complex 2 (mTORC2) can be found at the MAMs where it regulates MAM integrity via PACS-2 phosphorylation in a protein kinase B (PKB/Akt)-dependent way (Betz et al. 2013).

17.2.3.3 PERK

PERK, a kinase protein located in the ER membrane, functions as an ER stress sensor that controls the unfolded protein response (UPR), alongside IRE1 and activating transcription factor 6 (ATF6) (Sano and Reed 2013). The main effect of the UPR on the cell is the diminishment of mRNA translation to avoid further accumulation of improperly folded proteins, while at the same time, the transcription of chaperones is stimulated (Sano and Reed 2013). Furthermore, retrograde transport of unfolded proteins to the cytosol takes place, where they undergo ubiquitination and subsequent degradation (Sano and Reed 2013). PERK is mainly responsible for the halt of translation by virtue of phosphorylating eukaryotic translation initiation factor 2 (eIF2α), a protein needed for correct mRNA translation (Sano and Reed 2013). The phosphorylation disrupts recycling of eIF2α from its GTP-free form to its GTP-bound form (Sano and Reed 2013). Furthermore, eIF2α phosphorylation is responsible for the preferential translation of UPR-involved genes. In addition, PERK phosphorylates nuclear erythroid 2 p45-related factor 2 (Nrf2) inducing the expression of antioxidant genes to alleviate oxidative stress (Cullinan and Diehl 2004). When ER stress continues for a longer period, the cell will brace itself for apoptosis. In this process, PERK contributes by promoting the transcription of CCAAT-enhancer-binding protein homologous protein, a pro-apoptotic transcription factor (Oyadomari and Mori 2004; Sano and Reed 2013). Moreover, PERK was shown to be involved in autophagy regulation via ATF4-dependent transcription of autophagy-related (ATG) genes (Harding et al. 2000).

Additionally, PERK also serves as an ER-mitochondria tether at the MAMs, thereby facilitating the propagation of ROS signals between these two organelles (Verfaillie et al. 2012). Hence, PERK-knockout cells displayed significantly weaker ER-mitochondria contact sites, counteracting ROS-triggered apoptosis. This function of PERK to maintain ER-mitochondria juxtapositions was independent of its kinase activity but required its cytoplasmic domains (Verfaillie et al. 2012).

17.2.3.4 Foetal and Adult Testis Expressed 1 (FATE1)

FATE1 is a protein that belongs to the group of cancer-testis antigens (CTAs) (Dong et al. 2003), which is a heterogeneous group of proteins with limited expression in normal testis tissue. However, in several types of cancer, these CTAs are upregulated (Simpson et al. 2005; Whitehurst 2014). Recently, FATE1 was found to reside at the MAMs where it regulates ER-mitochondrial distance and ER-mitochondrial Ca^{2+} flux (cfr. infra) (Doghman-Bouguerra et al. 2016). Remarkably, silencing FATE1 led to an increased sensitivity towards paclitaxel, a chemotherapeutic drug, in non-small cell lung cancer cell lines (Whitehurst et al. 2007).

17.3 MAMs in Cancer

All types of cancer share certain stereotypical traits, called the hallmarks of cancer (Hanahan and Weinberg 2011; Giampazolias and Tait 2016). These features, acquired gradually during the development of tumours, include sustaining proliferative signalling, resisting cell death, activating invasion and metastasis, inducing angiogenesis and rewiring metabolism (Hanahan and Weinberg 2011; Giampazolias and Tait 2016). Importantly, MAMs and mitochondria play key roles in many cellular processes such as cell death, cell migration and energy production (Giampazolias and Tait 2016). Therefore, functioning of these cellular compartments is frequently altered and affected during acquisition of the hallmarks of cancer. In this section, we discuss the role of the proteins listed above in the various hallmarks of cancer.

17.3.1 Tumour Growth, Proliferation and Metastasis

17.3.1.1 The IP_3R

Since Ca^{2+} signalling controls a plethora of cellular functions that relate to cancer hallmarks, IP_3Rs have emerged as important regulators of tumour biology. A striking example of the importance of the receptor's function is the observation that lack of IP_3R-mediated Ca^{2+} signalling in thymocytes causes the development of malignancies in mice, resembling T-cell acute lymphoblastic leukaemia (Ouyang et al. 2014). Furthermore, there is a growing body of evidence that suggests IP_3R-mediated Ca^{2+} release plays a role in cancer cell migration (Wei et al. 2009, 2012; Huang et al. 2016). Also, migrating fibroblasts displayed cytosolic Ca^{2+} flickers mediated in part by IP_3R2 (Wei et al. 2009, 2012). Furthermore, IP_3R3 is overexpressed in glioblastoma cells, whereas reducing its expression via siRNA attenuated migration via inhibition of cytosolic Ca^{2+} signalling (Kang et al. 2010). Recently, it was also found that overexpression of ER protein 44, which negatively regulates Ca^{2+} release, prevented migration of A549 cells by suppressing IP_3R2-dependent Ca^{2+} release (Huang et al. 2016). Moreover, the IP_3R was shown to play a role in MCF-7 cell growth, since growth inhibition occurred upon blockage of the IP_3R (Szatkowski et al. 2010). Interestingly, 17-β-estradiol, which induces cell proliferation in MCF-7 cells, may do so by elevating IP_3R3 levels (Szatkowski et al. 2010). In the same cell line, a molecular and functional coupling between IP_3R3 and large-conductance Ca^{2+}- and voltage-dependent K^+ (BK_{Ca}) channels was responsible for ATP-induced proliferation in a cyclin-D1/cyclin-dependent kinase 4-dependent mechanism (Mound et al. 2013). Ablation of IP_3R3 or BK_{Ca} resulted in attenuated proliferation (Mound et al. 2013). Interestingly, the IP_3R is also implied in senescence (Wiel et al. 2014), protecting cells from tumour onset and progression (Ben-Porath and Weinberg 2004; Collado and Serrano 2010; Kang et al. 2011). It was shown that loss of IP_3R2 allowed cells to avoid oncogene-induced senescence (Wiel et al. 2014). This was also the case for the mitochondrial Ca^{2+} uniporter. This points to mitochondrial Ca^{2+} accumulation playing an important role in senescence through lowering the mitochondrial membrane potential and ROS (Wiel et al. 2014).

17.3.1.2 VDAC1

The expression levels of VDAC1 are correlated with tumour growth in different types of cancer. Zhang et al. showed that a decrease in miRNA-320a allowed for a high VDAC1 expression in non-small cell lung cancer cells and that this was correlated with the initiation and progression of cancer (Zhang et al. 2016b). Furthermore, cervical cancer tissues positive for VDAC1 showed an increased tumour size and deep stromal invasion compared to tissues negative for VDAC1 (Wu et al., 2016a). In the same study, VDAC1 knockdown inhibited cell proliferation and migration (Wu et al. 2016a), which was also shown in human papilloma virus-related cervical cancers (Zhang et al. 2016a). This evidence suggests that VDAC1 promotes tumour survival and invasion. Interestingly, knockout of VDAC1 in MEF cells increased proliferation rates under hypoxic conditions through activation of the extracellular signal-regulated protein kinase (ERK) 1/2 pathway (Brahimi-Horn et al. 2015).

Moreover, VDAC1's role in Ca^{2+} signalling has been linked to cell migration. Myeloid cell leukaemia sequence 1 (Mcl-1), an anti-apoptotic protein from the B-cell lymphoma 2 (Bcl-2) protein family, is able to bind VDAC1 with high affinity, thereby seemingly promoting mitochondrial Ca^{2+} uptake. Mcl-1 binding to VDAC1 promoted cell migration without affecting cell proliferation. The pro-migration effect of Mcl-1 could be antagonised by VDAC-based peptides that interfere with VDAC1/Mcl-1-complex formation (Huang et al. 2014). Also other anti-apoptotic Bcl-2 proteins, including Bcl-XL, inhibit VDAC1-mediated Ca^{2+} uptake in the mitochondria (Arbel et al. 2012; Monaco et al. 2015; Vervliet et al. 2016). The mechanism involved Bcl-XL's Bcl-2 homology 4 (BH4) domain and VDAC1's N-terminus (Monaco et al. 2015). The inhibitory impact of Bcl-2 proteins on VDAC1-mediated Ca^{2+} uptake in the mitochondria is consistent with the original papers that describe Bcl-2 proteins as negative regulators of VDAC1-mediated apoptosis (Shimizu et al. 1999, 2000).

17.3.1.3 The Chaperones

A study by Vilner et al. (1995) demonstrated that Sig-1R was overexpressed in a large range of cancer cell lines, both human and rodent. Later studies added that there might be a link between Sig-1R overexpression and metastasis. This was proposed by Aydar et al. since the highest expression levels were found in metastatic cell lines (Aydar et al. 2006). Sig-1R mRNA levels were also found to be higher in invasive breast cancer tissue derived from patients, compared to normal breast tissue (Wang et al. 2004); Sig-1R mRNA was overexpressed in colorectal cancer and colorectal cancer liver metastases (Skrzycki and Czeczot 2013), and overexpression of Sig-1R in hilar cholangiocarcinoma was linked to poor differentiation, lymph node metastasis and advanced disease stage (Xu et al. 2014).

Concerning the molecular mechanisms that are possibly involved, the ability of Sig-1R to interact with several ion channels seems to be important for various oncogenic features (Crottès et al. 2013). In K562 myeloid leukaemia cells, the link between Sig-1R and expression of the potassium ion channel human ether-à-go-g-related gene (hERG), which controls several processes like migration and adhesion (Pillozzi et al. 2007, 2011), was investigated (Crottès et al. 2011). It was observed that Sig-1R is important for hERG maturation by improving maturation efficiency and stabilisation of the α-subunit (Crottès et al. 2011). Other ion channels that interact with Sig-1R and that play a role in cancer include L-type voltage-gated Ca^{2+} channels, voltage-gated Na^{+} channels and Ca^{2+}-activated K^{+} channels (for extensive review, see Crottès et al. 2013). However, the interaction of Sig-1R with these ion channels is not necessarily related to its role at the MAMs.

Not only by directly interacting with other proteins, but also indirectly, Sig-1R is able to modify properties of ion channels: Palmer et al. (2007) found that Sig-1R can bind cholesterol and stabilise lipid rafts via the insertion of cholesterol. In turn, the cholesterol level of lipid rafts can impact the signalling molecules present in these domains (Gniadecki 2004; Palmer et al. 2007), thereby altering the activity of ion channels nearby.

In addition, GRP75 overexpression is correlated with tumour growth and invasion (Kaul et al. 1998; Yi et al. 2008; Jin et al. 2016). Also, in K562 cells high expression levels of GRP75, as well as other chaperone proteins, coincided with resistance towards the proteasome inhibitor bortezomib (Kliková et al. 2015), and inhibition of GRP75 reduced cisplatin resistance in ovarian cancer (Yang et al. 2013).

Also calnexin, as a chaperone, may play a role in tumoural growth in response to growth factors. Lakkaraju and van der Goot (2013) found that in squamous carcinoma cells, caspase-8-mediated cleavage of calnexin occurs upon stimulation of the cells with epidermal growth factor. This yields a calnexin fragment that inhibits protein inhibitor of activated STAT3 (PIAS3), an inhibitor of signal transducer and activator of transcription 3 (STAT3), which functions as an oncogenic transcription factor (Lakkaraju and van der Goot 2013). This, in turn, promotes STAT3-dependent transcription and possibly tumour growth (Lakkaraju and van der Goot 2013).

17.3.1.4 Proteins Defining MAM Structure

The importance of Mfn2 for cell proliferation is suggested by findings in vascular smooth muscle cells: overexpression of Mfn2 in cultured vascular smooth muscle cells inhibited proliferation by blocking the mitogen-activated protein kinase (MAPK)/ERK signalling pathway (Chen et al. 2004). This mechanism was found to be independent of its role in mitochondrial fusion (Chen et al. 2004; Guo et al. 2007). In concert with observations of Mfn2 acting in an anti-proliferative way, Zhang et al. (2013) showed that Mfn2 expression was lower in gastric tumours than in normal mucosal tissue and that expression levels were negatively correlated with tumour size, while Wu et al. (2016b) observed that poor overall survival in hepatocellular carcinoma patients correlated with low Mfn2 expression levels. Also in primary breast cancer, a loss of Mfn2 was detected (Kannan et al. 2016). Strikingly, knockdown of TMX1 in HeLa and A375P melanoma cells generates a similar phenotype as low-level Mfn2 expression (Raturi et al. 2016). This includes increased SERCA activity and altered MAM structure (Raturi et al. 2016). In the case of TMX1, Raturi et al. propose that the stimulatory effect on tumour growth upon TMX1 knockdown is due to an elevated Ca^{2+} retention capacity at the ER combined with an increased ER-mitochondrial distance (Raturi et al. 2016). This, in turn, leads to reduced ER-mitochondrial Ca^{2+} flux and impairment of mitochondrial metabolism, possibly contributing to the Warburg effect (cfr. infra) (Raturi et al. 2016).

Interestingly, another study revealed that Mfn2 deficiency decreased proliferation by blocking autophagy in HeLa cells (Ding et al. 2015). Similarly, A549 human lung adenocarcinoma cells showed disturbed cell proliferation and invasion upon Mfn2 knockdown (Lou et al. 2015). A recent bioinformatics study conducted on the same cell line revealed that Mfn2 knockdown resulted in repression of genes implicated in cell-cycle progression as well as DNA replication and MAPK signalling pathway (Lou et al. 2016). These opposing data suggest that Mfn2's role in cancer is highly context dependent. It may however also relate to a critical window for proper ER-mitochondria distance, in which too close apposition results in excessive apoptosis sensitivity, whereas too far apposition results in defective energetic and metabolic features, as discussed elsewhere (Naon and Scorrano 2014).

PERK, the ER stress sensor, also fulfils this double-edged function in relation to tumour growth. For example, PERK activity has been linked to cell-cycle arrest (Brewer and Diehl 2000; Hamanaka et al. 2005). Since PERK is involved in a general slowdown of the translation process, several proteins are impacted including the drivers of the cell cycle, which comprise the cyclins (Brewer and Diehl 2000; Hamanaka et al. 2005). In this case, cyclin D1's expression is severely altered due to its short half-life (Brewer and Diehl 2000; Hamanaka et al. 2005). This brings about a redistribution of p21, which blocks cyclin-dependent kinase 2, resulting in cell-cycle arrest in the G1 phase (Brewer and Diehl 2000; Hamanaka et al. 2005).

In addition, PERK causes an increase in p53 through decreased E3 ubiquitin ligase human double minute 2 (Hdm2)-dependent removal, promoting apoptosis when a certain threshold is reached (Li et al. 2006; Zhang et al. 2006). Another feature of this p53 accumulation is the induction of p21 and subsequent cell-cycle arrest, as stated above (Ono et al. 1997). Furthermore, PERK is involved in the upregulation of p47, an N-terminal truncated analogue of p53, which mediates cell-cycle arrest in the G2 phase (Bourougaa et al. 2010). Conversely, PERK has been proposed to improve the degradation of p53 in a glycogen synthase kinase 3 β (GSK3β)-dependent manner (Qu et al. 2004; Pluquet et al. 2005).

Also, PERK and its downstream signalling axis have been implicated in metastasis of several cancers, e.g. cervix cancer, breast cancer and head and neck squamous cell carcinoma (Nagelkerke et al. 2013, 2015; Mujcic et al. 2013). Epithelia to mesenchymal transition, which is an indication of the level of invasiveness, is also correlated with PERK signalling, which is underpinned by observations in primary breast cancer, colon cancer, gastric cancer and lung cancer (Feng et al. 2014). Furthermore, the human epidermal growth factor receptor 2 (HER2)/Neu protein is able to induce PERK activity, which allows for redox homeostasis via Nrf2 (Bobrovnikova-Marjon et al. 2010). Subsequently, loss of PERK in HER2/Neu-dependent mammary adenocarcinoma was responsible for growth attenuation and decreased metastasis (Bobrovnikova-Marjon et al. 2010).

About the role of PACS-2 in tumour progression, not much is known. However, PACS-2 was shown to be a regulator of ADAM17, a metalloproteinase that is involved in epithelial development, growth and tumour progression (Dombernowsky et al. 2015). More specifically, loss of PACS-2 diminished ADAM17 levels at the cell surface due to increased degradation (Dombernowsky et al. 2015).

17.3.2 Apoptosis

Another major hallmark of cancer cells is their ability to evade apoptosis (Hanahan and Weinberg 2011). As briefly indicated before, Ca^{2+} signalling is able to regulate the apoptotic process. While Ca^{2+} oscillations were found to be pro-survival signals due to stimulation of critical enzymes of the tricarboxylic acid cycle (TCA), high-amplitude Ca^{2+} signals that last for a longer time can cause apoptosis through mitochondrial Ca^{2+} overload (Hajnóczky et al. 1995; Orrenius et al. 2003; Joseph and Hajnóczky 2007; Roderick and Cook 2008; Denton 2009). This, in turn, causes the opening of the mitochondrial permeability transition pore, mitochondrial swelling and eventually the release of pro-apoptotic factors like cytochrome c in the cytosol (Halestrap 2014; Morciano et al. 2015; Jonas et al. 2015). In order to prevent Ca^{2+}-induced apoptosis and/or to promote Ca^{2+}-dependent bioenergetics, cells may rewire their Ca^{2+}-signalling toolkit (Capiod et al. 2007; Chen et al. 2013; Stewart et al. 2015).

17.3.2.1 The IP_3R

The IP_3R exerts a central role in ER-mitochondrial Ca^{2+} signalling, making it prone to the electrical rewiring of the cancer cell. IP_3R expression levels are altered in various cancers, supporting the critical role of the IP_3R in Ca^{2+} signalling from the ER. For instance, hormone-refractory prostate tumour cells showed increased levels of IP_3R1 (Boutin et al. 2015). This is thought to increase Ca^{2+} leakage from the ER, so that less Ca^{2+} is available for the induction of apoptosis by mitochondrial Ca^{2+} overload (Boutin et al. 2015). Interestingly, bladder cancer cells evade cell death by doing the opposite: treatment with cisplatin diminished IP_3R1 levels, provoking cisplatin resistance (Tsunoda et al. 2005). By lowering IP_3R expression levels, cancer cells prevent the event of toxic mitochondrial Ca^{2+} overload (Prevarskaya et al. 2014). Furthermore, some diffuse large B-cell lymphomas (DLBCLs) express high levels of IP_3R2 (Akl et al. 2013). The reason for this IP_3R2 elevation remains elusive, but one hypothesis is that in metabolically stressed cancer cells, low levels of ATP, a positive regulator of the IP_3R, are insufficient to provide the basal Ca^{2+} signalling needed to fuel mitochondrial bioenergetics (Akl et al. 2013; Akl et al. 2014). Hence, by upregulation of the IP_3R2, which is the IP_3R isoform most sensitive to IP_3, these cancer cells are able to survive (Akl et al. 2013, 2014).

In cancer cells, the process of apoptosis is not only influenced by modifying the expression levels of the IP_3R, but also by altering its Ca^{2+}-release properties. For instance, phosphorylation of the IP_3R dramatically changes its function (Vanderheyden et al. 2009). PKB/Akt is a serine-threonine kinase that phosphorylates the IP_3R C terminally via a substrate motive that is conserved in all isoforms (Khan et al. 2006). While ER Ca^{2+} levels remain unaffected in HeLa cells overexpressing PKB/Akt, IP_3R-dependent ER Ca^{2+} release was shown to be negatively affected (Szado et al. 2008; Marchi et al. 2008). Again, this mechanism may protect against mitochondrial Ca^{2+} overload and subsequent apoptosis (Marchi et al. 2008). Furthermore, it was shown that the protective effect of PKB/Akt overexpression was isoform specific: in COS7 cells, almost completely lacking IP_3R1, PKB/Akt activation led to a decreased IP_3-induced Ca^{2+} release and conferred a protective effect against apoptosis (Marchi et al. 2012). However, in SH-SY 5Y cells, lacking IP_3R3, ER Ca^{2+} release was not modified, while expressing the type 3 isoform in these cells restored the protective effect (Marchi et al. 2012). This suggests that the anti-apoptotic effect of PKB/Akt is mediated in an IP_3R3-dependent way (Marchi et al. 2012). The effect of PKB/Akt-mediated phosphorylation is thought to be directly counteracted by phosphatase and tensin homolog (PTEN), which also localises at the MAMs and dephosphorylates the IP_3R, thereby increasing again the IP_3R-mediated Ca^{2+} release (Bononi et al. 2013).

A recent study showed that extra-nuclear PML contributes to protection against Ca^{2+}-mediated apoptotic cell death via interaction with the IP_3R (Giorgi et al. 2010). PML located at the MAMs physically interacts with the IP_3R3. In cells expressing ER-targeted PML, apoptotic stimuli induced a higher cytosolic and mitochondrial Ca^{2+} response (Giorgi et al. 2010). Furthermore, it was revealed that PML-expressing cells displayed lower levels of phosphorylated IP_3R3 and phosphorylated, active PKB/Akt and higher levels of the phosphatase 2A (PP2A) compared to PML-negative cells (Giorgi et al. 2010). PML stimulates pro-apoptotic Ca^{2+} signalling at the MAMs by recruiting PP2A to IP_3R3-PKB/Akt complexes (Giorgi et al. 2010), resulting in a suppressed PKB/Akt-mediated IP_3R3 phosphorylation, since PP2A negatively regulates the activity of PKB/Akt at the ER (Pinton et al. 2011; Bittremieux et al. 2016).

IP_3R function can also be stimulated by phosphorylation (Gomez et al. 2016). During reperfusion injury of the heart, a fraction of the protein kinase GSK3β is localised at the sarco/endoplasmic reticulum (SR/ER) and the MAMs. There, it interacts with the IP_3R Ca^{2+}-channelling complex, regulating its protein composition and modulating Ca^{2+} transfer between the SR/ER and mitochondria. During hypoxia reoxygenation, GSK3β activity is augmented, resulting in increased IP_3R phosphorylation and IP_3R hyperactivity. Consequently, increased IP_3R-mediated SR/ER-mitochondria Ca^{2+} transfer leads to cardiomyocyte cell death. Therefore, inhibition of GSK3β may protect the heart from lethal reperfusion injury by cellular Ca^{2+} overload (Gomez et al. 2016).

The critical role of IP_3R3 underlying cellular apoptosis sensitivity has also been confirmed independently in two isogenic cell lines, one of which expresses oncogenic K-Ras and one in which this oncogenic allele was deleted via homologous recombination (Pierro et al. 2014). The presence of oncogenic K-Ras caused a lowering in the ER Ca^{2+}-store content, thereby decreasing the likelihood of pro-apoptotic Ca^{2+} transfer and thus decreasing apoptotic sensitivity (Pierro et al. 2014). This was due to the increased expression level of IP_3R1 relative to that of IP_3R3 in the cell line expressing oncogenic K-Ras, augmenting basal Ca^{2+} leak via IP_3R1 and suppressing pro-apoptotic Ca^{2+} transfer into the mitochondria via the IP_3R3 (Pierro et al. 2014).

Apart from phosphorylation and expression regulation, IP_3R-mediated Ca^{2+} release can be modulated directly by the binding of an increasing number of oncogenes and tumour suppressors (Akl and Bultynck 2013; Bittremieux et al. 2016). Notably, several members of the Bcl-2-protein family are known to interact with the IP_3R. First, there is the anti-apoptotic protein Bcl-2 itself, whereas its canonical function comprises the sequestration of pro-apoptotic Bcl-2-protein family members like Bik and Bid via its BH3 domain (Youle and Strasser 2008; Czabotar et al. 2014) evidence was found that Bcl-2 interacts via its BH4 domain with 20 amino acids in the central, modulatory part of the IP_3R (Rong et al. 2008, 2009; Monaco et al. 2012). This interaction protects cells against Ca^{2+}-mediated apoptotic cell death (Hanson et al. 2008; Rong et al. 2008). The importance of Bcl-2's complex formation with the IP_3R became clear in Bcl-2-dependent chronic lymphocytic leukaemia and in DLBCL cell lines (Zhong et al. 2011; Akl et al. 2013). Some subtypes of DLBCL display high levels of IP_3R2, the most sensitive isoform with respect to IP_3 (Akl et al. 2013). Therefore, Bcl-2 overexpression is needed in these cells to avoid Ca^{2+}-induced apoptosis triggered by the high expression levels of IP_3R2, making these cells balancing on the edge of apoptosis, a state that was coined "primed-for-death at the ER" (Akl et al. 2013; Akl et al. 2014). Following this concept, TAT-IDP, a peptide mimicking the IP_3R-binding site for Bcl-2, and its derivative BIRD-2, induced apoptosis by disrupting IP_3R/Bcl-2 interaction and eliciting spontaneous toxic Ca^{2+} signalling (Zhong et al. 2011; Akl et al. 2013; Akl et al. 2015; Lavik et al. 2015; Greenberg et al. 2015).

Apart from Bcl-2, the closely related anti-apoptotic Bcl-XL protein is also able to interact with the IP_3R, albeit not via its BH4 domain

(White et al. 2005; Monaco et al. 2012). Bcl-XL, through its hydrophobic cleft, binds the IP_3R at its C-terminal region (Eckenrode et al. 2010) by targeting two BH3-domain-like sequences (Yang et al. 2016). The binding between Bcl-XL and the IP_3R results in a reduction in ER [Ca^{2+}], increased Ca^{2+} oscillations in the cytosol and protection against apoptosis by sensitising the IP_3R to low basal levels of IP_3 (Li et al. 2007).

Bcl-XL does not only alter Ca^{2+} signalling in the cell via direct modulation of the IP_3R, but also via a nuclear factor of activated T-cells (NFAT)-dependent pathway, which modifies IP_3R expression levels (Li et al. 2007). Furthermore, Mcl-1, another Bcl-2-protein family member, was shown to interact directly with the IP_3R via the last transmembrane domain in its C-terminal tail (Eckenrode et al. 2010). Like Bcl-XL, Mcl-1 was found to sensitise the IP_3R, thereby increasing the frequency of Ca^{2+} oscillations in the cell and the number of oscillating cells (Eckenrode et al. 2010). Figure 17.2 displays the proteins that regulate ER-mitochondrial Ca^{2+} signalling via the IP_3R and VDAC1. For reasons of clarity, a distinction was made between the regulation of pro-survival and pro-apoptotic Ca^{2+} signalling.

Finally, inhibition of Bcl-2-family members has emerged as an attractive anticancer strategy, particularly by preventing the complex formation between the anti- and pro-apoptotic Bcl-2-family members (Davids and Letai 2012). However, the function of anti-apoptotic Bcl-2-family members appears to be involved in ER-mitochondrial contact sites as well. Indeed, recent studies showed that targeting the hydrophobic cleft using BH3 mimetics like ABT-737, a non-selective Bcl-2/Bcl-XL inhibitor, enhance anticancer treatments by increasing ER-mitochondria contact sites and stimulating ER-mitochondrial Ca^{2+} transfer (Fan et al. 2015; Xie et al. 2016). As such, ABT-737 could restore the sensitivity of cisplatin-resistant ovarian cancer cells to cisplatin treatment. This correlated with the ability of cisplatin to induce mitochondrial Ca^{2+} overload, an important feature of the successful induction of cell death by anticancer treatments (Bittremieux and Bultynck 2015; Bonora et al. 2015; Fan et al. 2015; Giorgi et al. 2015a; Xie et al. 2016).

17.3.2.2 VDAC1

As VDAC1 is the gateway for Ca^{2+} entry in the mitochondria, this protein's function may also be influenced in cancer cells to ensure their survival. Bcl-XL was found to bind to and block VDAC1 with its BH4 domain, thereby inhibiting Ca^{2+}-mediated apoptosis (Monaco et al. 2015). Intriguingly, cytosolic Ca^{2+} levels impact VDAC1 expression levels (Weisthal et al. 2014). An increase in cytosolic [Ca^{2+}], elicited by, for example, H_2O_2, induces a rise in VDAC1 expression, which at the same time correlates with the ability to form oligomers in the OMM, through which the pro-apoptotic protein cytochrome c is released from the mitochondria (Weisthal et al. 2014).

VDAC1 does not only contribute to Ca^{2+}-mediated apoptotic cell death, but also influences apoptosis occurring independently of Ca^{2+}. For instance, in human glioma cells subjected to hypoxia, VDAC1 has been implicated in the activation of mitophagy (Qiao et al. 2016). In these cells, the mitochondrial deacetylase sirtuin-3 (Sirt3) stimulates the association between VDAC1 and parkin, an E3 ubiquitin ligase, stimulating mitophagy (Qiao et al. 2016; Bernardini et al. 2017). Consequently, knockdown of Sirt3 inhibited mitophagy, rendering the cells prone to apoptotic cell death (Qiao et al. 2016). Thus, VDAC1 in concert with Sirt3 plays a role in protecting cancer cells through mitophagy. Interestingly, parkin also seems to play a role in the regulation of mitochondrial homeostasis and energy metabolism (Calì et al. 2013). Overexpression of parkin in HeLa and SH-SY5Y neuroblastoma cells increased physical as well as functional interactions between the ER and the mitochondria, whereas parkin silencing caused mitochondrial fragmentation and compromised mitochondrial Ca^{2+} transients due to reduced ER-mitochondria tethering (Calì et al. 2013).

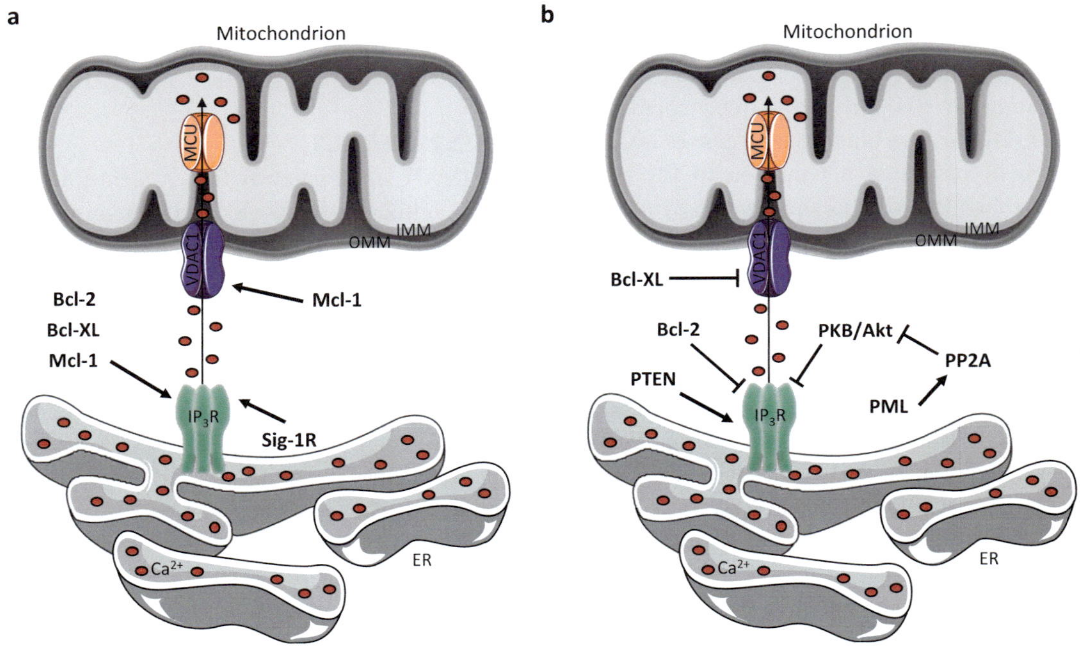

Fig. 17.2 Regulation of pro-survival and pro-apoptotic Ca^{2+} signalling at the MAMs. *Arrow-headed lines* indicate a stimulatory interaction, *while bar-headed lines* indicate an inhibitory interaction. (**a**) Regulation of pro-survival Ca^{2+} signalling at the MAMs. Bcl-2, Bcl-XL and Mcl-1 increase pro-survival Ca^{2+} oscillations and stimulate cell metabolism by interacting with the C-terminus of the IP_3R, which results in a sensitisation of the channel to basal IP_3 levels. Mcl-1 also enhances cell survival through binding to VDAC1, thereby increasing its activity and thus mitochondrial Ca^{2+} uptake. While these proteins directly impact the Ca^{2+}-flux properties of the IP_3R or VDAC1, Sig-1R indirectly promotes cell survival. Under conditions of ER stress, Sig-1R becomes active and stabilises the IP_3R, ensuring the transmission of pro-survival Ca^{2+} signalling into the mitochondria. (**b**) Regulation of pro-apoptotic Ca^{2+} signalling at the MAMs. Bcl-2 and Bcl-XL do not only support pro-survival Ca^{2+} signalling, but also inhibit pro-apoptotic Ca^{2+} signalling. Binding of Bcl-2 via its BH4 domain to the central region of the IP_3R diminishes the Ca^{2+} flux through the IP_3R, while Bcl-XL inhibits VDAC1 via its BH4 domain that targets the N-terminus of VDAC1. Ca^{2+} release from the ER is also decreased by PKB/Akt, which inhibits IP_3R function by phosphorylation. The phosphatase PTEN counteracts the function of PKB/Akt at the MAMs by dephosphorylating the IP_3R. PML indirectly influences the phosphorylation state of the IP_3R by recruiting the phosphatase PP2A, which negatively regulates PKB/Akt activity in the MAMs and counteracts PKB/Akt-mediated phosphorylation of IP_3R in the MAMs. Thus, PML alleviates the suppression of IP_3R-mediated Ca^{2+} flux from the ER to the mitochondria imposed by PKB/Akt (For more details, please see text)

Furthermore, VDAC1 provides a link between apoptosis and differentiation of cancer cells. In glioblastoma tumour cells, in which VDAC1 expression was silenced, a shift from pro-apoptotic proteins linked to cell proliferation, including avian myelocytomatosis virus oncogene cellular homolog (c-Myc) and nuclear factor κB (NF-κB), to pro-apoptotic proteins regulating cell differentiation, including p53, was observed (Arif et al. 2016). This led to differentiation of the glioblastoma cells into astrocyte- and neuron-like cells. Additionally, several studies showed that hexokinase (HK), the enzyme catalysing the first step of glycolysis (Wilson 2003), bound to VDAC1, confers protection from apoptosis in HEK and HeLa cells (Bryson et al. 2002). This protection is proposed to result from the inhibition of the interaction between VDAC1 and the pro-apoptotic Bcl-2 family member Bax by HK (Bryson et al. 2002). Interestingly, for HK to execute its anti-apoptotic effect, its binding to VDAC1 is needed (Arzoine et al. 2009; Abu-Hamad et al. 2009). Furthermore, evidence also suggests that the binding of HK to VDAC1 reduced mitochondrial ROS generation (da-Silva et al. 2004; Sun et al. 2008). Since ROS production is often elevated in cancer cells (Liou and Storz 2010; Panieri and Santoro 2016), this may be another mechanism by which HK protects tumour cells against cell death, as both the HK-I and -II isoforms conveyed protection towards apoptosis-inducing oxidants through their association with VDAC1 (Bryson et al. 2002; Ahmad et al. 2002).

17.3.2.3 The Chaperones

Also the chaperone proteins at the MAMs are able to modify the cell's apoptotic pathways. Similar to VDAC1, Sig-1R acts in a Ca^{2+}-dependent and Ca^{2+}-independent way. Apoptotic Ca^{2+} signalling is impacted by Sig-1R's ability to bind to the IP_3R. Knockdown of Sig-1R resulted in increased degradation of the IP_3R3 via the proteasome, suggesting that it has a stabilising function (Hayashi and Su 2007). Furthermore, in conditions of physiologically normal ER [Ca^{2+}], Sig-1R is in a resting state, bound to GRP78 (Hayashi and Su 2007). However, under conditions of depletion of the ER Ca^{2+} stores, the Sig-1R/GRP78 complex is disrupted, and Sig-1R obtains its chaperone activity (Hayashi and Su 2007). It targets the IP_3R to ensure that IP_3-mediated pro-survival Ca^{2+} signalling to the mitochondria occurs properly. Interestingly, while during short periods of ER Ca^{2+} depletion, Sig-1R remains localised at the MAMs it is redistributed throughout the ER upon longer ER Ca^{2+} depletion, again as a pro-survival mechanism under continued ER stress, which is often present in cancer cells (Hayashi and Su 2007). By stabilising the IP_3R, it sustains pro-survival Ca^{2+} signalling during ER stress. Alternatively, it was reported that Sig-1R supports cell survival during ER stress via another client protein: IRE1 (Mori et al. 2013). Activation of IRE1 triggers its endonuclease activity, needed for splicing X-box binding protein-1 (XBP1) mRNA. This transcription factor then promotes the transcription of various ER chaperones (Yoshida et al. 2001). It is thought that Sig-1R's stabilisation of IRE1 contributes to prolonged signalling along the IRE1-XBP1 axis, thereby supporting cell survival under conditions of augmented and prolonged ER stress (Mori et al. 2013).

GRP75, on the other hand, impacts apoptosis in cancer cells primarily through its alteration of MAPK/ERK signalling and influence on p53. In medullary thyroid carcinoma cells, apoptosis and inhibition of cell growth were caused by a depletion of GRP75 (Starenki et al. 2015). Investigations into the pathways involved revealed that a temporary activation of MAPK/ERK signalling was responsible for the growth arrest, while apoptosis was induced through mitochondrial dysfunction. This consisted of loss of the mitochondrial membrane potential, lowered oxygen consumption and an elevation of ROS levels. Furthermore, it was shown that these mitochondrial effects were linked to a decrease in Bcl-2 expression (Starenki et al. 2015). Similar results in different cancer cell lines confirmed GRP75 as a negative regulator of MAPK/ERK signalling (Wu et al. 2013). An additional contribution to the anti-apoptotic function of GRP75 is its capacity to bind the tumour suppressor p53 (Wadhwa et al. 1998; Kaul et al. 2001; Wadhwa et al. 2002b). The interaction prevents nuclear translocation of p53, abrogating its function as a transcription factor (Wadhwa et al. 2002b). Moreover, keeping p53 in the cytoplasm speeds up its proteasomal degradation (Kaul et al. 2005). Strikingly, applying the HSP90 inhibitor 17-AAG, which blocks other heat shock proteins but not GRP75, stimulated GRP75 expression and reinforced its binding to p53, weakening the effect of the HSP90 inhibitor in hepatocellular carcinoma (Guo et al. 2014).

Also calnexin has been implicated in the regulation of apoptotic cell death. However, it seems that calnexin may play an anti-apoptotic as well as a pro-apoptotic role dependently on the circumstances. Caspase-3 and Caspase-7 have both been shown to cleave calnexin in vitro, while overexpression of the cleavage product partially inhibited apoptosis (Takizawa et al. 2004). Indirectly, calnexin also fulfils a pro-apoptotic role: its cytosolic tail is able to recruit caspase-8, which is responsible for the cleavage of Bap31 (Breckenridge et al. 2002;

Delom et al. 2007). Bap31's cleaved form stimulates Ca^{2+} release from the ER and apoptosis ensues (Breckenridge et al. 2003). In the MCF-7 breast cancer cell line, which is resistant to tunicamycin-induced cell death, calnexin is able to sensitise the cells to tunicamycin, independently of its chaperone function (Delom et al. 2007).

17.3.2.4 Proteins Defining MAM Structure

In MCF-7 cells, Mfn2 mediated apoptosis via the phosphoinositide 3-kinase (PI3K)/PKB/Akt signalling pathway (Ma et al. 2015). These results mimic the findings in vascular smooth muscle cells. There, Mfn2 was shown to trigger mitochondrial apoptosis by inhibiting the GTPase Ras, resulting in decreased PKB/Akt signalling along the Ras-PI3K-PKB/Akt axis (Guo et al. 2007). However, it was not established whether lowered PKB/Akt signalling was linked to an increased Ca^{2+} signalling through reduced phosphorylation of the IP_3R. In this regard, it is interesting to note that Wang et al. (2015) found that overexpression of Mfn2 induced Ca^{2+}-dependent apoptosis in hepatocellular carcinoma.

Additionally, several studies have linked PACS-2 with the apoptotic process, both via the intrinsic pathway as through the extrinsic one. For example, PACS-2 has been shown to interact with the Bcl-2-family protein Bid (Simmen et al. 2005): upon the addition of apoptotic stimuli, PACS-2 is responsible for the translocation of Bid to the mitochondria, where cytochrome c release and caspase activation ensue (Simmen et al. 2005). Furthermore, PACS-2 is involved in tumour necrosis factor-related apoptosis inducing ligand (TRAIL)-triggered apoptosis, more specifically in lysosomal permeabilisation (Werneburg et al. 2012). PACS-2 recruits Bim and Bax, two other members of the Bcl-2 family, to the lysosomal membrane to bring about cathepsin B release and subsequent apoptosis (Werneburg et al. 2012). On this note, it is interesting that cellular inhibitor of apoptosis protein-1 and -2 (cIAP-1/cIAP-2) repress expression levels of PACS-2 by promoting its ubiquitinylation (Guicciardi et al. 2014). In this way, these cIAPs confer resistance to TRAIL-induced apoptosis in hepatobiliary cancer cell lines (Guicciardi et al. 2014). Curiously, PKB/Akt-mediated phosphorylation of Ser437 serves as a switch to shift from PACS-2's trafficking function to its function as a promoter of apoptosis (Aslan et al. 2009). When phosphorylated, PACS-2 is bound to the 14-3-3 scaffold protein, which inhibits its role in apoptosis (Aslan et al. 2009), in a similar way as the Bcl-2-protein family Bad (Zha et al. 1996). Interestingly, phosphorylation of PACS-2 does not merely serve to repress apoptosis, but is also required for polycystin-2 localisation to the ER (Aslan et al. 2009).

Apart from engaging Bcl-2-protein family members, PACS-2 also influences the regulation of p53. Sirt1 deacetylates p53, but upon DNA damage, PACS-2 is shuttled to the nucleus, where it interacts with Sirt1, preventing the deacetylation of p53 and inducing p21-dependent cell-cycle arrest (Atkins et al. 2014). Contrasting with its pro-apoptotic function, however, a recent study has found that PACS-2 was necessary for NF-κB-dependent Bcl-XL induction in response to DNA damage (Barroso-González et al. 2016).

Lastly, PERK has been shown to play a role in the survival of c-Myc-dependent cancer cells via its involvement in autophagy (Hart et al. 2012). As an oncogene, c-Myc regulates ribosome expression and biogenesis, increasing protein synthesis (van Riggelen et al. 2010). This increased synthesis load is accompanied by elevated PERK activity, which promotes autophagy as a survival mechanism through the upregulation of unc-51-like kinase 1 and ATG5 (Hart et al. 2012). This is underpinned by the observation that upon loss of PERK in these cancer cells, the balance tips from survival to apoptosis due to decreased autophagy (Hart et al. 2012).

Furthermore, PERK seems to play an important role in tumour cell survival under hypoxic conditions. Hypoxia was shown to trigger PERK signalling in xenograft models, while a dominant-negative PERK or eIF2α was linked with an increase in apoptotic cells in hypoxic regions of the tumour (Bi et al. 2005). Delving into the molecular mechanisms responsible for this protection against hypoxia, an increase in ATG5 expression was found, indicating that the induction of autophagy provides the protection for tumour cells (Kouroku et al. 2007; Rouschop et al. 2010).

Recently, the CTA FATE1 has been discovered at the MAMs, where it controls ER-mitochondrial distance (Doghman-Bouguerra et al. 2016). In fact, FATE1 functions as an anti-tether: it diminishes ER-mitochondrial contact sites and decreases mitochondrial Ca^{2+} uptake. Hence, it regulates the sensitivity towards pro-apoptotic stimuli that elicit apoptosis via Ca^{2+} signalling (Doghman-Bouguerra et al. 2016). Another anti-apoptotic function of FATE1 is its role in the prevention of accumulation of Bik, a pro-apoptotic Bcl-2-protein family member (Maxfield et al. 2015). The underlying molecular mechanism consists of the recruitment of the E3 ligase RNF183 by FATE1 and subsequent stimulation of Bik degradation (Maxfield et al. 2015), allowing cells to survive even in the presence of apoptotic stimuli.

17.3.3 Cellular Energetics and Biochemical Pathways

17.3.3.1 The IP_3R

Ca^{2+} plays an important role in controlling mitochondrial bioenergetics, since it stimulates ATP production and mitochondrial respiration as the α-ketoglutarate, isocitrate and pyruvate dehydrogenases are Ca^{2+}-dependent rate-limiting enzymes of the TCA (Cárdenas et al. 2010; Kaufman and Malhotra 2014). These dehydrogenases are inhibited in the absence of constitutive low-level Ca^{2+} transfer from ER to mitochondria, which turns on AMP-activated kinase (AMPK). This results in an increase in basal autophagic flux that is independent of mTOR (Cárdenas et al. 2010; Cárdenas and Foskett 2012). The pro-survival low-level IP_3R-mediated Ca^{2+} signalling can be modified by the activity of several anti-apoptotic Bcl-2-family proteins, which are often upregulated in cancer (Bittremieux et al. 2016). Bcl-2, Bcl-XL as well as Mcl-1 have been reported to interact with IP_3Rs (White et al. 2005; Li et al. 2007; Eckenrode et al. 2010). The anti-apoptotic proteins target the C-terminal region of the IP_3R (a.a. 2570-2749) (White et al. 2005), resulting in a sensitisation of the IP_3Rs to basal IP_3 levels, thereby enhancing IP_3R-dependent Ca^{2+} oscillations and stimulating mitochondrial bioenergetics. Bcl-XL is also present in the MAMs, targeting and stimulating IP_3Rs and driving mitochondrial metabolism (Williams et al. 2016). Bcl-XL is recruited to the MAMs during non-apoptotic ER stress induction, augmenting mitochondrial bioenergetics through interaction with IP_3Rs. IP_3R sensitisation by Bcl-XL occurs via its hydrophobic cleft, which binds two BH3-like domains in the C-terminus of IP_3Rs, although the BH4 domain of Bcl-XL also contributes by targeting the central, modulatory domain of the IP_3R (Yang et al. 2016; Williams et al. 2016). Therefore, BH3 mimetic drugs that target Bcl-XL may also antagonise Bcl-XL's ability to sensitise IP_3Rs and thus may suppress Ca^{2+}-driven mitochondrial metabolism. This is very important, since cancer cells are particularly addicted to these basal Ca^{2+}-signalling events to sustain adequate TCA cycling (Cárdenas et al. 2016; Bultynck 2016). This process provides mitochondrial substrates, like nucleosides, that are essential for proper cell-cycle progression and cell division. In the absence of these Ca^{2+} fluxes, non-tumorigenic cells tune down their cell cycle, while tumorigenic cells progress through the cell cycle irrespective of their energetic state, resulting in a mitotic catastrophe (Cárdenas et al. 2016; Bultynck 2016).

As described above, the phosphorylation state of the IP_3R has a determining role in its Ca^{2+}-release properties, and PML is able to alter this phosphorylation state. Recently, it has been discovered that MAM-localised PML inhibits autophagy via the control of ER-mitochondrial Ca^{2+} signalling (Missiroli et al. 2016). By stimulating ER-mitochondrial Ca^{2+} flux, the cell's metabolism is being stimulated (Cárdenas et al. 2010; Kaufman and Malhotra 2014). p53 is acting as a molecular bridge to keep PML at its place in the MAMs (Missiroli et al. 2016). Upon

PML loss, however, metabolic stimulation via Ca^{2+} is not present anymore, and this turns on AMPK signalling and subsequent autophagy (Missiroli et al. 2016). Additionally, also mTORC2 may control metabolism via IP_3R3 phosphorylation, in a similar way it contributes to MAM integrity via PACS-2 phosphorylation (Betz et al. 2013). Activation of PKB/Akt in an mTORC2-dependent manner reduces IP_3R-mediated Ca^{2+} signalling and hence might impact the rate of oxidative phosphorylation (Betz et al. 2013).

17.3.3.2 VDAC1

As a gateway connecting the mitochondria to their environment, VDAC1 is not only a regulator of apoptosis: its localisation at the OMM of the mitochondrion also allows for regulation of cellular bioenergetics. For a start, it has been reported that the Ca^{2+}-flux properties of VDAC1 are stimulated by Mcl-1, which binds with high affinity to the anion channel in the OMM (Huang et al. 2014). This results in an increased mitochondrial Ca^{2+} uptake, thereby promoting ATP production and stimulating cell survival, as described above (Huang et al. 2014).

Furthermore, VDAC1 is thought to play a major role in constituting the Warburg effect through its interaction with HK (Bustamante and Pedersen 1977; Azoulay-Zohar et al. 2004; Pedersen 2008). A common characteristic of cancer cells is that they show high levels of glycolysis, even though they are oxygenated (Vander Heiden et al. 2009; Liberti and Locasale 2016). This aerobic glycolysis is commonly referred to as the Warburg effect (Vander Heiden et al. 2009; Liberti and Locasale 2016). It is proposed that, by binding VDAC1, ATP produced by oxidative phosphorylation is readily accessible for HK-I and -II to fuel the conversion of glucose to glucose-6-phosphate (G-6-P) (Pedersen 2008). This provides a functional coupling of glycolysis with the TCA. Interestingly, evidence suggests that HK is less sensitive to product inhibition by G-6-P through its interaction with VDAC1 (Bustamante and Pedersen 1977; Azoulay-Zohar et al. 2004). Another clue to VDAC1's role in the Warburg effect is that especially the VDAC1-bound isoforms, HK-I and HK-II were found to show higher expression levels in several types of cancer, e.g. lymphoma, prostate and breast cancer (Pedersen 2008). Recently, mTORC2 was found to impact the binding of HK-II to VDAC1, again through PKB/Akt-dependent phosphorylation of HK-II, which stabilises its binding to VDAC1 (Betz et al. 2013).

The Warburg effect also comprises the suppression of oxidative phosphorylation in the mitochondria (Zheng 2012; Lu et al. 2015). A long unidentified player in this suppression is dimeric tubulin (Rostovtseva et al. 2008). Dimeric tubulin at concentrations in the nanomolar range was shown to reversibly block VDAC reconstituted into planar phospholipid membranes (Rostovtseva et al. 2008). Furthermore, this block of VDAC was demonstrated to decrease oxygen consumption in isolated mitochondria (Rostovtseva et al. 2008). In addition to this, it was shown that in HepG2 cells, an increase in dimeric tubulin resulted in mitochondrial depolarisation, while a decrease in dimeric tubulin was associated with mitochondrial hyperpolarisation (Maldonado et al. 2010). These results suggest that mitochondrial metabolism in cancer cells is attenuated by the tubulin-mediated blockage of VDAC (Maldonado et al. 2010). It is noteworthy that the effect of free tubulin on cellular metabolism was not found in primary hepatocytes: microtubule depolymerisation decreased the mitochondrial membrane potential, as in cancer cells, but inducing polymerisation did not increase it (Maldonado et al. 2010). This leaves the question whether the observations concerning free tubulin blocking VDAC are specific for cancer cells (Maldonado et al. 2010; Rostovtseva and Bezrukov 2012).

Apart from a shift towards aerobic glycolysis, other biochemical pathways like cholesterologenesis may be altered in cancer cells. Also in this process, a link to VDAC1 can be found: the channel is part of the polyprotein complex called the transduceosome, which is responsible for import of cholesterol into the mitochondria (McEnery et al. 1992; Liu et al. 2006; Rone et al. 2012). In this complex, VDAC1 is proposed to interact with the translocator protein, anchoring the transduceosome to the OMM and facilitating the binding and import of the steroidogenic acute regulatory protein (Hauet et al. 2005). It is proposed that HK bound to VDAC1 may influence the amount of cholesterol synthesis and its import in mitochondria in cancer cells (Campbell and Chan 2007), while at the same time, the channel properties of VDAC1 itself may be influenced by the augmented levels of cholesterol in the OMM (Pastorino and Hoek 2008).

17.4 Conclusions

The MAMs and their various components, including Ca^{2+}-transport systems, chaperones and structural components, establish an important Ca^{2+}-signalling domain between the ER, the main intracellular Ca^{2+}-storage organelle and the mitochondria, the main organelle controlling cell death and survival processes, including cellular bioenergetics and autophagy, apoptosis sensitivity, growth and proliferation. Dysregulation of these processes is a hallmark of cancer. Hence, alterations and perturbations in the structural organisation and functional properties of the MAMs have emerged as an important nexus that underlies oncogenesis, tumour growth and metastasis and responses to chemotherapy. Moreover, several oncogenes and tumour suppressors are localised at the MAMs. Thus, changes in MAMs can drive oncogenesis, while cancer cells at a later stage could remodel MAMs to favour tumour growth, proliferation and metastatic behaviour.

Acknowledgements GB and JBP are supported by grants from the Research Foundation - Flanders (FWO) (grants G.0571.12N, G.0819.13N, G.0C91.14N, G.0927.15N and G.0A34.16N) and the Research Council - KU Leuven (OT14/101). MB and MK are holders of a Ph.D. Fellowship from the FWO. P.P. is grateful to Camilla degli Scrovegni for continuous support. P.P. is supported by Telethon (GGP15219/B), the Italian Association for Cancer Research (AIRC) (IG-14442), the Italian Cystic Fibrosis Research Foundation (19/2014), the Italian Ministry of Education, University and Research (COFIN no. 20129JLHSY_002, FIRB no. RBAP11FXBC_002 and Futuro in Ricerca no. RBFR10EGVP_001), local funds from the University of Ferrara and the Italian Ministry of Health. C.G is supported by AIRC (MFAG-13521), the Italian Ministry of Health, Cariplo and local funds from the University of Ferrara.

References

Abu-Hamad S, Arbel N, Calo D, Arzoine L, Israelson A, Keinan N, Ben-Romano R, Friedman O, Shoshan-Barmatz V (2009) The VDAC1 N-terminus is essential both for apoptosis and the protective effect of anti-apoptotic proteins. J Cell Sci 122:1906–1916

Ahmad A, Ahmad S, Schneider BK, Allen CB, Chang LY, White CW (2002) Elevated expression of hexokinase II protects human lung epithelial-like A549 cells against oxidative injury. Am J Physiol Lung Cell Mol Physiol 283:573–584

Akl H, Bultynck G (2013) Altered Ca^{2+} signaling in cancer cells: Proto-oncogenes and tumor suppressors targeting IP_3 receptors. Biochim Biophys Acta 1835:180–193

Akl H, Monaco G, La Rovere R, Welkenhuyzen K, Kiviluoto S, Vervliet T, Molgó J, Distelhorst CW, Missiaen L, Mikoshiba K et al (2013) IP_3R2 levels dictate the apoptotic sensitivity of diffuse large B-cell lymphoma cells to an IP_3R-derived peptide targeting the BH4 domain of Bcl-2. Cell Death Dis 4:e632

Akl H, Vervloessem T, Kiviluoto S, Bittremieux M, Parys JB, De Smedt H, Bultynck G (2014) A dual role for the anti-apoptotic Bcl-2 protein in cancer: mitochondria versus endoplasmic reticulum. Biochim Biophys Acta 1843:2240–2252

Akl H, La Rovere RM, Janssens A, Vandenberghe P, Parys JB, Bultynck G (2015) HA14-1 potentiates apoptosis in B-cell cancer cells sensitive to a peptide disrupting IP_3 receptor/Bcl-2 complexes. Int J Dev Biol 59:391–398

Arbel N, Ben-Hail D, Shoshan-Barmatz V (2012) Mediation of the antiapoptotic activity of Bcl-xL protein upon interaction with VDAC1 protein. J Biol Chem 287:23152–23161

Arif T, Krelin Y, Shoshan-Barmatz V (2016) Reducing VDAC1 expression induces a non-apoptotic role for pro-apoptotic proteins in cancer cell differentiation. Biochim Biophys Acta 1857:1228–1242

Arzoine L, Zilberberg N, Ben-Romano R, Shoshan-Barmatz V (2009) Voltage-dependent anion channel 1-based peptides interact with hexokinase to prevent

its anti-apoptotic activity. J Biol Chem 284:3946–3955
Ashby MC, Tepikin AV (2001) ER calcium and the functions of intracellular organelles. Semin Cell Dev Biol 12:11–17
Aslan JE, You H, Williamson DM, Endig J, Youker RT, Thomas L, Shu H, Du Y, Milewski RL, Brush MH et al (2009) Akt and 14-3-3 control a PACS-2-homeostatic switch that integrates membrane traffic with TRAIL-induced apoptosis. Mol Cell 34:497–509
Assefa Z, Bultynck G, Szlufcik K, Nadif Kasri N, Vermassen E, Goris J, Missiaen L, Callewaert G, Parys JB, De Smedt H (2004) Caspase-3-induced truncation of type 1 inositol trisphosphate receptor accelerates apoptotic cell death and induces inositol trisphosphate-independent calcium release during apoptosis. J Biol Chem 279:43227–43236
Atkins KM, Thomas LL, Barroso-González J, Thomas L, Auclair S, Yin J, Kang H, Chung JH, Dikeakos JD, Thomas G (2014) The multifunctional sorting protein PACS-2 regulates SIRT1-mediated deacetylation of p53 to modulate p21-dependent cell-cycle arrest. Cell Rep 8:1545–1557
Aydar E, Onganer P, Perrett R, Djamgoz MB, Palmer CP (2006) The expression and functional characterization of sigma (sigma) 1 receptors in breast cancer cell lines. Cancer Lett 242:245–257
Azoulay-Zohar H, Israelson A, Abu-Hamad S, Shoshan-Barmatz V (2004) In self-defence: hexokinase promotes voltage-dependent anion channel closure and prevents mitochondria-mediated apoptotic cell death. Biochem J 377:347–355
Barroso-González J, Auclair S, Luan S, Thomas L, Atkins KM, Aslan JE, Thomas LL, Zhao J, Zhao Y, Thomas G (2016) PACS-2 mediates the ATM and NF-κB-dependent induction of anti-apoptotic Bcl-xL in response to DNA damage. Cell Death Differ 23:1448–1457
Báthori G, Csordás G, Garcia-Perez C, Davies E, Hajnóczky G (2006) Ca^{2+}-dependent control of the permeability properties of the mitochondrial outer membrane and voltage-dependent anion-selective channel (VDAC). J Biol Chem 281:17347–17358
Ben-Porath I, Weinberg RA (2004) When cells get stressed: an integrative view of cellular senescence. J Clin Invest 113:8–13
Bernardini JP, Lazarou M, Dewson G (2017) Parkin and mitophagy in cancer. Oncogene 36(10):1315–1327
Berridge MJ, Lipp P, Bootman MD (2000) The versatility and universality of calcium signalling. Nat Rev Mol Cell Biol 1:11–21
Berridge MJ, Bootman MD, Roderick HL (2003) Calcium signalling: dynamics, homeostasis and remodelling. Nat Rev Mol Cell Biol 4:517–529
Betz C, Stracka D, Prescianotto-Baschong C, Frieden M, Demaurex N, Hall MN (2013) Feature Article: mTOR complex 2-Akt signaling at mitochondria-associated endoplasmic reticulum membranes (MAM) regulates mitochondrial physiology. Proc Natl Acad Sci USA 110:12526–12534
Betzenhauser MJ, Wagner LE, Iwai M, Michikawa T, Mikoshiba K, Yule DI (2008) ATP modulation of Ca^{2+} release by type-2 and type-3 inositol (1, 4, 5)-triphosphate receptors. Differing ATP sensitivities and molecular determinants of action. J Biol Chem 283:21579–21587
Bezprozvanny I, Watras J, Ehrlich BE (1991) Bell-shaped calcium-response curves of Ins(1,4,5)P3- and calcium-gated channels from endoplasmic reticulum of cerebellum. Nature 351:751–754
Bi M, Naczki C, Koritzinsky M, Fels D, Blais J, Hu N, Harding H, Novoa I, Varia M, Raleigh J et al (2005) ER stress-regulated translation increases tolerance to extreme hypoxia and promotes tumor growth. EMBO J 24:3470–3481
Bittremieux M, Bultynck G (2015) p53 and Ca^{2+} signaling from the endoplasmic reticulum: partners in anti-cancer therapies. Oncoscience 2:233–238
Bittremieux M, Parys JB, Pinton P, Bultynck G (2016) ER functions of oncogenes and tumor suppressors: Modulators of intracellular Ca^{2+} signaling. Biochim Biophys Acta 1863:1364–1378
Blackshaw S, Sawa A, Sharp AH, Ross CA, Snyder SH, Khan AA (2000) Type 3 inositol 1,4,5-trisphosphate receptor modulates cell death. FASEB J Off Publ Fed Am Soc Exp Biol 14:1375–1379
Bobrovnikova-Marjon E, Grigoriadou C, Pytel D, Zhang F, Ye J, Koumenis C, Cavener D, Diehl JA (2010) PERK promotes cancer cell proliferation and tumor growth by limiting oxidative DNA damage. Oncogene 29:3881–3895
Bononi A, Bonora M, Marchi S, Missiroli S, Poletti F, Giorgi C, Pandolfi PP, Pinton P (2013) Identification of PTEN at the ER and MAMs and its regulation of Ca^{2+} signaling and apoptosis in a protein phosphatase-dependent manner. Cell Death Differ 20:1631–1643
Bonora M, Giorgi C, Pinton P (2015) Novel frontiers in calcium signaling: A possible target for chemotherapy. Pharmacol Res 99:82–85
Bourougaa K, Naski N, Boularan C, Mlynarczyk C, Candeias M, Marullo S, Fåhraeus R (2010) Endoplasmic reticulum stress induces G2 cell-cycle arrest via mRNA translation of the p53 isoform p53/47. Mol Cell 38:78–88
Boutin B, Tajeddine N, Monaco G, Molgo J, Vertommen D, Rider M, Parys JB, Bultynck G, Gailly P (2015) Endoplasmic reticulum Ca^{2+} content decrease by PKA-dependent hyperphosphorylation of type 1 IP_3 receptor contributes to prostate cancer cell resistance to androgen deprivation. Cell Calcium 57:312–320
Brahimi-Horn MC, Giuliano S, Saland E, Lacas-Gervais-S, Sheiko T, Pelletier J, Bourget I, Bost F, Féral C, Boulter E et al (2015) Knockout of Vdac1 activates hypoxia-inducible factor through reactive oxygen species generation and induces tumor growth by

promoting metabolic reprogramming and inflammation. Cancer Metab 3:8

Breckenridge DG, Nguyen M, Kuppig S, Reth M, Shore GC (2002) The procaspase-8 isoform, procaspase-8L, recruited to the BAP31 complex at the endoplasmic reticulum. Proc Natl Acad Sci USA 99:4331–4336

Breckenridge DG, Stojanovic M, Marcellus RC, Shore GC (2003) Caspase cleavage product of BAP31 induces mitochondrial fission through endoplasmic reticulum calcium signals, enhancing cytochrome c release to the cytosol. J Cell Biol 160:1115–1127

Brewer JW, Diehl JA (2000) PERK mediates cell-cycle exit during the mammalian unfolded protein response. Proc Natl Acad Sci USA 97:12625–12630

Bryson JM, Coy PE, Gottlob K, Hay N, Robey RB (2002) Increased hexokinase activity, of either ectopic or endogenous origin, protects renal epithelial cells against acute oxidant-induced cell death. J Biol Chem 277:11392–11400

Bultynck G (2016) Onco-IP_3Rs feed cancerous cravings for mitochondrial Ca^{2+}. Trends Biochem Sci 41:390–393

Bustamante E, Pedersen PL (1977) High aerobic glycolysis of rat hepatoma cells in culture: role of mitochondrial hexokinase. Proc Natl Acad Sci USA 74:3735–3739

Calì T, Ottolini D, Negro A, Brini M (2013) Enhanced parkin levels favor ER-mitochondria crosstalk and guarantee Ca^{2+}transfer to sustain cell bioenergetics. Biochim Biophys Acta 1832:495–508

Campbell AM, Chan SHP (2007) The voltage dependent anion channel affects mitochondrial cholesterol distribution and function. Arch Biochem Biophys 466:203–210

Capiod T, Shuba Y, Skryma R, Prevarskaya N (2007) Calcium signalling and cancer cell growth. Subcell Biochem 45:405–427

Carafoli E (2012) The interplay of mitochondria with calcium: an historical appraisal. Cell Calcium 52:1–8

Cárdenas C, Foskett JK (2012) Mitochondrial Ca^{2+} signals in autophagy. Cell Calcium 52:44–51

Cárdenas C, Miller RA, Smith I, Bui T, Molgó J, Müller M, Vais H, Cheung KH, Yang J, Parker I et al (2010) Essential regulation of cell bioenergetics by constitutive $InsP_3$ receptor Ca^{2+} transfer to mitochondria. Cell 142:270–283

Cárdenas C, Müller M, McNeal A, Lovy A, Jaňa F, Bustos G, Urra F, Smith N, Molgó J, Diehl JA, Ridky TW, Foskett JK (2016) Selective vulnerability of cancer cells by inhibition of Ca^{2+} transfer from endoplasmic reticulum to mitochondria. Cell Rep 14:2313–2324

Chen AH, Silver PA (2012) Designing biological compartmentalization. Trends Cell Biol 22:662–670

Chen K-H, Guo X, Ma D, Guo Y, Li Q, Yang D, Li P, Qiu X, Wen S, Xiao R-P et al (2004) Dysregulation of HSG triggers vascular proliferative disorders. Nat Cell Biol 6:872–883

Chen Y-F, Chen Y-T, Chiu W-T, Shen M-R (2013) Remodeling of calcium signaling in tumor progression. J Biomed Sci 20:23

Choe C-U, Ehrlich BE (2006) The inositol 1,4,5-trisphosphate receptor (IP_3R) and its regulators: sometimes good and sometimes bad teamwork. Sci STKE 2006:re15

Chu UB, Ruoho AE (2016) Biochemical pharmacology of the sigma-1 receptor. Mol Pharmacol 89:142–153

Clapham DE (2007) Calcium signaling. Cell 131:1047–1058

Collado M, Serrano M (2010) Senescence in tumours: evidence from mice and humans. Nat Rev Cancer 10:51–57

Colombini M (2012) VDAC structure, selectivity, and dynamics. Biochim Biophys Acta 1818:1457–1465

Cosson P, Marchetti A, Ravazzola M, Orci L (2012) Mitofusin-2 independent juxtaposition of endoplasmic reticulum and mitochondria: an ultrastructural study. PloS One 7:e46293

Crottès D, Martial S, Rapetti-Mauss R, Pisani DF, Loriol C, Pellissier B, Martin P, Chevet E, Borgese F, Soriani O (2011) Sig1R protein regulates hERG channel expression through a post-translational mechanism in leukemic cells. J Biol Chem 286:27947–27958

Crottès D, Guizouarn H, Martin P, Borgese F, Soriani O (2013) The sigma-1 receptor: a regulator of cancer cell electrical plasticity? Membr Physiol Membr Biophys 4:175

Csordás G, Thomas AP, Hajnóczky G (1999) Quasi-synaptic calcium signal transmission between endoplasmic reticulum and mitochondria. EMBO J 18:96–108

Csordás G, Renken C, Várnai P, Walter L, Weaver D, Buttle KF, Balla T, Mannella C, Hajnóczky G (2006) Structural and functional features and significance of the physical linkage between ER and mitochondria. J Cell Biol 174:915–921

Csordás G, Várnai P, Golenár T, Roy S, Purkins G, Schneider T, Balla T, Hajnóczky G (2010) Imaging interorganelle contacts and local calcium dynamics at the ER-mitochondrial interface. Mol Cell 39:121–132

Cullinan SB, Diehl JA (2004) PERK-dependent activation of Nrf2 contributes to redox homeostasis and cell survival following endoplasmic reticulum stress. J Biol Chem 279:20108–20117

Czabotar PE, Lessene G, Strasser A, Adams JM (2014) Control of apoptosis by the BCL-2 protein family: implications for physiology and therapy. Nat Rev Mol Cell Biol 15:49–63

da Silva WS, Gómez-Puyou A, de Gómez-Puyou MT, Moreno-Sanchez R, De Felice F, de Meis L, Oliveira MF, Galina A (2004) Mitochondrial bound hexokinase activity as a preventive antioxidant defense: steady-state ADP formation as a regulatory mechanism of membrane potential and reactive oxygen species generation in mitochondria. J Biol Chem 279:39846–39855

Davids MS, Letai A (2012) Targeting the B-cell lymphoma/leukemia 2 family in cancer. J Clin Oncol 30:3127–3135

de Brito OM, Scorrano L (2008) Mitofusin 2 tethers endoplasmic reticulum to mitochondria. Nature 456:605–610

De Stefani D, Bononi A, Romagnoli A, Messina A, De Pinto V, Pinton P, Rizzuto R (2012) VDAC1 selectively transfers apoptotic Ca^{2+} signals to mitochondria. Cell Death Differ 19:267–273

Decuypere J-P, Monaco G, Bultynck G, Missiaen L, De Smedt H, Parys JB (2011) The IP_3 receptor-mitochondria connection in apoptosis and autophagy. Biochim Biophys Acta 1813:1003–1013

Delom F, Emadali A, Cocolakis E, Lebrun JJ, Nantel A, Chevet E (2007) Calnexin-dependent regulation of tunicamycin-induced apoptosis in breast carcinoma MCF-7 cells. Cell Death Differ 14:586–596

Denton RM (2009) Regulation of mitochondrial dehydrogenases by calcium ions. Biochim Biophys Acta 1787:1309–1316

Diekmann Y, Pereira-Leal JB (2013) Evolution of intracellular compartmentalization. Biochem J 449:319–331

Ding Y, Gao H, Zhao L, Wang X, Zheng M (2015) Mitofusin 2-deficiency suppresses cell proliferation through disturbance of autophagy. PLoS One 10: e0121328

Doghman-Bouguerra M, Granatiero V, Sbiera S, Sbiera I, Lacas-Gervais S, Brau S, Fassnacht M, Rizzuto R, Lalli E (2016) FATE1 antagonizes calcium- and drug-induced apoptosis by uncoupling ER and mitochondria. EMBO Rep 17:1264–1280

Dombernowsky SL, Samsøe-Petersen J, Petersen CH, Instrell R, Hedegaard AM, Thomas L, Atkins KM, Auclair S, Albrechtsen R, Mygind KJ et al (2015) The sorting protein PACS-2 promotes ErbB signalling by regulating recycling of the metalloproteinase ADAM17. Nat Commun 6:7518

Dong X-Y, Su Y-R, Qian X-P, Yang X-A, Pang X-W, Wu H-Y, Chen W-F (2003) Identification of two novel CT antigens and their capacity to elicit antibody response in hepatocellular carcinoma patients. Br J Cancer 89:291–297

Eckenrode EF, Yang J, Velmurugan GV, Foskett JK, White C (2010) Apoptosis protection by Mcl-1 and Bcl-2 modulation of inositol 1,4,5-trisphosphate receptor-dependent Ca^{2+} signaling. J Biol Chem 285:13678–13684

Fan Z, Yu H, Cui N, Kong X, Liu X, Chang Y, Wu Y, Sun L, Wang G (2015) ABT737 enhances cholangiocarcinoma sensitivity to cisplatin through regulation of mitochondrial dynamics. Exp Cell Res 335:68–81

Fedorenko OA, Popugaeva E, Enomoto M, Stathopulos PB, Ikura M, Bezprozvanny I (2014) Intracellular calcium channels: inositol-1,4,5-trisphosphate receptors. Eur J Pharmacol 739:39–48

Feng Y-X, Sokol ES, Del Vecchio CA, Sanduja S, Claessen JH, Proia TH, Jin DX, Reinhardt F, Ploegh HL, Wang Q, Gupta PB (2014) Epithelial-to-mesenchymal transition activates PERK-eIF2α and sensitizes cells to endoplasmic reticulum stress. Cancer Discov 4:702–715

Filadi R, Greotti E, Turacchio G, Luini A, Pozzan T, Pizzo P (2015) Mitofusin 2 ablation increases endoplasmic reticulum-mitochondria coupling. Proc Natl Acad Sci USA 112:2174–2181

Finch EA, Turner TJ, Goldin SM (1991) Calcium as a coagonist of inositol 1,4,5-trisphosphate-induced calcium release. Science 252:443–446

Flachbartová Z, Kovacech B (2013) Mortalin - a multipotent chaperone regulating cellular processes ranging from viral infection to neurodegeneration. Acta Virol 57:3–15

Foskett JK, White C, Cheung K-H, Mak D-OD (2007) Inositol trisphosphate receptor Ca^{2+} release channels. Physiol Rev 87:593–658

Giampazolias E, Tait SWG (2016) Mitochondria and the hallmarks of cancer. FEBS J 283:803–814

Gincel D, Silberberg SD, Shoshan-Barmatz V (2000) Modulation of the voltage-dependent anion channel (VDAC) by glutamate. J Bioenerg Biomembr 32:571–583

Gincel D, Zaid H, Shoshan-Barmatz V (2001) Calcium binding and translocation by the voltage-dependent anion channel: a possible regulatory mechanism in mitochondrial function. Biochem J 358:147–155

Giorgi C, Ito K, Lin H-K, Santangelo C, Wieckowski MR, Lebiedzinska M, Bononi A, Bonora M, Duszynski J, Bernardi R, Rizzuto R et al (2010) PML regulates apoptosis at endoplasmic reticulum by modulating calcium release. Science 330:1247–1251

Giorgi C, Baldassari F, Bononi A, Bonora M, De Marchi E, Marchi S, Missiroli S, Patergnani S, Rimessi A, Suski JM et al (2012) Mitochondrial Ca^{2+} and apoptosis. Cell Calcium 52:36–43

Giorgi C, Bonora M, Sorrentino G, Missiroli S, Poletti F, Suski JM, Galindo Ramirez F, Rizzuto R, Di Virgilio F, Zito E et al (2015a) p53 at the endoplasmic reticulum regulates apoptosis in a Ca^{2+}-dependent manner. Proc Natl Acad Sci USA 112:1779–1784

Giorgi C, Missiroli S, Patergnani S, Duszynski J, Wieckowski MR, Pinton P (2015b) Mitochondria-associated membranes: composition, molecular mechanisms, and physiopathological implications. Antioxid Redox Signal 22:995–1019

Gniadecki R (2004) Depletion of membrane cholesterol causes ligand-independent activation of Fas and apoptosis. Biochem Biophys Res Commun 320:165–169

Gomez L, Thiebaut P-A, Paillard M, Ducreux S, Abrial M, Crola Da Silva C, Durand A, Alam MR, Van Coppenolle F, Sheu S-S et al (2016) The SR/ER-mitochondria calcium crosstalk is regulated by GSK3β during reperfusion injury. Cell Death Differ 23:313–322

Greenberg EF, McColl KS, Zhong F, Wildey G, Dowlati A, Distelhorst CW (2015) Synergistic killing of human small cell lung cancer cells by the Bcl-2-inositol 1,4,5-trisphosphate receptor disruptor BIRD-2 and the BH3-mimetic ABT-263. Cell Death Dis 6: e2034

Guicciardi ME, Werneburg NW, Bronk SF, Franke A, Yagita H, Thomas G, Gores GJ (2014) Cellular inhibitor of apoptosis (cIAP)-mediated ubiquitination of phosphofurin acidic cluster sorting protein 2 (PACS-2) negatively regulates tumor necrosis factor-related apoptosis-inducing ligand (TRAIL) cytotoxicity. PloS One 9:e92124

Guo X, Chen K-H, Guo Y, Liao H, Tang J, Xiao R-P (2007) Mitofusin 2 triggers vascular smooth muscle cell apoptosis via mitochondrial death pathway. Circ Res 101:1113–1122

Guo W, Yan L, Yang L, Liu X, E Q, Gao P, Ye X, Liu W, Zuo J (2014) Targeting GRP75 improves HSP90 inhibitor efficacy by enhancing p53-mediated apoptosis in hepatocellular carcinoma. PloS One 9: e85766

Gutstein DE, Marks AR (1997) Role of inositol 1,4,5-trisphosphate receptors in regulating apoptotic signaling and heart failure. Heart Vessels Suppl 12:53–57

Hajnóczky G, Robb-Gaspers LD, Seitz MB, Thomas AP (1995) Decoding of cytosolic calcium oscillations in the mitochondria. Cell 82:415–424

Hajnóczky G, Csordás G, Yi M (2002) Old players in a new role: mitochondria-associated membranes, VDAC, and ryanodine receptors as contributors to calcium signal propagation from endoplasmic reticulum to the mitochondria. Cell Calcium 32:363–377

Halestrap AP (2014) The C ring of the F_1F_o ATP synthase forms the mitochondrial permeability transition pore: a critical appraisal. Front Oncol 4:234

Hamanaka RB, Bennett BS, Cullinan SB, Diehl JA (2005) PERK and GCN2 contribute to eIF2α phosphorylation and cell cycle arrest after activation of the unfolded protein response pathway. Mol Biol Cell 16:5493–5501

Hanahan D, Weinberg RA (2011) Hallmarks of cancer: the next generation. Cell 144:646–674

Hanson CJ, Bootman MD, Distelhorst CW, Wojcikiewicz RJ, Roderick HL (2008) Bcl-2 suppresses Ca^{2+} release through inositol 1,4,5-trisphosphate receptors and inhibits Ca^{2+} uptake by mitochondria without affecting ER calcium store content. Cell Calcium 44:324–338

Harding HP, Zhang Y, Bertolotti A, Zeng H, Ron D (2000) Perk is essential for translational regulation and cell survival during the unfolded protein response. Mol Cell 5:897–904

Hart LS, Cunningham JT, Datta T, Dey S, Tameire F, Lehman SL, Qiu B, Zhang H, Cerniglia G, Bi M (2012) ER stress-mediated autophagy promotes Myc-dependent transformation and tumor growth. J Clin Invest 122:4621–4634

Hauet T, Yao Z-X, Bose HS, Wall CT, Han Z, Li W, Hales DB, Miller WL, Culty M, Papadopoulos V (2005) Peripheral-type benzodiazepine receptor-mediated action of steroidogenic acute regulatory protein on cholesterol entry into leydig cell mitochondria. Mol Endocrinol Baltim Md 19:540–554

Hayashi T, Su T-P (2007) Sigma-1 receptor chaperones at the ER-mitochondrion interface regulate Ca^{2+} signaling and cell survival. Cell 131:596–610

Hebert DN, Foellmer B, Helenius A (1995) Glucose trimming and reglucosylation determine glycoprotein association with calnexin in the endoplasmic reticulum. Cell 81:425–433

Hodge T, Colombini M (1997) Regulation of metabolite flux through voltage-gating of VDAC channels. J Membr Biol 157:271–279

Huang H, Shah K, Bradbury NA, Li C, White C (2014) Mcl-1 promotes lung cancer cell migration by directly interacting with VDAC to increase mitochondrial Ca^{2+} uptake and reactive oxygen species generation. Cell Death Dis 5:e1482

Huang X, Jin M, Chen Y-X, Wang J, Zhai K, Chang Y, Yuan Q, Yao K-T, Ji G (2016) ERP44 inhibits human lung cancer cell migration mainly via IP_3R2. Aging 8:1276–1286

Iino M (1990) Biphasic Ca^{2+} dependence of inositol 1,4,5-trisphosphate-induced Ca release in smooth muscle cells of the guinea pig taenia caeci. J Gen Physiol 95:1103–1122

Jayaraman T, Marks AR (1997) T cells deficient in inositol 1,4,5-trisphosphate receptor are resistant to apoptosis. Mol Cell Biol 17:3005–3012

Jin H, Ji M, Chen L, Liu Q, Che S, Xu M, Lin Z (2016) The clinicopathological significance of Mortalin overexpression in invasive ductal carcinoma of breast. J Exp Clin Cancer Res CR 35:42

Jonas EA, Porter GA, Beutner G, Mnatsakanyan N, Alavian KN (2015) Cell death disguised: the mitochondrial permeability transition pore as the c-subunit of the F_1F_O ATP synthase. Pharmacol Res 99:382–392

Joseph SK, Hajnóczky G (2007) IP_3 receptors in cell survival and apoptosis: Ca^{2+} release and beyond. Apoptosis Int J Program Cell Death 12:951–968

Kang SS, Han K-S, Ku BM, Lee YK, Hong J, Shin HY, Almonte AG, Woo DH, Brat DJ, Hwang EM et al (2010) Caffeine-mediated inhibition of calcium release channel inositol 1,4,5-trisphosphate receptor subtype 3 blocks glioblastoma invasion and extends survival. Cancer Res 70:1173–1183

Kang T-W, Yevsa T, Woller N, Hoenicke L, Wuestefeld T, Dauch D, Hohmeyer A, Gereke M, Rudalska R, Potapova A et al (2011) Senescence surveillance of pre-malignant hepatocytes limits liver cancer development. Nature 479:547–551

Kannan A, Wells RB, Sivakumar S, Komatsu S, Singh KP, Samten B, Philley JV, Sauter ER, Ikebe M, Idell S et al (2016) Mitochondrial reprogramming regulates breast cancer progression. Clin Cancer Res 22:3348–3360

Kaufman RJ, Malhotra JD (2014) Calcium trafficking integrates endoplasmic reticulum function with mitochondrial bioenergetics. Biochim Biophys Acta 1843:2233–2239

Kaul SC, Duncan EL, Englezou A, Takano S, Reddel RR, Mitsui Y, Wadhwa R (1998) Malignant transformation of NIH3T3 cells by overexpression of mot-2 protein. Oncogene 17:907–911

Kaul SC, Reddel RR, Mitsui Y, Wadhwa R (2001) An N-terminal region of mot-2 binds to p53 *in vitro*. Neoplasia N Y N 3:110–114

Kaul SC, Aida S, Yaguchi T, Kaur K, Wadhwa R (2005) Activation of wild type p53 function by its mortalin-binding, cytoplasmically localizing carboxyl terminus peptides. J Biol Chem 280:39373–39379

Khan MT, Wagner L, Yule DI, Bhanumathy C, Joseph SK (2006) Akt kinase phosphorylation of inositol 1,4,5-trisphosphate receptors. J Biol Chem 281:3731–3737

Kliková K, Štefaniková A, Pilchová I, Hatok J, Chudý P, Chudej J, Dobrota D, Račay P (2015) Differential impact of bortezomib on HL-60 and K562 cells. Gen Physiol Biophys 34:33–42

Kouroku Y, Fujita E, Tanida I, Ueno T, Isoai A, Kumagai H, Ogawa S, Kaufman RJ, Kominami E, Momoi T (2007) ER stress (PERK/eIF2α phosphorylation) mediates the polyglutamine-induced LC3 conversion, an essential step for autophagy formation. Cell Death Differ 14:230–239

Krols M, Bultynck G, Janssens S (2016) ER-Mitochondria contact sites: a new regulator of cellular calcium flux comes into play. J Cell Biol 214:367–370

Kronidou NG, Oppliger W, Bolliger L, Hannavy K, Glick BS, Schatz G, Horst M (1994) Dynamic interaction between Isp45 and mitochondrial hsp70 in the protein import system of the yeast mitochondrial inner membrane. Proc Natl Acad Sci USA 91:12818–12822

La Rovere RM, Roest G, Bultynck G, Parys JB (2016) Intracellular Ca^{2+} signaling and Ca^{2+}microdomains in the control of cell survival, apoptosis and autophagy. Cell Calcium 60:74–87

Lakkaraju AKK, van der Goot FG (2013) Calnexin controls the STAT3-mediated transcriptional response to EGF. Mol Cell 51:386–396

Lamriben L, Graham JB, Adams BM, Hebert DN (2016) N-Glycan-based ER molecular chaperone and protein quality control system: the Calnexin binding cycle. Traffic Cph Den 17:308–326

Lavik AR, Zhong F, Chang M-J, Greenberg E, Choudhary Y, Smith MR, McColl KS, Pink J, Reu FJ, Matsuyama S et al (2015) A synthetic peptide targeting the BH4 domain of Bcl-2 induces apoptosis in multiple myeloma and follicular lymphoma cells alone or in combination with agents targeting the BH3-binding pocket of Bcl-2. Oncotarget 6:27388–27402

Li J, Lee B, Lee AS (2006) Endoplasmic reticulum stress-induced apoptosis: multiple pathways and activation of p53-up-regulated modulator of apoptosis (PUMA) and NOXA by p53. J Biol Chem 281:7260–7270

Li C, Wang X, Vais H, Thompson CB, Foskett JK, White C (2007) Apoptosis regulation by Bcl-x_L modulation of mammalian inositol 1,4,5-trisphosphate receptor channel isoform gating. Proc Natl Acad Sci USA 104:12565–12570

Li G, Mongillo M, Chin K-T, Harding H, Ron D, Marks AR, Tabas I (2009) Role of ERO1-α-mediated stimulation of inositol 1,4,5-triphosphate receptor activity in endoplasmic reticulum stress-induced apoptosis. J Cell Biol 186:783–792

Liberti MV, Locasale JW (2016) The Warburg effect: how does it benefit cancer cells? Trends Biochem Sci 41:211–218

Liou G-Y, Storz P (2010) Reactive oxygen species in cancer. Free Radic Res 44:479–496

Liu J, Rone MB, Papadopoulos V (2006) Protein-protein interactions mediate mitochondrial cholesterol transport and steroid biosynthesis. J Biol Chem 281:38879–38893

Lou Y, Li R, Liu J, Zhang Y, Zhang X, Jin B, Liu Y, Wang Z, Zhong H, Wen S et al (2015) Mitofusin-2 over-expresses and leads to dysregulation of cell cycle and cell invasion in lung adenocarcinoma. Med Oncol 32:132

Lou Y, Zhang Y, Li R, Gu P, Xiong L, Zhong H, Zhang W, Han B (2016) Transcriptional profiling revealed the anti-proliferative effect of MFN2 deficiency and identified risk factors in lung adenocarcinoma. Tumor Biol 37:8643–8655

Lu J, Tan M, Cai Q (2015) The Warburg effect in tumor progression: mitochondrial oxidative metabolism as an anti-metastasis mechanism. Cancer Lett 356:156–164

Lynes EM, Bui M, Yap MC, Benson MD, Schneider B, Ellgaard L, Berthiaume LG, Simmen T (2012) Palmitoylated TMX and calnexin target to the mitochondria-associated membrane. EMBO J 31:457–470

Lynes EM, Raturi A, Shenkman M, Ortiz Sandoval C, Yap MC, Wu J, Janowicz A, Myhill N, Benson MD, Campbell RE et al (2013) Palmitoylation is the switch that assigns calnexin to quality control or ER Ca^{2+} signaling. J Cell Sci 126:3893–3903

Ma LI, Chang Y, Yu L, He W, Liu Y (2015) Pro-apoptotic and anti-proliferative effects of mitofusin-2 via PI3K/Akt signaling in breast cancer cells. Oncol Lett 10:3816–3822

Maldonado EN, Patnaik J, Mullins MR, Lemasters JJ (2010) Free tubulin modulates mitochondrial membrane potential in cancer cells. Cancer Res 70:10192–10201

Marchi S, Pinton P (2014) The mitochondrial calcium uniporter complex: molecular components, structure and physiopathological implications. J Physiol 592:829–839

Marchi S, Pinton P (2016) Alterations of calcium homeostasis in cancer cells. Curr Opin Pharmacol 29:1–6

Marchi S, Rimessi A, Giorgi C, Baldini C, Ferroni L, Rizzuto R, Pinton P (2008) Akt kinase reducing endoplasmic reticulum Ca^{2+} release protects cells from Ca^{2+}-dependent apoptotic stimuli. Biochem Biophys Res Commun 375:501–505

Marchi S, Marinello M, Bononi A, Bonora M, Giorgi C, Rimessi A, Pinton P (2012) Selective modulation of subtype III IP_3R by Akt regulates ER Ca^{2+} release and apoptosis. Cell Death Dis 3:e304

Marchi S, Giorgi C, Oparka M, Duszynski J, Wieckowski MR, Pinton P (2014a) Oncogenic and oncosuppressive signal transduction at mitochondria-associated endoplasmic reticulum membranes. Mol Cell Oncol 1:956469

Marchi S, Patergnani S, Pinton P (2014b) The endoplasmic reticulum-mitochondria connection: one touch, multiple functions. Biochim Biophys Acta 1837:461–469

Mavlyutov TA, Epstein M, Guo L-W (2015) Subcellular localization of the sigma-1 receptor in retinal neurons - an electron microscopy study. Sci Rep 5:10689

Maxfield KE, Taus PJ, Corcoran K, Wooten J, Macion J, Zhou Y, Borromeo M, Kollipara RK, Yan J, Xie J et al (2015) Comprehensive functional characterization of cancer-testis antigens defines obligate participation in multiple hallmarks of cancer. Nat Commun 6:8840

McEnery MW, Snowman AM, Trifiletti RR, Snyder SH (1992) Isolation of the mitochondrial benzodiazepine receptor: association with the voltage-dependent anion channel and the adenine nucleotide carrier. Proc Natl Acad Sci USA 89:3170–3174

Mendes CCP, Gomes DA, Thompson M, Souto NC, Goes TS, Goes AM, Rodrigues MA, Gomez MV, Nathanson MH, Leite MF (2005) The type III inositol 1,4,5-trisphosphate receptor preferentially transmits apoptotic Ca^{2+} signals into mitochondria. J Biol Chem 280:40892–40900

Messina A, Reina S, Guarino F, De Pinto V (2012) VDAC isoforms in mammals. Biochim Biophys Acta 1818:1466–1476

Mikoshiba K (2007) The IP_3 receptor/Ca^{2+} channel and its cellular function. Biochem Soc Symp:9–22

Missiroli S, Bonora M, Patergnani S, Poletti F, Perrone M, Gafà R, Magri E, Raimondi A, Lanza G, Tacchetti C, Kroemer G, Pandolfi PP, Pinton P, Giorgi C (2016) PML at mitochondria-associated membranes is critical for the repression of autophagy and cancer development. Cell Rep 16:2415–2427

Miyakawa T, Maeda A, Yamazawa T, Hirose K, Kurosaki T (1999) Encoding of Ca^{2+} signals by differential expression of IP_3 receptor subtypes. EMBO J 18:1303–1308

Monaco G, Decrock E, Akl H, Ponsaerts R, Vervliet T, Luyten T, De Maeyer M, Missiaen L, Distelhorst CW, De Smedt H et al (2012) Selective regulation of IP_3-receptor-mediated Ca^{2+} signaling and apoptosis by the BH4 domain of Bcl-2 versus Bcl-Xl. Cell Death Differ 19:295–309

Monaco G, Decrock E, Arbel N, van Vliet AR, La Rovere RM, De Smedt H, Parys JB, Agostinis P, Leybaert L, Shoshan-Barmatz V et al (2015) The BH4 domain of anti-apoptotic Bcl-XL, but not that of the related Bcl-2, limits the voltage-dependent anion channel 1 (VDAC1)-mediated transfer of pro-apoptotic Ca^{2+} signals to mitochondria. J Biol Chem 290:9150–9161

Morciano G, Giorgi C, Bonora M, Punzetti S, Pavasini R, Wieckowski MR, Campo G, Pinton P (2015) Molecular identity of the mitochondrial permeability transition pore and its role in ischemia-reperfusion injury. J Mol Cell Cardiol 78:142–153

Mori T, Hayashi T, Hayashi E, Su T-P (2013) Sigma-1 receptor chaperone at the ER-mitochondrion interface mediates the mitochondrion-ER-nucleus signaling for cellular survival. PloS One 8:e76941

Mound A, Rodat-Despoix L, Bougarn S, Ouadid-Ahidouch H, Matifat F (2013) Molecular interaction and functional coupling between type 3 inositol 1,4,5-trisphosphate receptor and BKCa channel stimulate breast cancer cell proliferation. Eur J Cancer Oxf Engl 1990 49:3738–3751

Mujcic H, Nagelkerke A, Rouschop KMA, Chung S, Chaudary N, Span PN, Clarke B, Milosevic M, Sykes J, Hill RP et al (2013) Hypoxic activation of the PERK/eIF2α arm of the unfolded protein response promotes metastasis through induction of LAMP3. Clin Cancer Res 19:6126–6137

Myhill N, Lynes EM, Nanji JA, Blagoveshchenskaya AD, Fei H, Carmine Simmen K, Cooper TJ, Thomas G, Simmen T (2008) The subcellular distribution of calnexin is mediated by PACS-2. Mol Biol Cell 19:2777–2788

Nagelkerke A, Bussink J, Mujcic H, Wouters BG, Lehmann S, Sweep FC, Span PN (2013) Hypoxia stimulates migration of breast cancer cells via the PERK/ATF4/LAMP3-arm of the unfolded protein response. Breast Cancer Res BCR 15:R2

Nagelkerke A, Sweep FC, Stegeman H, Grénman R, Kaanders JH, Bussink J, Span PN (2015) Hypoxic regulation of the PERK/ATF4/LAMP3-arm of the unfolded protein response in head and neck squamous cell carcinoma. Head Neck 37:896–905

Naon D, Scorrano L (2014) At the right distance: ER-mitochondria juxtaposition in cell life and death. Biochim Biophys Acta 1843:2184–2194

Naon D, Zaninello M, Giacomello M, Varanita T, Grespi F, Lakshminaranayan S, Serafini A, Semenzato M, Herkenne S, Hernández-Alvarez MI et al (2016) Critical reappraisal confirms that Mitofusin 2 is an endoplasmic reticulum-mitochondria tether. Proc Natl Acad Sci USA 113:11249–11254

Newton CL, Mignery GA, Südhof TC (1994) Co-expression in vertebrate tissues and cell lines of multiple inositol 1,4,5-trisphosphate (InsP3) receptors with distinct affinities for $InsP_3$. J Biol Chem 269:28613–28619

Ono Y, Tamiya T, Ichikawa T, Matsumoto K, Furuta T, Ohmoto T, Akiyama K, Seki S, Ueki K, Louis DN (1997) Accumulation of wild-type p53 in astrocytomas is associated with increased p21 expression. Acta Neuropathol (Berl) 94:21–27

Orrenius S, Zhivotovsky B, Nicotera P (2003) Regulation of cell death: the calcium-apoptosis link. Nat Rev Mol Cell Biol 4:552–565

Ouyang K, Leandro Gomez-Amaro R, Stachura DL, Tang H, Peng X, Fang X, Traver D, Evans SM, Chen

J (2014) Loss of IP_3R-dependent Ca^{2+} signalling in thymocytes leads to aberrant development and acute lymphoblastic leukemia. Nat Commun 5:4814
Oyadomari S, Mori M (2004) Roles of CHOP/GADD153 in endoplasmic reticulum stress. Cell Death Differ 11:381–389
Palmer CP, Mahen R, Schnell E, Djamgoz MB, Aydar E (2007) Sigma-1 receptors bind cholesterol and remodel lipid rafts in breast cancer cell lines. Cancer Res 67:11166–11175
Panieri E, Santoro MM (2016) ROS homeostasis and metabolism: a dangerous liaison in cancer cells. Cell Death Dis 7:e2253
Parys JB, De Smedt H (2012) Inositol 1,4,5-trisphosphate and its receptors. Adv Exp Med Biol 740:255–279
Parys JB, Sernett SW, DeLisle S, Snyder PM, Welsh MJ, Campbell KP (1992) Isolation, characterization, and localization of the inositol 1,4,5-trisphosphate receptor protein in Xenopus laevis oocytes. J Biol Chem 267:18776–18782
Pastorino JG, Hoek JB (2008) Regulation of hexokinase binding to VDAC. J Bioenerg Biomembr 40:171–182
Pedersen PL (2008) Voltage dependent anion channels (VDACs): a brief introduction with a focus on the outer mitochondrial compartment's roles together with hexokinase-2 in the "Warburg effect" in cancer. J Bioenerg Biomembr 40:123–126
Pierro C, Cook SJ, Foets TC, Bootman MD, Roderick HL (2014) Oncogenic K-Ras suppresses IP_3-dependent Ca^{2+} release through remodelling of the isoform composition of IP_3Rs and ER luminal Ca^{2+} levels in colorectal cancer cell lines. J Cell Sci 127:1607–1619
Pillozzi S, Brizzi MF, Bernabei PA, Bartolozzi B, Caporale R, Basile V, Boddi V, Pegoraro L, Becchetti A, Arcangeli A (2007) VEGFR-1 (FLT-1), β_1 integrin, and hERG K^+ channel for a macromolecular signaling complex in acute myeloid leukemia: role in cell migration and clinical outcome. Blood 110:1238–1250
Pillozzi S, Masselli M, De Lorenzo E, Accordi B, Cilia E, Crociani O, Amedei A, Veltroni M, D'Amico M, Basso G, Becchetti A, Campana D, Arcangeli A (2011) Chemotherapy resistance in acute lymphoblastic leukemia requires hERG1 channels and is overcome by hERG1 blockers. Blood 117:902–914
Pinton P, Giorgi C, Pandolfi PP (2011) The role of PML in the control of apoptotic cell fate: a new key player at ER–mitochondria sites. Cell Death Differ 18:1450–1456
Pluquet O, Qu L-K, Baltzis D, Koromilas AE (2005) Endoplasmic reticulum stress accelerates p53 degradation by the cooperative actions of Hdm2 and glycogen synthase kinase 3β. Mol Cell Biol 25:9392–9405
Prevarskaya N, Ouadid-Ahidouch H, Skryma R, Shuba Y (2014) Remodelling of Ca^{2+} transport in cancer: how it contributes to cancer hallmarks? Philos Trans R Soc Lond B Biol Sci 369:20130097
Qiao A, Wang K, Yuan Y, Guan Y, Ren X, Li L, Chen X, Li F, Chen AF, Zhou J et al (2016) Sirt3-mediated mitophagy protects tumor cells against apoptosis under hypoxia. Oncotarget 7:43390–43400
Qu L, Huang S, Baltzis D, Rivas-Estilla AM, Pluquet O, Hatzoglou M, Koumenis C, Taya Y, Yoshimura A, Koromilas AE (2004) Endoplasmic reticulum stress induces p53 cytoplasmic localization and prevents p53-dependent apoptosis by a pathway involving glycogen synthase kinase-3β. Genes Dev 18:261–277
Raffaello A, Mammucari C, Gherardi G, Rizzuto R (2016) Calcium at the center of Cell signaling: interplay between endoplasmic reticulum, mitochondria, and lysosomes. Trends Biochem Sci 41 (12):1035–1049
Ran Q, Wadhwa R, Kawai R, Kaul SC, Sifers RN, Bick RJ, Smith JR, Pereira-Smith OM (2000) Extramitochondrial localization of mortalin/mthsp70/PBP74/GRP75. Biochem Biophys Res Commun 275:174–179
Ranieri M, Brajkovic S, Riboldi G, Ronchi D, Rizzo F, Bresolin N, Corti S, Comi GP (2013) Mitochondrial fusion proteins and human diseases. Neurol Res Int 2013:293893
Rapizzi E, Pinton P, Szabadkai G, Wieckowski MR, Vandecasteele G, Baird G, Tuft RA, Fogarty KE, Rizzuto R (2002) Recombinant expression of the voltage-dependent anion channel enhances the transfer of Ca^{2+} microdomains to mitochondria. J Cell Biol 159:613–624
Raturi A, Gutiérrez T, Ortiz-Sandoval C, Ruangkittisakul A, Herrera-Cruz MS, Rockley JP, Gesson K, Ourdev D, Lou P-H, Lucchinetti E (2016) TMX1 determines cancer cell metabolism as a thiol-based modulator of ER-mitochondria Ca^{2+} flux. J Cell Biol 214:433–444
Rieusset J, Fauconnier J, Paillard M, Belaidi E, Tubbs E, Chauvin MA, Durand A, Bravard A, Teixeira G, Bartosch B (2016) Disruption of calcium transfer from ER to mitochondria links alterations of mitochondria-associated ER membrane integrity to hepatic insulin resistance. Diabetologia 59:614–623
Rizzuto R, Brini M, Murgia M, Pozzan T (1993) Microdomains with high Ca^{2+} close to IP_3-sensitive channels that are sensed by neighboring mitochondria. Science 262:744–747
Rizzuto R, Pinton P, Carrington W, Fay FS, Fogarty KE, Lifshitz LM, Tuft RA, Pozzan T (1998) Close contacts with the endoplasmic reticulum as determinants of mitochondrial Ca^{2+} responses. Science 280:1763–1766
Rizzuto R, Bernardi P, Pozzan T (2000) Mitochondria as all-round players of the calcium game. J Physiol 529 (Pt 1):37–47
Rizzuto R, De Stefani D, Raffaello A, Mammucari C (2012) Mitochondria as sensors and regulators of calcium signalling. Nat Rev Mol Cell Bio 13:566–578
Roderick HL, Cook SJ (2008) Ca^{2+} signalling checkpoints in cancer: remodelling Ca^{2+} for cancer cell proliferation and survival. Nat Rev Cancer 8:361–375
Roderick HL, Lechleiter JD, Camacho P (2000) Cytosolic phosphorylation of calnexin controls intracellular Ca^{2+}

oscillations via an interaction with SERCA2b. J Cell Biol 149:1235–1248

Rone MB, Midzak AS, Issop L, Rammouz G, Jagannathan S, Fan J, Ye X, Blonder J, Veenstra T, Papadopoulos V (2012) Identification of a dynamic mitochondrial protein complex driving cholesterol import, trafficking, and metabolism to steroid hormones. Mol Endocrinol Baltim Md 26:1868–1882

Rong Y-P, Aromolaran AS, Bultynck G, Zhong F, Li X, McColl K, Matsuyama S, Herlitze S, Roderick HL, Bootman MD et al (2008) Targeting Bcl-2-IP_3 receptor interaction to reverse Bcl-2's inhibition of apoptotic calcium signals. Mol Cell 31:255–265

Rong Y-P, Bultynck G, Aromolaran AS, Zhong F, Parys JB, De Smedt H, Mignery GA, Roderick HL, Bootman MD, Distelhorst CW (2009) The BH4 domain of Bcl-2 inhibits ER calcium release and apoptosis by binding the regulatory and coupling domain of the IP_3 receptor. Proc Natl Acad Sci USA 106:14397–14402

Rostovtseva TK, Bezrukov SM (2012) VDAC inhibition by tubulin and its physiological implications. Biochim Biophys Acta 1818:1526–1535

Rostovtseva TK, Kazemi N, Weinrich M, Bezrukov SM (2006) Voltage gating of VDAC is regulated by nonlamellar lipids of mitochondrial membranes. J Biol Chem 281:37496–37506

Rostovtseva TK, Sheldon KL, Hassanzadeh E, Monge C, Saks V, Bezrukov SM, Sackett DL (2008) Tubulin binding blocks mitochondrial voltage-dependent anion channel and regulates respiration. Proc Natl Acad Sci USA 105:18746–18751

Rouschop KMA, van den Beucken T, Dubois L, Niessen H, Bussink J, Savelkouls K, Keulers T, Mujcic H, Landuyt W, Voncken JW et al (2010) The unfolded protein response protects human tumor cells during hypoxia through regulation of the autophagy genes MAP1LC3B and ATG5. J Clin Invest 120:127–141

Rowland AA, Voeltz GK (2012) Endoplasmic reticulum-mitochondria contacts: function of the junction. Nat Rev Mol Cell Biol 13:607–625

Sano R, Reed JC (2013) ER stress-induced cell death mechanisms. Biochim Biophys Acta 1833:3460–3470

Schein SJ, Colombini M, Finkelstein A (1976) Reconstitution in planar lipid bilayers of a voltage-dependent anion-selective channel obtained from paramecium mitochondria. J Membr Biol 30:99–120

Scherer PE, Manning-Krieg UC, Jenö P, Schatz G, Horst M (1992) Identification of a 45-kDa protein at the protein import site of the yeast mitochondrial inner membrane. Proc Natl Acad Sci USA 89:11930–11934

Schneider HC, Westermann B, Neupert W, Brunner M (1996) The nucleotide exchange factor MGE exerts a key function in the ATP-dependent cycle of mt-Hsp70-Tim44 interaction driving mitochondrial protein import. EMBO J 15:5796–5803

Shimizu S, Narita M, Tsujimoto Y (1999) Bcl-2 family proteins regulate the release of apoptogenic cytochrome c by the mitochondrial channel VDAC. Nature 399:483–487

Shimizu S, Konishi A, Kodama T, Tsujimoto Y (2000) BH4 domain of antiapoptotic Bcl-2 family members closes voltage-dependent anion channel and inhibits apoptotic mitochondrial changes and cell death. Proc Natl Acad Sci USA 97:3100–3105

Shioda N, Ishikawa K, Tagashira H, Ishizuka T, Yawo H, Fukunaga K (2012) Expression of a truncated form of the endoplasmic reticulum chaperone protein, $\sigma 1$ receptor, promotes mitochondrial energy depletion and apoptosis. J Biol Chem 287:23318–23331

Shoshan-Barmatz V, Gincel D (2003) The voltage-dependent anion channel: characterization, modulation, and role in mitochondrial function in cell life and death. Cell Biochem Biophys 39:279–292

Shoshan-Barmatz V, De Pinto V, Zweckstetter M, Raviv Z, Keinan N, Arbel N (2010) VDAC, a multi-functional mitochondrial protein regulating cell life and death. Mol Aspects Med 31:227–285

Simmen T, Aslan JE, Blagoveshchenskaya AD, Thomas L, Wan L, Xiang Y, Feliciangeli SF, Hung C-H, Crump CM, Thomas G (2005) PACS-2 controls endoplasmic reticulum-mitochondria communication and Bid-mediated apoptosis. EMBO J 24:717–729

Simpson AJ, Caballero OL, Jungbluth A, Chen Y-T, Old LJ (2005) Cancer/testis antigens, gametogenesis and cancer. Nat Rev Cancer 5:615–625

Skrzycki M, Czeczot H (2013) Altered expression level of Sigma1 receptor gene in human colorectal cancer. J Recept Signal Transduct Res 33:313–318

Starenki D, Hong S-K, Lloyd RV, Park J-I (2015) Mortalin (GRP75/HSPA9) upregulation promotes survival and proliferation of medullary thyroid carcinoma cells. Oncogene 34:4624–4634

Stewart TA, Yapa KTDS, Monteith GR (2015) Altered calcium signaling in cancer cells. Biochim Biophys Acta 1848:2502–2511

Su TP (1982) Evidence for sigma opioid receptor: binding of [^{3}H]SKF-10047 to etorphine-inaccessible sites in guinea-pig brain. J Pharmacol Exp Ther 223:284–290

Su T-P, Hayashi T, Maurice T, Buch S, Ruoho AE (2010) The sigma-1 receptor chaperone as an inter-organelle signaling modulator. Trends Pharmacol Sci 31:557–566

Sun L, Shukair S, Naik TJ, Moazed F, Ardehali H (2008) Glucose phosphorylation and mitochondrial binding are required for the protective effects of hexokinases I and II. Mol Cell Biol 28:1007–1017

Szabadkai G, Bianchi K, Várnai P, De Stefani D, Wieckowski MR, Cavagna D, Nagy AI, Balla T, Rizzuto R (2006) Chaperone-mediated coupling of endoplasmic reticulum and mitochondrial Ca^{2+} channels. J Cell Biol 175:901–911

Szado T, Vanderheyden V, Parys JB, De Smedt H, Rietdorf K, Kotelevets L, Chastre E, Khan F, Landegren U, Söderberg O et al (2008) Phosphorylation of inositol 1,4,5-trisphosphate receptors by

protein kinase B/Akt inhibits Ca^{2+} release and apoptosis. Proc Natl Acad Sci USA 105:2427–2432
Szatkowski C, Parys JB, Ouadid-Ahidouch H, Matifat F (2010) Inositol 1,4,5-trisphosphate-induced Ca^{2+} signalling is involved in estradiol-induced breast cancer epithelial cell growth. Mol Cancer 9:156
Tagashira H, Bhuiyan MS, Shioda N, Fukunaga K (2014) Fluvoxamine rescues mitochondrial Ca^{2+} transport and ATP production through σ(1)-receptor in hypertrophic cardiomyocytes. Life Sci 95:89–100
Takizawa T, Tatematsu C, Watanabe K, Kato K, Nakanishi Y (2004) Cleavage of calnexin caused by apoptotic stimuli: implication for the regulation of apoptosis. J Biochem (Tokyo) 136:399–405
Tsai S-YA, Chuang J-Y, Tsai M-S, Wang X-F, Xi Z-X, Hung J-J, Chang W-C, Bonci A, Su T-P (2015) Sigma-1 receptor mediates cocaine-induced transcriptional regulation by recruiting chromatin-remodeling factors at the nuclear envelope. Proc Natl Acad Sci USA 112:6562–6570
Tsunoda T, Koga H, Yokomizo A, Tatsugami K, Eto M, Inokuchi J, Hirata A, Masuda K, Okumura K, Naito S (2005) Inositol 1,4,5-trisphosphate (IP_3) receptor type1 (IP_3R1) modulates the acquisition of cisplatin resistance in bladder cancer cell lines. Oncogene 24:1396–1402
Tu H, Wang Z, Bezprozvanny I (2005) Modulation of mammalian inositol 1,4,5-trisphosphate receptor isoforms by calcium: a role of calcium sensor region. Biophys J 88:1056–1069
Urra H, Dufey E, Avril T, Chevet E, Hetz C (2016) Endoplasmic reticulum stress and the hallmarks of cancer. Trends Cancer 2:252–262
van Riggelen J, Yetil A, Felsher DW (2010) MYC as a regulator of ribosome biogenesis and protein synthesis. Nat Rev Cancer 10:301–309
van Vliet AR, Verfaillie T, Agostinis P (2014) New functions of mitochondria associated membranes in cellular signaling. Biochim Biophys Acta 1843:2253–2262
Vance JE (2014) MAM (mitochondria-associated membranes) in mammalian cells: Lipids and beyond. Biochim Biophys Acta 1841:595–609
Vander Heiden MG, Chandel NS, Li XX, Schumacker PT, Colombini M, Thompson CB (2000) Outer mitochondrial membrane permeability can regulate coupled respiration and cell survival. Proc Natl Acad Sci USA 97:4666–4671
Vander Heiden MG, Li XX, Gottleib E, Hill RB, Thompson CB, Colombini M (2001) Bcl-xL promotes the open configuration of the voltage-dependent anion channel and metabolite passage through the outer mitochondrial membrane. J Biol Chem 276:19414–19419
Vander Heiden MG, Cantley LC, Thompson CB (2009) Understanding the Warburg effect: the metabolic requirements of cell proliferation. Science 324:1029–1033
Vanderheyden V, Devogelaere B, Missiaen L, De Smedt H, Bultynck G, Parys JB (2009) Regulation of inositol 1,4,5-trisphosphate-induced Ca^{2+} release by reversible phosphorylation and dephosphorylation. Biochim Biophys Acta 1793:959–970
Várnai P, Balla A, Hunyady L, Balla T (2005) Targeted expression of the inositol 1,4,5-triphosphate receptor (IP_3R) ligand-binding domain releases Ca^{2+} via endogenous IP_3R channels. Proc Natl Acad Sci USA 102:7859–7864
Verfaillie T, Rubio N, Garg AD, Bultynck G, Rizzuto R, Decuypere JP, Piette J, Linehan C, Gupta S, Samali A, Agostinis P (2012) PERK is required at the ER-mitochondrial contact sites to convey apoptosis after ROS-based ER stress. Cell Death Differ 19:1880–1891
Vervliet T, Parys JB, Bultynck G (2016) Bcl-2 proteins and calcium signaling: complexity beneath the surface. Oncogene 35:5079–5092
Vervloessem T, Yule DI, Bultynck G, Parys JB (2015) The type 2 inositol 1,4,5-trisphosphate receptor, emerging functions for an intriguing Ca^{2+}-release channel. Biochim Biophys Acta 1853:1992–2005
Vilner BJ, John CS, Bowen WD (1995) Sigma-1 and sigma-2 receptors are expressed in a wide variety of human and rodent tumor cell lines. Cancer Res 55:408–413
Voos W, Röttgers K (2002) Molecular chaperones as essential mediators of mitochondrial biogenesis. Biochim Biophys Acta 1592:51–62
Wadhwa R, Pereira-Smith OM, Reddel RR, Sugimoto Y, Mitsui Y, Kaul SC (1995) Correlation between complementation group for immortality and the cellular distribution of mortalin. Exp Cell Res 216:101–106
Wadhwa R, Takano S, Robert M, Yoshida A, Nomura H, Reddel RR, Mitsui Y, Kaul SC (1998) Inactivation of tumor suppressor p53 by mot-2, a hsp70 family member. J Biol Chem 273:29586–29591
Wadhwa R, Taira K, Kaul SC (2002a) An Hsp70 family chaperone, mortalin/mthsp70/PBP74/Grp75: what, when, and where? Cell Stress Chaperones 7:309–316
Wadhwa R, Yaguchi T, Hasan MK, Mitsui Y, Reddel RR, Kaul SC (2002b) Hsp70 family member, mot-2/mthsp70/GRP75, binds to the cytoplasmic sequestration domain of the p53 protein. Exp Cell Res 274:246–253
Wagner LE, Joseph SK, Yule DI (2008) Regulation of single inositol 1,4,5-trisphosphate receptor channel activity by protein kinase A phosphorylation. J Physiol 586:3577–3596
Wang B, Rouzier R, Albarracin CT, Sahin A, Wagner P, Yang Y, Smith TL, Meric-Bernstam F, Marcelo Aldaz C, Marcelo AC et al (2004) Expression of sigma 1 receptor in human breast cancer. Breast Cancer Res Treat 87:205–214
Wang W, Xie Q, Zhou X, Yao J, Zhu X, Huang P, Zhang L, Wei J, Xie H, Zhou L et al (2015) Mitofusin-2 triggers mitochondria Ca^{2+} influx from the endoplasmic reticulum to induce apoptosis in

hepatocellular carcinoma cells. Cancer Lett 358:47–58

Wei C, Wang X, Chen M, Ouyang K, Song L-S, Cheng H (2009) Calcium flickers steer cell migration. Nature 457:901–905

Wei C, Wang X, Zheng M, Cheng H (2012) Calcium gradients underlying cell migration. Curr Opin Cell Biol 24:254–261

Weisthal S, Keinan N, Ben-Hail D, Arif T, Shoshan-Barmatz V (2014) Ca^{2+}-mediated regulation of VDAC1 expression levels is associated with cell death induction. Biochim Biophys Acta 1843:2270–2281

Werneburg NW, Bronk SF, Guicciardi ME, Thomas L, Dikeakos JD, Thomas G, Gores GJ (2012) Tumor necrosis factor-related apoptosis-inducing ligand (TRAIL) protein-induced lysosomal translocation of proapoptotic effectors is mediated by phosphofurin acidic cluster sorting protein-2 (PACS-2). J Biol Chem 287:24427–24437

White C, Li C, Yang J, Petrenko NB, Madesh M, Thompson CB, Foskett JK (2005) The endoplasmic reticulum gateway to apoptosis by Bcl-X_L modulation of the $InsP_3R$. Nat Cell Biol 7:1021–1028

Whitehurst AW (2014) Cause and consequence of cancer/testis antigen activation in cancer. Annu Rev Pharmacol Toxicol 54:251–272

Whitehurst AW, Bodemann BO, Cardenas J, Ferguson D, Girard L, Peyton M, Minna JD, Michnoff C, Hao W, Roth MG et al (2007) Synthetic lethal screen identification of chemosensitizer loci in cancer cells. Nature 446:815–819

Wiel C, Lallet-Daher H, Gitenay D, Gras B, Le Calvé B, Augert A, Ferrand M, Prevarskaya N, Simonnet H, Vindrieux D et al (2014) Endoplasmic reticulum calcium release through ITPR2 channels leads to mitochondrial calcium accumulation and senescence. Nat Commun 5:3792

Williams A, Hayashi T, Wolozny D, Yin B, Su T-C, Betenbaugh MJ, Su T-P (2016) The non-apoptotic action of Bcl-xL: regulating Ca^{2+} signaling and bioenergetics at the ER-mitochondrion interface. J Bioenerg Biomembr 48:211–225

Wilson JE (2003) Isozymes of mammalian hexokinase: structure, subcellular localization and metabolic function. J Exp Biol 206:2049–2057

Wu P-K, Hong S-K, Veeranki S, Karkhanis M, Starenki D, Plaza JA, Park J-I (2013) A mortalin/HSPA9-mediated switch in tumor-suppressive signaling of Raf/MEK/extracellular signal-regulated kinase. Mol Cell Biol 33:4051–4067

Wu C-H, Lin Y-W, Wu T-F, Ko J-L, Wang P-H (2016a) Clinical implication of voltage-dependent anion channel 1 in uterine cervical cancer and its action on cervical cancer cells. Oncotarget 7:4210–4225

Wu Y, Zhou D, Xu X, Zhao X, Huang P, Zhou X, Song W, Guo H, Wang W, Zheng S (2016b) Clinical significance of mitofusin-2 and its signaling pathways in hepatocellular carcinoma. World J Surg Oncol 14:179

Xie Q, Su J, Jiao B, Shen L, Ma L, Qu X, Yu C, Jiang X, Xu Y, Sun L (2016) ABT737 reverses cisplatin resistance by regulating ER-mitochondria Ca^{2+} signal transduction in human ovarian cancer cells. Int J Oncol

Xu D, Yi W, Chen Y, Ma L, Wang J, Yu G (2014) Overexpression of Sig1R is closely associated with tumor progression and poor outcome in patients with hilar cholangiocarcinoma. Med Oncol Northwood Lond Engl 31:261

Yang L, Li H, Jiang Y, Zuo J, Liu W (2013) Inhibition of mortalin expression reverses cisplatin resistance and attenuates growth of ovarian cancer cells. Cancer Lett 336:213–221

Yang J, Vais H, Gu W, Foskett JK (2016) Biphasic regulation of $InsP_3$ receptor gating by dual Ca^{2+} release channel BH3-like domains mediates Bcl-xL control of cell viability. Proc Natl Acad Sci USA 113:1953–1962

Yi X, Luk JM, Lee NP, Peng J, Leng X, Guan X-Y, Lau GK, Beretta L, Fan S-T (2008) Association of mortalin (HSPA9) with liver cancer metastasis and prediction for early tumor recurrence. Mol Cell Proteomics 7:315–325

Yoshida H, Matsui T, Yamamoto A, Okada T, Mori K (2001) XBP1 mRNA is induced by ATF6 and spliced by IRE1 in response to ER stress to produce a highly active transcription factor. Cell 107:881–891

Yoshikawa F, Iwasaki H, Michikawa T, Furuichi T, Mikoshiba K (1999) Trypsinized cerebellar inositol 1,4,5-trisphosphate receptor. Structural and functional coupling of cleaved ligand binding and channel domains. J Biol Chem 274:316–327

Youker RT, Shinde U, Day R, Thomas G (2009) At the crossroads of homoeostasis and disease: roles of the PACS proteins in membrane traffic and apoptosis. Biochem J 421:1–15

Youle RJ, Strasser A (2008) The BCL-2 protein family: opposing activities that mediate cell death. Nat Rev Mol Cell Biol 9:47–59

Zha J, Harada H, Yang E, Jockel J, Korsmeyer SJ (1996) serine phosphorylation of death agonist bad in response to survival factor results in binding to 14-3-3 not BCL-XL. Cell 87:619–628

Zhang F, Hamanaka RB, Bobrovnikova-Marjon E, Gordan JD, Dai M-S, Lu H, Simon MC, Diehl JA (2006) Ribosomal stress couples the unfolded protein response to p53-dependent cell cycle arrest. J Biol Chem 281:30036–30045

Zhang G-E, Jin H-L, Lin X-K, Chen C, Liu X-S, Zhang Q, Yu J-R (2013) Anti-tumor effects of Mfn2 in gastric cancer. Int J Mol Sci 14:13005–13021

Zhang C, Ding W, Liu Y, Hu Z, Zhu D, Wang X, Yu L, Wang L, Shen H, Zhang W et al (2016a) Proteomics-based identification of VDAC1 as a tumor promoter in cervical carcinoma. Oncotarget 7:52317–52328

Zhang G, Jiang G, Wang C, Zhong K, Zhang J, Xue Q, Li X, Jin H, Li B, Zhang G et al (2016b) Decreased expression of microRNA-320a promotes proliferation

and invasion of non-small cell lung cancer cells by increasing VDAC1 expression. Oncotarget 7:49470–49480
Zheng J (2012) Energy metabolism of cancer: Glycolysis versus oxidative phosphorylation (Review). Oncol Lett 4:1151–1157
Zhong F, Harr MW, Bultynck G, Monaco G, Parys JB, De Smedt H, Rong Y-P, Molitoris JK, Lam M, Ryder C et al (2011) Induction of Ca^{2+} –driven apoptosis in chronic lymphocytic leukemia cells by peptide-mediated disruption of Bcl-2-IP_3 receptor interaction. Blood 117:2924–2934

Index

M. Tagaya, T. Simmen (eds.), *Organelle Contact Sites*, Advances in Experimental Medicine and Biology 997, DOI 10.1007/978-981-10-4567-7

MIX
Papier aus verantwortungsvollen Quellen
Paper from responsible sources
FSC® C105338

If you have any concerns about our products,
you can contact us on
ProductSafety@springernature.com

In case Publisher is established outside the EU,
the EU authorized representative is:
Springer Nature Customer Service Center GmbH
Europaplatz 3, 69115 Heidelberg, Germany

Printed by Libri Plureos GmbH
in Hamburg, Germany